International Max Planck Research School (IMPRS)
for Maritime Affairs
at the University of Hamburg

More information about this series at: http://www.springer.com/series/6888

Hamburg Studies on Maritime Affairs
Volume 35

Solène Guggisberg

The Use of CITES for Commercially-exploited Fish Species

A Solution to Overexploitation and Illegal, Unreported and Unregulated Fishing?

Springer

Solène Guggisberg
International Court of Justice
Peace Palace
The Hague
The Netherlands

The views and opinions expressed in this book are those of the author and do not necessarily reflect the views of any former of current employer.

Die vorliegende Arbeit wurde im Januar 2015 an der Bucerius Law School – Hochschule für Rechtswissenschaft – als Dissertation angenommen.
vorgelegt von Solène Guggisberg
Erstgutachter: Prof. Dr. Doris König, M.C.L.
Zweitgutachter: Prof. Dr. Dr. h.c. Peter Ehlers
Tag der mündlichen Prüfung: 10. April 2015

ISSN 1614-2462 ISSN 1867-9587 (electronic)
Hamburg Studies on Maritime Affairs
ISBN 978-3-319-23701-5 ISBN 978-3-319-23702-2 (eBook)
DOI 10.1007/978-3-319-23702-2

Library of Congress Control Number: 2015955393

Springer Cham Heidelberg New York Dordrecht London

Printed on acid-free paper

Springer International Publishing AG Switzerland is part of Springer Science+Business Media (www.springer.com)

Acknowledgments

Most people who met me in the last four years know that neither starting, nor pursuing or finishing this thesis was an easy task. Writing these words means that what appeared so terribly unreachable has actually been achieved. Persevering through hard times has been made possible by the passion I had and still have for the subject of my research and, more generally, for the questions related to the marine environment and sustainability upon which it touches. But even more, what kept me going were the people I was blessed to have around me who trusted my abilities whenever I doubted them.

Amongst such people are first of all the members of my family. I want to thank you, Maman and Roger for your support and empathy, Tony for your availability now and always, Coline for your unusual wisdom, which often reminds me of what really matters, as well as my Granny and Mariette, who unfortunately did not get to see this publication. As blood is not all that matters, I would like to extend these special thanks to Alexia, for your longstanding friendship: you definitely made it in the family group by now. As for the Strasbourg quartet, Caroline, Jean-Sébastien, Suzanne and Noémi, you offered me some of the greatest joys of these last years as well as structured weekends from which I always came more productive, if not always rested. Last but not least, in this first round of naming and expressing gratitude, I would like to thank you Garth for the understanding—and useful—ear with which you listened to my substantive questions and to footnotes-related frustrations.

Then, my appreciation goes to my friends, all over Europe, for their support at various stages of this adventure. Amongst many others, I would like to thank by name Mike, Hannah and Frank for their precious pieces of advice at an early stage, as well as my office colleagues in Hamburg, Lief and Lina for their positive and supportive attitude. A special thanks to Sarah for generously reading a full draft and providing me with valuable comments and encouragements. Then, my gratitude to Tess, Maja, Johannes and Ole for the refreshing off-time moments we shared in Hamburg. In Cambridge, through good work and hard times, my appreciative thoughts go to Mariangela, David, Laura and Pia for their friendship and the great memories we shared.

After these personal remarks, I would like to express my more formal thanks to the International Max Planck Research School for Maritime Affairs program that financed my scholarship and provided me with a comfortable situation, to Professor Doris König (Bucerius Law School, Hamburg) who acted as my supervisor and to Barbara Krah-Schröder, our program coordinator, who successfully dealt with my academic moods, administrative problems and serious doubts. Needless to say, I would not be writing the words prefacing this book had it not been for her patient and understanding presence.

Finally, a last group I wish to thank is composed of the many interesting people who listened to the development of my research, intellectually challenged me, and offered me a time 'off-thesis' with shorter-time professional gratification or achievements. In this category, I would in particular like to express my gratitude to the two academics who, by teaching me respectively the law of the sea and international environmental law at the University of Edinburgh, sent me on the path that led to this dissertation: these people are James Harrison, who provided me with some very much appreciated support in the difficult early months of settling in the Ph.D. routine, and Alan Boyle, who showed interest in my progress throughout these years. As for the time I spent at the FAO Fisheries and Aquaculture Department, it was not only an opportunity to undertake fundamental research for my thesis but also a chance to discover other issues in the fisheries field, in particular those related to small-scale fisheries—I would like to thank Rolf Willmann, Nicole Franz, Gail Lugten, Carlos Fuentevilla, Monica Barone and Johanne Fischer for their warm welcome.

To get started with the serious part now, a Babylonian proverb has it that the gods do not deduct from a (wo)man's allotted span the hours spent fishing; I do sincerely hope that they extend such generosity to the years spent writing a thesis about this activity!

Abbreviations

AB	Appellate Body (of the World Trade Organization)
CCAMLR	Commission on the Conservation of Antarctic Marine Living Resources
CCSBT	Commission for the Conservation of Southern Bluefin Tuna
CCRF	Code of Conduct for Responsible Fisheries
CIL	Customary International Law
CITES	Convention on International Trade in Endangered Species of Wild Fauna and Flora
CDS	Catch Documentation Scheme
CLCS	Commission on the Limits of the Continental Shelf
CMMs	Conservation and Management Measures
CMS	Convention on the Conservation of Migratory Species of Wild Animals
COFI	Committee on Fisheries
CoP	Conference of the Parties
DSB	Dispute Settlement Body (of the World Trade Organization)
EEZ	Exclusive Economic Zone
EU	European Union
FAO	Food and Agriculture Organization
GATT	General Agreement on Tariffs and Trade
GFCM	General Fisheries Commission for the Mediterranean
IFS	Introduction from the Sea
IATTC	Inter-American Tropical Tuna Commission
ICCAT	International Commission for the Conservation of Atlantic Tunas
ICES	International Council for the Exploration of the Sea
ICJ	International Court of Justice
ICRW	International Convention for the Regulation of Whaling
ILA	International Law Association
ILC	International Law Commission

ILC Draft articles of 2011	ILC draft articles on the responsibility of international organizations adopted in 2011
IMO	International Maritime Organization
IPOA-Sharks	International Plan of Action for the Conservation and Management of Sharks
IPOA-IUU	International Plan of Action to Prevent, Deter and Eliminate Illegal, Unreported and Unregulated Fishing
IOTC	Indian Ocean Tuna Commission
ITLOS	International Tribunal for the Law of the Sea
ITTA	International Tropical Timber Agreement
ITTO	International Tropical Timber Organization
IUCN	International Union for the Conservation of Nature
IUU	Illegal, Unreported and Unregulated
IWC	International Whaling Commission
MEA	Multilateral Environmental Agreement
MoC	Memorandum of Cooperation
MoU	Memorandum of Understanding
MSY	Maximum Sustainable Yield
NAFO	Northwest Atlantic Fisheries Organization
NAMMCO	North Atlantic Marine Mammal Commission
NASCO	North Atlantic Salmon Conservation Organization
NDF	Non-Detriment Finding
NEAFC	North East Atlantic Fisheries Commission
NGO	Non-Governmental Organization
Nm	Nautical miles
PCIJ	Permanent Court of International Justice
RFMO	Regional Fisheries Management Organization
SCRS	Standing Committee on Research and Statistics (of ICCAT)
SEAFO	South East Atlantic Fisheries Organisation
SIOFA	South Indian Ocean Fisheries Agreement
SOFIA	State of World Fisheries and Aquaculture
SRFC	Sub-Regional Fisheries Commission
TACs	Total Allowable Catches
TDS	Trade Documentation Scheme
UN	United Nations
UNCLOS	United Nations Convention on the Law of the Sea
UNFSA	United Nations Agreement for the Implementation of the Provisions of the United Nations Convention on the Law of the Sea of 10 December 1982 relating to the Conservation and Management of Straddling Fish Stocks and Highly Migratory Fish Stocks
UNGA	United Nations General Assembly
UNODC	United Nations Office on Drugs and Crime
USA	United States of America

VCLT	Vienna Convention on the Law of Treaties
WCPFC	Western and Central Pacific Fisheries Commission
WTO	World Trade Organization
WWF	World Wild Fund for Nature

Contents

Chapter 1
Introduction

> *"[T]he oceans of the world continue to suffer from the survival of the philosophy of the commons. Maritime nations still respond automatically to the shibboleth of the 'freedom of the seas.' Professing to believe in the 'inexhaustible resources of the oceans,' they bring species after species of fish and whales closer to extinction."*
> (G. Hardin, 'The Tragedy of the Commons' (1968) 3859 Science 1243, 1245)

The current situation of many commercially-exploited fish species is worrying with nearly 30 % of all stocks around the globe qualifying as overfished.[1] In particular, the biomass of big predatory species, such as tunas, is severely reduced. As for collapsed stocks, for instance the cod in the Northwest Atlantic, they will need several decades to recover from overexploitation.[2] Overfishing, caused by excessive—and yet often disregarded—quotas, driven by fleet overcapacity and aggravated by illegal, unreported and unregulated (IUU) fishing, dangerously depletes stocks and threatens to drive some species to extinction. The resource management regimes in charge have consequently been widely criticized; the institutions involved, their interactions and balance sheets are questioned, but with no easy one-size-fits-all solution in sight.

The current fishing crisis is a serious challenge on the social, economic and of course ecological levels, particularly since "nearly two-thirds of humanity inhabit coastal areas and depend on coastal and marine environments for their livelihoods."[3] The fisheries and aquaculture sector employed, in 2012, more than 58 million people directly and several hundred millions if ancillary (post-harvest) jobs and dependents are counted.[4] A billion people's protein intake mainly comes

[1] FAO, 'The State of the World Fisheries and Aquaculture' (Rome 2012) 11; FAO, 'The State of the World Fisheries and Aquaculture' (Rome 2014) 7.

[2] On the cod disaster, see C. Clover, *The end of the line* (Ebury Press 2005) chapter 8 'After the gold rush'.

[3] B. K. Sovacool, 'A Game of Cat and Fish: How to Restore the Balance in Sustainable Fisheries Management' (2009) 40 Ocean Development and International Law 97, 98.

[4] In 2008, 44.9 million people's livelihood directly depended upon fishing and 180 million people did so if the secondary – post-harvest – jobs are counted (FAO, 'The State of the World Fisheries

S. Guggisberg, *The Use of CITES for Commercially-exploited Fish Species*,
Hamburg Studies on Maritime Affairs 35, DOI 10.1007/978-3-319-23702-2_1

from fish[5], which represents more than 15 % of the total animal protein consumed in the world.[6] Also, "the estimated annual landed value of fish globally is around USD 90 billion",[7] that of fish and fishery products exports nearly reached USD 130 billion in 2011, making it "[f]or developing countries [. . .] by far the most valuable of agricultural commodities [traded internationally]".[8] As for the ecological impact of bad management leading to stocks' depletion, the disappearance of one—or more—of an ecosystem's constituent parts has obvious consequences on the whole trophic chain and even on the physical components' equilibrium.[9]

The fast-growing industry of aquaculture might appear to be the solution to the current fishing crisis. Undeniably, aquaculture is promising, but it also creates problems in relation to the sustainable management of wild fish stocks. This is particularly true if it has a farm-ranching component based on the fattening, in a controlled environment, of wild-caught juveniles. Nearly half of the fish products consumed nowadays comes from aquaculture sources.[10] It is hence impossible to discuss fisheries issues without looking at the effects of aquaculture on wild capture and at the implications of general conservation measures on such practices.

---

and Aquaculture' (Rome 2010) 6, 26). In 2010, 54.8 million people were engaged directly in fish production while it was estimated that the livelihoods of 660 to 820 million people depended upon it indirectly (FAO, 'The State of the World Fisheries and Aquaculture 2012' (n 1) 10, 41). In 2012, 58.3 million people were engaged directly in capture fisheries and aquaculture and 10 to 12 percent of the world's population depended on those sectors for their livelihood (FAO, 'The State of the World Fisheries and Aquaculture 2014' (n 1) 27, 31–32).

[5] T. Bostock and S. Walmsley, 'Enough to Eat? Fisheries and Food Security' in R. Bourne and M. Collins (eds), *From Hook to Plate: The State of Marine Fisheries: A Commonwealth Perspective* (2009) 105.

[6] FAO, 'The State of the World Fisheries and Aquaculture 2012' (n 1) 5; FAO, 'The State of the World Fisheries and Aquaculture 2014' (n 1) 66, 105.

[7] World Ocean Review, *Living with the oceans* (Maribus, Future Ocean 2010) 120.

[8] Bostock and Walmsley (n 5) 107; FAO, 'The State of the World Fisheries and Aquaculture 2014' (n 1) 7.

[9] N. Matz, *Wege zur Koordinierung völkerrechtlicher Verträge: Völkervertragsrechtliche und institutionelle Ansätze* (Springer 2005) 136. A well-known example is that of sea otters (see infra Chapter 4 A. II. 2.). Depletion can also lead to changes in the ecosystem balance (C. Mullon, P. Fréon and P. Curry 'The dynamics of collapse in world fisheries' (2006) 6 Fish and Fisheries 111, 112; J. B. C. Jackson and others, 'Historical overfishing and the recent collapse of coastal ecosystems' (2001) 293 Science 629; WWF, 'On the Brink – Mediterranean Bluefin Tuna – The Consequences of Collapse' http://assets.panda.org/downloads/onthebrinktunacollapse.pdf accessed 4 July 2015, 3–4). The heavily fished seas also become mostly populated by smaller fishes, jellyfish and microbes (K. M. Gjerde, 'High Seas Fisheries Governance: Prospects and Challenges in the 21st Century' in D. Vidas and P. J. Schei (eds), *The World Ocean in Globalisation: Climate Change, Sustainable Fisheries, Biodiversity, Shipping, Regional Issues* (Martinus Nijhoff Publishers 2011) 224; P. A. Larkin, 'Concepts and issues in marine ecosystem management' (1996) 6 Reviews of Fish Biology and Fisheries 139, 152).

[10] Nearly 46 % in 2008 as reported in FAO, 'The State of the World Fisheries and Aquaculture 2010' (n 4) 3, 18; 47 % in 2010 as reported in FAO, 'The State of the World Fisheries and Aquaculture 2012' (n 1) 26. Percentage reviewed down to 42.2 % for 2012 (FAO, 'The State of the World Fisheries and Aquaculture 2014' (n 1) 19).

The purpose of the present study is to consider possible ways to attain sustainable and legal fishing, thus avoiding further depletion of stocks or even extinction of species. The adequacy of using the Convention on International Trade in Endangered Species of Wild Fauna and Flora (CITES) for commercially-exploited fish species will be assessed and the cooperation between relevant institutions examined. In particular, the recent partnership between the Food and Agriculture Organization (FAO) and CITES will be analyzed, as will be the role of regional fisheries management organizations (RFMOs) in relation to a listing of fish species under CITES.

The organizations studied here, the FAO, RFMOs and CITES, adopt different approaches to the protection of marine species. A regime focusing on conservation is indeed proposed by CITES while rather exploitation-oriented regimes are based on the FAO for global principles and on RFMOs for the practical management of resources. RFMOs are usually the entities with primary responsibility for the management of stocks and they attempt—with variable results—to implement, at the regional level, the goals of optimum but sustainable utilization of fisheries promoted by the FAO.

In the current context where some fish species are depleted and international trade is a major factor in the market for fish products,[11] CITES, as a treaty regime regulating trade in endangered species, can be expected to play a beneficial role. Indeed, using the CITES structure to implement the FAO's main goal of sustainable fisheries as well as to provide an alternative or additional tool for fisheries matters when regional management proves inadequate is an attractive option. The same can be said of taking advantage of the RFMOs' and FAO's expertise to provide CITES with the best available scientific data and necessary technical information.

While a partnership between CITES and the FAO was undertaken already at the end of the 1990s, although informally at first, some concerns have been expressed about CITES' suitability regarding fisheries management. Highlighted as problematic were its lack of expertise in the marine field, its allegedly preservationist tendencies and the perceived threat it poses to RFMOs' mandates. The existence of a substantive overlap regarding the sustainable management of commercially-exploited and endangered fish species also appeared to be a major issue in the decision not to afford protection to certain fish species under CITES. However, such overlap does not necessarily mean that the institutions are or should be in confrontation. Cooperation has indeed the potential, under certain circumstances, to improve the situation and, more generally, is a practical way to face fragmentation of international law.

Two main concrete questions underlie this research: the first refers to whether CITES can and should be used for commercially-exploited fish species and the other examines whether the existing institutional cooperation is efficient. In addition to the practical contribution made by clarifying the value of using CITES and recommending ways to improve the system, this particular case-study provides an

[11] Sovacool (n 3) 98.

interesting lens to approach wider international law issues. Indeed, finding ways to achieve effective governance of transboundary or global natural resources is central for the peaceful use of oceans and land. Furthermore, the role of science in advising decision-makers is a sensitive issue which deserves scrutiny and is similar in many regimes. Finally, the complex problem of fragmentation of international law is acute in various fields of environmental law, as in all rapidly developing areas of international regulations. This has been recognized in 2012 in *The Future We Want*, when the heads of states and other representatives declared: "[w]e encourage parties to MEAs [i.e. multilateral environmental agreements] to consider further measures, in these and other clusters, as appropriate, to promote policy coherence at all relevant levels, improve efficiency, reduce unnecessary overlap and duplication, and enhance coordination and cooperation among MEAs".[12] Hence, looking at successes—and failures—in regimes' cooperation and analyzing the causes and risks that fragmentation gives rise to are aspects of the present research that are transferrable to most legal regimes in need of increased coordination and coherency.

In order to determine ways to improve the conservation and management of fish stocks, it is important to start by identifying the problems which stocks of commercially-exploited fish species are facing. The overview chapter introduces the issue of stock depletion through the most recent statistics available, before briefly describing the causes of such problem and the impact of aquaculture on the general picture (Chapter 2: Fishing Crisis and Aquaculture). In the following section, the treaties and institutions dealing with these issues are presented (Chapter 3: Global and Regional Legal Regimes Dealing with Commercially-exploited Marine Species). Fragmentation, lack of political will, inadequate measures, poor implementation and enforcement are a few of the numerous institutional and structural challenges which permit and/or cause overfishing as well as IUU fishing; these issues ought to be examined in parallel with the characteristics of what would represent or could facilitate appropriate governance of fisheries (Chapter 4: Structural and Governance Issues).

In the fifth section of this research, the legality, adequacy and expected efficacy of the protection of commercially-exploited fish species under CITES are discussed (Chapter 5: Protection of Commercially-exploited Fish Species Under CITES). The cooperation of CITES with other institutions in the listing of commercially-exploited fish species is the topic of the following chapter: the several situations in which partnerships exist and/or are needed are presented, their frameworks, strengths and weaknesses analyzed and compared with some of CITES' other partnerships (Chapter 6: CITES Cooperation with Other Institutions in Relation to Commercially-exploited Fish Species).

Finally, a concluding part addresses the desirability of using CITES for commercially-exploited fish species under the current partnerships. It also provides

[12] *The Future We Want* – endorsed in UNGA Resolution A/RES/66/288 27 July 2012 (Document adopted at Rio + 20) §89.

some recommendations as to the conditions under which CITES should be used for such species as well as to desirable changes both in the regime and in the collaborations (Chapters 7 and 8: Desirability of Using CITES and Recommendations).

The terminology 'commercially-exploited aquatic species' stems from the CITES-related activities of the FAO, but has not been formally defined. The gentlemen and ladies' agreement within CITES is to use the FAO understanding of the words, namely as a reference to all marine and freshwater fish and invertebrate species—mammals, birds, reptiles and plants are consequently excluded. This is wider than the ambit of the present research, which concentrates on marine species and tends to focus on fish (sharks included), with invertebrate used only as illustration of implementation aspects. The protection of marine mammals serves for comparison purposes. Hence the preferred wording will be 'commercially-exploited fish species' when referring to fish species only and 'commercially-exploited marine species' when referring to both marine fishes and mammals.

Part I
Fishing Crisis, Regulations and Structural Issues

Chapter 2
Fishing Crisis and Aquaculture

Many stocks of commercially-exploited fishes are overexploited with collapse as a possible consequence, especially in the case of particularly vulnerable species.[1] This is the overarching problem of today's fisheries. While the validity of the most alarming reports is questioned by some, the large majority of scientists agree that the current level of exploitation of many fisheries is not sustainable in the long-term[2] (A).

The problem of fish species' depletion is mostly caused by overfishing, IUU fishing and incidental by-catch.[3] However, other factors such as habitat alteration[4] or climatic variations have an important effect on the ability of stocks to replenish and thus on the level at which sustainable fishing can be set (B).

---

[1] On more vulnerable species, see W. W. L. Cheung and others, 'Intrinsic vulnerability in the global fish catch' (2007) 333 Marine Ecology Progress Series 1; J. A. Hutchings and J. D. Reynolds, 'Marine fish population collapses: consequences for recovery and extinction risk' (2004) 54 BioScience 297. Multiple examples of species extinction in marine species are provided in N. K. Dulvy, Y. Sadovy and J. D. Reynolds, 'Extinction vulnerability in marine populations' (2003) 4 Fish and Fisheries 25, 28–35.

[2] V. Christensen and others, 'Hundred-year decline of North Atlantic predatory fishes' (2003) 4 Fish and Fisheries 1; S. M. Garcia and C. Newton, 'Current situation, trends, and prospects in world capture fisheries' in E. K. Pikitch, D. D. Huppert and M. P. Sissenwine (eds), *Global Trends: Fisheries Management: Proceedings from the symposium held in Seattle, Washington, USA, June 14–16, 1994* (American Fisheries Society 1997); R. Hilborn and others, 'State of World Fisheries' (2003) 28 Annual Review of Environment and Resources 359; D. Pauly and others, 'Towards sustainability in world fisheries' (2002) 418 Nature 689; FAO, 'The State of the World Fisheries and Aquaculture' (Rome 2010) 8.

[3] These are also the main factors contributing to collapse of stocks (D. Freestone, 'Problems of High Seas Governance' in D. Vidas and P. J. Schei (eds), *The World Ocean in Globalisation: Climate Change, Sustainable Fisheries, Biodiversity, Shipping, Regional Issues* (Martinus Nijhoff Publishers 2011) 110; C. Mullon, P. Fréon and P. Curry 'The dynamics of collapse in world fisheries' (2006) 6 Fish and Fisheries 111, 119; Pauly and others (n 2)).

[4] N. K. Dulvy and others, 'Methods of assessing extinction risk in marine fishes' (2004) 4 Fish and Fisheries 255, 256.

S. Guggisberg, *The Use of CITES for Commercially-exploited Fish Species*,
Hamburg Studies on Maritime Affairs 35, DOI 10.1007/978-3-319-23702-2_2

Aquaculture and farm-ranching are presented by some of their supporters as the solutions to the fishing crisis. These practices have clearly some potential as ersatz to wild-capture, but they also bring problems of their own, particularly with regard to catching juveniles as spawning stock and to the use of wild-caught fish as food for predatory aquaculture stocks (C).

A. State of Wild Stocks

I. Decreasing Stocks

1. Fishes: Statistics

Without fishing pressure, stocks benefit from a biological equilibrium level, as the number of juveniles of one generation recruited in the mature stock and the biomass growth balance the losses both in number and biomass caused by natural mortality.

Fishing represents an additional withdrawal factor, but it is still possible to keep stable stocks in the long term if the levels harvested are below or at the maximum sustainable yield (MSY).[5] Indeed, after an sharp initial decrease, a stock that is exploited for the first time re-equilibrates at a new level, with fewer old individuals and more young ones which, all things being equal, have a faster growth rate.[6] A cycle of faster growth of stocks is hence put in place by removing the older fish which grow only marginally. In some general models, it seems accepted that a decline of 50 % of the pre-exploited stock quantities represents a sustainable utilization level, therefore a desirable target.[7]

Nowadays, the amount of fish caught is in many cases over that equilibrium point "where withdrawals through catch are exactly replaced by natural additions".[8] Statistical data such as the FAO State of World Fisheries and Aquaculture (SOFIA) reports non-sustainable fishing practices. According to the most recent edition, nearly 30 % of fish stocks are overexploited. Furthermore, 57 % of stocks are fully exploited and should consequently not be envisioned for expansion. This leaves only about 13 % of worldwide stocks which could produce more than their current level of exploitation, against 40 % in the mid-1970s.[9] In contrast, the values

[5] International Encyclopedia of the Social Sciences, 'Overfishing' (2008) http://www.encyclopedia.com/doc/1G2-3045301846.html accessed 4 July 2015.

[6] D. R. Rothwell and T. Stephens, *The International Law of the Sea* (Hart Publishing 2010) 295.

[7] K. L. Cochrane, 'A fishery manager's guidebook. Management measures and their application' FAO Fisheries Technical Paper 424 (Rome 2002) 9.

[8] B. K. Sovacool, 'A Game of Cat and Fish: How to Restore the Balance in Sustainable Fisheries Management' (2009) 40 Ocean Development and International Law 97, 106.

[9] FAO, 'The State of the World Fisheries and Aquaculture' (Rome 2012) 11–12; FAO, 'The State of the World Fisheries and Aquaculture' (Rome 2014) 7.

of overexploited stocks have steadily increased from 10 % in 1974 to 30 % in 2012, at the expense of stocks that are under- or moderately fished.

After the dramatic increase witnessed in marine capture fisheries' catches between 1950 and 1990—from 20 to 80 million tons—the overall quantity of marine fish and seafood caught remained constant[10] or slightly diminished according to other studies.[11] Usually, the main sign of overfishing is a decrease in catches for similar fishing effort, as less fish are available in the water. However, while the near stability observed here seems, at first glance, to suggest sustainability of the current levels of catches, it is most probably not the case.

Indeed, the fishing industry has increased its efforts and moved to new fishing grounds to catch similar quantities—no fishing ground is now left unexploited.[12] Also, after depleting one stock, fishing fleets have turned to others, a process often known as 'fishing down marine food webs',[13] as fishers shift to "less desirable species as the traditional stocks are exhausted."[14] The trophic chain has been consequently modified globally, with the levels of bigger and more vulnerable fishes steeply decreasing. A strong illustration thereof is that "the ocean has become increasingly dominated by small-bodied, fast-growing, less vulnerable species."[15] Furthermore, stable levels of catches can hide the danger of a sudden 'plateau-shaped' collapse, if depensatory mechanisms are at work:[16] "[o]bserving persistence at the level of catches can give the false impression that the fisheries have reached equilibrium and that effort has stabilized, when in fact the fishery is about to collapse."[17] Finally, it appears that harvested stocks are more prone to high

[10] FAO, 'The State of the World Fisheries and Aquaculture 2012' (n 9) 3; FAO, 'The State of the World Fisheries and Aquaculture 2014' (n 9) 4; World Ocean Review, *Living with the oceans* (Maribus, Future Ocean 2010) 120; World Ocean Review, *The Future of Fish – The Fisheries of the Future* (Maribus, Future Ocean 2013) 46.

[11] R. Watson and D. Pauly, 'Systematic distortion in world fisheries catch trends' (2001) 414 Nature 534, 536.

[12] R. Sumaila and W. W. L. Cheung, 'Vulnerability and Sustainability of Fish Stocks Worldwide: With Emphasis on Fish Stocks within the Commonwealth' in R. Bourne and M. Collins (eds), *From Hook to Plate: The State of Marine Fisheries: A Commonwealth Perspective* (2009) 206; World Ocean Review, *The Future of Fish – The Fisheries of the Future* 2013 (n 10) 46.

[13] Freestone, 'Problems of High Seas Governance' (n 3) 101; D. Pauly and others, 'Fishing down marine food webs' (1998) 279 Science 860; Mullon, Fréon and Curry (n 3) 119.

[14] A. Powers, 'Farming the Ocean' (2007–2008) 22 Natural Resources and Environment 45, 45; see also K. I. Stergiou and V. Christensen, 'Fishing down food webs' in V. Christensen and J. MacLean (eds), *Ecosystem Approaches to Fisheries: A Global Perspective* (Cambridge University Press 2011). Catch composition can thus be used as a source of information on the overall sustainability of fishing (Christensen and others (n 2) 8–9).

[15] Sumaila and Cheung (n 12) 204 (parenthesis omitted).

[16] "[D]epensatory mechanisms such as reduced probability of fertilization, impaired group dynamics, predator saturation and conditioning of the environment" (Mullon, Fréon and Curry (n 3) 118–119).

[17] Ibid. 119.

fluctuations due to the changed demographic distribution in the stock: the lower proportion of older and hence more efficient spawners indeed increases their vulnerability.[18]

A number of commercially-exploited fish species are thus facing a grim fate;[19] the International Union for the Conservation of Nature's (IUCN) list of animals threatened with extinction, the Red List, includes fishes such as the Southern Bluefin Tuna, Atlantic cod or Atlantic halibut.[20] Studies furthermore show that collapse of stocks and extinction of fish species are no myth and have happened several times in the recent past. According to one such piece of research, due to overexploitation, 366 fisheries have collapsed in the last 50 years (one in four fisheries).[21] For example, the Northwest Atlantic cod stocks crashed at the end of the twentieth century, leading to the fishery's closure in 1992 and will need at least until 2020 to recover the pre-1950 levels.[22]

2. Sharks: Statistics

Although data is partly missing, it appears well accepted that sharks are decreasing in numbers;[23] it is estimated that declines in sharks "may be of similar or greater magnitude than those of bony fishes."[24] In general, 20 % of the species of sharks assessed are threatened[25] but this number increases to 52 % for pelagic sharks.[26]

In certain regions decline is likely to be even more acute: in the Mediterranean for example, of the 71 shark species assessed by the regional program of the IUCN, 13 are critically endangered, eight endangered and nine vulnerable. This means that

[18] C. N. K. Anderson and others, 'Why fishing magnifies fluctuations in fish abundance' (2008) 452 Nature 835.

[19] A. Gillespie, *Conservation, biodiversity and international law* (Edward Elgar 2011) 72–73.

[20] Dulvy and others 2004 (n 4) 256.

[21] Mullon, Fréon and Curry (n 3) 111.

[22] Sovacool (n 8) 105 n 48. Some scientists fear that it will never recover (World Ocean Review, *Living with the oceans* 2010 (n 10) 124).

[23] H. Edwards, 'When predators become prey: the need for international shark conservation' (2006–2007) 12 Ocean and Coastal Law Journal 305, 310–311; Gillespie, *Conservation, biodiversity and international law* (n 19) 73; E. J. Techera and N. Klein, 'Fragmented governance: Reconciling legal strategies for shark conservation and management' (2011) 35 Marine Policy 73, 73; FAO, 'The State of the World Fisheries and Aquaculture 2014' (n 9) 121–122.

[24] R. Myers and B. Worm, 'Rapid worldwide depletion of predatory fish communities' (2003) 423 Nature 280, 282 (references omitted).

[25] R. D. Cavanagh and C. Gibson, 'Overview of the Conservation Status of Cartilaginous Fishes (Chondrichthyans) in the Mediterranean Sea' (IUCN 2007) 9.

[26] N. K. Dulvy and others, 'You can swim but you can't hide: the global status and conservation of oceanic pelagic sharks and rays' (2008) Aquatic Conservation: Marine and Freshwater Ecosystems. http://www.iotc.org/documents/you-can-swim-you-cant-hide-global-status-and-conservation-oceanic-pelagic-sharks-and-rays accessed 4 July 2015, 459, 466.

42 % are at the very least threatened.[27] As for the Gulf of Mexico, the oceanic whitetip declined of more than 99 % in the four decades before the 1990s.[28] In the Northwest Atlantic, in average, abundance in all species of sharks, except the mako shark—which on the other hand is critically endangered in the Mediterranean[29]—has been reduced by more than 50 %, within a timeframe of eight to 15 years.[30]

These declines are not the acceptable rate of decline related to the start of exploitation as such stocks were already under fishing pressure. Also, shark stocks do not react with much of the compensatory mechanism described above because of their life characteristics.[31] Hence, "[t]he magnitude of the declines estimated [. . .] suggests that several sharks may also now be at risk of large-scale extirpation."[32]

3. Particular Vulnerabilities

Some species, because of their physical characteristics, migrating habits and/or specific habitats are more vulnerable to the pressure exerted by fishing than others. Large-bodied, long-lived, late-maturing (such as deep-water fishes) as well as predatory species, which usually have a low fecundity rate,[33] are "intrinsically more vulnerable to exploitation and have less capacity to recover from exploitation."[34] Hence, the level of fishing which can be considered as sustainable is lower and the consequences of overfishing on such species can be quite dramatic.

For example, tunas' biomass has notably decreased in the last 50 years.[35] Similarly, the population of orange roughy—a very long-living and slow growing deep-sea species—has declined by 70 % at least after being commercially exploited for only 10 years[36] and stocks will be long to recover, if

---

[27] Cavanagh and Gibson (n 25) 9, and even higher percentages according to other studies (California Environmental Associates, 'Charting a Course to Sustainable Fisheries' (16 January 2012) http://www.chartingacourse.org/downloads/ accessed 4 July 2015, 27–28).

[28] F. Ferretti and others, 'Loss of Large Predatory Sharks from the Mediterranean Sea' (2008) 22 Conservation Biology 952, 953.

[29] Dulvy and others 2008 (n 26) 470.

[30] J. K. Baum and others, 'Collapse and Conservation of Shark Populations in the Northwest Atlantic' (2003) 299 Science 389, 389–390.

[31] Edwards (n 23) 311–312; on this, see infra Chapter 2 A. I. 3.

[32] Baum and others (n 30) 390.

[33] M. Clark, 'Are deepwater fisheries sustainable? The example of orange roughy (*Hoplostethus atlanticus*) in New Zealand' (2001) 51 Fisheries Research 123, 124; T. Morato and others, 'Fishing down the deep' (2006) 7 Fish and Fisheries 24, 25; R. Warner, *Protecting the Oceans Beyond National Jurisdiction* (Martinus Nijhoff Publishers 2009) 16; World Ocean Review, *The Future of Fish – The Fisheries of the Future* 2013 (n 10) 62–63.

[34] Dulvy and others 2004 (n 4) 259 (references omitted).

[35] Myers and Worm (n 24); Sovacool (n 8) 98.

[36] Sovacool (n 8) 105.

ever.[37] In the North Atlantic, the biomass of fishes that rank high in the trophic chain, the big predatory species, has decreased of two-thirds in the second half of the twentieth century.[38] Shark species also share these intrinsic vulnerabilities, since many of them mature late, have low reproduction rates and long lifespans.[39]

Additionally, in general, straddling and highly migratory species, being caught on the high seas, face protection gaps.[40] This can be a factor to explain that "more than half of the stocks for which information is available [is...] listed as overexploited or depleted."[41]

II. Difficulty in Stocks Assessments

Criticisms arose with regard to statistics estimating the level of stocks, both on the method used by the FAO for its general overview[42] and on the particular results reached in species-specific studies. It is certainly important, but equally difficult, to inquire into the meaning of the numbers publicized and into the assumptions on which the estimates are based.

Appraisals of stocks' health in SOFIA proceed from national catches' reports made available to the organization by member states. These evaluations usually "assume that the level of captures bears some direct relationship to the relative size of the stock biomass remaining in the sea".[43] The rationale behind this assumption is that with equal or increasing fishing effort, similar catches should be reported year after year if the level of catches is sustainable; "reduction in catch [...thus] indicates that the majority of the world's fisheries resources have already reached their maximum sustainable levels."[44]

While this relationship is widely accepted[45] and the usefulness of trends in catch is rarely questioned, some authors highlight that it should not be the only figure

[37] Clark (n 33) 124–126, 133–134; D. Freestone, 'Protecting Our Oceans: New Challenges, New Solutions An Overview of This Issue' (2006–2007) 7 Sustainable Development Law and Policy Review 2, 3; M. Lack, K. Short and A. Willock, 'Managing risk and uncertainty in deep-sea fisheries: lessons from Orange Roughy' (2003) http://www.mar-eco.no/learning-zone/__data/page/89/WWF-orange_roughy.pdf accessed 4 July 2015, 1, 6, 32–33, 42; B. Weeber, K. Thomas and C. Dorey, 'New Zealand orange roughy (*Hoplostethus atlanticus*)' (September 2010) http://www.greenpeace.org/new-zealand/Global/new-zealand/P3/publications/oceans/2010/Orange%20Roughy.pdf accessed 4 July 2015.

[38] Christensen and others (n 2) 17; Sumaila and Cheung (n 12) 206.

[39] Cavanagh and Gibson (n 25) 2; Ferretti and others (n 28) 953.

[40] On this, see infra Chapter 4 A. II.

[41] FAO, 'The State of the World Fisheries and Aquaculture' (Rome 2008) 35.

[42] World Ocean Review, *The Future of Fish – The Fisheries of the Future* 2013 (n 10) 44–46.

[43] A. Longhurst, *Mismanagement of Marine Fisheries* (Cambridge University Press 2010) 177.

[44] Sumaila and Cheung (n 12) 197 (parenthesis omitted).

[45] Christensen and others (n 2) 17.

used to indicate sustainability levels.[46] Other methods, such as ecosystem modeling, biomass modification as well as changes in the average depths of fishing in relation to global landings, have been utilized and also conclude that some fish species are overfished and others already depleted.[47] Also, data distortions have been pointed out, but not only towards underreporting, a tendency which can logically be expected from the fishing industry. Over-reporting mistakes in the Chinese reports have been found and were not insignificant. Indeed, once corrected, trends show decreases in the global catches instead of a slight growth.[48] Hence, criticisms which assert that fish stocks are in a better state than announced by the official statistics—and consequently that conservation warnings are groundless—are not the only ones to be noted. Quite to the contrary, overly optimistic estimates sometimes (seem to) back up an unfounded impression of sustainability and are questioned by the scientific community.[49]

A particularly acute debate arose about the level of decrease in tunas' biomass. While Myers and Worm published a paper stating that it had possibly been reduced, in 50 years, to 90 % of their pre-industrial equivalents, the majority of scientists criticized the methodology employed.[50] It is quite symptomatic of the complexity of making estimates in the fisheries field. On the one hand, the main study uses the catch-per-unit-effort of the Japanese longline-fishery to reach the worrying conclusion that 15 years of exploitation reduced the biomass of large pelagic fish by 80 %, and that tuna is currently at 10 % of pre-industrial times.[51] On the other hand, these results have been criticized due to the method and choice of data, in particular since the relevant fishing behavior has changed over time.[52] An alternative study consequently affirms that "biomass decline and fishing impacts are much less severe".[53] It also seems that while the 90 % decline was a miscalculation, independent

[46] Mullon, Fréon and Curry (n 3) 119.

[47] For example, Christensen and others ((n 2) 3) used a methodology "rel[ying] on a combination of ecosystem modelling, information from hydrographic databases, statistical analysis and GIS modelling." Myers and Worm ((n 24) 280) have mainly based their results on catch and biomass modifications. Morato and others ((n 33) 25) "analysed changes in the mean depth of fishing to test if the predicted expansion into deeper-waters can be detected in global landings datasets."

[48] Watson and Pauly (n 11) 535–536.

[49] Mullon, Fréon and Curry (n 3) 119.

[50] Pelagic Fisheries Research Program of the Joint Institute for Marine and Atmospheric Research, 'Changes in the Biomass of Large Pelagic Predators' http://www.soest.hawaii.edu/PFRP/large_pelagics/large_pelagic_predators.html accessed 4 July 2015; Longhurst (n 43) 158.

[51] Myers and Worm (n 24).

[52] R. Hilborn, 'Faith-based Fisheries' (2006) 31 Fisheries Research 554, 554; T. Polacheck, 'Tuna longline catch rates in the Indian Ocean: did industrial fishing result in a 90 % rapid decline in the abundance of large predatory fish?' (2006) 30 Marine Policy 470.

[53] J. Hampton and others, 'Decline of Pacific tuna populations exaggerated?' (2005) 434 Nature, E1.

research following this controversy acknowledged a decline of 65–80 %,[54] an already very high proportion.

A strong advocate of subtle and less alarmists conclusions points out that focusing on stocks' abundance instead of sustainable yield can be misleading as some species' MSY is at 20–40 % of their unfished biomass.[55] However, he also admits that lower fishing pressure is desirable[56] and that "a number [of the world's major fisheries] ha[s] been severely overfished, and many more stocks appear to be heading towards depletion."[57]

When it comes to sharks, data is in general deficient[58] and the probably high occurrence of by-catch and discards further makes it difficult to provide reliable estimates. For some regions, such as the Indo-Pacific Ocean, the data is totally missing.[59] Also, the tendency is to report global shark catches, without differentiating between species, which complicates assessments of stocks.[60] Some studies based on the Hong Kong shark fin market put the actual level of sharks caught three to four times higher than that recorded by the FAO.[61]

In conclusion, it seems hard for specialists in fish biology and statistics to agree on stocks assessments, let alone for non-specialists to determine where truth lies. Nevertheless, it is the exact extent and not the issue of decrease in stocks and biomass that is questioned. For the start of this study,[62] it is possible to affirm that many fisheries are overexploited leading to declining catches and biomass, with very serious consequences to be expected—and at times already witnessed—for vulnerable species.

---

[54] J. Jackson and J. Jacquet, 'The shifting baselines syndrome: perception, deception, and the future of our oceans' in V. Christensen and J. MacLean (eds), *Ecosystem Approaches to Fisheries: A Global Perspective* (Cambridge University Press 2011) 133.

[55] R. Hilborn, 'Reinterpreting the State of Fisheries and their Management' (2007) 10 Ecosystems 1362, 1365.

[56] Ibid. 1367.

[57] Hilborn and others (n 2) 359.

[58] Baum and others (n 30) 389; Ferretti and others (n 28) 953.

[59] Dulvy and others 2008 (n 26) 472.

[60] Cavanagh and Gibson (n 25) 3; M. Lack and G. Sant, 'The Future of Sharks: A Review of Action and Inaction' (TRAFFIC January 2011) 19.

[61] S. C. Clarke and others, 'Identification of Shark Species Composition and Proportion in the Hong Kong Shark Fin Market Based on Molecular Genetics and Trade Records' (2006) 20 Conservation Biology 201; S. C. Clarke and others, 'Global estimates of shark catches using trade records from commercial markets' (2006) 9 Ecology Letters 1115, both publications as interpreted by Dulvy and others 2008 (n 26) 461.

[62] The question of threat of extinction will be further discussed infra in Chapter 5 C. I. 1. b) aa), with the difference made between extinction and commercial extinction. As to the problem of scientific uncertainty and/or lack of data, it is examined infra in Chapter 4 A. III. 1.

B. Causes of Depletion

I. Increased Mortality due to Fishing Activities

1. Overfishing

While it is clear that many factors influence the current crisis, stock depletion is nevertheless most importantly caused by unsustainable fishing practices.[63] By overfishing one understands fishing over the MSY, in other words over the long-term sustainability levels. In lay terms, it can be defined as catching fish in such large quantities that the remaining adult population is not able to reproduce in sufficient abundance in order to replenish the stock.[64] Such excessive fishing is partly due to the important overcapacity of fishing fleets but also to ineffective management[65] and to non-compliance of both vessels and states with the relevant regulations.[66]

An important aspect to keep in mind is that overfishing can actually be legal, in particular when quotas are set too high compared to scientific recommendations or when scientific committees lack reliable data to convincingly advise decision makers. As legality does not mean efficiency, such practices are unsustainable and possibly lead to depletion of stocks.[67]

In the International Commission for the Conservation of Atlantic Tunas (ICCAT), total allowable catches (TACs) for Bluefin Tuna were for many years set well above the scientific recommendations. For example in 1999, the agreed TACs were at 32,000 tons whereas the ICCAT scientific committee, the Standing Committee on Research and Statistics (SCRS), had previously recommended 25,000 tons.[68] In 2010 finally, this RFMO dropped in line with the scientific advice, with TACs set at 12,900 tons. Hence, until recently, it was legal under this resource management regime to fish over the level of catch recommended by its own scientific committee.

This is particularly serious for two reasons. First, these recommendations are usually already high compared to the estimates from conservation bodies: when advising the 1999 TACs of 25,000, the SCRS itself noted that such a level would "halt the decline in spawning biomass" but not permit to restore historical levels.[69] Also, there is another percentage of illegal and unreported catch on top of this legal

---

[63] Gillespie, *Conservation, biodiversity and international law* (n 19) 194.

[64] J. K. Murphy, 'FAO's general fisheries commission for the Mediterranean: a new role, a new role model' (2007) 12 Drake Journal of Agricultural Law 391, 394–395.

[65] Sumaila and Cheung (n 12) 195.

[66] Several of these governance issues are addressed infra in Chapter 4 A.

[67] World Ocean Review, *Living with the oceans* 2010 (n 10) 121.

[68] WWF, 'ICCAT's four-decade record of failure: Media backgrounder' (2010) http://awsassets.panda.org/downloads/tunatimelinenov2010.pdf accessed 4 July 2015, 2.

[69] Ibid.

overfishing:[70] within ICCAT, in 2004–2005, IUU fishing was estimated to represent more than 40 % of over-quota fishing, this number being confirmed by the SCRS itself.[71]

Sharks are caught by traditional fisheries, for meat and increasingly for their fins.[72] In the last three decades, directed shark exploitation has indeed augmented,[73] for example from 300,000 pounds of shark being landed in the United States of America (USA) in 1979 to 16 million pounds ten years later.[74] Only a few coastal states actually regulate shark fishing in their waters.[75]

2. Illegal, Unreported and Unregulated Fishing

IUU fishing is defined as follows in the FAO International Plan of Action on IUU Fishing:

1. **Illegal fishing** refers to activities:
1.1 conducted by national or foreign vessels in waters under the jurisdiction of a State, without the permission of that State, or in contravention of its laws and regulations;
1.2 conducted by vessels flying the flag of States that are parties to a relevant regional fisheries management organization but operate in contravention of the conservation and management measures adopted by that organization and by which the States are bound, or relevant provisions of the applicable international law; or
1.3 in violation of national laws or international obligations, including those undertaken by cooperating States to a relevant regional fisheries management organization.
2. **Unreported fishing** refers to fishing activities:
2.1 which have not been reported, or have been misreported, to the relevant national authority, in contravention of national laws and regulations; or

---

[70] Gillespie, *Conservation, biodiversity and international law* (n 19) 201.

[71] WWF, 'On the Brink – Mediterranean Bluefin Tuna – The Consequences of Collapse' http://assets.panda.org/downloads/onthebrinktunacollapse.pdf accessed 4 July 2015, 2; as for 2005, the quotas were set at 29,500 tons, while the SCRS recommended 15,000 to avoid the stock's collapse. So, the TACs were already set 14,500 tons above the scientific advice. Then, Turkey and Libya decided to object to their allocated quotas and set their own limits. Finally, instead of 15,000, it is estimated that 50,000 tons were caught.

[72] Edwards (n 23) 313–318; S. Fordham and C. Dolan, 'A Case Study in International Shark Conservation: The Convention on International Trade in Endangered Species and the Spiny Dogfish' (2004) 34 Golden Gate University Law Review 531, 553; Techera and Klein (n 23) 73; M. A. Young, *Trading Fish, Saving Fish: The Interaction between Regimes and International Law* (Cambridge Studies in International and Comparative Law, Cambridge University Press 2011) 135–136.

[73] Baum and others (n 30) 389; J. D. Stevens and others, 'The effects of fishing on sharks, rays, and chimaeras (chondrichthyans), and the implications for marine ecosystems' (2000) 57 ICES Journal of Marine Science 476, 477.

[74] SeaWeb, 'Shark-Finning' http://www.seaweb.org/resources/briefings/shark.php accessed 4 July 2015.

[75] Stevens and others (n 73) 478; C. A. Ward-Paige and others, 'Recovery potential and conservation option for elasmobranchs' (2012) 80 Journal of Fish Biology 1844, 1855–1856.

2.2 undertaken in the area of competence of a relevant regional fisheries management organization which have not been reported or have been misreported, in contravention of the reporting procedures of that organization.
3. **Unregulated fishing** refers to fishing activities:
3.1 in the area of application of a relevant regional fisheries management organization that are conducted by vessels without nationality, or by those flying the flag of a State not party to that organization, or by a fishing entity, in a manner that is not consistent with or contravenes the conservation and management measures of that organization; or
3.2 in areas or for fish stocks in relation to which there are no applicable conservation or management measures and where such fishing activities are conducted in a manner inconsistent with State responsibilities for the conservation of living marine resources under international law.[76]

IUU practices are a major problem in the management of fisheries,[77] an important financial loss[78] and an aggravating factor in the fishing crisis. Indeed, while they do not automatically encompass unsustainable fishing, they often contribute to it. It has been said that IUU fishing puts sustainability at risk, creates an illegitimate competition towards legal fishermen and tarnishes RFMOs' reputation.[79] IUU fishing undeniably increases catches—landed or discarded—which, not being reported, are not taken into account by the authorities in the count of TACs. Being at best an estimated figure in reports, it also adds to the uncertainty related to fisheries data.[80] As such, it represents an unpredictable extra-pressure on stocks, which, according to some assessments, amounts to a third of all landings.[81]

[76] FAO, International Plan of Action to prevent, deter and eliminate illegal, unreported and unregulated fishing (IPOA-IUU) 2001 §3.1–3.3 (bold added and numbering simplified).

[77] For an overview of the recognition of IUU fishing as a major issue, see for example K. M. Gjerde, 'High-seas Fisheries Management under the Convention on the Law of the Sea' in D. Freestone, R. Barnes and D. Ong (eds), *The Law of the Sea – Progress and Prospects* (Oxford University Press 2006) 289–290; C.-C. Schmidt, 'Addressing Illegal, Unreported and Unregulated (IUU) Fishing: Paper prepared for the International Fisheries Compliance 2004 Conference, 29–30 September 2004, Brussels' (2004) http://www.oecd.org/greengrowth/fisheries/34029751.pdf accessed 4 July 2015. On the global measures taken to fight IUU practices, see J. Swan, 'International Action and Responses by Regional Fishery Bodies or Arrangements to Prevent, Deter and Eliminate Illegal, Unreported and Unregulated Fishing' FAO Fisheries and Aquaculture Circular 996 (Rome 2004) 1–11.

[78] D. J. Agnew and others, 'Estimating the Worldwide Extent of Illegal Fishing' (2009) http://www.plosone.org/article/info:doi/10.1371/journal.pone.0004570 accessed 4 July 2015, 1–2; The World Bank and FAO, *The Sunken Billions – The Economic Justification for Fisheries Reform* (2009).

[79] K. M. Gjerde, 'High Seas Fisheries Governance: Prospects and Challenges in the 21st Century' in D. Vidas and P. J. Schei (eds), *The World Ocean in Globalisation: Climate Change, Sustainable Fisheries, Biodiversity, Shipping, Regional Issues* (Martinus Nijhoff Publishers 2011) 228.

[80] J. Beer-Gabel and V. Lestang, *Les Commissions de Pêche et leur Droit: La Conservation et la Gestion des Ressources Marines Vivantes* (Bruylant 2003) 135–136.

[81] World Ocean Review, *Living with the oceans* 2010 (n 10) 129; World Ocean Review, *The Future of Fish – The Fisheries of the Future* 2013 (n 10) 70.

Problems of IUU fishing are well known for expensive species such as toothfish or tuna.[82]

Regarding unregulated fishing, most pelagic sharks are not managed under any regional or international plan with catch limits in place.[83] Indeed, except for a few sporadic measures, such fisheries are not regulated, or at least insufficiently so, which of course increases the risk of depletion for these species.[84]

3. By-Catch and Discarding

It is nearly impossible to catch only the targeted species when fishing; this activity hence involves a level of unwanted catches[85] known as by-catch, which are frequently discarded. The extent of the problem caused by the disposal of these unwanted catches (unwanted because they form part of protected, untargeted or low value species),[86] thrown back dead into the water, is not exactly known.

It is nevertheless estimated to have serious consequences on the actual mortality rates. Discards should be calculated in the catches,[87] but it is likely that these fish (or other marine animals such as mammals or birds) will be forgotten in the log books. The FAO estimated in 2010 the amount of discards in marine fisheries to seven million tons,[88] much lower than the 20 million tons at the end of the twentieth century, but still a significant and problematic quantity.[89] Indeed, if a legal quota is already set at the estimated MSY—when not above—, the actual catches including IUU fishing and discards will bring the withdrawal caused by harvesting well over the sustainability level.[90]

Although direct fishing is also a threat, many shark species are particularly vulnerable to high levels of by-catch and discarding, which remain largely

[82] K. W. Riddle, 'Illegal, Unreported, and Unregulated Fishing: Is International Cooperation Contagious?' (2006) 37 Ocean Development and International Law 265, 267–269; Schmidt (n 77) 4; Swan, 'International Action and Responses by Regional Fishery Bodies or Arrangements to Prevent, Deter and Eliminate Illegal, Unreported and Unregulated Fishing' (n 77) 29–30; N. Yagi, 'High Seas Fisheries Today: Challenges and Remedies Under the Global Economy' in D. Vidas and P. J. Schei (eds), *The World Ocean in Globalisation: Climate Change, Sustainable Fisheries, Biodiversity, Shipping, Regional Issues* (Martinus Nijhoff Publishers 2011) 212–213.

[83] Dulvy and others 2008 (n 26) 473–474; on some recent RFMOs' measures on sharks, see infra Chapter 3 B. II. 2. b).

[84] Dulvy and others 2008 (n 26) 460.

[85] W. T. Burke, *The New International Law of Fisheries* (Clarendon Press 1994) 27; generally on by-catch, see Gillespie, *Conservation, biodiversity and international law* (n 19) 306–326.

[86] Hilborn and others (n 2) 370–371.

[87] D. Zeller and D. Pauly, 'Good news, bad news: Global fisheries discards are declining, but so are total catches' (2005) 6 Fish and Fisheries 156, 157.

[88] FAO, 'The State of the World Fisheries and Aquaculture 2010' (n 2) 12.

[89] Zeller and Pauly (n 87) 157.

[90] O. S. Stokke, 'Trade Measures and the Combat of IUU Fishing: Institutional Interplay and Effective Governance in the Northeast Atlantic' (2009) 33 Marine Policy 339, 341–342.

unregulated.[91] In relation to the former, in the Mediterranean for example, sharks are often caught in the longline fisheries which are directed at swordfish or tuna.[92] All over the world, this is one of the main threats for sharks.[93] Another current practice—increasingly prohibited—is the discarding of sharks caught only for their fins; in that case the shark, after having been finned, is thrown back into water, where it dies.[94] The value of fins being higher than that of meat, the incentive is strong to discard the shark's body in order to keep space available on board for more of the high-value body parts.[95]

II. Habitat Destruction, Pollution and Climate Change

While fishing is estimated to have caused 55 % of the global, local and regional extinctions, 37 % are attributed to habitat destruction and the remaining 8 % to climate variations, pollution, disease and invasive species.[96] Even when they do not directly cause a collapse in stocks, these factors influence the oceans' resilience and hence species' ability to replenish stocks.[97]

Habitat destruction is indeed an additional factor which increases the vulnerability of some species. For example, sharks in the Mediterranean have seen their spawning areas and nurseries shrink with coastal communities continuing to spread and modifying the ecosystem.[98] Bottom trawling, the fishing practice where a net is dragged on the sea-floor, is also an extreme example of habitat destruction[99] and its use has increased in the second half of the twentieth century.[100]

---

[91] Australia, 'CoP12 Doc. 41.1 – Conservation and management of sharks' (3–15 November 2002) 1–2 §4; Edwards (n 23) 315–316, 318–319.

[92] Cavanagh and Gibson (n 25) 3.

[93] Ibid. 10.

[94] S. van Osch, 'Save our Sharks: Using International Fisheries Law Within Regional Fisheries Management Organizations to Improve Shark Conservation' (2011–2012) 33 Michigan Journal of International Law 383, 390–391; Ward-Paige and others (n 75) 1856.

[95] Dulvy and others 2008 (n 26) 461.

[96] Dulvy and others 2004 (n 4) 258; Hilborn and others (n 2) 363–367; A. Rieser, 'International Fisheries Law, Overfishing and Marine Biodiversity' (1996–1997) 9 Georgetown International Environmental Law Review 251, 253.

[97] Gjerde, 'High Seas Fisheries Governance: Prospects and Challenges in the 21st Century' (n 79) 225–226.

[98] Cavanagh and Gibson (n 25) 3.

[99] It damages vulnerable and important ecosystems (D. S. Calley, *Market Denial and International Fisheries Regulation: The Targeted and Effective Use of Trade Measures Against the Flag of Convenience Fishing Industry* (Martinus Nijhoff Publishers 2012) 63–71; Gillespie, *Conservation, biodiversity and international law* (n 19) 307–308; Murphy, 'FAO's general fisheries commission for the Mediterranean: a new role, a new role model' (n 64) 395; Rothwell and Stephens, *The International Law of the Sea* (n 6) 313).

[100] Sumaila and Cheung (n 12) 207.

Various ecological features influence the habitat's and ecosystem's health.[101] Marine currents and other external factors have a major and cyclical influence on stocks' growth.[102] Also, the temperature, salinity, oxygen concentration or water quality are central and can change, for example due to pollution. Additionally, qualitatively unsustainable practices—such as trawling for fish or nodules mining—impact habitats and thus affect marine species' regeneration abilities. Furthermore, following a species' extinction, the interactions amongst the remaining species and consequently the whole ecosystem's balance can be disturbed.

Moreover, climate change, with increasing levels of carbon dioxide and global warming, will most probably have an effect on fish stocks' behavior and response to fishing. However, it is not easy to determine how this effect will be felt, as climate change will influence habitat, as well as water temperature and pH level. Multiple interrelated factors having both direct and indirect impacts on fish (through the habitat and the production of plankton for example) make it very difficult to predict how much additional stress climate change will generate for marine species.[103]

In sum, it should be kept in mind that fishing is not the only element influencing the balance of fish stocks. Be it from natural or human induced causes, various factors transform the environment where fish live and add to the complexity of projecting how species will cope with fishing pressure. So, levels of uncertainty should be taken into account "when assessing the current state of sea fish populations, and when predicting their ability to maintain present landings."[104]

However, the questions related to these other factors will not be directly discussed here, because of their scientific nature and because the present study focuses on the regulation of international trade of fish, in order to decrease fishing pressure on endangered stocks. Also, while climate variations, pollution, diseases and invasive species have influenced stocks for centuries, these aspects are only of relative interest as the current crisis appears to stem mainly from the recent increase in fishing pressure and as this is the activity which can be most easily regulated.[105]

---

[101] World Ocean Review, *The Future of Fish – The Fisheries of the Future* 2013 (n 10) 17, 19.

[102] R. T. Carson and others, 'Fisheries Management Under Cyclical Population Dynamics' (2009) 42 Environmental and Resource Economics 379, 380; Longhurst (n 43) 97–107.

[103] World Ocean Review, *Living with the oceans* 2010 (n 10) 102, 117.

[104] Longhurst (n 43) 161.

[105] Hilborn and others (n 2) 368.

C. Aquaculture

Aquaculture has been defined as "[t]he propagation and rearing of aquatic organisms in controlled or selected environments".[106] The main difference with wild-capture is the input from human beings, who produce and grow aquatic animals instead of catching them.[107]

I. An Increasing Practice: Statistics

Aquaculture represents 42 % of total food fish supply[108] and "[i]t is set to overtake capture fisheries as a source of food fish."[109] Although inland aquaculture is more important in volume than marine aquaculture, with approximately 42 million tons against 25 million tons, marine aquaculture nevertheless already corresponds to nearly a quarter of the total marine fish production—that is 25 million tons of the 105 million tons harvested and produced.[110]

It is also the industry with the most important rate of increase in production, not only in aquatic food supply, but in all the animal food producing sectors;[111] it has grown from 32.4 million tons in 2000 to 66.6 million tons in 2008 and is estimated to attain 79 million tons in 2021.[112] It can be expected, that the increasing demand for fish will be met in the future—as has already been the case in the last decade—mainly by aquaculture production, especially as the levels of capture fisheries are either stagnating or decreasing.[113]

---

[106] U.S. National Aquaculture Act of 1980, Public Law 96–362, Stat. 1198, 16 U.S.C. 2801, as found in J. L. Anderson, 'Sustainable Aquaculture: What Does It Mean and How Do We Get There?' in P. Leung, C.-S. Lee and P. J. O'Bryen (eds), *Species and System Selection for Sustainable Aquaculture* (Blackwell Publishing 2007) 11 (italics omitted).

[107] M. Ariji, Y. Senda and A. Takahara, 'Achievements and Problems of Aquaculture in Japan' in P. Leung, C.-S. Lee and P. J. O'Bryen (eds), *Species and System Selection for Sustainable Aquaculture* (Blackwell Publishing 2007) 148.

[108] FAO, 'The State of the World Fisheries and Aquaculture 2014' (n 9) 19.

[109] FAO, 'The State of the World Fisheries and Aquaculture 2010' (n 2) 6.

[110] FAO, 'The State of the World Fisheries and Aquaculture 2014' (n 9) 4.

[111] FAO, 'The State of the World Fisheries and Aquaculture 2010' (n 2) 5–6; FAO, 'The State of the World Fisheries and Aquaculture 2012' (n 9) 188.

[112] FAO, 'The State of the World Fisheries and Aquaculture 2010' (n 2) 154; FAO, 'The State of the World Fisheries and Aquaculture 2012' (n 9) 188; FAO, 'The State of the World Fisheries and Aquaculture 2014' (n 9) 9–10.

[113] FAO, 'The State of the World Fisheries and Aquaculture 2010' (n 2) 68; FAO, 'The State of the World Fisheries and Aquaculture 2012' (n 9) 188; FAO, 'The State of the World Fisheries and Aquaculture 2014' (n 9) 24, 107; World Ocean Review, *The Future of Fish – The Fisheries of the Future* 2013 (n 10) 80.

While aquaculture is so far limited to coastal areas, which are mostly under national jurisdiction, such practices are increasingly likely to be developed on the high seas.[114]

II. Main Issues

Aquaculture is seen by some as the one solution to the fishing crisis; its proponents argue that it "is the only way to meet the food demands of a relentlessly growing world population".[115] While it surely has strong advantages, such as more regularity in quantity and quality of production,[116] as well as hopefully the ability to reduce pressure on wild stocks, it is nevertheless not the panacea and that for several reasons.

1. Consequences for the Environment

First, aquaculture brings some environmental problems of its own, whether through the release of feeding and medication products in the wild or the destruction of coastal ecosystems.[117] Also, disease transmissions, competition for food or genetic contamination are some of the harmful effects felt by wild species and caused by specimens' escapes from the farms or by eggs releases.[118]

Indeed, escape of some non-native species of farmed fish can have negative consequences,[119] arguably similar to those of introducing alien and invasive

[114] Powers (n 14) 45–46; J.-F. de Pulvenis Séligny, 'The marine living resources and the evolving law of the sea' (2010) 1 Aegean Review of the Law of the Sea and Maritime Law 61, 66.

[115] P. Greenberg, 'Green to the Gills' *The New York Times* (18 June 2006) http://www.nytimes.com/2006/06/18/magazine/18fish.html accessed 4 July 2015; R. S. V. Pullin, 'Aquaculture up and down the food web' in V. Christensen and J. MacLean (eds), *Ecosystem Approaches to Fisheries: A Global Perspective* (Cambridge University Press 2011) 94.

[116] OECD, 'Globalisation in Fisheries and Aquaculture: Opportunities and Challenges' (Paris 2010) 44–45; R. Hannesson, 'Aquaculture and Fisheries' (2003) 27 Marine Policy 169, 169. Even in the Code of Conduct for Responsible Fisheries (FAO 1995) aquaculture is referred to as to be developed, while capture fisheries is to be managed (respectively articles 7 and 9).

[117] F. Asche and F. Khatun, 'Aquaculture: Issues and Opportunities for Sustainable Production and Trade' ICTSD Natural Resources, International Trade and Sustainable Development Series Issue Paper 5 (Geneva 2006) 29–30; E. R. Englebrecht, 'Can Aquaculture continue to circumvent the regulatory net of the Magnuson-Stevens Fishery Conservation and Management Act?' (2002) 51 Emory Law Journal 1187, 1188, 1193–1194; Hannesson (n 116) 169; F. Ottolenghi and others, 'Capture-based aquaculture: The fattening of eels, groupers, tunas and yellowtails' (FAO 2004) 200–207; R. Vita and A. Marin, 'Environmental impact of capture-based bluefin tuna aquaculture on benthic communities in the western Mediterranean' (2007) 38 Aquaculture Research 331; FAO, 'Aquaculture development – 6. Use of wild fishery resources for capture-based aquaculture' FAO Technical Guidelines for Responsible Fisheries 5 Suppl. 6 (Rome 2011) 23–24.

[118] Powers (n 14) 46.

[119] Greenberg (n 115); World Ocean Review, *The Future of Fish – The Fisheries of the Future* 2013 (n 10) 84–86, 89.

species.[120] Also, farmed fish such as salmon is known to have developed diseases that can be transmitted to wild fish and are dangerous for them, especially when the wild stocks are not extremely healthy themselves.[121] Both types of fish moreover feed on similar forage species, which means that wild fish will compete with extra-voracious farmed-fish for the same prey. Furthermore, mix-breeding can cause changes in wild species' genetic composition, which has a potentially detrimental effect on them.

2. Bred-in-Captivity v. Catches of Juveniles

Moreover, a difference must be made between aquaculture of bred-in-captivity specimens—where the new generation is obtained by eggs and fingerlings from fish spawning in a hatchery—and farm-ranching of wild-captured juveniles further fattened in human-supervised conditions—where the new generation is constituted of wild-caught juveniles.[122]

While the former type has a strong potential to decrease the pressure on wild-capture of similar species, as it 'creates' both specimens and biomass, the second is not so clearly efficient in that regard. Indeed, the juveniles caught as stocking material, for example tuna or eel, will not breed hence not contribute to the further generation of wild stocks.[123] Depending on the circumstances, farm-ranching can even be detrimental to the state of a species, for example when the development of such practices increases the demand on wild juveniles.[124] In the case of Bluefin Tuna in the Mediterranean, it has been said that "[t]he introduction of bluefin tuna farming activities in the Mediterranean in 1997 exacerbated the problems with management of the fisheries."[125]

The existence of farms using wild-caught juveniles is based on the fact that some fishes have not yet reproduced outside their natural wild surroundings or that commercial hatchings are still at trial stages. Unlike salmon, most marine species

[120] R. L. Naylor and others, 'Effect of aquaculture on world fish supplies' (2000) 405 Nature 1021.

[121] Englebrecht (n 117) 1194–1197; Y. Liu, J. O. Olaussen and A. Skonhoft, 'Wild and farmed salmon in Norway – A review' (2011) 35 Marine Policy 413, 415–416.

[122] Ariji, Senda and Takahara (n 107) 148; Ottolenghi and others (n 117) 3–4. The nomenclature has only recently started to be clarified and most studies still refer to aquaculture as including both aquaculture and farm-ranching. However, for the present contribution, it is important to differentiate what Ottolenghi named culture-based fisheries from capture-based aquaculture, or, in CITES terms, bred-in-captivity, from fish caught in the wild and fattened in captivity.

[123] OECD, 'Globalisation in Fisheries and Aquaculture: Opportunities and Challenges' (n 116) 53.

[124] Ariji, Senda and Takahara (n 107) 152; Murphy, 'FAO's general fisheries commission for the Mediterranean: a new role, a new role model' (n 64) 399; Ottolenghi and others (n 117) 6–7, 192–194; Yagi (n 82) 217; FAO, 'Aquaculture development – 6. Use of wild fishery resources for capture-based aquaculture' (n 117) 1.

[125] Monaco, 'CoP15 Prop. 19 – Proposal to include Atlantic Bluefin Tuna (*Thunnus thynnus* (Linnaeus, 1758)) on Appendix I of CITES in accordance with Article II 1 of the Convention' (13–25 March 2010) 17 §7.2.

indeed hatch small and fragile eggs that are complicated to nourish and hence that can be brought to maturity in an artificial environment only with difficulty.[126] Research is nevertheless on its way to change the situation: cod is now successfully hatched in captivity[127] and tuna, for example, has been spawned in captivity for the first time recently.[128] Hence, for big predatory fish, the breakthrough might be at the doorstep, even though fish farmers have not yet been able to hatch specimens on a commercial basis.[129]

3. Fishmeal and Fish Oil Feed

Finally, another issue must be highlighted for both aquaculture and farm-ranching: many species have to be fed with fishmeal and fish oil, produced from wild-captured specimens. With the increase in aquaculture production the demand for forage fish has logically also risen.[130] It is assumed that for each kilo of farm-ranched fish (for example salmon), three kilos of smaller fish are needed.[131] The ratio for tuna is much higher, in several cases estimated at more than ten kilos of forage fish needed for each kilo of farmed species.[132]

Hence, risks of overfishing the wild species that serve as forage fish, especially juveniles, are to be taken into account when assessing sustainability;[133] "[n]atural limitations in the marine resource base could lead to global shortage of fishmeal and oil for fish feed production."[134] Even though forage species are small and fast reproducing species, most of them already range, according to some studies, from fully fished to overexploited[135] and are known to be particularly sensitive to

[126] Greenberg (n 115).

[127] M. Klinkhardt and B. Myrseth, 'New aquaculture candidates' in R. Arthur and J. Nierentz (eds), *Global Trade Conference on Aquaculture,* FAO Fisheries Proceedings 9 (Rome 2007) 179–183.

[128] Greenberg (n 115); Ottolenghi and others (n 117) 124.

[129] S. F. Locke, 'Breeding the Overfished Bluefin Tuna' (17 March 2008) http://www.livescience.com/4862-breeding-overfished-bluefin-tuna.html accessed 4 July 2015; T. Wall, 'Tuna Ranch Claims Spawning Success' (9 August 2011) http://news.discovery.com/animals/tuna-spawning-naturally-in-captivity-110809.htm accessed 4 July 2015.

[130] Asche and Khatun (n 117) 31; World Ocean Review, *The Future of Fish – The Fisheries of the Future* 2013 (n 10) 86–87.

[131] Naylor and others (n 120) 1019; Powers (n 14) 46.

[132] Ottolenghi and others (n 117) 134–135.

[133] OECD, 'Globalisation in Fisheries and Aquaculture: Opportunities and Challenges' (n 116) 52; A. Garric, 'Les petits poissons menacés par les fermes aquacoles' (12 April 2012) http://ecologie.blog.lemonde.fr/2012/04/12/les-petits-poissons-menaces-par-les-fermes-aquacoles/ accessed 4 July 2015; Pullin (n 115) 101–105.

[134] OECD, 'Globalisation in Fisheries and Aquaculture: Opportunities and Challenges' (n 116) 52.

[135] R. L. Naylor and others, 'Feeding aquaculture in an era of finite resources' (2009) 106 Proceedings of the National Academy of Science 15103, 15104, making reference to J. Adler and others, 'Forage fish: from ecosystems to markets' (2008) 33 Annual Review of Environmental Resources 153.

external factors such as El Niño.[136] Also, the more forage fish is caught for aquaculture—and other farming uses[137]—the less prey is left available to wild marine species with a predatory diet;[138] "increased catches of feed fish that are necessary for a further expansion of aquaculture seem bound to have an adverse effect on the production of edible fish in the sea."[139]

The industry has lately been trying to address these issues, by reducing feed conversion ratios—in other words the quantity of wild-fish input needed per unit of farmed-fish output—[140] and by developing non-fish based food for the fish.[141] This latter option will nevertheless only be possible for some sorts of fish, as carnivorous species such as salmon or tuna will unlikely feed exclusively on grain.

D. Conclusion

In conclusion, some fish species are endangered; overexploitation of stocks has caused collapse in the past and the current amount of legal overfishing and additional estimated IUU fishing is likely to provoke further depletion of species, maybe even some occurrences of extinctions, particularly for vulnerable species. Scientists sometimes disagree on the extent, but rarely on the fact that current practices are unsustainable.

Aquaculture is a growing practice, with the potential to decrease pressure on wild stocks in certain cases, but is also an additional source of withdrawals for forage fish and for juveniles when species cannot—yet—be bred in captivity. Due to the increasing share of this practice in the production of fish and the important part of farmed fish entering trade, aquaculture needs to be discussed together with wild-capture in considering ways to improve sustainability.

---

[136] T. Huntington and M. R. Hasan, 'Fish as feed inputs for aquaculture – practices, sustainability and implications: a global synthesis' in M. R. Hasan and M. Halwart (eds), *Fish as feed inputs for aquaculture: Practices, sustainability and implications,* FAO Fisheries and Aquaculture Technical Paper 518 (Rome 2009) 12.

[137] Competition indeed exists with other sectors, see ibid. 18–22.

[138] Englebrecht (n 117) 1197–1198; Greenberg (n 115); R. L. Naylor and G. H. Leonard, 'Aquaculture made safe' *Los Angeles Times* (15 February 2010) http://articles.latimes.com/2010/feb/15/opinion/la-oe-naylor15-2010feb15 accessed 4 July 2015.

[139] Hannesson (n 116) 170.

[140] Naylor and others (n 135) 15103.

[141] OECD, 'Globalisation in Fisheries and Aquaculture: Opportunities and Challenges' (n 116) 52; Huntington and Hasan (n 136) 46; M. Klinkhardt, 'The blue revolution – feed alternatives for aquaculture' in R. Arthur and J. Nierentz (eds), *Global Trade Conference on Aquaculture,* FAO Fisheries Proceedings 9 (Rome 2007) 263.

Chapter 3
Global and Regional Legal Regimes Dealing with Commercially-exploited Marine Species

Several sets of rules have been developed in international law, which are either directly meant to regulate fisheries or which can be used to do so. In some cases, institutions have also been established to coordinate and oversee their implementation by states.

For the purpose of this research, it is necessary to look at these different regimes in order to clarify the legal framework applicable to fisheries, and this for several reasons. The highly debated use of CITES for commercially-exploited fish species does indeed not fall into a vacuum field. Different sets of rules apply to fisheries, rules which are important to introduce in order to later identify governance issues,[1] to assess the added value of bringing CITES into play[2] and finally to discuss the types of relations between CITES and other institutions.[3] Also, knowing the rules applicable to fishing in different areas helps in the determination of the legality of catches, a necessary condition for a CITES permit to be issued.[4]

The relevant global and regional legal regimes will be presented: the framework regime (A), the management regimes (B), the conservation regimes (C) and the trade regime of the World Trade Organization (WTO) (D). Many management plans are adopted at the national level for waters under national jurisdiction and are important with regard to fishing and aquaculture, but the focus will stay here on regional and global regulations.

[1] See infra Chapter 4 A.

[2] See infra Chapter 5 D. II.

[3] See infra Chapter 6.

[4] See infra Chapter 5 C. I. 4.

S. Guggisberg, *The Use of CITES for Commercially-exploited Fish Species*,
Hamburg Studies on Maritime Affairs 35, DOI 10.1007/978-3-319-23702-2_3

A. Framework Regime for the Law of the Sea

I. 1982 United Nations Convention on the Law of the Sea

The United Nations Convention on the Law of the Sea (UNCLOS) was adopted with the aim, amongst others, to promote "the equitable and efficient utilization of the [... oceans'] resources [and] the conservation of their living resources".[5] Different rules and ways of implementing these goals have been developed for the different maritime zones.

1. Waters Under National Sovereignty

In its internal waters, territorial sea and archipelagic waters the coastal state exercises its sovereignty[6] and has the benefit of exclusive access over the fish stocks. It can thus take any measure it wishes with regard to fisheries up to 12 nautical miles (nm). Indeed, UNCLOS' relevant provisions "refrain from giving any indication as to which policy should be pursued by the Coastal States".[7]

The coastal state can adopt laws and regulations to prevent the infringement of its fisheries laws.[8] In the contiguous zone, up to 24 nm, it might take measures to punish violations of its laws in the territorial sea.[9]

UNCLOS Part XII on the protection of the marine environment applies to fisheries even though it is not explicitly stated. The Third Committee of the Third United Nations Conference on the Law of the Sea agreed that "the term 'marine environment' included 'marine life.'"[10] Also, the International Tribunal for the Law of the Sea (ITLOS) made that point clear in the provisional measures indicated in the *Southern Bluefin Tuna* case by stating "that the conservation of the living resources of the sea is an element in the protection and preservation of the marine environment".[11] Relevant provisions offer guidance as to the development of fishery policies applicable to waters under national jurisdiction and national

---

[5] 'United Nations Convention on the Law of the Sea' (UNCLOS) 1982, 1833 *United Nations Treaty Series* 397 preamble.

[6] Ibid. article 2.

[7] R. Wolfrum and N. Matz, 'The Interplay of the United Nations Convention on the Law of the Sea and the Convention on Biological Diversity' (2000) 4 Max Planck Yearbook of United Nations Law 445, 448.

[8] UNCLOS (n 5) article 21(1)(e).

[9] Ibid. article 33(1)(b).

[10] M. H. Nordquist and others (eds), *United Nations Convention on the Law of the Sea 1982: A Commentary: Volume IV* (Martinus Nijhoff Publishers 1991) 43 (parentheses omitted).

[11] *Southern Bluefin Tuna cases (New Zealand v. Japan; Australia v. Japan), Provisional Measures,* 27 August 1999, International Tribunal for the Law of the Sea (ITLOS) cases n°3 & 4 §70; Wolfrum and Matz, 'The Interplay of the United Nations Convention on the Law of the Sea and the Convention on Biological Diversity' (n 7) 451.

control—hence these remarks similarly apply to the exclusive economic zone (EEZ).[12] Particularly, articles 192 and 194 (5) respectively emphasize the obligation of states to protect the marine environment and the obligation to take measures to protect as well as to preserve fragile ecosystems and depleted, threatened or endangered species.

These provisions arguably limit the discretion of coastal states in designing their fisheries laws and regulations. However, while reference is explicitly made to fragile ecosystems,[13] the ecosystem approach in itself is not mentioned[14] and thus its application is not required. Nevertheless, the protection of fragile ecosystems could be relevant for aquaculture practices in waters under national sovereignty, for example by requiring states to ensure that such practices do not put them at risk.

2. Waters Under National Control

One of the main changes brought about by UNCLOS was the establishment of a functional and 'special legal regime', that of the EEZ up to 200 nm.[15] There, the coastal state has sovereign rights, inter alia for the purpose of conserving and managing living resources.[16] Articles 61 to 73 provide the framework applicable to conservation and utilization of living resources within a coastal state's EEZ. Taking into account the best scientific evidence available to them,[17] these states must determine TACs.[18] Subject to the obligation to ensure that living resources are not overexploited,[19] they must promote their optimum utilization.[20]

---

[12] Wolfrum and Matz, 'The Interplay of the United Nations Convention on the Law of the Sea and the Convention on Biological Diversity' (n 7) 448, 451.

[13] UNCLOS (n 5) article 194(5).

[14] Wolfrum and Matz, 'The Interplay of the United Nations Convention on the Law of the Sea and the Convention on Biological Diversity' (n 7) 451. The ecosystem approach is generally understood in the sense of determining measures applicable to the target species while taking into account all the interactions which are relevant within the ecosystem, both between species and with the physical environment – it is also referred to as the ecosystem-based fishery management, in opposition to the ecosystem-based management which holistically examines all activities (pollution, shipping, fisheries, etc.) in an area. For more details on this question, see infra Chapter 4 A. II. and B. III. 1. b).

The opinion of J. Morishita ('What is the ecosystem approach for fisheries management?' (2008) 32 Marine Policy 19, 20) that articles 61(4) and 119(1) of UNCLOS, without naming it, require an ecosystem approach for the conservation of living resources, is not convincing. Indeed, independent and associated species would only be some of the factors to take into account, even for the narrower ecosystem-based fishery management.

[15] UNCLOS (n 5) articles 55, 57.

[16] Ibid. article 56(1)(a).

[17] Ibid. article 61(2).

[18] Ibid. article 61(1).

[19] Ibid. article 61(2).

[20] Ibid. article 62(1).

On the one hand, coastal states must adopt and implement conservation measures to avoid overexploitation, to allow restoration of species if needed, as well as take into account associated or dependent species.[21] This is "a first step towards an ecosystem approach",[22] but the interactions with the ecosystem's physical components are still not mentioned.[23] The coastal state has discretionary powers in the choice of measures and no control is envisioned in the Convention to overview them. On the other hand, the coastal state must promote optimum utilization of living resources within its EEZ. If it cannot harvest the entirety of the set quotas, it must give access to other countries.[24] The MSY is hence no upper limit, but a goal in itself. Since the coastal state is the one setting the TACs in the first place, it also benefits on the exploitative side of its obligations of wide discretionary powers.

When more than one state is involved in the protection and management of a stock which migrates through an area under national control, UNCLOS requires cooperation. Hence, for fish stocks that move between two or more EEZs, the coastal states shall seek to agree on measures to conserve and develop stocks, either directly or through an appropriate organization.[25] Then, for fish stocks that are harvested both in an EEZ and in the adjacent area of the high seas (the so-called straddling stocks), the coastal state and the states fishing there shall seek to agree on measures to conserve stocks in the adjacent area—though not in the EEZ—either directly or through an appropriate organization.[26]

Finally, for fish stocks that are highly migratory, the coastal state and other states fishing this species in the region shall cooperate directly or through an organization, in order to ensure conservation and optimum utilization of the species.[27] If an appropriate organization does not exist, states shall cooperate towards its establishment.[28] This provision is relevant for a number of tuna species and oceanic sharks, which are listed as highly migratory species in Annex I of UNCLOS.

In conclusion, the coastal state's freedom is restricted since it must take some measures regarding conservation and it must cooperate with other concerned states when the fish stocks are transboundary, straddling or highly migratory.[29] However, whether the coastal state is managing its EEZ's fish stocks in a manner that respects

[21] Ibid. article 61(3)–(4).

[22] Wolfrum and Matz, 'The Interplay of the United Nations Convention on the Law of the Sea and the Convention on Biological Diversity' (n 7) 450.

[23] Ibid.

[24] UNCLOS (n 5) article 62.

[25] Ibid. article 63(1).

[26] Ibid. article 63(2).

[27] Ibid. article 64(1).

[28] Ibid.

[29] Wolfrum and Matz, 'The Interplay of the United Nations Convention on the Law of the Sea and the Convention on Biological Diversity' (n 7) 449.

the Convention is difficult for third parties to scrutinize and assess.[30] Indeed, this is part of the restrictions to compulsory dispute settlement set in article 297 (3).[31] This provision "qualifies the respective powers of Coastal States as being of a discretionary nature."[32] At the most, if a state has manifestly failed to avoid overexploitation or is refusing to set TACs, it could be forced into a compulsory conciliation.[33] However, since UNCLOS' binding settlement of disputes mechanisms apply to disputes regarding the marine environment, practices which do not comply with Part XII could arguably be reviewed by such mechanisms.

3. High Seas

UNCLOS Part VII Section 2 provides the framework applicable to fishing activities in the high seas. The freedom of fishing for nationals of all states is restated, qualified by the obligation for states to respect their treaty obligations, the interests of coastal states when relevant—for straddling and highly migratory species—and the specific provisions developed in Part VII.[34] While rules of implementation—by the flag state instead of by the coastal state—differ from the regime applicable in EEZs, the provisions' content is mostly similar.[35]

When their nationals fish in the same area and/or the same resources states shall cooperate in the conservation and management of living resources. They can do so directly or through a fisheries organization. Though subject to an 'as appropriate' qualification, the establishment of such organization is encouraged in UNCLOS.[36]

In determining the conservation and management measures (CMMs), states must, on the basis of the best scientific evidence available, aim at maintaining or restoring populations at levels where they can produce the MSY.[37] They also have to consider the effects on associated and dependent species.[38] States shall

[30] R. R. Churchill and A. V. Lowe, *The law of the sea* (3rd edn Juris Publishing – Manchester University Press 1999) 289.

[31] UNCLOS (n 5) article 297(3); R. Barnes, 'The Convention on the Law of the Sea: An Effective Framework for Domestic Fisheries Conservation?' in D. Freestone, R. Barnes and D. Ong (eds), *The Law of the Sea – Progress and Prospects* (Oxford University Press 2006) 239; M. A. Young, *Trading Fish, Saving Fish: The Interaction between Regimes and International Law* (Cambridge Studies in International and Comparative Law, Cambridge University Press 2011) 38.

[32] Wolfrum and Matz, 'The Interplay of the United Nations Convention on the Law of the Sea and the Convention on Biological Diversity' (n 7) 450.

[33] Churchill and Lowe (n 30) 289, 455.

[34] UNCLOS (n 5) article 116.

[35] Wolfrum and Matz, 'The Interplay of the United Nations Convention on the Law of the Sea and the Convention on Biological Diversity' (n 7) 452.

[36] UNCLOS (n 5) article 118.

[37] Ibid. article 119(1)(a).

[38] Ibid. article 119(1)(b).

furthermore "tak[e] into account [...] any generally recommended international minimum standards, whether subregional, regional or global".[39]

The discretion of states in taking and implementing CMMs is high and arguably even increased by the existence of RFMOs. Indeed, in the *Southern Bluefin Tuna* case, the Arbitral Tribunal judged that the UNCLOS dispute settlement does not take precedence over review mechanisms developed under such regional organizations, even if the latter are non-binding.[40]

4. Overview of Implementation

The United Nations (UN) Secretariat annually organizes the meeting of states parties to UNCLOS; this competence is based on article 319 (2) (e). UNCLOS meeting of the parties is occupied with administrative matters such as dealing with ITLOS's budget as well as with the election of its judges and that of the Commission on the Limits of the Continental Shelf's members.[41] Hence, the UN Secretariat performs the functions of the Convention's Secretariat.[42] The UN Secretary General is also the depositary of the Convention and shall, in addition to this role, "report to all States Parties, the [International Seabed] Authority and competent international organizations on issues of a general nature that have arisen with respect to this Convention".[43]

Two United Nations General Assembly (UNGA) Resolutions are adopted every year on the law of the sea, one related directly to fisheries and one more generally to oceans' governance. Hence, the UNGA debates fisheries questions every year in the general discussion on the law of the sea or in thematic Resolutions and, since 2003, in the separate Resolution focused on sustainable fisheries.[44] These resolutions

[39] Ibid. article 119(1)(a).

[40] *Southern Bluefin Tuna (New Zealand v. Japan, Australia v. Japan), Arbitral Award,* 4 August 2000, Arbitral Tribunal, (2006) XXIII Reports of International Arbitral Awards 1 §53–59, 72; T. Tyler, 'Saving Fisheries on the High-seas: The Use of Trade Sanctions To Force Compliance with Multilateral Fisheries Agreements' (2006–2007) 20 Tulane Environmental Law Journal 43, 60–61; Young, *Trading Fish, Saving Fish* (n 31) 44–45. See also infra Chapter 4 A. IV. 3. b).

[41] Division for Ocean Affairs and the Law of the Sea, 'Meetings of States Parties to the 1982 United Nations Convention on the Law of the Sea' http://www.un.org/depts/los/meeting_states_parties/meeting_states_parties.htm accessed 4 July 2015.

[42] H. Corell, 'Oversight of the Implementation of the Global Ocean Regime: The Role of the United Nations' in D. Vidas and W. Ostreng (eds), *Order for the oceans at the turn of the century* (Kluwer Law International 1999) 339–340.

[43] UNCLOS (n 5) article 319(1)(2)(a).

[44] J. Harrison, *Making the Law of the Sea: A Study in the Development of International Law* (Cambridge University Press 2011) 201, 203. At first, questions related to fisheries were dealt with in the general part of the Law of the Sea Resolution, but in 2003, two years after the entry into force of the UNFSA, a specific resolution on sustainable fisheries started to be issued.

have addressed various issues, such as large-scale pelagic driftnets,[45] by-catch and discard,[46] or fishing overcapacity.[47] They look, in a comprehensive way, at the issues faced by fisheries, make recommendations and reaffirm applicable principles. Since they are negotiated as thoroughly as a binding instrument would be and they have undoubtedly a considerable impact on states and international organizations, it is worth examining them.[48]

Any recent *Resolution on Sustainable Fisheries* is typical of this coordinating role, where the UNGA recalls the rules applicable as well as calls upon different actors, be they states, RFMOs or the FAO to take more effective measures,[49] ratify treaties,[50] increase cooperation,[51] or improve practices.[52] It refers to all major issues and, for example, calls upon the three sets of entities above mentioned to cooperate in achieving sustainable aquaculture.[53] With regard to sharks, the 2008 Resolution recognized the need for measures to promote the long-term conservation, management and sustainable use of shark populations, given their vulnerability and the fact that some are threatened with extinction.[54]

---

[45] United Nations General Assembly (UNGA), Large-scale pelagic driftnet fishing and its impact on the living and marine resources of the world's oceans and seas 22 December 1989, A/RES/44/225.

[46] UNGA, Resolution adopted by the General Assembly on 19 December 1994 – Fisheries by-catch and discards and their impact on the sustainable use of the world's living marine resources 2 February 1995, A/RES/49/118.

[47] UNGA, Resolution adopted by the General Assembly on 7 December 2010 – Sustainable fisheries, including through the 1995 Agreement for the Implementation of the Provisions of the United Nations Convention on the Law of the Sea of 10 December 1982 relating to the Conservation and Management of Straddling Fish Stocks and Highly Migratory Fish Stocks, and related instruments 30 March 2011, A/RES/65/38 part VI.

[48] L. de La Fayette, 'The Role of the UN in International Oceans Governance' in D. Freestone, R. Barnes and D. Ong (eds), *The Law of the Sea – Progress and Prospects* (Oxford University Press 2006) 69.

[49] UNGA, Resolution adopted by the General Assembly on 7 December 2010, A/RES/65/38 (n 47) §62, with regard to the traceability of fish products.

[50] Ibid. §5, 36.

[51] Ibid. §101 between RFMOs.

[52] Ibid. §105, 107 urges the RFMOs and states to respectively improve transparency, rely on the best scientific information available, incorporate the precautionary approach and ecosystem approaches etc. and to undertake performance reviews of the RFMOs they are members of.

[53] Ibid. §116.

[54] UNGA, Resolution adopted by the General Assembly on 5 December 2008 – Sustainable fisheries, including through the 1995 Agreement for the Implementation of the Provisions of the United Nations Convention on the Law of the Sea of 10 December 1982 relating to the Conservation and Management of Straddling Fish Stocks and Highly Migratory Fish Stocks, and related instruments 24 February 2009, A/RES/63/112 preamble.

5. Applicability and Appropriateness of UNCLOS Rules

Most of UNCLOS' substantive provisions have entered customary international law (CIL).[55] Even though doubts were voiced as to whether articles 61 and 62 had passed into this body of general international law, due to the lack of uniform reference to their content in national laws,[56] those rights and obligations linked to fisheries are nowadays usually considered a reflection of CIL.[57]

The legal regime provided by UNCLOS gives some insights as to the rights and obligations of states towards fisheries in the different maritime zones. So, not only is UNCLOS of major importance because it codified—or at least now reflects—CIL, but it also has been widely ratified and its obligations are linked to a compulsory dispute settlement.

However, its articles related to fisheries are not detailed provisions,[58] thus "fall[ing] woefully short of providing the muscle, direction, and vision needed for stewardship of marine species and their ecosystems."[59] Indeed, these are of a very general nature and vaguely termed. For example, even the obligation to take into account international minimum standards in setting quotas on the high seas is formulated in a non-prescriptive manner; 'taking into account' is indeed an obligation of conduct and not one of result. In addition to the uncertainty as to the actual state of fish stocks, the vagueness of this obligation would make it "very difficult, if not impossible, to tell whether in any particular case the duties had been observed."[60] All in all, the relevant provisions give the coastal and flag states, in the EEZ and on the high seas respectively, wide discretionary powers as to the measures to take. Additionally, some aspects of fisheries are not addressed: for example, no framework is provided for aquaculture activities, neither in waters under national jurisdiction—which is understandable and where national laws would apply—nor in waters beyond national jurisdiction[61]—which is rather more problematic.

---

[55] W. T. Burke, *The New International Law of Fisheries* (Clarendon Press 1994) 40, 99, 104.

[56] Churchill and Lowe (n 30) 290; F. Orrego Vicuña, *The exclusive economic zone: regime and legal nature under international law* (Cambridge University Press 1989) 244–246.

[57] M. Bowman, P. Davies and C. Redgwell, *Lyster's International Wildlife Law* (2nd edn Cambridge University Press 2010) 125; P. Sands and others, *Principles of International Environmental Law* (3rd edn Cambridge University Press 2012) 403. The International Court of Justice (ICJ), without commenting on the fisheries aspect of it, nevertheless explicitly stated that the establishment of EEZs is part of CIL (*Case concerning the Continental Shelf (Libyan Arab Jamahiriya/Malta), Judgment,* 3 June 1985, I.C.J. Reports 1985 13, 33 §34).

[58] Barnes (n 31) 237, 239; J.-F. de Pulvenis Séligny, 'The marine living resources and the evolving law of the sea' (2010) 1 Aegean Review of the Law of the Sea and Maritime Law 61, 85–86; Tyler (n 40) 79–80.

[59] W. M. von Zharen, 'Ocean Ecosystem Stewardship' (1998) 23 William and Mary Environmental Law and Policy Review 1, 33.

[60] Churchill and Lowe (n 30) 290.

[61] Pulvenis Séligny (n 58) 90.

Practically speaking, the rules developed in UNCLOS are based on a division of oceans into jurisdictional spaces where states have different rights and obligations.[62] This is totally disconnected from the physical reality of ecosystems. The consequent need for cooperation between relevant states when stocks cross political borders is faced, although not sufficiently, by UNCLOS' general obligation to cooperate. Also, the extension of the coastal states' sovereign rights was seen as a major step towards a better management of stocks; it was believed that coastal states would take greater care of resources once they were theirs.[63] However, this hope was partially unfounded and unsustainable practices have flourished rather than decreased.[64] Furthermore, the lack of an ecosystem approach is highly problematic. Finally, the MSY is not unanimously accepted as the best concept to manage fisheries anymore;[65] taking into account the lack of exact data and the complex processes, some precaution should be preferred to a management at the margin.[66]

[62] J. Ellis, 'The Straddling Stocks Agreement and the Precautionary Principle as Interpretive Device and Rule of Law' (2001) 32 Ocean Development and International Law 289, 290; L. Juda, 'The 1995 United Nations agreement on straddling fish stocks and highly migratory fish stocks: A critique' (1997) 28 Ocean Development and International Law 147, 160.

[63] While pointing out to expected inefficiencies for straddling stocks and potential difficulties with enforcement, R. D. Eckert (*The Enclosure of Ocean Resources: Economics and the Law of the Sea* (Hoover Institution Press 1979) 129–131) was of the opinion that the EEZ can, by creating an economic rent for the resource owner, reduce unsustainable practices; see also R. Rayfuse, 'The Interrelationship between the Global Instruments of International Fisheries Law' in E. Hey (ed), *Developments in International Fisheries Law* (Kluwer Law International 1999) 110–111.

[64] P. Birnie, A. Boyle and C. Redgwell, *International Law and the Environment* (3rd edn Oxford University Press 2009) 705; Bowman, Davies and Redgwell (n 57) 148; D. R. Christie, 'The Conservation and Management of Stocks Located Solely within the Exclusive Economic Zone' in E. Hey (ed), *Developments in International Fisheries Law* (Kluwer Law International 1999) 397; Pulvenis Séligny (n 58) 72; Sands and others (n 57) 397.

[65] The MSY is traditionally used for a management at the margin and does not take into account the relationships with other species and their habitat (Barnes (n 31) 243–244; Birnie, Boyle and Redgwell (n 64) 591; Burke, *The New International Law of Fisheries* (n 55) 206; K. Rigg, R. Parmentier and D. Currie, 'Halting IUU Fishing: Enforcing International Fisheries Agreements' (OECD Directorate for Food, Agriculture and Fisheries 2004) 5; G. J. Schram and A. Tahindro, 'Developments in Principles for the Adoption of Fisheries Conservation and Management Measures' in E. Hey (ed), *Developments in International Fisheries Law* (Kluwer Law International 1999) 257–258). This concept has however evolved and now tends to consider uncertainties and interactions between species (World Ocean Review, *The Future of Fish – The Fisheries of the Future* (Maribus, Future Ocean 2013) 101).

[66] The precautionary approach is to be understood generally as a method to deal with uncertainties: at the very least scientific uncertainty should not justify a delay in action, but more generally it should be counter-balanced by measures such as buffer-zones, lower quotas, etc. On this concept, see infra Chapter 4 B. II. 2.

6. Conclusions

UNCLOS provides a general framework to fishing activities. Coastal states have exclusive rights and responsibilities up to 200 nm in their EEZs. They have to avoid overexploitation while aiming at an optimum utilization. As for fish stocks that are transboundary, straddling or highly migratory, cooperation between relevant states is required, be it directly or through the appropriate organizations. On the high seas, the flag states are the ones responsible for their fishing fleets' actions. There, UNCLOS links the freedom of fishing to the obligation to cooperate in the conservation of living resources.

The framework provided by UNCLOS is applicable to all states since it reflects CIL, but these rules are very general and, as such, arguably not opposable to states. It is indeed hard to verify compliance with 'soft' obligations and article 297 (3) on limitations to compulsory dispute settlement removes from the competence of binding mechanisms most questions on fisheries policies up to 200 nm. The avenues to scrutinize whether decisions of the coastal state fulfill a correct balancing act are thus very limited, although making reference to obligations found in Part XII might be an option.

Also, the relevant rules are, with the exception of Part XII, very much oriented towards exploitation, particularly in the EEZ. The traditional concept of the MSY does not reflect the precautionary principle, quite the opposite. Finally, even though ecosystems as well as dependent and related species are mentioned, no ecosystem approach is required in the Convention. While the normativity of the precautionary principle and the ecosystem approach are still subject to discussion, it is nevertheless accepted that they should, in the field of fisheries, at least be taken into account in decision-making processes.[67]

II. 1995 United Nations Fish Stock Agreement

1. Main Changes

In order to meet the challenges faced by straddling and highly migratory fish stocks, and as a response to the call for sustainable fisheries in Agenda 21,[68] an Implementation Agreement to UNCLOS was adopted in 1995 for straddling and highly migratory fish species (UNFSA). This instrument builds upon the obligations found in UNCLOS and, in 50 articles, provides a detailed framework with some directly

---

[67] For more on the precautionary approach/principle, see infra Chapter 4 B. II. 2. and on the ecosystem approach, see infra Chapter 4 B. III. 1. b).

[68] United Nations, Agenda 21 – The United Nations Programme of Action from Rio (1992) chapter 17 §17.46; this call was further repeated by the UNGA in the Resolution adopted by the General Assembly on 22 December 1992 – United Nations Conference on Straddling Fish Stocks and Highly Migratory Fish Stocks 29 January 1993, A/RES/47/192 preamble.

applicable provisions for the sustainable use of such stocks. The UNFSA "seeks to ensure a harmonious development of coherent conservation and management measures for the high seas and the exclusive economic zone, based upon cooperation."[69]

It aims mainly at activities on the high seas, but the precautionary approach and the requirement to adopt measures compatible inside and outside of an EEZ are also to be applied to areas under national jurisdiction.[70] Indeed, states involved must agree on measures regarding fishing stocks straddling or migrating between areas within and outside national jurisdiction. If coastal states and distant water fishing nations fail to cooperate towards this compatibility of measures the dispute settlement mechanisms will be triggered.[71]

The novelties of the UNFSA are first the obligation to apply a precautionary approach in the management of fisheries.[72] In this context, it means that lack of scientific information is not to be used as a justification for not taking conservation and management measures. States must show caution in their decisions and actions when information is missing.[73] However, the burden of proof is not totally shifted towards the users, in which case the soundness of a harvesting project would have to be demonstrated before being allowed. This would de facto stop an important number of fishing activities.[74]

Second, an approach closer to that of an ecosystem-based one is also required, particularly in the protection of the marine environment's biodiversity[75] and references to the 'biological unit' are made.[76] Also, contracting parties are obliged to minimize catches of non-target species and impacts on associated or dependent species, in particular the ones that are endangered.[77] This is important for sharks which are often caught incidentally in fisheries directed at other species and for species which are not considered as straddling and highly migratory, and hence do

[69] Wolfrum and Matz, 'The Interplay of the United Nations Convention on the Law of the Sea and the Convention on Biological Diversity' (n 7) 453, on the 'Agreement for the Implementation of the Provisions of the United Nations Convention on the Law of the Sea of 10 December 1982 Relating to the Conservation and Management of Straddling Fish Stocks and Highly Migratory Fish Stocks' (UNFSA) 1995, 2167 *United Nations Treaty Series* 3 articles 7–8.

[70] UNFSA (n 69) article 3(1).

[71] Ibid. article 7.

[72] Ibid. article 6 and annex II, which develops guidelines to the application of the precautionary approach.

[73] Ibid. article 6(2).

[74] Ellis (n 62) 298–299.

[75] UNFSA (n 69) article 5(d)-(g).

[76] Ibid. article 7(2)(d); Wolfrum and Matz, 'The Interplay of the United Nations Convention on the Law of the Sea and the Convention on Biological Diversity' (n 7) 453.

[77] UNFSA (n 69) article 5(f).

not fall directly within the ambit of the convention. However, while the UNFSA "is more clearly based on the ecosystem approach still, it mitigates the situation for specific species or stocks only."[78]

Third, while the MSY terminology has been kept, it appears that, rather than as a goal to reach, "[t]his reference point is to be considered a limiting reference point which constraints utilization within safe biological limits."[79] It also nowadays takes into account factors such as uncertainty and interactions between species; hence, the wording remained but the concept somehow evolved.[80]

Fourth, an important clarification brought about by the UNFSA is the way in which the duty to cooperate must be implemented. A special role is given to RFMOs,[81] which are to be open to states with a real interest in the fishery. RFMOs, or similar arrangements, are regarded as the appropriate means of cooperation. If none exists, states shall cooperate to establish one.[82] Obligation is made to join the relevant organization or at least to cooperate with it in order to be allowed to fish in the managed area. This requirement brings about a paradigm shift in comparison to the traditional regime of freedom of fishing on the high seas.[83] Indeed, states involved in harvesting stocks that are managed by an RFMO must become member or cooperate with the organization, or else they have to refrain from fishing in the high seas areas under the RFMO's jurisdiction.[84]

Fifth, regarding implementation, the duties of the flag state are listed,[85] as well as various progressive methods of compliance and enforcement. While the flag state remains the main actor in enforcement,[86] should it not take its responsibilities, other players, such as members of RFMOs fishing on the high seas and port states may move in and act.[87] Non-flag state enforcement on the high seas is envisioned for vessels of states parties to the UNFSA or, if it is so provided in the RFMO, members

---

[78] Wolfrum and Matz, 'The Interplay of the United Nations Convention on the Law of the Sea and the Convention on Biological Diversity' (n 7) 451.

[79] Ibid. 453–454 and n 24; see also Barnes (n 31) 247.

[80] World Ocean Review, *The Future of Fish – The Fisheries of the Future* 2013 (n 65) 101.

[81] UNFSA (n 69) article 8; E. Franckx, 'Fisheries Enforcement: Related legal and institutional issues: national, subregional or regional perspectives' FAO Legislative Study 71 (Rome 2001) 5.

[82] UNFSA (n 69) article 8(5).

[83] J. R. Mack, 'International Fisheries Management: How the U.N. Conference on Straddling and Highly Migratory Fish Stocks Changes the Law of Fishing on the High Seas' (1995–1996) 26 California Western International Law Journal 313, 326; P. Vincent, *Droit de la mer* (Larcier 2008) 218.

[84] UNFSA (n 69) articles 8(4), 17; E. DeLone, 'Improving the Management of the Atlantic Tuna: The Duty to Strengthen the ICCAT in Light of the 1995 Straddling Stocks Agreement' (1998) 6 New York University Environmental Law Journal 656, 663–664; P. Oerebech, K. Sigurjonsson and T. L. McDorman, 'The 1995 United Nations Straddling and Highly Migratory Fish Stocks Agreement: Management, Enforcement and Dispute Settlement' (1998) 13 International Journal of Marine and Coastal Law 119, 122.

[85] UNFSA (n 69) article 18.

[86] Ibid. article 19.

[87] Juda (n 62) 157–158.

thereof.[88] A progressive provision inserted in the Agreement even sets up a boarding and inspection scheme applicable to RFMOs which have not developed their own within the two years following the UNFSA's adoption.[89]

Accordingly, the procedure for boarding and inspection includes the obligation to inform the flag state of the inspection,[90] but does not require its authorization. Once on board, inspection of all relevant areas, equipment and documents "which the inspector deems necessary to verify the compliance with the measures established [by the relevant RFMO]"[91] takes place. Following inspection, if there are 'clear grounds for believing'—a situation which is not defined—that a violation of CMMs has taken place, the inspecting state must notify the flag state, which has to respond within three days to say whether it will take measures itself and inform the inspecting state of its findings and the measures taken, or whether it authorizes the inspecting state to further investigate the alleged violations—often after having brought the vessel back to the nearest port.[92] Should the flag state fail to respond to notification or fail to take action, such failure also gives the inspecting state the right to bring the vessel to port to investigate, if the situation is one of 'serious violations', as defined in article 21 (11).[93] From the bypass of flag state authorization to the possibility to take measures beyond inspection should the flag state not react within three days this provision is very innovative.

Measures to be taken by a port state such as inspections and prohibition of landing or transshipment are also mentioned, although in less detail.[94] Both sets of non-flag state measures have the potential to improve the verification of the activities undertaken by non-compliant vessels and/or made possible by flags of convenience and hence to reduce the free-riding problem. This issue is indeed caused by the lack of willingness of certain flag states to join RFMOs and hence to be bound by the CMMs;[95] when a large number of vessels take advantage of such a loophole, regulatory schemes are deemed to fail.

Finally, the UNFSA acknowledges the importance and particularity of small-scale fisheries. Artisanal and subsistence fisheries represent a significant percentage of the global landings, usually suffer from overfishing but are crucial for coastal communities.[96] According to the UNFSA, account must be taken of these special interests.[97]

---

[88] UNFSA (n 69) article 21(1)(2).

[89] Ibid. articles 21(3)–(18), 22.

[90] Ibid. article 22(1)(b).

[91] NEAFC Scheme of Control and Enforcement article 18(2). For more on inspection schemes, see infra Chapter 3 B. II. 3.

[92] UNFSA (n 69) article 21(5)–(7).

[93] Ibid. article 21(8) and (11); Oerebech, Sigurjonsson and McDorman (n 84) 132.

[94] UNFSA (n 69) article 23.

[95] On this, see infra Chapter 4 A. IV. 1.

[96] Schram and Tahindro (n 65) 270–271.

[97] UNFSA (n 69) article 5(i).

2. UNFSA Review Procedure

According to article 36 of the UNFSA, four years after the entry into force of this treaty, a review conference shall be organized, to "[assess] the effectiveness of this Agreement in securing the conservation and management of straddling fish stocks and highly migratory fish stocks." This process is to take place with states parties, but also non-parties, non-governmental organizations (NGOs) and intergovernmental organizations. After examining how adequate the treaty is in order to conserve and manage highly migratory and straddling fish stocks, the Review Conference also makes suggestions as to improvement ways.

The first meeting took place in 2006 and highlighted further ameliorations needed, such as establishing new RFMOs where there were none, starting the drafting of a binding instrument on port state measures, reviewing the performance of RFMOs, or reducing fishing capacity.[98] Another conference was called upon in 2010, which noted positive developments since 2006, such as the creation of new RFMOs, the adoption of a treaty on port state measures, the "application of the ecosystem and precautionary approaches", the "completion of performance reviews by some regional fisheries management organizations and arrangements" as well as "progress in cooperative activities among regional fisheries management organizations and arrangements."[99] However, concerns were still voiced in relation to the sustainability of fishing. The need to continue strengthening the requirements placed on states and improving RFMOs performance was expressed.[100] The tentative use of CITES for shark species was mentioned, alongside with the consequences such listing proposals had in putting pressure on RFMOs to improve their management.[101]

3. Applicability and Appropriateness of the UNFSA

a) In General

The UNFSA makes several positive clarifications and/or modifications to the international law of fisheries, particularly through the emphasis it puts on the importance of science in decision-making and the precautionary principle.

[98] UNFSA, 'Annex: Outcome of the Review Conference' New York 26 May 2006, A/CONF.210/2006/15', as presented in D. A. Balton and H. R. Koehler, 'Reviewing the United Nations Fish Stocks Treaty' (2006–2007) 7 Sustainable Development Law and Policy Review 5, 7.

[99] UNFSA, 'Report of the resumed Review Conference on the Agreement for the Implementation of the Provisions of the United Nations Convention on the Law of the Sea of 10 December 1982 relating to the Conservation and Management of Straddling Fish Stocks and Highly Migratory Fish Stocks' 27 July 2010, A/CONF.210/2010/7 §24, see also 39–40.

[100] Ibid. §26, 33.

[101] Ibid. §27.

However, the requirement to apply an ecosystem approach is still missing from this framework.

Also, one should not forget that the UNFSA only applies to straddling and highly migratory fish stocks and their dependent and related species—although these terms are not defined. UNCLOS is consequently still the relevant framework agreement for all other stocks and species.[102] Finally, while the UNFSA principles are mainly to be implemented through RFMOs, there is no mechanism set up in the treaty to review whether RFMOs indeed comply with the UNFSA.[103]

b) Applicability of the UNFSA

As for the applicability of this instrument, one must keep in mind that UNCLOS and the UNFSA's memberships do not overlap: the former benefits from 167 ratifications, whereas the latter only obtained 82. Also, some parties to the UNFSA have not ratified UNCLOS—the USA being the main example.[104]

The nature of the UNFSA is quite debated, particularly as to whether it reached CIL status, whether it is an interpretative agreement implementing/giving effect to UNCLOS obligations, or whether it is simply changing the law.[105] Determining the nature of this instrument is crucial to assessing the appropriateness of the response it offers to issues facing commercial fisheries. Particularly, the innovative semi-enclosure of the high seas by RFMOs is a useful means to control and conserve exploited species. However, this is very unlikely to solve the free-riding problem if the UNFSA is not applicable to all flag states. Indeed, flags of non-compliance will obviously not join the Agreement and voluntarily submit themselves to its provisions.

---

[102] Ellis (n 62) 296–297.

[103] E. J. Molenaar, 'Non-Participation in the Fish Stocks Agreement: Status and Reasons' (2011) 26 International Journal of Marine and Coastal Law 195, 222.

[104] Division for Ocean Affairs and the Law of the Sea, 'Chronological lists of ratifications of, accessions and successions to the Convention and the related Agreements as at 3 October 2014' (2014) http://www.un.org/depts/los/reference_files/chronological_lists_of_ratifications.htm#Agreement%20for%20the%20implementation%20of%20the%20provisions%20of%20the%20Convention%20relating%20to%20the%20conservation%20and%20management%20of%20straddling%20fish%20stocks%20and%20highly%20migratory%20fish%20stocks accessed 4 July 2015.

[105] Barnes (n 31) 248–249; Oerebech, Sigurjonsson and McDorman (n 84) 123–124; J. Ziemer, *Das gemeinsame Interesse an einer Regelung der Hochseefischerei: Dargestellt am Beispiel des Fish Stocks Agreement* (Veröffentlichungen des Walther-Schücking-Instituts für Internationales Recht an der Universität Kiel vol 128, Duncker & Humblot 2000) 187–291.

In international law, according to the *pacta tertiis* principle, third parties are generally not bound by the obligations found in a treaty that they have not ratified.[106] Indeed, as they have not agreed to them, they do not have to comply with them even if their conduct seems to imply acceptance.[107] Hence, if the UNFSA is merely treaty law and creating new law, it only applies *inter partes*.[108] But if this

[106] 'Vienna Convention on the Law of Treaties' (VCLT) 1969, 1155 *United Nations Treaty Series* 331 article 34; *North Sea Continental Shelf Case (Federal Republic of Germany/Denmark; Federal Republic of Germany/Netherlands), Judgment,* 20 February 1969, I.C.J. Reports 1969 3, 24–25 §26–27; A. Aust, *Modern Treaty Law and Practice* (Cambridge University Press 2000) 207.

An exemption could be that of objective regimes: these regimes address issues of general interest for the international community and have effects on all states, so-called *erga omnes* rights and obligations, hence also towards non-parties (C. F. de Casadevante Romani, 'Objective Regime' Max Planck Encyclopedia of Public International Law www.mpepil.com §1–2) and might be applicable to some fisheries treaties (F. Salerno, 'Treaties Establishing Objective Regimes' in E. Cannizzaro (ed), *The Law of Treaties Beyond the Vienna Convention* (Oxford University Press 2011) 228). Sir Waldock stated that "[a] treaty establishes an objective régime when it appears from its terms and from the circumstances of its conclusion that the intention of the parties is to create in the general interest general obligations and rights relating to a particular region, State, territory, locality, river, waterway, or to a particular area of sea, sea-bed, or air-space; provided that the parties include among their number any State having territorial competence with reference to the subject-matter of the treaty, or that any such State has consented to the provisions in question" (H. Waldock (United Nations International Law Commission Special Rapporteur), 'Third Report on the Law of Treaties (1964)' 1964, UN Doc A/CN.4/167 and Add.1–3 §26).

The legal framework surrounding them is highly controversial (Casadevante Romani (n 106) §17; for earlier arguments see I. Sinclair, *The Vienna Convention on the Law of Treaties* (2nd edn Manchester University Press 1984) 104–106) but it also appears that obligations would still need to be acquiesced to by third parties (B. Vukas, 'Treaties, Third-Party Effect' Max Planck Encyclopedia of Public International Law www.mpepil.com §15). Hence it seems to depart only slightly from the traditional law of treaties and will not be examined in detail here. In general on objective regimes and the example of the Antarctic Treaty System, see Aust (n 106) 208–209; M. Fitzmaurice, 'Third Parties and the Law of Treaties' (2002) 6 Max Planck Yearbook of United Nations Law 37, 66–83, 121–136; Salerno (n 106).

[107] VCLT (n 106) article 35 (obligation must be accepted in writing); A. Mahiou, *Le droit international ou la dialectique de la rigueur et de la flexibilité: Cours général de droit international public* (Recueil des Cours vol 337, Martinus Nijhoff Publishers 2008) 314.

[108] Burke considers that views to the contrary are non-defendable ('Compatibility and Precaution in the 1995 Straddling Stock Agreement' in H. N. Scheiber (ed), *Law of the Sea: The Common Heritage and Emerging Challenges* (Martinus Nijhoff Publishers 2000) 110); Hayashi argues that the UNFSA has not become CIL and hence is not opposable to third parties ('Regional Fisheries Management Organisations and Non-Members' in T. M. Ndiaye and R. Wolfrum (eds), *Law of the sea, environmental law and settlement of disputes: Liber amicorum Judge Thomas A. Mensah* (Nijhoff 2007) 761–762).

instrument (or some of its provisions) reached CIL status then the rules contained in it can be applied to all states.[109] Also, or alternatively, if the UNFSA is interpreting or implementing rather than amending UNCLOS rules, its provisions are to be applicable to all states, whether they are contracting parties to either treaty or not, through the customary nature of UNCLOS.

It is a defendable view to say that most of the UNFSA provisions reflect CIL, as both *opinio juris* and practice can be found to justify this conclusion. Indeed, these treaty provisions were negotiated by the international community as a whole and were adopted by consensus.[110] Most articles, except the ones on enforcement and on dispute settlement, are directed to all states, not only to states parties—and this is the wording the whole international community has agreed upon during the negotiations.

Also, the UNGA, in several resolutions, called upon all states, not only states parties, to ensure that their vessels respect measures adopted by RFMOs, consistently with some provisions of the UNFSA.[111] Then, already at the Review Conference in 2006 "virtually all governments represented [...] – even those not yet party to the UNFSA – indicated their acceptance of the treaty as an expression of basic standards for the management of ocean fisheries",[112] with the exception of the enforcement provisions.[113] The outcome of the Review Conference in 2010 repeated that UNCLOS and the UNFSA constitute the legal framework applicable to straddling and highly migratory fish stocks.[114]

---

[109] VCLT (n 106) article 38; *North Sea Continental Shelf Case (Federal Republic of Germany/Denmark; Federal Republic of Germany/Netherlands)* (n 106) 41 §71; Aust (n 106) 210–211. According to R. Rayfuse, some of the provisions of the UNFSA are CIL ('To Our Children's Children's Children: From Promoting to Achieving Compliance in High-seas Fisheries' (2005) 20 International Journal of Marine and Coastal Law 509, 525).

[110] Harrison (n 44) 108.

[111] For example in UNGA, Resolution adopted by the General Assembly on 12 December 2002 – Agreement for the Implementation of the Provisions of the United Nations Convention on the Law of the Sea of 10 December 1982 relating to the Conservation and Management of Straddling Fish Stocks and Highly Migratory Fish Stocks 26 February 2003, A/RES/57/143 §8; UNGA, Resolution adopted by the General Assembly on 17 November 2004 – Sustainable fisheries, including through the 1995 Agreement for the Implementation of the Provisions of the United Nations Convention on the Law of the Sea of 10 December 1982 relating to the Conservation and Management of Straddling Fish Stocks and Highly Migratory Fish Stocks, and related instruments 17 January 2005, A/RES/59/25 §9.

[112] Balton and Koehler (n 98) 7.

[113] Enforcement on the high seas against non-flag state vessels has from the start been very controversial. See for example M. Hayashi, 'Enforcement by Non-Flag States on the High Seas Under the 1995 Agreement on Straddling and Highly Migratory Fish Stocks' (1996–1997) 9 Georgetown International Environmental Law Review 1.

[114] UNFSA, 'Report of the resumed Review Conference on the Agreement', A/CONF.210/2010/7 (n 99) annex §1.

Moreover, several states, both parties and non-parties, have taken steps to implement this instrument.[115] The relevant articles are mainly the provisions which develop or give content to some of UNCLOS general rules, particularly the obligation to cooperate in conservation.[116]

Furthermore, RFMOs seem, in practice, to be the main vehicle for states' cooperation and they are referred to as such in other internationally adopted documents.[117] Finally, many RFMOs declare that they base their decisions on science, make mention of the precautionary principle and try to close the high seas fisheries to non-members, for example by refusing landing of catches.[118] In the older agreements, an evolution towards these principles can be witnessed, such as within NAFO, CCAMLR, or ICCAT, which, already at the end of the twentieth century, implemented a program against non-members fishing in its area, in the form of trade sanctions.[119] As for the RFMOs that were created or updated after the

[115] G. Vigneron, 'The Most Recent Efforts in the International Community to Implement the 1995 United Nations Straddling Fish Stocks Agreement' (1999) 10 Colorado Journal of International Environmental Law and Policy 225, 231–233.

[116] F. Orrego Vicuña, *The Changing International Law of High Seas Fisheries* (Cambridge University Press 1999) 205.

[117] For example, the Code of Conduct for Responsible Fisheries (FAO 1995) article 6(12); FAO, International Plan of Action to prevent, deter and eliminate illegal, unreported and unregulated fishing (IPOA-IUU) 2001 §22 and 28 in particular.

[118] R. Barston, 'The Law of the Sea and Regional Fisheries Organisations' (1999) 14 International Journal of Marine and Coastal Law 333, 346–347; M. W. Lodge and others, 'Recommended Best Practices for Regional Fisheries Management Organizations: Report of an independent panel to develop a model for improved governance by Regional Fisheries Management Organizations' (Chatham House 2007) 21, 133–139; G. Lugten, 'The role of international fishery organizations and other bodies in the conservation and management of living aquatic resources' FAO Fisheries and Aquaculture Circular 1054 (Rome 2010) 21–22; F. Orrego Vicuña, 'Le droit de la pêche en haute mer: développement et conservation des ressources' in Mélanges offerts à Laurent Lucchini et Jean-Pierre Quéneudec (ed), *La mer et son droit* (Pedone 2003) 473–475. For more on these aspects, see infra Chapter 3 B. II. 1.-3.

[119] Balton and Koehler (n 98) 7–8; P. A. Nickler, 'A tragedy of the commons in coastal fisheries: contending prescriptions for conservation, and the case of the Atlantic Bluefin Tuna' (1998–1999) 26 Environmental Affairs Law Review 549, 557–558. More generally on the evolution following the UNFSA's adoption, see Committee on Fisheries, 'Strengthening regional fisheries management organizations and their performances including the outcome of the 2007 RFMOs meeting', presented at the Twenty-seventh Session of the Committee on Fisheries (Rome, 5–9 March 2007) COFI/2007/9 Rev.1 (Rome 2007) 4 §14.

adoption of the UNFSA, the legal framework they implement reflects many of its provisions.[120]

It is true that the CIL nature of the UNFSA might be rejected, on the basis that several states did not join or cooperate with RFMOs, that a number of RFMOs do not feel they have to adapt to the UNFSA,[121] and that decisions taken in these organizations are in fact not always based on a precautionary approach. However, even if such a view is favored, it is nowadays accepted by several commentators that many provisions of the UNFSA implement the relevant articles of UNCLOS and hence represent their authoritative interpretation, or at least are a subsequent instrument to use for means of interpretation.[122] In contrast to other parts of the UNFSA which further develop the law of the sea or even bring in new concepts and principles,[123] articles 5, 6 and arguably 8 can be considered provisions that

---

[120] Balton and Koehler (n 98) 7; 'Convention on the Conservation and Management of Highly Migratory Fish Stocks in the Western and Central Pacific Ocean' (WCPFC Convention) 2000, 2275 *United Nations Treaty Series* 43; 'Convention on the Conservation and Management of Fishery Resources in the South East Atlantic Ocean' (SEAFO Convention) 2001, 2221 *United Nations Treaty Series* 189, on which see A. Jackson, 'The Convention on the Conservation and Management of Fishery Resources in the South East Atlantic Ocean, 2001: an introduction' (2002) 17 International Journal of Marine and Coastal Law 33; NAFO is being amended (Bowman, Davies and Redgwell (n 57) 138–139); NEAFC has undergone major transformations from 1995 to 2005; T. Henriksen, G. B. Hønneland and A. K. Sydnes, *Law and Politics in Ocean Governance: The UN Fish Stocks Agreement and Regional Fisheries Management Regimes* (Publications on Ocean Development, Martinus Nijhoff Publishers 2006) 117–118, 129–130; Lugten (n 118) 18–19; Orrego Vicuña, 'Le droit de la pêche en haute mer: développement et conservation des ressources' (n 118) 474–475.

[121] Lugten (n 118) 18.

[122] D. H. Anderson, 'The Straddling Stocks Agreement of 1995: An Initial Assessment' (1996) 45 International and Comparative Law Quarterly 463, 468; D. Freestone, 'Implementing Precaution Cautiously: the Precautionary Approach in the Straddling and Highly Migratory Fish Stocks Agreement' in E. Hey (ed), *Developments in International Fisheries Law* (Kluwer Law International 1999) 300; Harrison (n 44) 107; Henriksen, Hønneland and Sydnes (n 120) 15; T. Henriksen, 'Revisiting the Freedom of Fishing and Legal Obligations on States Not Party to Regional Fisheries Management Obligations' (2009) 40 Ocean Development and International Law 80, 81; Schram and Tahindro (n 65) 284.

[123] Hayashi lists for example the non-flag state enforcement measures, the precautionary approach and the principle of compatibility, but also quite confusingly the compulsory participation in RFMOs, which he similarly inserts in his enumeration of measures facilitating the implementation of UNCLOS (M. Hayashi, 'The 1995 Agreement on the Conservation and Management of Straddling and Highly Migratory Fish Stocks: Significance for the Law of the Sea Convention' (1995) 29 Ocean and Coastal Management 51, 53–54, 56–62). See also E. Franckx, 'Pacta Tertiis and the Agreement for the Implementation of the Straddling and Highly Migratory Fish Stocks Provisions of the United Nations Convention on the Law of the Sea' (2000) 8 Tulane Journal of International and Comparative Law 49, 61–62.

facilitate the implementation of UNCLOS "by setting forth specific measures to be taken."[124] As such, these articles would be applicable to some extent to all states since these are bound by obligations under UNCLOS or the CIL it reflects. It means that, if article 8 indeed provides the interpretation and/or implementation method for the obligation to cooperate, all states are bound to join/cooperate with RFMOs or must refrain from fishing on the high seas.[125]

However, some arguments can be developed against the global applicability of the UNFSA.[126] It is said to have been "adopted to supplement the regime of the Convention"[127] and that some provisions go beyond interpretation or implementation of UNCLOS,[128] "br[oke] new ground in international fisheries law",[129] or have not yet reached CIL status.[130] As seen above, some commentators have tried to categorize provisions. Reference is also made to the rather narrow membership of the UNFSA and its consequently limited applicability, thus considering it as normal treaty law.[131] Furthermore, it has been said that article 8 (4) is novel and, hence,

---

[124] Hayashi, 'The 1995 Agreement on the Conservation and Management of Straddling and Highly Migratory Fish Stocks: Significance for the Law of the Sea Convention' (n 123) 53–54; F. Orrego Vicuña, 'The International Law of High Seas Fisheries: From Freedom of Fishing to Sustainable Use' in O. S. Stokke (ed), *Governing High Seas Fisheries: The Interplay of Global and Regional Regimes* (Oxford University Press 2001) 40–42.

[125] The question is always how much interpretation is needed or allowed and how to avoid that interpretation becomes substitution. On this, see *Gabcikovo-Nagymaros Project (Hungary/Slovakia), Judgment,* 25 September 1997, I.C.J. Reports 1997 7, Separate Opinion of Judge Bedjaoui §12; Harrison (n 44) 108; Henriksen (n 122) 81.

[126] For example, A. Serdy argued that it does not create an objective regime, since its quasi-closure of the commons is more in the interest of the states already fishing in the region than in the general interest ('Postmodern International Fisheries Law, or We Are All Coastal States Now' (2011) 60 International and Comparative Law Quarterly 387, 389–390).

[127] H. Tuerk, 'The Waning Freedom of the Seas' in R. Casado Raigón and G. Cataldi (eds), *L'évolution et l'état actuel du droit international de la mer: Mélanges de droit de la mer offerts à Daniel Vignes* (Bruylant 2009) 932.

[128] C. Kojima and V. S. Vereshchetin, 'Implementation Agreements' Max Planck Encyclopedia of Public International Law www.mpepil.com §9, 22; Orrego Vicuña, 'Le droit de la pêche en haute mer: développement et conservation des ressources' (n 118) 471.

[129] Henriksen, Hønneland and Sydnes (n 120) 1; R. Bratspies, 'Finessing King Neptune: Fisheries Management and the Limits of International Law' (2001) 25 Harvard Environmental Law Review 213, 237–241.

[130] T. Bjorndal, 'Overview, roles, and performance of the North East Atlantic fisheries commission (NEAFC)' (2009) 33 Marine Policy 685, 687; Vincent, *Droit de la mer* (n 83) 225.

[131] S. Beslier, 'The exercise of jurisdiction over vessels: new developments in the sector of fisheries' in E. Franckx and P. Gautier (eds), *The Exercise of Jurisdiction over Vessels: New Developments in the Fields of Pollution, Fisheries, Crimes at Sea and Trafficking of Weapons of Mass Destruction* (Bruylant 2010) 51; D. S. Calley, *Market Denial and International Fisheries Regulation: The Targeted and Effective Use of Trade Measures Against the Flag of Convenience Fishing Industry* (Martinus Nijhoff Publishers 2012) 114; Henriksen, Hønneland and Sydnes (n 120) 12; Rayfuse, 'The Interrelationship between the Global Instruments of International Fisheries Law' (n 63) 113–114.

cannot break the *pacta tertiis* rule.[132] It nevertheless appears that objection to the content of article 8 was not, so far, presented as a reason not to ratify the treaty.[133] In any case, no global consensus seems to have been reached by the doctrine at this point in time as to the nature of the UNFSA content.

It should also be noted that, while the practice of the states parties to RFMOs is logically one of protecting 'their' resources against non-parties, the practice of non-members is not always one of cooperation.[134] Such an attitude from non-members does not seem to be criticized as a violation of international law. Solutions are found in a softer manner: for example in the 2006 UN Review Conference, "members of regional fisheries management organizations should continue to *encourage* non-member States and entities fishing in areas under the competence of those organizations to participate".[135] Encouragement would probably not be the term used if it was beyond doubt that non-members are de jure obliged to participate.

Finally, in the *Southern Bluefin Tuna* case, the Arbitral Tribunal judged that the situation could have been different, both in terms of substance and competence of the dispute settlement mechanism under UNCLOS, for states parties to the UNFSA, if it had been in force.[136] This could well support the view that at least parts of the UNFSA are to be treated as new law since "[t]he substantive provisions of the Straddling Stocks Agreement are more detailed and far-reaching than the pertinent provisions of UNCLOS or even of the CCSBT."[137]

4. Conclusions

The UNFSA requires a precautionary approach and gets closer to an ecosystem approach. It also affirms the pre-eminent role of RFMOs and the obligation to either join them or refrain from fishing in the relevant areas. Furthermore, although it confirms the primary role of the flag state and its duties, it develops non-flag state compliance and enforcement mechanisms, particularly on the high seas and through port states.

Whether the content of this treaty reached CIL or is the authoritative interpretation of UNCLOS is still subject to some debate, hence leaving the scope of applicability of this instrument as well as the status of the freedom of fishing

[132] Hayashi, 'The 1995 Agreement on the Conservation and Management of Straddling and Highly Migratory Fish Stocks: Significance for the Law of the Sea Convention' (n 123) 66, as presented in Franckx, 'Pacta Tertiis and the Agreement for the Implementation of the Straddling and Highly Migratory Fish Stocks Provisions of the United Nations Convention on the Law of the Sea' (n 123) 63 n 63.

[133] Molenaar, 'Non-Participation in the Fish Stocks Agreement: Status and Reasons' (n 103) 227.

[134] UNFSA, 'Annex: Outcome of the Review Conference' (n 98) 16–17 §76, 78.

[135] Ibid. 16 §76 (italics added).

[136] *Southern Bluefin Tuna Arbitral Award (New Zealand v. Japan, Australia v. Japan)* (n 40) §71.

[137] Ibid.

non-clarified.[138] However, while the innovative nature of some provisions cannot be denied—in particular the ones related to non-flag state enforcement—and the conduct of some non-parties to RFMOs could hint to an applicability of the UNFSA limited to its members, most provisions, drafted by the international community and adopted by consensus, are literally directed to all states. Also, that some states disregard their obligations of control over their vessels and of cooperation in the conservation of marine resources, for example by serving as flags of convenience, is not representative of the behavior of the international community in its entirety. Moreover, the widespread application of the precautionary principle, of the ecosystem approach and the role of RFMOs have all increased after the adoption of the UNFSA; this Agreement hence appears to have clarified the content of states' obligations in the field of fisheries. One can reasonably consider the provisions of the UNFSA, at the exception of the enforcement ones, to be a reflection of CIL and/or to be the authoritative interpretation of UNCLOS.

B. Management Regimes

I. Food and Agriculture Organization

1. Institutional Structure

The FAO, created in 1945, is a UN specialized agency. Its main functions are "to assist member states in the utilization of their agriculture, including fisheries, by providing information and recommending research and concerted action with respect to, inter alia, the conservation of natural resources."[139] It does not hold any mandate to actually manage fisheries and refrains from doing so.

a) Fisheries and Aquaculture Department

The Fisheries and Aquaculture Department aims at ensuring food security and improving conservation and utilization of aquatic resources, by strengthening the

[138] This matter was not directly discussed by ITLOS in the advisory opinion it issued following the request of the Sub-Regional Fisheries Commission (SRFC), received on 28 March 2013 (SRFC, 'Request for an advisory opinion' (27 March 2013) http://www.itlos.org/fileadmin/itlos/documents/cases/case_no.21/Request_eng.pdf accessed 6 July 2015), but ITLOS did examine the nature of the duty to cooperate in conservation and management in the EEZs. See infra Chapter 4 A. IV. 3. c) aa).

[139] Young, *Trading Fish, Saving Fish* (n 31) 47, on the 'Constitution of the Food and Agriculture Organization of the United Nations' 1945 article I.

global governance as well as the more local capacities.[140] It does so by developing targeted plans of action, as well as by supporting knowledge dissemination through various databases, interfaces, statistics and publications.[141] Technical assistance is also provided through field projects and capacity building. As for its advisory and coordination role, the FAO is to provide support to RFMOs established under its auspices and it promotes collaboration between all RFMOs.[142]

b) Committee on Fisheries

The Committee on Fisheries (COFI) is the main body dealing with fisheries matters within the FAO. This subsidiary body of the FAO Council[143] was set up in 1965. It is meant to examine major fisheries and aquaculture issues, make recommendations to the relevant institutions and stakeholders as well as review the programs established by the FAO and their implementation.[144] Two sub-committees have been established to examine specific issues, one on trade and one on aquaculture.

Membership is open to members and non-members of the FAO.[145] Representatives of UN bodies, international organizations, NGOs and RFMOs are also welcome as participants to the biennial meetings but they cannot vote.[146]

COFI is the major forum where governments discuss fisheries and aquaculture problems, where recommendations are made to all major relevant stakeholders—e.g. governments, RFMOs, FAO—and where some fisheries instruments are negotiated.[147] COFI is said to "supplemen[t] rather than supplan[t] other organizations working in the field of fisheries and aquaculture."[148] As such it is one of the places where all fisheries matters can be addressed with a global overview. Various issues

---

[140] FAO, 'About us – Fisheries and Aquaculture Department' http://www.fao.org/fishery/about/en accessed 4 July 2015.

[141] For example the Fisheries Global Information System, FishStat and SOFIA.

[142] FAO, 'FAO and Regional Fishery Bodies' http://www.fao.org/fishery/topic/16918/en accessed 4 July 2015.

[143] The Conference is the supreme governing body of the FAO, is composed of a representative of each member state and meets every two years. The Council is a non-plenary body of 49 representatives chosen to ensure geographical distribution and which serves of governing body between the Conference meetings as well as undertakes the activities it is entrusted with in the FAO Constitution. It is supported by a number of Committees, including the COFI. See K. Mechlem, 'Food and Agriculture Organization of the United Nations (FAO)' Max Planck Encyclopedia of Public International Law www.mpepil.com §9–13; P. Sands, P. Klein and D. W. Bowett, *Bowett's law of international institutions* (6th edn Sweet & Maxwell 2009) 85.

[144] FAO, 'Committee on Fisheries (COFI) – Fisheries and Aquaculture Department' http://www.fao.org/fishery/about/cofi/en accessed 4 July 2015; see in detail 'General Rules of the Organization', *Basic texts of the FAO* (2013 edn) rule XXX.

[145] 'Rules of Procedures of the Committee on Fisheries', *Basic texts of the FAO* (2013 edn) rule III.

[146] Ibid; 'General Rules of the Organization' (n 144) rule XVII.

[147] FAO, 'Committee on Fisheries (COFI) – Fisheries and Aquaculture Department' (n 144).

[148] Ibid.

are brought up at the meetings. For example, at the COFI Sub-Committee on Fish Trade in February 2012, several key issues were examined such as traceability, market access requirements, CITES and monitoring of the implementation of the article on trade of the Code of Conduct for Responsible Fisheries.[149] The update on CITES-related activities focused on the application of CITES criteria, examined the newly proposed Memorandum of Cooperation between FAO and CITES and considered the priorities in the CITES-related work.[150]

COFI cannot take decisions on its own and must refer to the FAO Council or Conference therefor. As most international organizations' acts, FAO Resolutions and Decisions are not directly binding on the member states, particularly since the FAO Constitution describes such decisions as "recommendations [...] for consideration by the [member states] with a view to implementation by national action."[151]

2. Binding Agreements Developed Under the FAO

Under article XIV of its Constitution, the FAO has a role to play in treaty-making. Indeed, the FAO Conference and the FAO Council can adopt conventions, respectively with a global reach and regional or implementing treaties.[152] Two binding agreements regarding fisheries were developed under the FAO auspices: the Compliance Agreement and the Port State Measures Agreement.

a) 1993 Compliance Agreement

The Agreement to Promote Compliance with International Conservation and Management Measures by Fishing Vessels on the High Seas was adopted in 1993. It was developed—and so was the 1995 Code of Conduct presented here below[153]—as an answer to the Cancun Declaration issued at the 1992 International Conference on Responsible Fishing. Just after this conference, the UN Conference on Environment and Development was held in Rio and Agenda 21 was adopted.[154] As made evident by these meetings and declarations, the awareness as to problems related to high

---

[149] Committee on Fisheries – Sub-Committee on Fish Trade, 'Provisional Agenda and Time Table of the thirteenth session of the Sub-Committee on Fish Trade (Hyderabad, 20–24 February 2012)' COFI:FT/XIII/2012/1 (2012); Code of Conduct for Responsible Fisheries (n 117) article 11.

[150] Committee on Fisheries – Sub-Committee on Fish Trade, 'Update on CITES related activities' presented at the Thirteenth Session of the Sub-Committee on Fish Trade (Hyderabad, 20–24 February 2012) COFI:FT/XIII/2012/8 (2012).

[151] Constitution of the FAO (n 139) article IV(3).

[152] Ibid. article XIV(1)(2).

[153] See infra Chapter 3 B. I. 3 a).

[154] Churchill and Lowe (n 30) 302.

seas fisheries was rising, particularly in what concerned the use of flags of convenience by fishing vessels.[155]

The Compliance Agreement clarifies the responsibilities of flag states towards their vessels on the high seas.[156] Such states must take necessary measures to ensure that vessels flying their flag are not engaging in any activity undermining CMMs' effectiveness. Authorization for fishing must be given by the relevant national authorities only if the flag state is able to exercise effective control on its vessels. Limitations to the liberties of flag shopping are also provided.

This instrument promotes information gathering and sharing:[157] flag states have to maintain records of all vessels authorized to fish under their flag and share this information with the FAO. Information on vessels undermining CMMs also has to be forwarded.

Nevertheless, some gaps exist in this treaty, such as that vessels under 24 meters of length can be exempted from the obligation to be put on record and to receive an authorization to fish.[158] Also, there is no compulsory dispute settlement mechanism. Hence, the discretionary powers of flag states continue to de facto apply. This is particularly relevant since the exclusive jurisdiction of the flag states is not questioned in this instrument.[159] Finally, and mainly, the Compliance Agreement, which entered into force 10 years after its adoption, suffers from a low membership of 40 parties.[160] While some general rules are a reflection of CIL, the detailed provisions were only meant to apply between parties and certainly do so.[161]

b) 2009 Port State Measures Agreement

The Agreement on Port State Measures to Prevent, Deter and Eliminate Illegal, Unreported and Unregulated Fishing is the most recent fishing-related agreement adopted under the FAO auspices. It clearly builds on the FAO Port State Scheme,

---

[155] D. A. Balton, 'The Compliance Agreement' in E. Hey (ed), *Developments in International Fisheries Law* (Kluwer Law International 1999) 34–35; G. Moore, 'The FAO Compliance Agreement' in M. H. Nordquist and G. Moore (eds), *Current Fisheries Issues and the Food and Agriculture Organization of the United Nations* (Kluwer Law International 2000) 78. On general comments on flags of convenience, see for example Y. Tanaka, *The International Law of the Sea* (Cambridge University Press 2012) 157–159.

[156] 'FAO Agreement to Promote Compliance with International Conservation and Management Measures by Fishing Vessels on the High Seas: Compliance Agreement' (Compliance Agreement) 1993, 2221 *United Nations Treaty Series* 91 article III.

[157] Ibid. articles IV-VI.

[158] Ibid. article II(2).

[159] Harrison (n 44) 209.

[160] FAO, 'Parties to the Agreement to Promote Compliance with International Conservation and Management Measures by Fishing Vessels on the High Seas' http://www.fao.org/fileadmin/user_upload/legal/docs/012s-e.pdf accessed 4 July 2015.

[161] Harrison (n 44) 212.

which was adopted in 2005.[162] Its purpose is to fight IUU fishing by promoting the use of port state measures. These states are indeed required to take a number of measures, such as designating ports which fishing vessels can enter,[163] requesting information from vessels before entry,[164] and then refusing entry into ports to vessels suspected of IUU activities.[165] Inspections in ports are also to be undertaken, followed, if necessary, by notifications to the flag state which is the only entity entitled to take enforcement actions.[166] In such cases, the port state must nevertheless deny landing, transshipment, or also port services, which is a form of sanction.[167]

Unfortunately, there is no provision for a binding compulsory dispute settlement.[168] On the positive side, while mechanisms to check port states' compliance with such measures are currently lacking within RFMOs,[169] the FAO agreement requires "the regular and systematic monitoring and review of the implementation of this Agreement".[170] This could be the basis for more accountability and cooperation. Finally, port state measures, by their nature, are applicable to both parties and non-parties to the Port State Measures Agreement. Indeed, should enough coastal states implement them, the benefits of flags of convenience could be importantly decreased.

3. Non-binding Codes and Plans of Action

Several non-binding codes and plans of action were developed under the FAO auspices. Of major importance here are the 1995 Code of Conduct for Responsible Fisheries, the 1999 International Plan of Action on Sharks and the 2001 International Plan of Action on Illegal, Unreported and Unregulated Fishing. A number of

[162] E. J. Molenaar, 'Port State Jurisdiction' Max Planck Encyclopedia of Public International Law www.mpepil.com §33.

[163] 'Agreement on Port State Measures to Prevent, Deter and Eliminate Illegal, Unreported and Unregulated Fishing' (Port State Measures Agreement) 2009 article 7.

[164] Ibid. article 8.

[165] Ibid. article 9.

[166] Ibid. articles 12, 15, 18, 20.

[167] Ibid. article 18(1)(b).

[168] Ibid. article 22.

[169] S. Flothmann and others, 'Closing Loopholes: Getting Illegal Fishing Under Control' (2010) 328 Science 1235, 1236.

[170] Port State Measures Agreement (n 163) article 24; FAO, *Groundbreaking treaty on illegal fishing approved: Port state measures broaden the fight against IUU fishing* (25 November 2009, Rome) http://www.fao.org/news/story/en/item/37627/icode/ accessed 4 July 2015.

(technical) guidelines have also been developed on several aspects of fisheries and aquaculture management,[171] but since they are not adopted by the states or a political body,[172] and hence have no binding, nature they will not be described here.

a) 1995 Code of Conduct for Responsible Fisheries

The 1995 Code of Conduct for Responsible Fisheries (CCRP) is a voluntary instrument[173] which "provides principles and standards applicable to the conservation, management and development of all fisheries."[174] These standards can be implemented at the national or regional level. The Code covers a much broader set of topics than other instruments such as the UNFSA. This comprehensive approach would have been difficult to attain, had the Code been of a binding nature.[175] The Code is applicable to all aspects of fisheries, be it "capture, processing and trade of fish and fishery products, fishing operations, aquaculture, fisheries research and the integration of fisheries into coastal area management."[176] It provides guidance to all actors in fisheries[177] to attain responsible fisheries and applies to all areas of jurisdiction.[178]

The Code acknowledges the need for a (quasi-)ecosystem approach,[179] for science-based management decisions,[180] for the precautionary principle in case of uncertainty,[181] as well as the obligation of states to cooperate,[182] and the necessity to protect biodiversity.[183] Responsibilities of the coastal states, flag states and port states are listed.[184]

Article 9 develops coastal states' duties with regard to aquaculture, inter alia, the obligation to promote responsible development and management, to evaluate the

---

[171] For example the FAO, 'International Guidelines for the Management of Deep-sea Fisheries in the High Seas' (Rome 2009); FAO, 'Technical Guidelines on Aquaculture Certification' (Rome 2011).

[172] J. Friedrich, 'Legal Challenges of Nonbinding Instruments: The Case of the FAO Code of Conduct for Responsible Fisheries' (2008) 9 German Law Journal 1539, 1550.

[173] Code of Conduct for Responsible Fisheries (n 117) article 1.1.

[174] Ibid. article 1.3.

[175] W. R. Edeson, 'Current Legal Developments – Food and Agriculture Organization of the UN: The Code of Conduct for Responsible Fisheries: An Introduction' (1996) 11 International Journal of Marine and Coastal Law 233, 233–234; Friedrich (n 172) 1561.

[176] Code of Conduct for Responsible Fisheries (n 117) article 1.3.

[177] Ibid. article 1.2.

[178] Friedrich (n 172) 1547–1548.

[179] Code of Conduct for Responsible Fisheries (n 117) article 6.1.

[180] Ibid. articles 6.4, 7.3.1.

[181] Ibid. articles 6.5, 7.5.

[182] Ibid. articles 6.12, 6.15, 7.1.3.

[183] Ibid. article 7.2.2(d).

[184] Ibid. article 8.1–8.3.

effects on genetic diversity and ecosystem integrity, to ensure that aquaculture development is ecologically sustainable and to cooperate with other states to protect transboundary aquatic ecosystems.[185] The Code then goes on to mention relationships between fisheries and international trade, by emphasizing the need for responsible international trade, which should not compromise the sustainable development of fisheries and responsible utilization of aquatic living resources. Hence, it recognizes that, under certain conditions, restrictions to trade are possible.[186]

b) 1999 International Plan of Action for the Conservation and Management of Sharks

The 1999 International Plan of Action for the Conservation and Management of Sharks (IPOA-Sharks) is a voluntary instrument designed within the framework of the CCRP.[187] It applies to states in the waters of which sharks are caught by their own or foreign vessels and to states the vessels of which catch sharks on the high seas. All states concerned are encouraged to participate in sharks' management.[188]

"The objective of the IPOA-[Sharks] is to ensure the conservation and management of sharks and their long-term sustainable use."[189] The IPOA-Sharks "include[s] all species of sharks, skates, rays and chimaeras" and targets all types of catches (directed, by-catch, commercial, recreational or others).[190]

This instrument lists the actions needed for the management and conservation of sharks to be efficient, thus allowing their long-term sustainable use. These include sharks' status assessments and the development of national plans of actions.[191] The aims of a Shark-Plan are inter alia to ensure the sustainability of shark catches, to evaluate threats to shark populations, to conserve critical habitats, to minimize waste and discards, and to protect biodiversity.[192] To keep catches within sustainable levels, management and conservation should apply a precautionary

---

[185] Ibid. article 9.

[186] Ibid. article 11.

[187] Ibid. article 2(d); FAO, International Plan of Action for the Conservation and Management of Sharks (IPOA-Sharks) 2001 §10.

[188] Ibid. §13. It must be noted that the adoption and implementation of national plans of actions have been slow, incomplete and with debated efficiency (Australia, 'CoP12 Doc. 41.1 – Conservation and management of sharks' (3–15 November 2002) 3–4 §13; M. Lack and G. Sant, 'The Future of Sharks: A Review of Action and Inaction' (TRAFFIC January 2011) 16), but are progressing (J. Fischer and others, 'Review of the Implementation of the International Plan of Action for the Conservation and Management of Sharks' FAO Fisheries and Aquaculture Circular 1076 (Rome 2012)).

[189] IPOA-Sharks (n 187) §16.

[190] Ibid. §11.

[191] Ibid. §18, 21.

[192] Ibid. §22.

approach.[193] States are also urged to take effective CMMs for transboundary, straddling, highly migratory and high seas stocks.[194]

c) 2001 International Plan of Action Against IUU Fishing

As for the 2001 International Plan of Action against IUU fishing (IPOA-IUU), its focus, indicated by its name, is on IUU fishing. It was also developed under article 2 (d) of the CCRP. It provides a broadly recognized definition of IUU fishing.[195] As a voluntary document, it is a 'tool box' for states and RFMOs, where they can find ideas on how to tackle IUU.[196] Each country which desires can notify the FAO of its national plan of action (NPOA).[197]

The international plan of action covers most of the strategies and governance aspects relevant in the fight against IUU, such as market-based measures, the identification of responsibilities and duties of various groups of states (all states, flag states, coastal states, port states and states of nationality), the importance of research and the role of RFMOs.[198] This document reaffirms a number of rules and principles already developed in other non-binding or binding instruments. It was innovative in the sense that, at the time of its adoption, it was further developing the measures applicable, particularly these based on trade, port state controls and the control of states over their nationals.[199]

4. Conclusions

The Fisheries and Aquaculture Department of the FAO is a key actor in promoting and advancing the fisheries and aquaculture policies and more generally the relevant legal framework. Although it is not an overarching institution for fisheries law,[200] it brings together all states and applies a holistic approach to the field.[201] It plays a coordinating role through information gathering and sharing, being both

[193] Ibid. §14.

[194] Ibid. §26.

[195] IPOA-IUU (n 117) §3.1–3.3.

[196] W. Edeson, 'The International Plan of Action on Illegal, Unreported and Unregulated Fishing: The Legal Context of a Non-Legally Binding Instrument' (2001) 16 International Journal of Marine and Coastal Law 603, 623.

[197] Young, *Trading Fish, Saving Fish* (n 31) 48–49.

[198] IPOA-IUU (n 117) part IV.

[199] Edeson (n 196) 605; IPOA-IUU (n 117) §18–19, 52–76.

[200] Bowman, Davies and Redgwell (n 57) 123.

[201] Harrison (n 44) 234.

"the principal repositor[y] and disseminato[r] of technical expertise"[202] and it promotes national and international actions. It also has helped developing relevant international regulations[203] in particular by adopting binding and non-binding agreements of utmost importance in international fisheries law. The role of these documents "lies in their ability to provide a framework for the development of fisheries conservation and management measures and policies by other actors such as fishing states, other fishing entities and regional fishing bodies."[204]

Also, the procedure to design and adopt both binding and non-binding instruments is very inclusive. Indeed, usually, either the Conference, the Council, or COFI mandates the Secretariat to undertake some actions regarding a particular topic. Then, an expert consultation (i.e. a gathering of experts working in their personal capacities) is organized, followed by a technical consultation (where member states' representatives meet and where consensus is sought for decisions). The results of this second consultation are endorsed, either by COFI or one of the Sub-Committees. The report of COFI is further endorsed by the Council (for administrative matters) or the Conference (for policy matters). Hence, the soft-law guidelines are the result of a text negotiated by the parties, after it was put together during an expert consultation. As for the binding instruments that have been negotiated under the auspices of the FAO (such as the Port State Measures Agreement), they have further been ratified by a certain number of countries before entering into force.

Nevertheless, it must be noted that the binding instruments suffer from a low ratification rate and that voluntary instruments have not always been implemented as widely as could be hoped for.[205]

---

[202] A. Boyle and C. Chinkin, *The Making of International Law* (Oxford University Press 2007) 125.

[203] Ibid. 126.

[204] Harrison (n 44) 234–235.

[205] For example, the IPOA-Sharks has not had for consequence a widespread adoption of national plans of action (see supra footnote 188) and, even when such plan was developed, it can suffer from major inconsistencies with the international requirements (on the Canadian example, see B. Davis and B. Worm, 'The International Plan of Action for Sharks: How does national implementation measure up?' (2013) 38 Marine Policy 312).

II. Regional Fisheries Management Organizations

1. General Characteristics

Regional fisheries bodies have been developed since the second part of the twentieth century and now cover most of the high seas.[206] While a majority of them encompass both the high seas and adjacent waters under the national jurisdiction of their members,[207] a few adopt measures only directed to the high seas.[208] Some are deciding bodies, the RFMOs, while others only have an advisory or scientific role without the power to adopt binding measures or to directly manage fisheries, an example being the International Council for the Exploration of the Sea.[209]

The focus of this research is on the first group which includes about 20 organizations.[210] Some have a regional basis such as NEAFC,[211] NAFO, GFCM or CCAMLR. Others are rather species-specific, such as the tuna RFMOs: IOTC, IATTC, WCPFC, CCSBT and ICCAT.

Most RFMOs are international organizations with legal personality,[212] a specific budget and staff, but this ultimately depends on their founding treaty or on their need for legal personality to perform the functions entrusted to them, as established

[206] For a full typology, see A. K. Sydnes, 'Regional Fishery Organizations: How and Why Organizational Diversity Matters' (2001) 32 Ocean Development and International Law 349, 353–356; and for a general presentation, see E. Meltzer, 'Global Overview of Straddling and Highly Migratory Fish Stocks: Maps and Charts Detailing RFMO Coverage and Implementation' (2005) 20 International Journal of Marine and Coastal Law 571.

[207] CCAMLR, CCSBT, GFCM, IATTC, ICCAT, IOTC, NAFO, NEAFC, WCPFC.

[208] SEAFO, SIOFA, SPRFMO.

[209] Henriksen, Hønneland and Sydnes (n 120) 3; Sydnes (n 206) 354.

[210] For an overview of the major RFMOs which will be referred to, see tables in Annexes 1 and 2 as well as illustrative maps on the FAO website ('Regional Fishery Bodies Map Viewer' http://www.fao.org/figis/geoserver/factsheets/rfbs.html accessed 6 July 2015).

[211] For a general presentation of NEAFC, see Bjorndal (n 130).

[212] International organizations are defined as "forms of cooperation (1) founded on an international agreement; (2) having at least one organ with a will of its own; and (3) established under international law" (H. G. Schermers and N. Blokker, *International institutional law: Unity within diversity* (5th edn Nijhoff 2011) §33); see also the similar definition in International Law Commission (ILC), 'Draft Articles on the Responsibility of International Organizations, with Commentaries' 2011 article 2(a).

As for international personality, it is said that "[h]aving international legal personality for an international organization means possessing rights, duties, powers and liabilities etc. as distinct from its members or its creators on the international plane and in international law" (C. F. Amerasinghe, *Principles of the Institutional Law of International Organization* (Cambridge Studies in International and Comparative Law, 2nd edn Cambridge University Press 2005) 78).

in the doctrine of implied powers.[213] Some have been established under the auspices of the FAO,[214] but the majority of them are independent from such links.[215] Due to the negotiations that led to UNCLOS and the new attribution of large portions of the seas to national EEZs, several RFMOs' founding treaties were revised to take these changes into account.[216]

RFMOs are probably the most important institutions in conservation and management of fish stocks, or at least the ones that are the most practically involved therein.[217] They implement the obligation under UNCLOS to cooperate in dealing with transboundary, straddling or highly migratory stocks and are referred to, in the UNFSA, as of primary significance in that regard. Membership depends on each

[213] RFMOs the legal personality of which is explicitly declared include: CCAMLR ('Convention on the Conservation of Antarctic Marine Living Resources' (CCAMLR Convention) 1980, 1329 *United Nations Treaty Series* 47 article VIII); CCSBT ('Convention for the Conservation of Southern Bluefin Tuna' (CCSBT Convention) 1993, 1819 *United Nations Treaty Series* 360 article 6(9)); IATTC ('Convention for the Strengthening of the Inter-American Tropical Tuna Commission Established by the 1949 Convention Between the United States of America and the Republic of Costa Rica' (Antigua Convention) 2003 article VI(3)); NAFO ('Convention on Future Multilateral Cooperation in the Northwest Atlantic Fisheries' (NAFO Convention) 1978, 1135 *United Nations Treaty Series* 369 article II(3)); NEAFC ('Convention on the Future Multilateral Cooperation in North-East Atlantic Fisheries' (NEAFC Convention) 1980, 1285 *United Nations Treaty Series* 129 (revised 2006, 'New Convention' http://www.neafc.org/system/files/Text-of-NEAFC-Convention-04.pdf accessed 4 July 2015) article 3(2)); SEAFO (SEAFO Convention (n 120) article 5(3)); SPRFMO ('Convention on the Conservation and Management of High Seas Fishery Resources in the South Pacific Ocean' (SPRFMO Convention) 2009 article 6(3)); WCPFC (WCPFC Convention (n 120) article 9(6)). See also Henriksen, Hønneland and Sydnes (n 120) 2; J. Beer-Gabel and V. Lestang, *Les Commissions de Pêche et leur Droit: La Conservation et la Gestion des Ressources Marines Vivantes* (Bruylant 2003) 59.

The implied powers doctrine provides that an international organization may benefit from a wider range of competences than expressly provided for in its founding treaty, if such additional powers are "essential to the performance of its duties" (*Reparation for injuries suffered in the service of the United Nations, Advisory Opinion*, 11 April 1949, I.C.J. Reports 1949 174, 182). On this, see for example J. Klabbers, *An introduction to international institutional law* (Cambridge University Press 2009) 59–66; Sands, Klein and Bowett (n 143) 475–476; G. Ulfstein, 'Treaty Bodies' in D. Bodansky, J. Brunnée and E. Hey (eds), *Oxford Handbook of International Environmental Law* (Oxford University Press 2007) 880–881. On the international legal personality of international organizations, see for example M. Bettati, 'Création et personnalité juridique des organisations internationales' in R.-J. Dupuy (ed), *Manuel sur les organisations internationales – A Handbook on International Organizations* (2nd edn Martinus Nijhoff Publishers 1998) 48–59; T. Gazzini, 'Personality of international organizations' in J. Klabbers and A. Wallendahl (eds), *Research Handbook on the Law of International Organizations* (Edward Elgar 2011) 33–55.

[214] These founded under the Constitution of the FAO (n 139) article VI are of an advisory nature, while those founded under its article XIV can have management functions (Barston (n 118) 343–344).

[215] Beer-Gabel and Lestang (n 213) 53.

[216] Henriksen, Hønneland and Sydnes (n 120) 9, for example NAFO and NEAFC which were re-established respectively in 1978 and 1980.

[217] Harrison (n 44) 226.

RFMO; some are totally open to new members, but in other cases membership is explicitly restricted to coastal states and/or states with a (real) interest in fishing in the region, or entry is otherwise limited by the agreement of existing members.[218]

RFMOs regulate fishing through CMMs adopted by the conference of the parties, by a vote or by consensus,[219] on the basis of scientific advice, usually stemming from recommendations made by their scientific committee.[220] In most cases, it is possible for a member state to opt out from a CMM which it does not want to be bound by.[221] In recent years, some RFMOs such as ICCAT, GFCM and CCSBT have started regulating aquaculture practices, in addition to their traditional focus on capture fisheries.

---

[218] The conditions differ from one RFMO to the next. Unqualified openness of the RFMO to new entrants: CCSBT (CCSBT Convention (n 213) article 18); ICCAT ('International Convention for the Conservation of Atlantic Tunas' (ICCAT Convention) 1966, 673 *United Nations Treaty Series* 63 article XIV); unqualified for members of the FAO in IOTC ('Agreement for the Establishment of the Indian Ocean Tuna Commission' (IOTC Agreement) 1993, 1927 *United Nations Treaty Series* 329 articles IV and XVII); NAFO (NAFO Convention (n 213) article XXII(4)); SEAFO (SEAFO Convention (n 120) article 26); SIOFA ('Southern Indian Ocean Fisheries Agreement' (SIOFA Agreement) 2006 article 23); SPRFMO (SPRFMO Convention (n 213) article 37).

Some of these RFMOs nevertheless require an interest and/or fishing activities in the region: CCSBT, IATTC, IOTC, SEAFO, SIOFA, SPRFMO, and WCPFC. As for CCAMLR, it also includes states that have a real interest in research in the region CCAMLR Convention (n 213) article XXIX(1).

Vote and/or acceptance of existing members: two-third majority in GFCM ('Agreement for the establishment of the General Fisheries Commission for the Mediterranean' (GFCM Agreement) 1949, 126 *United Nations Treaty Series* 237 article XIII); various options including a consensus based one in IATTC (Antigua Convention (n 213) article XXX); three-quarter majority in NEAFC (NEAFC Convention (n 213) article 20(4)). Invitation by the contracting parties deciding by consensus: WCPFC (WCPFC Convention (n 120) article 35(2)).

[219] Consensus: CCAMLR for substantive matters (CCAMLR Convention (n 213) article XII(1)); IATTC (Antigua Convention (n 213) article IX); SEAFO (SEAFO Convention (n 120) article 17 (3)); SIOFA for substantive matters (SIOFA Agreement (n 218) article 8(1)); SPRFMO (SPRFMO Convention (n 213) article 16); WCPFC (WCPFC Convention (n 120) article 20(1)). Also to be noted is the CCSBT procedure of adoption by unanimity (CCSBT Convention (n 213) article 7).

Majority vote: GFCM (GFCM Agreement (n 218) article II(2)); ICCAT (ICCAT Convention (n 218) article VIII(1)(b)); IOTC (IOTC Agreement (n 218) article VI(2)); NAFO (NAFO Convention (n 213) article V(2)); NEAFC (NEAFC Convention (n 213) article 3(9)).

[220] CCAMLR (CCAMLR Convention (n 213) articles IX(1)(f) and (4)); CCSBT (CCSBT Convention (n 213) article 8(3)); GFCM (GFCM Agreement (n 218) article III); IATTC (Antigua Convention (n 213) article VII); ICCAT (ICCAT Convention (n 218) article VIII(1)); IOTC (IOTC Agreement (n 218) article V(1) and (2)(c)); NAFO (NAFO Convention (n 213) article XI); NEAFC (NEAFC Convention (n 213) articles 4(2), 5–6); SEAFO (SEAFO Convention (n 120) articles 3(a), 6); SIOFA (SIOFA Agreement (n 218) articles 4(a), 6(1)); SPRFMO (SPRFMO Convention (n 213) articles 3(1)(a), 8); WCPFC (WCPFC Convention (n 120) article 5(a)(b)).

[221] CCAMLR (CCAMLR Convention (n 213) article IX(6)(c)); GFCM (GFCM Agreement (n 218) article V(3)); ICCAT (ICCAT Convention (n 218) article VIII(3)); IOTC (IOTC Agreement (n 218) article IX(5)); NAFO (NAFO Convention (n 213) article XII(1)); NEAFC (NEAFC Convention (n 213) article 12(2)); SEAFO (SEAFO Convention (n 120) article 23(1)(c)-(g)); SPRFMO (SPRFMO Convention (n 213) article 17(2)).

One RFMO, CCAMLR, is particularly well-known for its innovative and environmentally friendly measures.[222] It applies an ecosystem approach to management,[223] the purpose of which is to conserve the marine life of the Southern Ocean.[224] This means that it does not only regulate harvesting and associated activities in the area, but that it also aims at conserving the ecosystem in its entirety.[225] While many RFMOs now make reference to the ecosystem approach,[226] few bodies other than CCAMLR have taken, in practice, such a strong view regarding ecosystem protection. Most management regimes are still species-based and nearly all fail to address the impacts of fishing on the physical environment. At the best indeed, the effects on "species belonging to the same ecosystem or dependent upon or associated with the target stocks" are taken into account.[227]

CCAMLR also applies a precautionary approach to set TACs, meaning that "the extent and effect of the uncertainties and gaps in [. . .data are weighed up] before making a management decision."[228] Other RFMOs, such as WCPFC, NEAFC, SEAFO and IATTC, have a reference to the precautionary principle in their

[222] J. P. Croxall and S. Nicol, 'Management of Southern Ocean fisheries: global forces and future sustainability' (2004) 16 Antarctic Science 569, 569.

[223] CCAMLR Convention (n 213) articles II(3)(c), IX(2)(i); CCAMLR, 'CCAMLR Ecosystem Monitoring Program' http://www.ccamlr.org/en/science/ccamlr-ecosystem-monitoring-program-cemp accessed 4 July 2015; CCAMLR, 'CCAMLR's Management of the Antarctic' (Hobart, Australia 2001) http://archive.ccamlr.org/pu/E/e_pubs/am/man-ant/e-management.pdf accessed 4 July 2015, 5–6; D. Freestone, 'Problems of High Seas Governance' in D. Vidas and P. J. Schei (eds), *The World Ocean in Globalisation: Climate Change, Sustainable Fisheries, Biodiversity, Shipping, Regional Issues* (Martinus Nijhoff Publishers 2011) 124–125; K.-H. Kock, 'Understanding CCAMLR's Approach to Management' (2000) www.ccamlr.org/en/system/files/am-all.pdf accessed 4 July 2015, 7–10, 24–27.

[224] CCAMLR, 'About CCAMLR' http://www.ccamlr.org/en/organisation/about-ccamlr accessed 4 July 2015.

[225] CCAMLR Convention (n 213) article II(3); G. Parkes, 'Precautionary fisheries management: the CCAMLR approach' (2000) 24 Marine Policy 83, 83.

[226] A. Fabra and V. Gascón, 'The Convention on the Conservation of Antarctic Marine Living Resources (CCAMLR) and the Ecosystem Approach' (2008) 23 International Journal of Marine and Coastal Law 567, 572; S. M. Garcia, 'The Ecosystem Approach to Fisheries: on the Way to Implementation' in M. H. Nordquist and others (eds), *Law, Science & Ocean Management* (Martinus Nijhoff Publishers 2007) 214; GFCM (Lugten (n 118) 38); IATTC (Antigua Convention (n 213) article VII(1)(f)); NEAFC (NEAFC Convention (n 213) article 4(2)(c)); SEAFO (SEAFO Convention (n 120) article 3(c)(d)); SPRFMO (SPRFMO Convention (n 213) articles 2, 3(2), 20); WCPFC (WCPFC Convention (n 120) article 5(d)).

[227] WCPFC (WCPFC Convention (n 120) article 5(d)); on the ecosystem approach, see infra Chapter 4 B. III. 1. b).

[228] CCAMLR, 'CCAMLR's Management of the Antarctic' (n 223) 6. CCAMLR is known for its precautionary approach (CCAMLR Convention (n 213) article II) although the reference is not explicit.

founding document or refer to it in Resolutions.[229] Established after the adoption of the UNFSA, these organizations indeed see the MSY as a limit not to overpass, not as a goal in itself. Some older RFMOs, such as ICCAT, still have not included the precautionary principle in their management.[230]

2. Measures

a) General Conservation and Management Measures

CMMs take various forms, but have all the same declared objective, namely to ensure the maximum but sustainable use of the resources under their mandate.[231] Quotas are the most widely used measure: the states parties set a TAC, supposedly on the basis of scientific information. Then either states negotiate for a national share of this TAC[232] or the fishery grounds are open until the TAC is reached—the so-called Olympic fishery method. Other measures include prohibition/regulation

---

[229] GFCM (GFCM Agreement (n 218) article III(2) and its biological reference points, as presented in S. Marr, *The Precautionary Principle in the Law of the Sea: Modern Decision Making in International Law* (Martinus Nijhoff Publishers 2003) 157–158); IATTC (Antigua Convention (n 213) article IV); IOTC (Resolution 12/01 On the implementation of the precautionary principle 2012 (IOTC)); NEAFC (NEAFC Convention (n 213) article 4(2)(b). Since 1997 the precautionary principle was introduced in the advice provided by ICES, see Henriksen, Hønneland and Sydnes (n 120) 111); SEAFO (SEAFO Convention (n 120) article 7); SIOFA (SIOFA Agreement (n 218) article 4(c)); SPRFMO (SPRFMO Convention (n 213) articles 2, 3(2), 20, 22); WCPFC (WCPFC Convention (n 120) article 6). As to within NAFO, the precautionary approach is made use of but is not found in the Convention or Scheme (NAFO, 'Precautionary Approach' http://www.nafo.int/fisheries/frames/precautionary.html accessed 4 July 2015); NAFO rather uses buffer zones as presented in Marr (n 229) 154 and exemplified in Kock (n 223) 27–28. CCBST also makes reference to it (CCSBT, 'Total Allowable Catch' http://www.ccsbt.org/site/total_allowable_catch.php accessed 4 July 2015).

[230] On the precautionary approach, see infra Chapter 4 B. II. 2.

[231] CCAMLR (CCAMLR Convention (n 213) article II); CCSBT (CCSBT Convention (n 213) article 3); GFCM (GFCM Agreement (n 218) preamble, article III); IATTC (Antigua Convention (n 213) preamble, article II); ICCAT (ICCAT Convention (n 218) preamble, article IV(2)(b)); IOTC (IOTC Agreement (n 218) article V(1)); NAFO (NAFO Convention (n 213) preamble, article II(1)); NEAFC (NEAFC Convention (n 213) preamble, article 2); SEAFO (SEAFO Convention (n 120) article 2); SIOFA (SIOFA Agreement (n 218) article 2); SPRFMO (SPRFMO Convention (n 213) article 2); WCPFC (WCPFC Convention (n 120) article 2); Beer-Gabel and Lestang (n 213) 50.

[232] Beer-Gabel and Lestang (n 213) 100–102.

of certain types of fishing gears,[233] lists of vessels allowed (or inversely prohibited) to fish in the managed area,[234] designation of open and closed seasons for

[233] GFCM (Recommendation on the mesh size of trawl nets exploiting demersal resources 2007, Rec. GFCM/31/2007/3; Resolution on the 40 mm square mesh size in codend of trawl nets exploiting demersal resources 2007, Res. GFCM/31/2007/3; Recommendation on a minimum Mesh Size in the codend of demersal trawls nets 2009 Rec. GFCM/33/2009/2); IATTC (Resolution prohibiting fishing on data buoys 2011, Resolution C-11-03); IOTC (Resolution 12/12 To prohibit the use of large-scale driftnets on the high seas in the IOTC area 2012); NAFO (Conservation and Enforcement Measures 2014, NAFO/FC Doc. 14/1 article 13); NEAFC (Minimum mesh size when fishing for capelin 1984, Recommendation 1; Minimum mesh size when fishing for blue whiting 1986, Recommendation 2; and Recommendation to Temporarily Prohibit the Use of Gillnets, Entangling Nets and Trammel Nets in the NEAFC Regulatory Area 2006, Recommendation III); CCAMLR (Conservation Measures Category 22 – Gear Regulation); SEAFO (Recommendation 1/2010 on Banning of gillnets 2010); SPRFMO (Conservation and Management Measure for Gillnets in the SPRFMO Convention Area 2013, CMM 1.02); WCPFC (Conservation and Management Measure to Prohibit the use of Large Scale Driftnets on the High Seas in the Convention Area 2008, CMM 2008-24; and Conservation and Management Measure Prohibiting Fishing on Data Buoys 2009, CMM 2009-05).

[234] CCAMLR (Scheme to promote compliance by Contracting Party vessels with CCAMLR conservation measures 2008, Scheme to promote compliance by non-Contracting Party vessels with CCAMLR conservation measures 2009); CCSBT (Resolution on amendment of the Resolution on "Illegal, Unregulated and Unreported Fishing (IUU) and Establishment of a CCSBT Record of Vessels over 24 meters Authorized to Fish for Southern Bluefin Tuna" 2008; Resolution on Establishing a List of Vessels Presumed to have Carried Out Illegal, Unreported and Unregulated Fishing Activities For Southern Bluefin Tuna (SBT) 2013); GFCM (Recommendation concerning the establishment of a GFCM record of vessels over 15 metres authorized to operate in the GFCM area amending the Recommendation GFCM/2005/2 2009, Rec. GFCM/33/2009/6; Recommendation on the establishment of a list of vessels presumed to have carried out IUU fishing in the GFCM Area, amending Recommendation GFCM/2006/4 2009, Rec. GFCM/33/2009/8); IATTC (Resolution to establish a list of vessels presumed to have carried out illegal, unreported and unregulated fishing activities in the Eastern Pacific Ocean 2005; Resolution (amended) on the establishment of a list of longline fishing vessels over 24 meters authorized to operate in the Eastern Pacific Ocean 2011, Resolution C-11-05); ICCAT (Concerning the Establishment of an ICCAT Record of Vessels 20 m or greater Authorized to Operate in the Convention Area 2011, Recommendation 11–12; various lists established per species; Recommendation by ICCAT Further Amending Recommendation 09–10 Establishing a List of Vessels Presumed to Have Carried out Illegal, Unreported, and Unregulated Fishing Activities in the ICCAT Convention Area 2011); IOTC (Resolution 13/02 Concerning the IOTC record of vessels authorised to operate in the IOTC area of competence 2013; Resolution 11/03 On establishing a list of vessels presumed to have carried out illegal, unreported and unregulated fishing in the IOTC area of competence 2011); NAFO (Conservation and Enforcement Measures 2014 (n 233) articles 53–54); NEAFC (NEAFC Scheme of Control and Enforcement (n 91) article 44); SEAFO (System of observation, inspection, compliance and enforcement 2013 article 28); SPRFMO (Establishment of the Commission record of Vessels authorised to fish in the Convention Area 2014, CMM 2.05; Establishing a List of Vessels presumed to have carried out Illegal, Unreported and Unregulated fishing activities in the SPRFMO Convention Area 2013); WCPFC (Conservation and Management Measure to establish a List of Vessels presumed to have carried out Illegal, Unreported and Unregulated fishing activities in the WCPO 2010; Record of Fishing Vessels and Authorization to Fish 2013, CMM 2013-10).

harvesting,[235] by-catch limits or reduction,[236] rules on bottom-fishing,[237] limits to fishing capacity,[238] etc.

---

[235] CCAMLR (General measure for the closure of all fisheries 2007, Conservation Measure 31-02); GFCM (Recommendation on the establishment of a closed season for the dolphin fish fisheries based on fishing aggregation devices (FADs) 2006, Rec. GFCM/2006/2; Recommendation concerning the recommendation by ICCAT amending recommendation [08-05] to establish a multiannual recovery plan for bluefin tuna in the eastern Atlantic and Mediterranean 2010, Rec. GFCM/34/2010/4(b)); IATTC (Multiannual program for the conservation of tuna in the Eastern Pacific Ocean during 2014–2016 2013, Resolution C-13-01); ICCAT (Recommendation by ICCAT Amending the Recommendation by ICCAT to Establish a Multi-Annual Recovery Plan for Bluefin tuna in the Eastern Atlantic and Mediterranean 2013, Recommendation 13-07 §21–25); NAFO (Conservation and Enforcement Measures 2014 (n 233) article 5(5)); NEAFC (Recommendation to adopt regulatory measures for the protection of blue ling in the NEAFC regulatory area (ICES Division XIV) from 2013 to 2016, Recommendation 5: 2013).

[236] CCAMLR (General measure for the closure of all fisheries (n 235); Limitation of By-Catch in New and Exploratory Fisheries in the 2013/14 Season 2013, Conservation Measure 33-03; Limitation of By-Catch in Statistical Division 58.5.2 in the 2013/14 Season 2013, Conservation Measure 33-02); GFCM (Recommendation on reducing incidental by-catch of seabirds in fisheries in the GFCM Competence Area 2011, Rec. GFCM/35/2011/3; Recommendation on the incidental by-catch of sea turtles in fisheries in the GFCM Competence Area 2011, Rec. GFCM/35/2011/4); IATTC (Consolidated Resolution on bycatch 2006, Resolution C-04-05 (Rev 2)); ICCAT (Recommendation by ICCAT on reducing incidental by-catch of seabirds in longline fisheries 2007, Recommendation 07-07; Recommendation by ICCAT on the by-catch of sea turtles in ICCAT fisheries 2010, Recommendation 10-09; Supplemental recommendation by ICCAT on reducing incidental by-catch of seabirds in longline fisheries 2011, Recommendation 11-09); IOTC (Resolution 12/06 On reducing the incidental bycatch of seabirds in longline fisheries 2012); NAFO (Conservation and Enforcement Measures 2014 (n 233) article 6); SEAFO (Conservation Measure 25/12 on Reducing Incidental By-catch of Seabirds in the SEAFO Convention Area 2012); SPRFMO (Minimising bycatch of seabirds in the SPRFMO Convention Area 2014, CMM 2.04); WCPFC (Resolution on the Incidental Catch of Seabirds 2005, Resolution 2005-01; Conservation and Management Measure to Address the Impact of Purse Seine Activity on Cetaceans 2012, Conservation and Management Measure on the protection of whale sharks from purse seine operations 2012, CMM 2012-04; Conservation and Management Measure for Mitigating Impacts of Fishing on Seabirds 2012).

[237] CCAMLR (Bottom fishing in the Convention Area 2012, Conservation Measure 22-06); NAFO (Conservation and Enforcement Measures 2014 (n 233) articles 15–24); NEAFC (Consolidated text of all NEAFC recommendations on regulating bottom fishing 2011); SEAFO (Conservation Measure 26/13 on Bottom Fishing Activities in the SEAFO Convention Area 2013); SPRFMO (Management of Bottom fishing in the SPRFMO Convention Area 2014, CMM 2.03).

[238] GFCM (Resolution on Guidelines on the management of fishing capacity in the GFCM area 2013, Res. GFCM/37/2013/2; Recommendation on the management of fishing capacity 2010, Rec. GFCM/34/2010/2); IATTC (Resolution on the capacity of the tuna fleet operating in the Eastern Pacific Ocean (revised) 2002); ICCAT (Recommendation by ICCAT Concerning the Limitation of Fishing Capacity on Northern Albacore 1998, Recommendation 98-8); IOTC (Resolution 12/11 On the implementation of a limitation of fishing capacity of contracting parties and cooperating non-contracting parties 2012); WCPFC (Resolution on Reduction of Overcapacity 2005, Resolution 2005-02).

Also, in an upward trend, RFMOs adopt and implement trade documentation schemes (TDS) and/or catch documentation schemes (CDS).[239] These documents, issued by the flag states, have to accompany the fish, respectively when it is traded internationally and from landing on, with information regarding its traceability.[240] The importing state or port state has to verify the document's existence and validity when fish subject to such a scheme is imported or landed. TDS/CDS are based on a reverse burden of proof for both contracting and non-contracting parties to the relevant RFMO: the documents must be presented in order for the fish to be landed, transshipped or traded. Thus, only participants in the scheme who can show that they have respected the relevant CMMs benefit from access to the port's facilities—this is however only applicable to the ports of states who are parties to or cooperating with the RFMO.

Hence, TDS/CDS "promote a way of keeping track of 'legal' catch."[241] Such schemes aim at avoiding that specimens be caught in violation of CMMs or, if harvested, enter international trade. If and when working, such measures make IUU fishing less profitable. To be efficient, TDS/CDS must be backed up by dedicated resources in the flag states and by a strong network of ports verifying the documentation before allowing landing/trade. How comprehensive in terms of participation and geographical area of application the scheme is also influences its actual effects,[242] which are generally considered very positive.[243] Nevertheless, it is true

[239] CDS: ICCAT (Recommendation by ICCAT Amending Recommendation 09–11 on an ICCAT Bluefin Tuna Catch Documentation Program 2011, Recommendation 11–20); CCAMLR (Catch Documentation Scheme for Dissostichus spp. 2013, Conservation Measure 10-05; on the CCAMLR CDS development, see D. J. Agnew, 'The illegal and unregulated fishery for toothfish in the Southern Ocean, and the CCAMLR catch documentation scheme' (2000) 24 Marine Policy 361); CCSBT (Resolution on the Implementation of a CCSBT Catch Documentation Scheme 2013). On the IOTC, ICCAT, IATTC and CCSBT early documentation schemes, usually TDS see M. Lack, 'Catching On?: Trade-related Measures as a Fisheries Management Tool' (TRAFFIC International 2007) 7–8; R. Rayfuse, 'Building sustainable high seas fisheries through certification processes: issues and perspectives' (2009) 35 Océanis 93, 97. In general, see S. Clarke, 'Best Practice Study of Fish Catch Documentation Schemes' (23 August 2010) http://www.m2cms.com.au/uploaded/5/Final%20CDS%20Report%20-August%2023.pdf accessed 4 July 2015; C. A. Roheim and J. G. Sutinen, 'Trade and Market-Related Instruments to Promote Sustainable Fishing Practices' International Trade and Sustainable Development Series Issue Paper 3 (Geneva 2006) 2–7.

[240] Clarke, 'Best Practice Study of Fish Catch Documentation Schemes' (n 239) 1–2; Rayfuse, 'Building sustainable high seas fisheries through certification processes: issues and perspectives' (n 239) 98; Lack, 'Catching On?' (n 239) 8–9.

[241] C.-C. Schmidt, 'Addressing Illegal, Unreported and Unregulated (IUU) Fishing: Paper prepared for the International Fisheries Compliance 2004 Conference, 29–30 September 2004, Brussels' (2004) http://www.oecd.org/greengrowth/fisheries/34029751.pdf accessed 4 July 2015, 7.

[242] Lack, 'Catching On?' (n 239) 11.

[243] M. Lack, 'Continuing CCAMLR's Fight Against IUU Fishing for Toothfish' (WWF Australia and TRAFFIC International 2008) 25.

that mislabeling and other document forgeries can decrease the efficiency of those trade measures which are based on documentation schemes.[244]

b) Measures on Sharks

Most RFMOs have adopted some—mostly similar—measures regarding sharks.[245] It must be noted that such measures are quite recent and that, for the species the fishing of which is not banned, no quotas are set[246] except in one occurrence—skates since 2004 in NAFO.[247]

Many RFMOs require a full use of the retained catches of sharks. Discard before the first point of offloading is prohibited, except for the head, guts and skin. Fishermen are also forbidden to have on board shark fins totaling more than 5 % of the weight of sharks on board. Fishing vessels are not to keep on board, transship or land the carcass of certain species and/or any fins harvested in contravention of the relevant measures.

Incidentally caught sharks, particularly specimens of certain species, juveniles and/or pregnant females, have to be released alive, whenever possible if they are not used for food or subsistence. As for parties to the RFMO, they have, when possible, to undertake research to make fishing gears more selective and to identify shark nursery areas.

Under several RFMOs, fishing certain species, for example the (bigeye) thresher sharks, oceanic whitetip sharks, hammerhead sharks, porbeagle, spiny dogfish, basking shark, and/or deep-sea sharks, has been banned. Within the CCAMLR area, fishing sharks for other reasons than scientific research is forbidden. As for NAFO, it adopted measures banning finning, as well as transshipment and landing of shark fins.

Some RFMOs require that their members adopt and implement a national plan of action on sharks, encourage their contracting parties to do so, or simply make reference to the IPOA-Sharks. Reporting is required from contracting parties, be it on catches, fishing effort directed at sharks, discards and releases, or the implementation of adopted measures and of the IPOA/NPOA.

---

[244] Roheim and Sutinen (n 239) 4.

[245] Lack and Sant (n 188) 11; on the measures adopted by RFMOs for sharks, see Annex 2.

[246] S. van Osch, 'Save our Sharks: Using International Fisheries Law Within Regional Fisheries Management Organizations to Improve Shark Conservation' (2011–2012) 33 Michigan Journal of International Law 383, 409–417.

[247] M. R. Simpson, C. M. Miri and C. Busby, 'Assessment of Thorny Skate (*Amblyraja radiata Donovan*, 1808) in NAFO Divisions 3LNO and Subdivision 3Ps' SCR Doc. 08/43 (June 2008) http://archive.nafo.int/open/sc/2008/scr08-043.pdf accessed 4 July 2015, 1; Shark Alliance, *World's only international ray fishing quota reduced: Northwest Atlantic thorny skate limit will soon be lower, yet still excessive* (2009) http://www.sharkalliance.org/content.asp?did=33850 accessed 4 July 2015.

c) Measures on Aquaculture

Fish farms are increasingly regulated at a regional level. In ICCAT, facilities farming Bluefin Tuna caught in the managed area have to be listed.[248] Farms must request catch documentation and provide it for any Bluefin Tuna arriving in or leaving the facility.[249] GFCM requires information from its members on aquaculture practices, on fish caught for such purposes, and on the minimum size of juveniles.[250] CCSBT also centralizes a list of authorized farms. Additionally, its "Members and Cooperating Non-Members will not validate trade documents for farms not on the list and will not accept imports of Southern Bluefin Tuna product from farms not on the list."[251] NASCO, which is responsible to manage salmon in the North Atlantic, regulates aquaculture, a major practice in the region for this species. The focus is put on minimizing adverse effects.[252]

It appears that the existing, and quite new, measures are subject to several sets of criticisms. First, compliance is low, misreporting is a normal practice[253] and the measures have not allowed curbing the important increase in catches of juveniles that aquaculture practices have initiated. These shortcomings are particularly problematic as aquaculture has taken very important proportions: "the current farming capacity in the Mediterranean is estimated by the SCRS to be around 64,000 t, more than double the Total Allowable Catch adopted for past years."[254] Then, the requirements applicable to Bluefin Tuna appear to be too soft, with optional tagging and unspecified timing for its implementation.[255] This does not take into account the very international nature of aquaculture and the related need for strict traceability. Indeed, juveniles are more often than not caught in waters under national jurisdiction or in the high seas, transferred to pens in third countries and then made to enter international trade, sometimes directly on board factory ships.[256]

---

[248] Recommendation by ICCAT on Bluefin Tuna Farming 2006, Recommendation 06–07.

[249] Recommendation by ICCAT Amending Recommendation 09–11 on an ICCAT Bluefin Tuna Catch Documentation Program, Recommendation 11–20 (n 239).

[250] Recommendation by GFCM on reporting of aquaculture data and information, amending Recommendation GFCM/33/2009/4 2011, Recommendation GFCM/35/2011/6; J. K. Murphy, 'FAO's general fisheries commission for the Mediterranean: a new role, a new role model' (2007) 12 Drake Journal of Agricultural Law 391, 404–405.

[251] CCSBT, 'CCSBT Record of Authorised Farms' http://www.ccsbt.org/site/authorised_farms.php accessed 4 July 2015.

[252] NASCO, 'Aquaculture, Introductions, Transfers & Transgenics' http://www.nasco.int/aquaculture.html accessed 4 July 2015.

[253] Monaco, 'CoP15 Prop. 19 – Proposal to include Atlantic Bluefin Tuna (*Thunnus thynnus* (Linnaeus, 1758)) on Appendix I of CITES in accordance with Article II 1 of the Convention' (13–25 March 2010) 17 §7.2.

[254] Ibid. 17 §7.2 (reference omitted).

[255] Ibid. 19 §8.3.1.

[256] Ibid.

3. Implementation

In all RFMOs, the flag state is the main implementer of CMMs, in particular enforcement-related ones.[257] It must indeed overview its vessels' activities through control, monitoring and surveillance measures, verify compliance with quotas, liaise with the port state when necessary, etc. Vessel monitoring systems are required under certain schemes to allow the continuous reporting of vessels' positions to the flag state.[258]

The port state has recently become a major actor in the enforcement of CMMs. Port state measures, initially developed in relation to shipping, are indeed increasingly used in the field of fisheries:[259] entry into port, provision of services or landing of fish can be/have to be refused to vessels suspected of IUU fishing, to

[257] Parkes (n 225) 88; see for example provisions and measures in CCAMLR (Licensing and inspection obligations of Contracting Parties with regard to their flag vessels operating in the Convention Area 2013); IATTC (Antigua Convention (n 213) article XX); NAFO (Conservation and Enforcement Measures 2014 (n 233)); NEAFC (NEAFC Scheme of Control and Enforcement (n 91)); WCPFC (WCPFC Convention (n 120) part V); SIOFA (SIOFA Agreement (n 218) article 11).

[258] CCAMLR (Automated satellite-linked Vessel Monitoring Systems (VMS) 2013, Conservation Measure 10-04); CCSBT (Resolution on establishing the CCSBT Vessel Monitoring System 2008); GFCM (Recommendation concerning minimum standards for the establishment of a Vessel Monitoring System (VMS) in the GFCM area 2009, Rec. GFCM/33/2009/7); IATTC (Resolution on the establishment of a vessel monitoring system (VMS) 2004, Resolution C-04-06); ICCAT (Recommendation by ICCAT concerning minimum standards for the establishment of a vessel monitoring system in the ICCAT convention area 2003, Recommendation 2003–14); IOTC (Resolution 06/03 On establishing a vessel monitoring system programme 2006); NAFO (Conservation and Enforcement Measures 2014 (n 233) article 29); NEAFC (NEAFC Scheme of Control and Enforcement (n 91) article 11); SEAFO (System of observation, inspection, compliance and enforcement (n 234) article 13); SPRFMO (Establishment of the Vessel Monitoring System in the SPRFMO Convention Area 2014, CMM 2.06); WCPFC (Commission vessel monitoring system 2007, CMM 2007-02).

[259] Measures based on port state control originated in the 1980s in relation to shipping; the ones in the field of fisheries are modelled on those earlier rules. See for example D. Anderson, 'Port States and Environmental Protection' in A. Boyle and D. Freestone (eds), *International Law and Sustainable Development – Past Achievements and Future Challenges* (Oxford University Press 1999); A. Gillespie, *Conservation, biodiversity and international law* (Edward Elgar 2011) 455–456; R. Lagoni, 'Ports' Max Planck Encyclopedia of Public International Law www.mpepil.com §21–27, 39–40; T. A. Mensah, 'Marine Pollution from Ships, Prevention of and Responses to' Max Planck Encyclopedia of Public International Law www.mpepil.com §32–37; Molenaar, 'Port State Jurisdiction' (n 162) on fisheries, see particularly §33; J. Swan, 'Port State Measures to Combat IUU Fishing: International and Regional Developments' (2006–2007) 7 Sustainable Development Law and Policy Review 38, 38.

vessels inspected and found in breach of CMMs, or to vessels not authorized to fish.[260] Usually, specific ports are designated for landing and transshipment.[261]

Then, some non-flag state measures have been developed in the last decade, such as boarding and inspection schemes. Non-flag states have indeed, in certain RFMOs, the right to board and inspect vessels flying the flag of other members.[262] Also, vessels of states parties to the UNFSA can be submitted to such verifications of compliance. Then, regional observer schemes are meant to promote compliance; according to the provisions of the particular scheme, an observer must be on board at all times, or for specific maneuvers such as transshipment.[263] Transshipment in port is also subject to specific rules.[264]

---

[260] CCAMLR (Port inspections of fishing vessels carrying Antarctic marine living resources 2013, Conservation Measure 10-03); GFCM (Recommendation on a regional scheme on port state measures to combat illegal, unreported and unregulated fishing in the GFCM area 2008, Rec. GFCM/2008/1); ICCAT (Recommendation by ICCAT for an ICCAT Scheme for Minimum Standards for Inspection in Port 2012, Recommendation 12-07); IOTC (Resolution 10/11 On port state measures to prevent, deter and eliminate illegal, unreported and unregulated fishing 2010); NAFO (Conservation and Enforcement Measures 2014 (n 233) chapters VII-VIII); NEAFC (NEAFC Scheme of Control and Enforcement (n 91) chapter V); SEAFO (System of observation, inspection, compliance and enforcement (n 234) articles 22–23); SIOFA (SIOFA Agreement (n 218) article 12).

[261] GFCM (Recommendation on a regional scheme on port state measures to combat illegal, unreported and unregulated fishing in the GFCM area (n 260)); ICCAT (Recommendation by ICCAT for an ICCAT Scheme for Minimum Standards for Inspection in Port (n 260)); IOTC (Resolution 10/11 On port state measures to prevent, deter and eliminate illegal, unreported and unregulated fishing (n 260)); NAFO (Conservation and Enforcement Measures 2014 (n 233) article 43(1)); NEAFC (NEAFC Scheme of Control and Enforcement (n 91) article 21); SEAFO (System of observation, inspection, compliance and enforcement (n 234) article 20).

[262] CCAMLR (CCAMLR Convention (n 213) article XXIV); ICCAT (Recommendation by ICCAT Amending the Recommendation by ICCAT to Establish a Multi-Annual Recovery Plan for Bluefin tuna in the Eastern Atlantic and Mediterranean (n 235) part V); NAFO (Conservation and Enforcement Measures 2014 (n 233) chapter VI); NEAFC (NEAFC Scheme of Control and Enforcement (n 91) chapter IV); SEAFO (System of observation, inspection, compliance and enforcement (n 234) chapter IV); WCPFC (WCPFC Boarding and Inspection Procedures 2006, CMM 2006–08).

[263] CCSBT (Resolution on Establishing a Program for Transshipment by Large-Scale Fishing Vessels 2008); IATTC (Amendment to resolution C-11-09 on establishing a program for transhipments by large-scale fishing vessels 2012, Resolution C-12-07); ICCAT (Recommendation by ICCAT on a Programme for Transhipment 2012, Recommendation 12-06; Recommendation by ICCAT Amending the Recommendation by ICCAT to Establish a Multi-Annual Recovery Plan for Bluefin tuna in the Eastern Atlantic and Mediterranean (n 235) §90-92); IOTC (Resolution 11/04 On a regional observer scheme 2011); NAFO (Conservation and Enforcement Measures 2014 (n 233) article 30); WCPFC (Conservation and Management Measure on Regulation of Transhipment 2009, CMM 2009-06).

[264] IATTC (Amendment to resolution C-11-09 on establishing a program for transhipments by large-scale fishing vessels (n 263)); ICCAT (Recommendation by ICCAT Amending the Recommendation by ICCAT to Establish a Multi-Annual Recovery Plan for Bluefin tuna in the Eastern Atlantic and Mediterranean (n 235) §65); IOTC (Resolution 12/05 On establishing a programme for transhipment by large-scale fishing vessels 2012); SEAFO (System of observation, inspection, compliance and enforcement (n 234) article 14).

A crucial and complex aspect of implementation is related to the question of non-contracting parties. As discussed above,[265] should the UNFSA be an authoritative interpretation of UNCLOS or should it have reached CIL, as considered likely in this study, both parties and non-parties to the 1995 Agreement have to join the relevant RFMO or to ensure that their vessels refrain from fishing in the managed area. Several RFMOs have specific rules applying to non-contracting parties, usually making their access to ports more difficult except for the vessels of states officially cooperating.[266] CCAMLR, IOTC, NAFO and NEAFC for example are of the opinion that non-contracting non-cooperating parties are undermining the effectiveness of CMMs, unless they can prove the contrary; as such, they are to be prohibited to use the ports of member states or have to be automatically inspected.[267]

Such a reversal of the burden of proof is not uncontroversial since it imposes double-standards (and stricter ones) to states which have not agreed to the measures they are judged upon.[268] Moreover, in the view of those who consider the UNFSA as treaty law binding *inter partes* only,[269] the conduct of RFMOs members who adopt measures applicable to third parties and enforce them is a violation of the *pacta tertiis* rule.[270]

Most RFMOs have a reference to dispute settlement procedures in their founding treaties:[271] some are explicitly of a non-binding nature,[272] others are related to the

[265] See supra Chapter 3 A. II. 3. b).

[266] Beer-Gabel and Lestang (n 213) 140–146.

[267] CCAMLR (Scheme to promote compliance by non-Contracting Party vessels with CCAMLR conservation measures (n 234)); IOTC (Resolution 01/03 Establishing a scheme to promote compliance by Non-Contracting Party vessels with Resolutions established by IOTC 2001; Resolution 05/03 Relating to the establishment of an IOTC programme of inspection in port 2005); NAFO (Conservation and Enforcement Measures 2014 (n 233) chapter VIII); NEAFC (NEAFC Scheme of Control and Enforcement (n 91) articles 37, 40, 44, 45); Y. Tanaka, 'The Changing Approaches to Conservation of Marine Living Resources in International Law' (2011) 71 Zeitschrift für ausländisches öffentliches Recht und Völkerrecht 291, 325–328; R. Rayfuse, *Non-Flag State Enforcement in High-seas Fisheries* (Martinus Nijhoff Publishers 2004) 174–180, 190–191, 220–222, 253–256, 278–282.

[268] A. J. Bederman, 'CCAMLR in crisis: a case study of marine management in the southern oceans' in H. N. Scheiber (ed), *Law of the Sea: The Common Heritage and Emerging Challenges* (Martinus Nijhoff Publishers 2000) 190; Tanaka, 'The Changing Approaches to Conservation of Marine Living Resources in International Law' (n 267) 324–328.

[269] On the question of the UNFSA's status, see supra Chapter 3 A. II. 3. b).

[270] Hayashi, 'Regional Fisheries Management Organisations and Non-Members' (n 108) 760–761.

[271] However, ICCAT and NAFO do not have such provisions, and NEAFC's new founding treaty only states that recommendations shall be made by the Commission "establishing procedures for the settlement of disputes" (NEAFC Convention (n 213) article 18 bis (not yet in force)).

[272] GFCM (GFCM Agreement (n 218) article XVII); IATTC (Antigua Convention (n 213) article XXV); IOTC for the first stage (IOTC Agreement (n 218) article XXIII); SEAFO for the first stage (SEAFO Convention (n 120) article 24(3)).

mechanisms envisioned in UNCLOS Part XV and/or UNFSA Part VIII,[273] while a third group refers to the possibility to use the International Court of Justice (ICJ) and/or arbitration, in most cases without providing for the competence of a member state to start proceedings unilaterally.[274]

4. Cooperation

Many RFMOs cooperate on an *ad hoc* basis, for example by referring to each other's IUU vessels lists; the tuna RFMOs and NAFO for example do so.[275] These lists are a useful source of information on which RFMO members can base their decision to take the relevant port state measures towards vessels and/or to undertake further controls.

Other forms of *ad hoc* cooperation can be noted mostly between RFMOs with a geographical overlap in the areas of competence (such as IATTC and WCPFC), or with a shared stock (such as ICCAT and NEAFC with regard to sharks, or NAFO and NEAFC). The most common form of cooperation remains through exchange of information and attendance at each other's meetings.[276]

The tuna RFMOs collaborate in a more organized and continuous way. They indeed established the 'tuna-org network' through which they share information and analyze issues of common interest in the areas of science, management, compliance and enforcement. At one of its recent meetings, this working group of the five tuna RFMOs recognized that most of the recommendations it had made in previous occasions such as on data sharing, provisions of scientific advice had been implemented by the RFMOs and that the harmonized IUU vessel lists between 4 of the 5 RFMOs was a positive move. It further put on the agenda the development of common minimum standards for stocks assessments, recommended the

---

[273] SEAFO (SEAFO Convention (n 120) article 24(4)); SIOFA (SIOFA Agreement (n 218) article 20(1)); SPRFMO (SPRFMO Convention (n 213) article 34(2)); WCPFC (WCPFC Convention (n 120) article 31).

[274] CCAMLR (CCAMLR Convention (n 213) article XXV(2)); CCSBT (CCSBT Convention (n 213) article 16(2)); IOTC (IOTC Agreement (n 218) article XXIII). Note however GFCM (GFCM Agreement (n 218) article XVII) which does not refer to the need for both parties to agree.

[275] Tuna-org, 'IUU Vessel Lists' http://www.tuna-org.org/vesselneg.htm accessed 4 July 2015; NAFO, 'Links to other RFMOs' http://www.nafo.int/fisheries/fishery/iuu/rfmo-links.html accessed 4 July 2015.

[276] For example, CCAMLR attends many meetings (CCAMLR, 'Cooperation with others' https://www.ccamlr.org/en/organisation/cooperation-others accessed 4 July 2015).

development of a common vessel database as coordinated by the FAO, and the establishment of a common format for assessing compliance. A Steering Committee was to be created to overview the implementation of the recommendations made through this so-called Kobe process.[277]

Then, some cooperation takes place through the FAO. A number of regional fishery bodies, including some RFMOs, have been created under the auspices of the FAO,[278] or have given the FAO some depository functions.[279] Even these RFMOs that are outside the framework of the FAO[280] collaborate in the Regional Fishery Body Secretariats Network. The FAO plays a facilitator role for the Network. This informal working group allows FAO and non-FAO RFMOs to meet and discuss common issues such as ways to integrate ecosystem-related aspects into management, overcapacity, methods to combat IUU, aquaculture or performance reviews.[281]

5. Applicability and Appropriateness

Many RFMOs are under strong criticism because they are perceived, in some cases rightly so, as associations of fishing states which distribute access and quotas to each other, with a strong bias towards exploitation,[282] and with overfishing as a

[277] Tuna-org, 'Chair's report of the third joint meeting of the tuna regional fisheries management organizations (Kobe III)' (12 July 2011) http://www.tuna-org.org/Documents/TRFMO3/REP-KOBE3-ENG.pdf accessed 5 July 2015.

[278] Young, *Trading Fish, Saving Fish* (n 31) 39–40; APFIC, GFCM, IOTC, RECOFI and CACFish under article XIV; CECAF, CIFAA, WECAFC, COPESCAALC, SWIOFC and EIFAAC under article VI.

[279] ICCAT (ICCAT Convention (n 218) articles XIII(2), XIV(2), XVI); SEAFO (SEAFO Convention (n 120) article 25(2)); and SIOFA (SIOFA Agreement (n 218) article 25).

[280] CCAMLR, IATTC, NAFO, NEAFC, WCPFC.

[281] FAO, 'Report of the First Meeting of Regional Fishery Body Secretariats Network' FAO Fisheries Report 837 (Rome 2007); FAO, 'FAO Report of the Second Meeting of Regional Fishery Body Secretariats Network' FAO Fisheries and Aquaculture Report 908 (Rome 2009); FAO, 'Statement of the third meeting of the regional fishery body secretariats network' (Rome 8 February 2011).

[282] Gillespie, *Conservation, biodiversity and international law* (n 259) 450; Young, *Trading Fish, Saving Fish* (n 31) 40; California Environmental Associates, 'Charting a Course to Sustainable Fisheries' (16 January 2012) http://www.chartingacourse.org/downloads/ accessed 4 July 2015, 23–24.

result.[283] It is correct that they struggle with, or are part of, many of the major governance problems that fisheries face, such as free-riding, lack of sufficient data, or non-compliance.[284] This is made worse by the fact that there is no formal oversight body to verify compliance of an RFMO with its founding treaty and with general international law.[285] However, in the last few years, and as a response to the call for accountability made by COFI[286] and the UNGA,[287] some RFMOs have decided to be audited. So far, NEAFC, SEAFO, CCSBT, ICCAT, IOTC and IATTC have undertaken performance reviews internally or by external experts.[288]

---

[283] For example, in the NEAFC area, out of five stocks, only one was considered, in 2009, to be exploited sustainably (Bjorndal (n 130) 696; for slightly different figures but with a similar conclusion see A. Longhurst, *Mismanagement of Marine Fisheries* (Cambridge University Press 2010) 178–180). S. Cullis-Suzuli and D. Pauly demonstrate that, in average, RFMOs perform poorly both in terms of best practices 'on paper' and in reality ('Failing the high seas: A global evaluation of regional fisheries management organizations' (2010) 34 Marine Policy 1036). Here are other examples of what has been called 'mismanagement' within ICCAT: banning of fishing during the spawning season had been advised but rejected and, in 2007, the SCRS advised not to set quotas at more than 15,000 tons, but ICCAT members within the deciding body overlooked this advice by setting the TACs at 22,000 tons for 2008. As a result, the estimated catches, including IUU, reached 61,000 tons in 2007, i.e. four times the advice of ICCAT's scientific body (Monaco, 'CoP15 Prop. 19 – Proposal to include Atlantic Bluefin Tuna' (n 253) 16 §6.4, 18–19 §8.1; Greenpeace, 'CITES – Last Chance for Atlantic Bluefin Tuna' (2010) http://www.greenpeace.org/international/en/publications/reports/cites-last-chance-for-bluefin/ accessed 4 July 2015, 2).

[284] Governance issues are discussed in more detail infra Chapter 4 A. On issues facing RFMOs, see for example J. Swan, 'Regional Fishery Bodies and Governance: Issues, Actions and Future Directions' FAO Fisheries and Aquaculture Circular 959 (Rome 2000).

[285] As international organizations, they are bound not only by their founding treaty but also by CIL (Sands, Klein and Bowett (n 143) 461–463). On this, see infra Chapter 4 A. IV. 2. c).

[286] Committee on Fisheries, 'Report of the Twenty-sixth Session of the Committee on Fisheries (Rome, 7–11 March 2005)' FAO Fisheries Report 780 (Rome 2005) 18 §111–112; Committee on Fisheries, 'Report of the Twenty-seventh Session of the Committee on Fisheries (Rome, 5–9 March 2007)' FAO Fisheries Report 830 (Rome 2007) 14 §86; Committee on Fisheries, 'Strengthening regional fisheries management organizations and their performances including the outcome of the 2007 RFMOs meeting' (n 119) 6–7 §29–35.

[287] In its Resolutions, the UNGA urges states parties to RFMOs to undertake performance reviews "initiated either by the organization or arrangement itself or with external partners, including in cooperation with the Food and Agriculture Organization of the United Nations, using transparent criteria based on the provisions of the Agreement and other relevant instruments" (UNGA, Resolution adopted by the General Assembly on 8 December 2006 – Sustainable fisheries, including through the 1995 Agreement for the Implementation of the Provisions of the United Nations Convention on the Law of the Sea of 10 December 1982 relating to the Conservation and Management of Straddling Fish Stocks and Highly Migratory Fish Stocks, and related instruments, 6 March 2007, A/RES/61/105 §73; UNGA, Resolution adopted by the General Assembly on 7 December 2010, A/RES/65/38 (n 47) §107).

[288] Tuna-org, 'Informal framework for sharing information from tuna Regional Fishery Management Organizations' http://www.tuna-org.org/ accessed 4 July 2015; Bjorndal (n 130) 694–695; M. Ceo and others, 'Performance Reviews by Regional Fishery Bodies: Introduction, summaries, synthesis and best practices, Volume I: CCAMLR, CCSBT, ICCAT, IOTC, NAFO, NASCO, NEAFC' FAO Fisheries and Aquaculture Circular 1072 (Rome 2012); Lugten (n 118) 6–7. See infra Chapter 4 A. IV. 2. c) cc).

Also, compliance by states is a problematic aspect.[289] As long as the high seas are considered by some actors as the commons, the possibility for states to stay outside of the legal framework of RFMOs remains, in practice, open. Also, even as parties, states can opt out of measures. Furthermore, in case of breaches, strong enforcement mechanisms are lacking. However, some compliance mechanisms are being put in place, such as the IOTC compliance committee[290] or the verification of member states and cooperating non-members by GFCM.[291]

Finally, fragmentation of both rules and space is a weakness.[292] Each organization acts more or less independently and ecosystems are not always under the management of one single organization. One must nevertheless note the existence of the Regional Fishery Body Secretariats' Network, of some (quasi-) institutionalized partnership initiatives, particularly between the five tuna RFMOs as well as several *ad hoc* cooperation mechanisms.

6. Conclusions

RFMOs are institutions through which states interested in a fishery cooperate in the management and conservation of a species or of several species in a particular area. Most parts of the high seas are nowadays covered. Since these areas are traditionally seen as free for all—though this is challenged by the UNFSA—RFMOs are struggling with free-riders. Increasingly, measures are taken against non-parties to reduce the attractiveness of fishing in the managed area; however, these are not uncontroversial.

Different measures have been developed to manage fisheries, varying principles have been followed and diverse implementation mechanisms put in place. RFMOs are indeed independent from each other and only loosely coordinated. Overall, quotas are used, some level of precautionary approach is declared applicable—if not always applied—, and flag states' primary jurisdiction is respected, although most organizations envision a strong role for non-flag states. Finally, documentation schemes are gradually implemented in most RFMOs.

---

[289] See infra Chapter 4 A. IV. 1.

[290] IOTC Resolution 10/09 Concerning the functions of the Compliance Committee 2010.

[291] GFCM Recommendation concerning the identification of the non-compliance 2010, Rec. GFCM/34/2010/3.

[292] See infra Chapter 4 A. II.

C. Conservation Regimes

I. 1973 Convention on the International Trade in Endangered Species

1. General Aspects

a) Rationale and Purpose

CITES is one of the oldest multilateral environmental agreements and deals specifically with trade in endangered species. It is based on the premise that "the prevention or reduction of international commerce in certain species will reduce harvest pressures."[293] Indeed, as international trade puts strain on wildlife survival, regulating this human activity is perceived as a crucial way to avoid the extinction of endangered species. Trade includes import, export, re-export and introduction from the sea actions.[294]

All species of fauna and flora can be listed. Hence, marine animals fall within the ambit of the Convention although special provisions apply to them, including 'introduction from the sea'.

b) System of Appendices

The Convention regime applies graded restrictions according to the level of protection needed by each species. Species that are granted protection under CITES are to be accompanied, when traded internationally, by a permit certifying, inter alia, that the trade of that specimen is not detrimental to the survival of the species.

Hence, for listed species, commercial trade is restricted (Appendices II, III), or forbidden (Appendix I), by an international system of permits delivered at the national level. Approximately 5,600 animal species and 30,000 plants are currently listed.[295]

[293] M. Giordano, 'The Internationalization of Wildlife and Efforts Towards its Management: A Conceptual Framework and the Historic Record' (2001–2002) 14 Georgetown International Environmental Law Review 607, 612.

[294] 'Convention on International Trade in Endangered Species of Wild Fauna and Flora' (CITES) 1973, 993 *United Nations Treaty Series* 243 article I(c).

[295] CITES, 'The CITES species' (2013) http://www.cites.org/eng/disc/species.php accessed 4 July 2015.

2. Processes and Institutions

a) Listing, Up – / Down – / De – Listing

At the biennial Conference of the Parties (CoP), the addition of a species to an Appendix, up-, down- or de-listing is decided by a two-third majority vote of the members present and voting. Secret ballot is possible, if the request is backed up by at least 10 parties.[296] While this rule of procedure has been questioned several times since it was adopted in 1994, all amendment proposals were clearly rejected.[297] Once an amendment is adopted it becomes binding on the parties except if they make a reservation within 90 days; this procedure is known as the tacit consent/opt out procedure.[298]

On the basis of biological and trade criteria, parties present proposals to amend the Appendices at the CoP. The criteria have evolved over time, and currently the Fort Lauderdale criteria are applied.[299] They are included in a CoP Resolution and consequently, formally speaking only amount to soft-law. However, they are considered binding by the parties who all use them without questioning their status.[300]

Down- or de-listing criteria were, already from the beginning, to be based on positive scientific evidence that the species is fit to deal with a lower level of protection or none.[301] This condition is further developed in Annex 4 of the current criteria, meaning that, now, the precautionary principle is applied to avoid unsustainable de-listing.

---

[296] R. Reeve, *Policing International Trade in Endangered Species* (Earthscan 2002) 42–43.

[297] The latest amendment proposals by the European Union on one hand (Denmark (on behalf of the EU), 'CoP16 Doc. 4.2 (Rev. 1) – Proposal to improve transparency of voting during meetings of the Conference of the Parties' (3–14 March 2013)) and by Mexico and Chile on the other hand (Chile and Mexico, 'CoP16 Doc. 4.3 (Rev. 1) – Proposed amendment to Rule 25 on Methods of Voting use of secret ballots' (3–14 March 2013)), which suggested to increase the number of parties necessary to support a secret ballot (respectively to a simple majority and to one third), were rejected at CoP16 (International Institute for Sustainable Development, 'Earth Negotiations Bulletin – Summary of the Sixteenth Meeting of the Conference of the Parties to the Convention on International Trade in Endangered Species' (18 March 2013) http://www.iisd.ca/vol21/enb2183e.html accessed 4 July 2015).

[298] CITES (n 294) article XV(3); D. König, 'Tacit Consent/Opting Out Procedures' Max Planck Encyclopedia of Public International Law www.mpepil.com §10.

[299] Resolution Conf. 9.24 (Rev. CoP16) – Criteria for amendment of Appendices I and II (Resolution Conf. 9.24 (Rev. CoP16) – Lauderdale criteria) 1994.

[300] The legal nature of CoP Resolutions is further discussed infra Chapter 3 C. I. 2. f).

[301] Resolution Conf. 1.2 – Criteria for the Deletion of Species and Other Taxa from Appendices I and II (Bern criteria) 1976 (repealed).

While aiming at making the listing process a scientific-based and neutral one, the Fort Lauderdale criteria, although they include quantitative guidelines,[302] "still fai[l] to address a fundamental truth about the listing process: a listing decision is a political decision, not merely a mechanical test."[303] States indeed vote and often do so defending their interests and values.[304] Favre notes that "[s]ome of the proposals in the initial treaty drafts of CITES would have turned the entire listing process to an independent scientific authority, but ultimately the decision was to rest, as it should, upon the judgment of the Parties."[305] Nevertheless, it has been demonstrated that the existing set of criteria and procedure for amending the Appendices facilitates decisions taken on the basis of scientific arguments rather than of parochial interests and bargaining powers.[306]

Indeed, while decision-making is still in the hands of the states, who will to some extent defend their interests and potentially use power instead of arguments, the CITES regime creates as efficient a forum as can be, through its general criteria, fact-finding procedures, the strong role given to the Secretariat and consultations with a wide range of actors.[307]

b) Trade in Appendix I

Appendix I is to include species which are or may be affected by trade and currently threatened by extinction, meaning either that their wild population is small, has a restricted area of distribution, is declining or will meet one of these criteria within five years if not added to Appendix I. These terms are defined numerically.

For these species, international trade is allowed only in exceptional circumstances, in which case it is strictly regulated. An import and an export permit are necessary. The import permit is to be issued by the Management Authority of the importing state, only if that specimen's trade is not for primarily commercial purposes and not detrimental to the survival of the species—this is determined by the Scientific Authority of the importing state. An export certificate must be issued by the Management Authority of the exporting state, ensuring that the specimen was legally obtained, that trade is non-detrimental to the survival of the species (so-called non-detriment finding or NDF, which is undertaken by the Scientific

[302] S. Young, 'Contemporary Issues of the Convention on International Trade in Endangered Species of Wild Fauna and Flora (CITES) and the Debate Over Sustainable Use' (2003) 14 Colorado Journal of International Environmental Law and Policy 167, 178.

[303] Ibid. 188.

[304] S. Dansky, 'The CITES "Objective" Listing Criteria: Are They Objective Enough to Protect the African Elephant?' (1999) 73 Tulane Law Review 961, 975.

[305] D. S. Favre, *International Trade in Endangered Species: A Guide to CITES* (Martinus Nijhoff Publishers 1989) 311.

[306] T. Gehring and E. Ruffing, 'When Arguments Prevail Over Power: The CITES Procedure for the Listing of Endangered Species' (2008) 8 Global Environmental Politics 123, 130–138.

[307] Ibid. 124, 130–131, 133–135, 145.

Authority of the exporting state) and only after an import permit has been issued.[308] For live animals, specific conditions regarding suitable housing, care and humane shipping are required.[309]

Specimens caught in waters not under the jurisdiction of any state need a certificate of introduction from the sea (IFS) to enter the state of introduction and then another permit (export or re-export) if they are to go further into international trade.[310] The certificate is issued by the Management Authority of the state of introduction, on the basis of an NDF and if it is satisfied that the specimen is not to be used for primarily commercial purposes.[311] No legality of catches verification is required.

Both legality verification and NDF are made under "[p]rotocols [. . .] established by individual range nations."[312] For the determination of the legality of catch, a positive decision means that "a Management Authority of the State of export is satisfied that the specimen was not obtained in contravention of the laws of that State for the protection of fauna and flora".[313]

As for the NDF, the Scientific Authority is responsible for this procedure. While NDF is a central feature of the CITES system, the treaty does not detail the ways to determine where the limit between sustainable use and overexploitation stands.[314] Overtime, guidance on NDF has been developed,[315] the Secretariat is available to

[308] CITES (n 294) article III(2)-(3).

[309] Ibid. article III(3)(b), (4)(b) (5)(b). These criteria will not be discussed here or for other Appendices, as fish are mostly traded dead.

[310] Whether an export or a re-export permit is to be issued remains to be clarified, see infra Chapter 5 C. I. 2. b) cc).

[311] CITES (n 294) article III(5).

[312] J. Grogan and P. Barreto, 'Big-Leaf Mahogany on CITES Appendix II: Big Challenge, Big Opportunity' (2005) 19 Conservation Biology 973, 974.

[313] CITES (n 294) article III(2)(b).

[314] A. Willock, 'Administrative and Monitoring Implications of listing and down-listing of commercially-exploited aquatic species, including the implications of Annex 4 of Resolution Conf. 9.24' Background paper to CITES Workshop on Introduction from the Sea Issues, 30 November – 2 December 2005, Geneva (Switzerland) http://www.cites.org/eng/news/meetings/ifs-05/IFS05-TRAFFIC-paper.pdf accessed 4 July 2015, 8. It does not appear clear to all parties whether the NDF for species listed under the look-alike criterion refers to the survival of the endangered species or the survival of the look-alike species (Bowman, Davies and Redgwell (n 57) 506–507).

[315] See Resolution Conf. 8.6 (Rev) – Role of the Scientific Authority, 1992 (repealed), which suggested elements to take into account when verifying the NDF. See also Resolution Conf. 10.3 – Designation and role of the Scientific Authorities 1997, which develops the role of the Scientific Authorities. On examples of NDF provided by CITES, see CITES (Chairs of Animals and Plants Committees), 'AC26/PC20 Doc. 8.4 – Draft guidance on the making of non-detriment findings' (22 March 2012); CITES, 'Resources for implementation: Sustainability' http://cites.org/eng/prog/shark/sustainability.php accessed 4 July 2015. For an independent research advising on methods to be further developed for NDF, see M. J. Smith and others, 'Assessing the impacts of international trade on CITES-listed species: Current practices and opportunities for scientific research' (2011) 144 Biological Conservation 82.

provide some support[316] and a Checklist to assist the Scientific Authorities in this process has been produced.[317] Most recently, at CoP16, a Resolution on NDF has been adopted, putting together and further developing concepts as well as non-binding principles to guide Scientific Authorities in NDF. It most prominently reminds states that NDF should be the result of a science-based assessment and that the best scientific available information should be used.[318]

As for 'primarily commercial', it is to be determined on the basis of the intended use in the country of import. The burden of proof to demonstrate that an import will predominately not be used for economic benefits lies with the person wanting to import the specimen.[319]

c) Trade in Appendix II

Appendix II comprises more than 30,000 species.[320] Appendix II species, although not currently threatened with extinction, could soon fulfill Appendix I criteria if trade was not regulated.[321] Some species which are not themselves under threat are also listed to allow the effective control of an endangered species—usually because they closely resemble an endangered species.[322]

Trade in those species is strictly regulated, so as to avoid utilization that could threaten their survival. To be exported, they must be accompanied by an export permit delivered by the Management Authority of the exporting state. The criteria to issue such a permit include a legality check and an NDF.[323] For live specimens, a humane way of transport is required.[324]

Specimens caught in waters not under the jurisdiction of any state need a certificate of IFS to enter the state of introduction and then a (re-)export permit if they are to go into international trade. The certificate is issued on the sole basis of an NDF.[325]

---

[316] A. Stroud, 'Essay: A Review of the Role of the CITES Secretariat in the Implementation of the Non-Detriment Finding Requirement' (2005–2006) 30 William and Mary Environmental Law and Policy Review 661.

[317] CITES, 'CoP11 Inf. 11.3 – Scientific Authorities' Checklist to assist in making Non-detriment Findings for Appendix II Exports' (10–20 April 2000).

[318] Resolution Conf. 16.7 – Non-detriment findings 2013.

[319] Resolution Conf. 5.10 (Rev. CoP15) – Definition of 'primarily commercial purposes' 1985.

[320] CITES 'The CITES species' (2013) (n 295).

[321] CITES (n 294) article II(2)(a).

[322] Ibid. article II(2)(b). Look-alike species inclusion was originally only planned in relation to Appendix II species but was broadened to deal with Appendix I species in the Resolution Conf. 9.24 (Rev. CoP16) – Lauderdale criteria (n 299) annex 2b; Gillespie, *Conservation, biodiversity and international law* (n 259) 97.

[323] CITES (n 294) article IV(2)(a)-(b).

[324] Ibid. article IV(2)(c).

[325] Ibid. article IV(6)(a).

d) Trade in Appendix III

Appendix III encompasses species "that in the eyes of the state, that has jurisdiction over their exploitation, need the cooperation of other states to prevent or restrict their exploitation."[326] So, species on this list are protected at least in one country, which requests help from other CITES parties in regulating international trade. A party can unilaterally list a species in Appendix III if this species is native to its country, but only after consulting other parties involved in relevant trade on the effects of such inclusion.[327]

Export from the state which included a species in Appendix III necessitates an export permit, issued on the basis of a legality check.[328] If the specimen is alive, conditions regarding humane transport are also to be respected.[329] If any other state exports such specimen, only a certificate of origin is needed; this document is based on neither an NDF nor a legality finding.

As Appendix III listing is unilateral and hence should target species that are not (highly) migratory—the need to involve all range states would be fulfilled only with difficulty—it will not be considered in this study.[330]

e) Exemptions and Reservations

According to the treaty, trade permits are not needed in the following circumstances: when the specimens have been acquired before the species has been listed; if they are in transit or being transshipped; if they are personal or household effects, non-commercial loans or exchanges between scientific establishments, part of travelling circuses and exhibitions; or if these specimens have been bred in captivity.[331] In this latter case, if the Management Authority is satisfied that the specimen has been bred in captivity, it can issue a certificate which will serve as an ersatz for the otherwise required trade permits.[332]

[326] E. Franckx, 'Applications of the term 'Introduction from the sea' http://www.cites.org/eng/news/meetings/ifs-05/term_IFS.pdf accessed 4 July 2015, 2 (reference omitted).

[327] Resolution Conf. 9.25 (Rev. CoP16) – Inclusion of species in Appendix III 1994.

[328] CITES (n 294) article V(2)(a).

[329] Ibid. article V(2)(b).

[330] For an interesting study of the use of CITES Appendix III for marine species, see A. Willock, M. Burgener and A. Sancho, 'First Choice or Fall Back? An Examination of Issues Relating to the Application of Appendix III of CITES to Marine Species' (TRAFFIC 2004).

[331] CITES (n 294) article VII; Resolution Conf. 9.7 (Rev. CoP15) – Transit and transhipment 1994; Resolution Conf. 12.3 (Rev. CoP16) – Permits and certificates 2002; Resolution Conf. 13.6 (Rev. CoP16) – Implementation of Article VII, paragraph 2, concerning 'pre-Convention' specimens 2004; Resolution Conf. 13.7 (Rev. CoP16) – Control of trade in personal and household effects 2004.

[332] CITES (n 294) article VII(5); Resolution Conf. 10.16 (Rev) – Specimens of animal species bred in captivity 1997.

Furthermore, reservations to a species' listing are possible: the state that enters into such reservation is to be treated as a non-party with regard to trade in that particular species.[333] In fact, this does not mean an exemption of control, as states parties are to require from non-parties "comparable documentation [...] which substantially conforms to the requirements of the present Convention for permits and certificates".[334] Hence, only trade between reserving or non-parties is exempt from CITES-related management; chances are high that of two trading partners, one at least will be a member without a reservation. The treaty has indeed, in its almost 40 years of being into force, gathered an impressive membership of 181 states.

f) Evolution of the Treaty Regime

The enhancement of the regime is possible through amendments of the treaty, but also through the adoption of documents by the CoP.[335] A few changes of the Convention have taken place through two amendments, but it is a lengthy and difficult process[336]—the proof being that one of the two amendments took 30 years before entering into force.[337]

Most developments and changes have been embraced through CoP Resolutions and Decisions which can be adopted by this body, as it is mandated "where appropriate, [to] make recommendations for improving the effectiveness of the present Convention."[338] The existence of an institutional structure which can adapt the Convention's regime without the lengthy process of a diplomatic conference

[333] CITES (n 294) article XXIII.

[334] Ibid. article X.

[335] Resolutions and Decisions are the types of documents adopted at CoPs; the former are meant to provide advice on the long-term, while the latter are shorter lived and provide for direct specific actions to be undertaken by a committee or the Secretariat (CITES, 'Decisions of the Conference of the Parties' http://www.cites.org/eng/dec/intro.php accessed 4 July 2015). Of the other documents presented at CoP meetings, the discussion documents are either abbreviated 'Doc.' for documents or 'Prop.' for proposals. Information documents, which are assigned the letters 'Inf.' are there for information purposes only ('Document numbering' *CITES World – Official Newsletter of the Parties* (July 2004) 6).

[336] According to CITES (n 294) article XVII, it is possible to amend the Convention if a third of the parties requests in writing an extraordinary meeting of the CoP to consider such amendment and if two-thirds at least vote in favor of the said amendment, which will then enter into force 60 days after two-thirds of the parties have deposited an instrument of acceptance to it.

[337] Of the two amendments adopted in almost 40 years, one entered into force in 1987, eight years after its adoption – the Bonn Amendment, which broadened the competence of the CoP – (Bonn amendment to the text of the Convention, http://www.cites.org/eng/disc/bonn.php accessed 4 July 2015), while the other – Gaborone Amendment, which aimed at allowing accession of regional economic integration organizations – finally entered into force on 29 November 2013, 30 years after its adoption and following the deposition of Costa Rica's instrument of acceptance (Gaborone amendment to the text of the Convention, http://www.cites.org/eng/disc/gaborone.php accessed 4 July 2015).

[338] CITES (n 294) article XI(3)(e).

and amendment is now the norm for MEAs but was quite new in the early 1970s.[339] Within the CITES regime, no formal 'opt-out' procedure exists in relation to Resolutions which are adopted by consensus or a two-third majority.[340]

The legal nature of such CoP documents is formally soft law,[341] but their effects are subject to discussion.[342] On the one hand indeed, it is true that "[a]lthough binding on the Secretariat, Conference resolutions and decisions, unless embodied into national enabling legislation, provide guidance only to Parties on a uniform implementation of the Convention."[343]

Nevertheless, on the other hand, the CoP does "unquestionably perform normative functions."[344] Resolutions adopted by consensus reflect the states parties' collective views[345] and might even represent later state practice.[346] In that respect, it is worth mentioning the ICJ's view on the International Whaling Commission's (IWC) resolutions: although not binding, "when they are adopted by consensus or

---

[339] R. R. Churchill and G. Ulfstein, 'Autonomous Institutional Arrangements in Multilateral Environmental Agreements: A Little-Noticed Phenomenon in International Law' (2000) 94 American Journal of International Law 623, 628–630.

[340] CITES Rules of Procedure of the Conference of the Parties, as amended at the 16th meeting, Bangkok, 2013 rules 21.1, 21.2 and 26.1.

[341] Birnie, Boyle and Redgwell (n 64) 690; Bowman, Davies and Redgwell (n 57) 488; Mahiou, *Le droit international ou la dialectique de la rigueur et de la flexibilité* (n 107) 349; D. Shelton ('Soft law' in D. Armstrong (ed), *Routledge Handbook of International Law* (Routledge 2009) 69) presents a definition of soft law being "any written international instrument, other than a treaty, containing principles, norms, standards, or other statements of expected behavior." However, Churchill and Ulfstein ((n 339) 642) seem to consider that only sets of guidelines are soft law documents in the CITES framework.

[342] Schermers and Blokker (n 212) §1217–1243, 1254; J. Werksman, 'The Conference of Parties to Environmental Treaties' in J. Werksman (ed), *Greening International Institutions* (Earthscan 1996) 63–64.

[343] R. W. G. Jenkins, 'An Overview of the Fundamental Principles of CITES as a Mechanism for Regulating Trade in Listed Species' http://www.cites.org/eng/news/meetings/ifs-05/IFS05-principle.pdf accessed 4 July 2015, 2.

[344] P. H. Sand, 'Enforcing CITES: The Rise and Fall of Trade Sanctions' (2013) 22 RECIEL 251, 255. On this question, see for example J. Brunnée, 'COPing with Consent: Law-making Under Multilateral Environmental Agreements' (2002) 15 Leiden Journal of International Law 1, 1–52; L. K. Camenzuli, 'The Development of International Law and the Multilateral Environmental Agreements' Conference of the Parties and Its Validity' (2007) http://cmsdata.iucn.org/downloads/cel10_camenzuli.pdf accessed 4 July 2015; J. Sommer, 'Environmental Law-Making by International Organizations' (1996) 56 Zeitschrift für ausländisches öffentliches Recht und Völkerrecht 628, 631–638; A. Wiersema, 'The New International Law-makers? Conferences of the Parties to Multilateral Environmental Agreements' (2009) 31 Michigan Journal of International Law 231, 232–87.

[345] Mahiou, *Le droit international ou la dialectique de la rigueur et de la flexibilité* (n 107) 352; Wiersema, 'The New International Law-makers? Conferences of the Parties to Multilateral Environmental Agreements' (n 344) 251.

[346] Ulfstein, 'Treaty Bodies' (n 213) 884, as in VCLT (n 106) article 31(3)(b).

by a unanimous vote, they may be relevant for the interpretation of the Convention or its Schedule."[347] States are also allowed to adopt authoritative interpretations of a treaty[348] and can further develop or modify a treaty by their practice, if so is their intent.[349]

It has been said that "an organization's legal output can help to shape expectations and constrain behavior."[350] Indeed, some of these documents with wordings such as 'agree', 'decide', or 'urge' have a binding content,[351] direct actions or establish an agreed understanding of terms. In some other Resolutions, the CoP 'invites', 'encourages' or 'recommends' a certain course of action, which can thus be considered non-mandatory. Consequently, the wording chosen for a Resolution and the drafting history are to be given a close look in order to determine the binding nature of its provisions.[352]

Also, except if otherwise decided, Resolutions are to "enter into effect 90 days after the meeting",[353] formulation which again might show a binding intent.[354] The compliance of states with obligations under the Convention is furthermore understood as interpreted in the Resolutions. Indeed, "[r]ecommendations have grown into a body of rules which, although not considered legally binding, transformed the regime in an unforeseeable way [...and n]on-compliance with them triggers

---

[347] *Whaling in the Antarctic (Australia v. Japan; New Zealand intervening), Judgment,* 31 March 2014, ICJ 24 §46.

[348] VCLT (n 106) article 31(3)(a); P. H. Sand, 'Endangered Species, International Protection' Max Planck Encyclopedia of Public International Law www.mpepil.com §10; Churchill and Ulfstein (n 339) 641.

[349] M. Akehurst, 'The Hierarchy of Sources in International Law' (1974–1975) 47 British Yearbook of International Law 273, 277–278.

[350] N. D. White, 'Separate but Connected: Inter-Governmental Organizations and International Law' (2008) 5 International Organizations Law Review 175, 183 (reference omitted).

[351] The terminology chosen is a crucial factor in determining the nature of an instrument as highlighted in the *Maritime Delimitation and Territorial Questions between Qatar and Bahrain (Qatar v Bahrain), Jurisdiction and Admissibility,* 1 July 1994, I.C.J. Reports 1994 112, 120–121 §23–25; see also Wiersema, 'The New International Law-makers? Conferences of the Parties to Multilateral Environmental Agreements' (n 344) 253–255.

[352] See for example the ICJ's view on how to consider UNGA Resolutions' normative value: "it is necessary to look at its content and the conditions of its adoption" (*Legality of the Threat or Use of Nuclear Weapons, Advisory Opinion,* 8 July 1996, I.C.J. Reports 1996 226, 254 §70). On a very critical view of international organizations' resolutions being considered somehow binding, see for example P. Weil, 'Towards relative normativity in international law?' (1983) 77 American Journal of International Law 413, 415–417.

[353] Resolution Conf. 4.6 (Rev. CoP16) – Submission of draft resolutions and other documents for meetings of the Conference of the Parties 1983.

[354] Sand refers to the fact that the French argument in the Bolivia Furskin Case (ECJ, *Case C-182/89 Commission v France (Bolivian Furskin Case)* 4344) that Resolutions are recommendations 'without legal effect' was rejected. *A contrario,* he argues that they have practical effects (Sand, 'Endangered Species, International Protection' (n 348) §10).

reactions under the compliance regime."[355] Non-compliance with some rules developed in the Resolutions is even explicitly considered to have similar consequences than non-compliance with the treaty itself.[356] Some examples such as trade restrictions decided on the basis of inadequate legislation or reporting, as understood in Resolutions Conf. 8.4 and 11.17, do support the binding force of those instruments.[357]

A few illustrations of major evolutions based on CoP Resolutions are the criteria for listing, the compliance and enforcement mechanisms, and several understandings of terms.

g) Review and Compliance

At the national level, states have to implement the Convention by adopting the necessary laws and designating a Management and a Scientific Authority.[358] Enforcement should take place in the form of verifications at the borders and sanctions in case of disrespect towards the permitting system. States are also under the obligation to prepare a report every year for the Secretariat, with information on traded specimens, quantities and trading partners.[359] Every second year, another report on "legislative, regulatory and administrative measures taken to enforce" the Convention is to be provided.[360] For states identified by the Secretariat as having problems in the development of adequate national legislation, technical

[355] C. Fuchs, 'Convention on International Trade in Endangered Species of Wild Fauna and Flora (CITES) – Conservation Efforts Undermine the Legality Principle' (2008) 9 German Law Journal 1565, 1575–1576 (reference omitted), P. H. Sand, 'Whither CITES? The Evolution of a Treaty Regime in the Borderland of Trade and Environment' (1997) 8 European Journal of International Law 29, 35. This is for example not totally accepted by V. Röben ('Conference (Meeting) of States Parties' Max Planck Encyclopedia of Public International Law www.mpepil.com §23) who states that this "would still fall short of imposing new substantive obligations".

[356] Resolution Conf. 14.3 – CITES compliance procedures 2007 annex §30 and footnote 1.

[357] Ibid. annex §1 (italics added) states that it "describes existing procedures in order to facilitate consistent and effective handling of compliance matters relating to obligations under the Convention, *taking into account relevant Resolutions and Decisions*". Sand, 'Enforcing CITES: The Rise and Fall of Trade Sanctions' (n 344) 256. The criteria used as benchmarks for compliance were developed respectively in Resolution Conf. 8.4 (Rev. CoP15) – National laws for implementation of the Convention 1992 and Resolution Conf. 11.17 (Rev. CoP16) – National reports 2000.

[358] CITES (n 294) article IX(1).

[359] Ibid. article VIII(6)–(7)(a).

[360] Ibid. article VIII (7)(b).

assistance can be offered.[361] Unfortunately, the obligation to report is poorly complied with.[362]

In addition to national reports, information about international trade is gathered by NGOs, particularly TRAFFIC, with which CITES has a Memorandum of Understanding. TRAFFIC, a joint program of the World Wild Fund for Nature (WWF) and IUCN, monitors trade around the world and provides CITES with information on illegal wildlife trade.[363] Intergovernmental organizations such as Interpol and the World Customs Organization are also involved in transmitting data on illegal trade.[364] Wildlife traffic is increasingly considered a serious criminal activity that ought to be fought in the same manner as other highly denounced transnational crimes.[365]

At the CITES level, the Secretariat reviews national reports. It uses this information, the one provided by NGOs, by international organizations and by on-site findings[366] to verify compliance with the Convention, for example, by discovering illegal trade through discrepancies in declared import/export of a species. The Secretariat also monitors trade, particularly through the Significant Trade Review.[367]

The Significant Trade Review procedure has been developed to monitor trade in Appendix II species which are traded in high numbers.[368] Based on article IV (2), (3) and (6), and as an answer to the poor implementation of these provisions, the Animals and Plants Committees are mandated "to review the biological, trade and

---

[361] Bowman, Davies and Redgwell (n 57) 520; see Reeve, *Policing International Trade in Endangered Species* (n 296) 134–147.

[362] CITES Secretariat, 'CoP12 Doc.22.1 – Report on national reports required under Article VIII, paragraph 7, of the Convention – Annual Reports' (3–15 November 2002) 1–2 §5, noting that levels of on-time submission ranges between 35 % and 60 % since the entry into force of the Convention. Bowman, Davies and Redgwell (n 57) 521–522; Gillespie, *Conservation, biodiversity and international law* (n 259) 425; Reeve, *Policing International Trade in Endangered Species* (n 296) 62–67, 147–152.

[363] Reeve, *Policing International Trade in Endangered Species* (n 296) 68; Sand, 'Whither CITES? The Evolution of a Treaty Regime in the Borderland of Trade and Environment' (n 355) 49–50.

[364] Fuchs (n 355) 1585; Reeve, *Policing International Trade in Endangered Species* (n 296) 227–230.

[365] J. E. Scanlon, 'CITES at Its Best: CoP16 as a 'Watershed Moment' for the World's Wildlife' (2013) 22 RECIEL 222, 223–225.

[366] Reeve, *Policing International Trade in Endangered Species* (n 296) 75–76.

[367] Resolution Conf. 12.8 (Rev. CoP13) – Review of Significant Trade in specimens of Appendix-II species 2002 operative part; see Bowman, Davies and Redgwell (n 57) 519; Reeve, *Policing International Trade in Endangered Species* (n 296) chapter 7.

[368] S. Aguilar, 'Regulatory Tools for the Management of Fish and Timber Species through CITES' (2013) 22 RECIEL 281, 284–285; Bowman, Davies and Redgwell (n 57) 522–525.

other relevant information on Appendix-II species subject to significant levels of trade, to identify problems and solutions concerning the implementation of Article IV".[369] This provides a second-layer species-specific verification of compliance with the Convention and the possibility to have solutions formulated. It counteracts the issue linked to trade in Appendix II, namely that "the implementation of Article IV relies entirely on the will and capacity of the exporting countries."[370]

No sanction mechanism was planned in the Convention and no compliance committee has been set up[371]—although this possibility was openly discussed in 2004[372]—but a system of voluntary yet coordinated non-compliance measures was developed.[373] Indeed, as states are allowed to apply stricter domestic measures regarding conditions to trade,[374] it is possible for them—but not compulsory—to suspend trade against non-compliant parties.[375] The Standing Committee, a non-plenary body created in 1979 which took over a number of functions of the CoP in between meetings, plays a central role in the application of sanctions. It recommends the suspension of trade and sets conditions to be fulfilled before trade ought to be resumed.[376]

Two types of sanctions exist: those undertaken under the Significant Trade Review process and those focusing on one country.[377] In the first case, sustainability of trade for a particular species is reviewed and the CITES Animals and Plants

---

[369] Resolution Conf. 12.8 (Rev. CoP13) – Review of Significant Trade in specimens of Appendix-II species (n 367).

[370] R. W. G. Jenkins, 'The Significant Trade Process: Making Appendix II Work' in J. Hutton and B. Dickson (eds), *Endangered Species-Threatened Convention: The Past, Present and Future of CITES* (Earthscan 2000) 48.

[371] R. Reeve, 'The Convention on International Trade in Endangered Species of Wild Fauna and Flora (CITES)' in G. Ulfstein (ed), *Making treaties work: Human Rights, Environment and Arms Control* (Cambridge University Press 2007) 136; G. Ulfstein, 'Dispute resolution, compliance control and enforcement in international environmental law' in G. Ulfstein (ed), *Making treaties work: Human Rights, Environment and Arms Control* (Cambridge University Press 2007) 125.

[372] V. Koester concluded that a compliance committee was not necessary nor would necessarily be beneficial ('Compliance Committees within MEAs and the Desirability and Feasibility of Establishing Special Compliance Bodies under CITES' 20 May 2004, SC54 Inf. 3 (CITES 2006)).

[373] Gillespie, *Conservation, biodiversity and international law* (n 259) 439–441; Reeve, *Policing International Trade in Endangered Species* (n 296) chapters 5–7; Sand, 'Enforcing CITES: The Rise and Fall of Trade Sanctions' (n 344) 251–262.

[374] CITES (n 294) article XIV(1).

[375] Even though the mechanism is not referred to as such in Resolution Conf. 11.3 (Rev. CoP16) – Compliance and enforcement 2000, it is quite regularly used against both parties and non-parties (Fuchs (n 355) 1587; Reeve, 'The Convention on International Trade in Endangered Species of Wild Fauna and Flora (CITES)' (n 371) 149–152; Sand, 'Endangered Species, International Protection' (n 348) §18). Aguilar considers that trade sanctions on the recommendation of the Standing Committee or the CoP are actually compulsory, as they implement article II(4) ((n 368) 285).

[376] Bowman, Davies and Redgwell (n 57) 518–519.

[377] Reeve, 'The Convention on International Trade in Endangered Species of Wild Fauna and Flora (CITES)' (n 371) 149.

Committees can recommend remedial measures. Failure to implement these recommendations can bring the Standing Committee to suspend trade with that country with regard to the reviewed species.[378] As for the suspension of trade for a country, it can be triggered by illegal trade, failure to develop domestic law and/or failure to submit annual reports for three consecutive years, after attempted communication with the party turns out to be insufficient to improve the situation.[379] It is also recommended to the parties by the CoP or the Standing Committee but the awareness of this committee is in turn usually triggered by the Secretariat.[380]

Dispute settlement with binding outcomes is envisioned in front of the Permanent Court of Arbitration, but it requires mutual consent of the parties involved.[381] It has not been used so far; interpretation issues are solved through the adoption of Resolutions.[382] Moreover, even though some states targeted by sanctions complained about the legality of such sanctions, they never moved towards third party dispute resolution.[383]

3. Applicability and Appropriateness

The CITES regime has the advantage of being quasi-universal. With 181 parties,[384] and also being indirectly applicable to non-members in their dealings with members, its measures are wide-reaching. Also, the enforcement measures developed to address non-compliant states' behavior or unsustainable significant trade are perceived by many commentators as quite efficient.[385] Indeed, suspension of trade

[378] Willock, 'Administrative and Monitoring Implications of listing' (n 314) 10.

[379] Resolution Conf. 11.17 (Rev. CoP16) – National reports (n 357).

[380] Reeve, 'The Convention on International Trade in Endangered Species of Wild Fauna and Flora (CITES)' (n 371) 140–141. The power of recommendation of the CoP is explicitly provided for in CITES (n 294) article XIII(3); on this, see M. Yeater and J. Vasquez, 'Demystifying the Relationship Between CITES and the WTO' (2001) 10 RECIEL 271, 275.

[381] CITES (n 294) article XVIII(2).

[382] W. Wijnstekers, *The Evolution of CITES* (9th edn CIC – International Council for Game and Wildlife Conservation 2011) 477.

[383] Reeve, 'The Convention on International Trade in Endangered Species of Wild Fauna and Flora (CITES)' (n 371) 158.

[384] Since 29 November 2013, regional economic integration organizations are also able to accede to the Convention, following the entry into force of the Gaborone amendment (CITES Secretariat, 'Notification to the parties n° 2013/045 – Entry into force of the amendment to Article XXI of the text of the Convention (Gaborone, 30 April 1983)' (4 October 2013)).

[385] See for example Bowman, Davies and Redgwell (n 57) 484; D. Harland, *Killing Game: International Law and the African Elephant* (Praeger 1994) 12; Reeve, *Policing International Trade in Endangered Species* (n 296) 6; Sand, 'Enforcing CITES: The Rise and Fall of Trade Sanctions' (n 344) 252, 255. However, on the other hand, see for example the nuanced assessment in K. Baakman, *Testing Times: The Effectiveness of Five International Biodiversity-Related Conventions* (Wolf Legal Publishers 2011) 264–265. For more on the regime's efficacy, see infra Chapter 5 D.

against a particular state regarding one or all species is a strong incentive to change one's behavior.

However, the Convention focuses only on international commerce, leaving domestic matters to be solely regulated by the national authorities. Furthermore, other important threats, mainly habitat destruction and introduction of alien species are also affecting species' health and recovery abilities[386] but are not included in this treaty's realm. Such non-holistic approach has been criticized as not taking into account the inherent complexity of biodiversity.[387] Since combatting habitat loss, for example, is crucial to species' survival chances,[388] "whether CITES regulation improves the conservation status of a species is likely to depend on how these many different factors combine to influence the incentives faced by actors at the local level."[389] It has also been said that regulating trade only is partly inefficient in promoting conservation of species.[390] The good news is that CITES parties have actually, at least to some extent, faced this issue by taking other threats into account in assessing a species' vulnerability.[391] Moreover, regimes do exist to protect habitats, a goal which CITES' drafters purposely did not include in the Convention.[392]

While the soft law nature of Resolutions has been identified by some commentators as a weakness of the regime[393] it can also be said that it has little impact on the effectiveness of these documents. Indeed, states parties adopt Resolutions to further develop the regime or clarify terms; since they agree, at the CoPs, to both content and wording, it can be expected that they will comply with them—and

---

[386] B. Dickson, 'CITES and the livelihoods of the poor' (2008) 42 Oryx 548, 549; Gillespie, *Conservation, biodiversity and international law* (n 259) 166–170; Sand, 'Whither CITES? The Evolution of a Treaty Regime in the Borderland of Trade and Environment' (n 355) 30–31.

[387] E. Couzens ('CITES at Forty: Never Too Late to Make Lifestyle Changes' (2013) 22 RECIEL 311, 311–323) expresses such concerns and suggests an innovative evolution of CITES, where blacklisting of endangered or threatened species is replaced by a whitelisting of species that can be traded.

[388] J. Copeland Nagle, 'Why Chinese Wildlife Disappears as CITES Spreads?' (1996–1997) 9 Georgetown International Environmental Law Review 435, 439–440; M. A. Du Plessis, 'CITES and the Causes of Extinction' in J. Hutton and B. Dickson (eds), *Endangered Species-Threatened Convention: The Past, Present and Future of CITES* (Earthscan 2000) 22; J. Hutton and B. Dickson, 'Introduction' in J. Hutton and B. Dickson (eds), *Endangered Species-Threatened Convention: The Past, Present and Future of CITES* (Earthscan 2000) xvi; M. Recharte Uscamaita and R. Bodmer, 'Recovery of the Endangered giant otter Pteronura brasiliensis on the Yavarí-Mirín and Yavarí Rivers: a success story for CITES' (2009) 44 Oryx 83, 86.

[389] Dickson, 'CITES and the livelihoods of the poor' (n 386) 549.

[390] Hutton and Dickson (n 388) xvi.

[391] According to the Resolution Conf. 9.24 (Rev. CoP16) – Lauderdale criteria (n 299) annex 5 extrinsic vulnerabilities include inter alia habitat loss, threats from alien species or rapid environmental change.

[392] M. Bowman, 'A Tale of Two CITES: Divergent Perspectives upon the Effectiveness of the Wildlife Trade Convention' (2013) 22 RECIEL 228, 235.

[393] Birnie, Boyle and Redgwell (n 64) 690.

according to some commentators, they are expected to do so, at the risk otherwise to violate the principle of good faith and to commit an estoppel.[394] Moreover, non-compliance with some particular Resolutions is treated in the same way as non-compliance with the treaty.[395] Then, as illustrations of the subsequent state practice and/or authoritative interpretation, their legal nature is generally strengthened.[396] Hence, more and more, the limit between hard and soft law is blurred. Last but not least, beyond the legal nature of the Resolutions, their efficacy is key.[397] In that regard, the flexibility offered by soft law instruments allows quicker decision-making, rhythm which is more in line with the environmental needs than the cautious speed of formal treaty-making.[398]

Then, even if the Fort Lauderdale criteria are supposed to facilitate science-based decisions, the decision-making process is more political than scientific. As was highlighted by Eldridge in 1995 the down-listing of some whale species was to be a test of how objective the new listing criteria were;[399] they certainly have not rated very well in that particular evaluation. Many species which should be listed are not or are under-listed and others stay in Appendices while they do not need it.

Secret ballot has also an impact on the usefulness of the newest criteria in promoting science-based decisions. Indeed, whether scientifically convincing or not, recommendations will be followed by states only if their interests tell them to do so, without any fear of criticisms or call for accountability. While it is true that it could also have the opposite effect, with states better able to sustain pressure from powerful lobbies or other states, the widespread use of secret ballots, in particular for controversial issues such as marine species or elephants and ivory trade matters, in fact threatens transparency, if not even the integrity of the Convention.[400] There

---

[394] Mahiou, *Le droit international ou la dialectique de la rigueur et de la flexibilité* (n 107) 349. However, states relying on the agreement would to have suffered a detriment or prejudice (*North Sea Continental Shelf Case (Federal Republic of Germany/Denmark; Federal Republic of Germany/Netherlands)* (n 106) 26 §30; T. Cottier and J. P. Müller 'Estoppel' Max Planck Encyclopedia of Public International Law www.mpepil.com §6–7).

[395] Resolution Conf. 14.3 – CITES compliance procedures (n 356) annex §30 and footnote 1; Sand, 'Enforcing CITES: The Rise and Fall of Trade Sanctions' (n 344) 256.

[396] Boyle and Chinkin (n 202) 216.

[397] Mahiou, *Le droit international ou la dialectique de la rigueur et de la flexibilité* (n 107) 350.

[398] Boyle and Chinkin (n 202) 214; Shelton (n 341) 75.

[399] K. Eldridge, 'Whale For Sale?: New Developments in the Convention on International Trade in Endangered Species of Wild Fauna and Flora' (1994–1995) 24 Georgia Journal of International and Comparative Law 549, 565.

[400] See for example Denmark (on behalf of the EU) 'CoP16 Doc. 4.2 (Rev. 1) – Proposal to improve transparency of voting during meetings of the Conference of the Parties' (n 297); Chile and Mexico 'CoP16 Doc. 4.3 (Rev. 1) – Proposed amendment to Rule 25 on Methods of Voting use of secret ballots' (n 297).

are moreover some allegations of vote-buying at CITES.[401] Other processes, such as NDF (by the national scientific authorities) and sometimes even Significant Trade Review (by CITES) are also not invariably based on science only, but rather appear to be influenced by political and economic interests.[402]

Additionally, the strength of the system is undermined by reservations. States have indeed the option to object to the addition or up-listing of a species in an Appendix. In the 90 days following the decision taken at the CoP, any party can enter into a reservation. For that party, it signifies that it will be treated as a non-party in relation to that species. There is no limit to the number of reservations a country can enter. Furthermore, exceptions which allow trade in Appendices I and II species create loopholes in the system and can consequently increase the risk of cheating.

Finally, implementation and enforcement measures are mainly to be taken by the parties, who do not always succeed in doing so efficiently. Issues of implementation include for example the failure to create national authorities to monitor the operations[403] or the inadequate delivery of permits by those authorities. As for enforcement, states are to "penalis[e] trade which violates the terms of the Convention, confiscate[e] illegally traded specimens, [can] designat[e] special ports of exit and entry for wildlife [and must maintain] records".[404] It appears that, in reality, trade bans have not always been enforced successfully, with illegal trade as a consequence.[405] Finally, however useful those requirements are, they are only checked through compulsory annual records, the required submission of which is not well complied with.

---

[401] A. Gillespie, *Whaling Diplomacy: Defining Issues in International Environmental Law* (New Horizons in Environmental Law, Edward Elgar 2005) 432–433; E. J. Molenaar, 'Marine Mammals: The Role of Ethics and Ecosystem Considerations' (2003) 6 Journal of International Wildlife Law and Policy 31, 40.

[402] See for example the NDF in the trade of bottlenose dolphins as shown in E. C. M. Parsons, N. A. Rose and T. M. Telecky, 'The trade in live Indo-Pacific bottlenose dolphins from Solomon Islands: A CITES decision implementation case study' (2010) 34 Marine Policy 384.

[403] As required in CITES (n 294) article IX.

[404] Ibid. article VIII(1), (3) and (6), as presented in S. Lyster, *International Wildlife Law* (Grotius Publications Limited 1985) 264, with precisions added in square brackets.

[405] Hutton and Dickson (n 388) xvi.

If non-compliance reaches certain levels, international enforcement is possible, but is limited to suspension of trade. The effectiveness of such sanctions is subject to various opinions.[406] They should in any event be kept for a limited number of cases; were they to be used against many actors at once, compliance with the system would be delegitimized and too many free-riders would find trade partners.

4. Conclusions

CITES aims at protecting endangered species by regulating the access to markets and hence reducing the harvest pressure put on internationally traded species. It uses a system of nationally delivered permits, with different requirements depending on the level of threat faced by the species. Sanctions mechanisms have been created over time and significantly strengthen the regime.

CITES applies to all species of wild fauna and flora, and hence is potentially relevant for marine species, although it has not yet been used for marine commercially-exploited fish species. Questions about the particular aspects of such a listing's legality, adequacy and efficacy will be looked into in detail in Chapter 5.

II. 1979 Convention on Migratory Species

1. Relevant Provisions

The 1979 Convention on Migratory Species (CMS) intends to protect species crossing national boundaries and/or living in areas beyond national jurisdiction. It "concentrates on the taking of migratory species by the range states that feature in relevant migration routes, and imposes obligations on these states to protect or restore habitat for certain species."[407] Range states include flag states of vessels engaged in activities in waters beyond national jurisdiction.[408]

---

[406] D. Brack ('Environmental Treaties and Trade: Multilateral Environmental Agreements and the Multilateral Trading System' in G. P. Sampson and W. B. Chambers (eds), *Trade, Environment and the Millennium* (United Nations University Press 2002) 334) refers to "an almost 100 percent success rate" and Sand ('Enforcing CITES: The Rise and Fall of Trade Sanctions' (n 344) 255) to "[t]he extraordinary effectiveness of the scheme", but R. Reeve ('Wildlife Trade, Sanctions and Compliance' (2006) 82 International Affairs 881, 892–895) presents a more balanced evaluation.

[407] Young, *Trading Fish, Saving Fish* (n 31) 61.

[408] 'Convention on the Conservation of Migratory Species of Wild Animals' (CMS) 1979, 1651 *United Nations Treaty Series* 333 article I(1)(h).

The CMS applies to marine species:[409] marine mammals, sharks and sturgeons are listed,[410] but major commercially-exploited species are not included in this instrument. It is worth noting that during the negotiations a number of states attempted, unsuccessfully, to explicitly exclude such species from the ambit of the CMS, to avoid putting at risk the effectiveness of other regimes.[411]

The CMS establishes general protective standards to be applied by member states, for endangered or 'of unfavourable conservation status' migratory species, listed respectively in Appendix I or II.[412] It thus serves as a framework for states' cooperation in the conservation of species and the protection of habitats.[413]

Appendix I species are those "facing a very high risk of extinction in the wild in the near future".[414] The species' status is examined by the CMS scientific body[415] which recognized that it would follow as much as possible the IUCN Red List system.[416] To be put in—or removed from—Appendix I, a migratory species must be considered endangered if "reliable evidence, including the best scientific evidence available,"[417] shows respectively that it is "in danger of extinction throughout all or a significant portion of its range"[418] or that it is no longer under such threat. Range states shall work towards conservation and, where feasible and appropriate, restoration of important habitats of those species, minimization of sources of obstacles on migratory routes, control of the introduction of exotic species[419] and they shall prohibit the taking of listed animals.[420] However, some exceptions exist, such as scientific permits or traditional subsistence catches.[421]

Migratory species that need or would significantly benefit from international cooperation are listed in Appendix II. They are not guaranteed any specific protection under the CMS.[422] Indeed, for these species, the CMS acts as a framework convention: it requires from states parties that they conclude global or regional

---

[409] Young, *Trading Fish, Saving Fish* (n 31) 62.

[410] CMS CoP, 'Appendices I and II of the Convention on the Conservation of Migratory Species of Wild Animals' (effective as of 8 February 2015) http://www.cms.int/manage/sites/default/files/document/Appendices_COP11_E_version5June2015.pdf accessed 4 July 2015.

[411] Bowman, Davies and Redgwell (n 57) 538 n 13.

[412] R. Caddell, 'International Law and the Protection of Migratory Wildlife: An Appraisal of Twenty-Five Years of the Bonn Convention' (2005) 16 Colorado Journal of International Environmental Law and Policy 113, 115–116.

[413] Birnie, Boyle and Redgwell (n 64) 681.

[414] CMS, Resolution 5.3 – Interpretation of certain terms of the Convention, 1997 §1.

[415] Established according to CMS (n 408) article VIII(1).

[416] CMS, Resolution 5.3, Interpretation of certain terms of the Convention (n 568) §2, as presented in Bowman, Davies and Redgwell (n 57) 541.

[417] CMS (n 408) article III(2).

[418] Ibid. article I(1)(e).

[419] Ibid. article III(4).

[420] Ibid. article III(5).

[421] Ibid.

[422] Bowman, Davies and Redgwell (n 57) 537.

AGREEMENTS,[423] which are open to all states.[424] Species can be listed under both Appendices at the same time.[425] Only four such AGREEMENTS have been developed yet and none of them relates to commercially-exploited aquatic species.[426]

Additionally, states are encouraged to develop and adopt other agreements for populations that periodically migrate through borders.[427] This can be done with regard to any species, whether it is listed on an Appendix or not. Such instruments are open to non-members and are not necessarily legally binding. More than 20 agreements have been adopted so far, many of them on marine mammals.[428]

Also, a non-legally binding Memorandum of Understanding (MoU) on the conservation of migratory sharks was adopted under the CMS in 2010, has 38 signatories and is applicable to seven species or populations of sharks.[429] Recognizing the importance of other institutions such as the FAO and RFMOs in the conservation and management of sharks,[430] the Signatories adopted at their first meeting in September 2012 a Conservation Plan which aims inter alia at improving the scientific understanding on the relevant species, ensuring the sustainability of both targeted and non-targeted shark fisheries and enhancing cooperation between the states and institutions involved.[431] An Advisory Committee composed of experts on science and management of shark species was created, which has, amongst other tasks, to assess the status of shark populations listed or which

---

[423] CMS (n 408) article IV(3).

[424] Ibid. article V(2).

[425] Ibid. article IV(2); Birnie, Boyle and Redgwell (n 64) 681.

[426] 'Agreement on the Conservation of Populations of European Bats' 1991, 1863 *United Nations Treaty Series* 101; 'Agreement on the Conservation of African-Eurasian Migratory Waterbirds' 1995; 'Agreement on the Conservation of Albatrosses and Petrels' 2001, 2258 *United Nations Treaty Series* 257; 'Agreement on the Conservation of Gorillas and Their Habitats' 2007, 2545 *United Nations Treaty Series* 55.

[427] CMS (n 408) article IV(4).

[428] Bowman, Davies and Redgwell (n 57) 558, 560–562.

[429] Memorandum of Understanding on the Conservation of Migratory Sharks, 'Signatories' http://sharksmou.org/list-of-signatories accessed 6 July 2015.

[430] 'Memorandum of Understanding on the Conservation of Migratory Sharks' (MoU on the Conservation of Migratory Sharks) 2010 preamble and section 3 (Fundamental Principles) §6–7; Lack and Sant (n 188) 15.

[431] Ibid. preamble and section 4 (Conservation Plan); Annex 3 to the MoU on the Conservation of Migratory Sharks: Conservation Plan 27 September 2012 http://cms.eaudeweb.ro/fr/document/conservation-plan accessed 4 July 2015, CMS/Sharks/Outcome 1.2 objectives A, B, E.

could be considered for listing.[432] The MoU is not only open for signature to all range states, but also to organizations who, as cooperating partners, can develop joint work plans.[433]

2. Applicability and Appropriateness

The CMS presents some complementarity to CITES[434] since it concerns harvesting for both domestic and international trade. It also applies to species that are not traded but threatened in a different way and arguably promotes the protection of habitat, at least to some extent.[435] Some of the species under its focus are also listed under CITES.

As for most fisheries-related treaties, even if this Convention is quite widely ratified,[436] it would need universal membership to really fulfill its goals.[437] Indeed "heavy exploitation at just one stage of [those species'] migratory route can undermine the most stringent protective measures at all other stages."[438] Here, no fish species except sharks and sturgeons has been listed (yet), but a call for research and dialogue regarding fisheries has been voiced, including on the by-catch of migratory species.[439]

Also, except in relation to the prohibition of harvesting, the language of the obligations is qualified with 'feasible and appropriate' or 'endeavour', thus leaving leeway in national implementation.[440] Then, beside a non-compulsory dispute settlement mechanism and national reports, the CMS does not benefit from a strong

---

[432] MoU on the Conservation of Migratory Sharks (n 430) section 7 (Advisory Committee); Terms of References of the Advisory Committee 24 September 2012 http://www.cms.int/huemul/sites/default/files/document/Outcome_1_1_AC_ToR_E_0.pdf accessed 4 July 2015.

[433] Authorization of Cooperating Organizations to Sign the MoU 24 September 2012, CMS/Sharks/MOS1/Doc.6.2 http://sharksmou.org/MOS1 accessed 4 July 2015.

[434] N. Matz, 'Chaos or Coherence? – Implementing and Enforcing the Conservation of Migratory Species through Various Legal Instruments' (2005) 65 Zeitschrift für ausländisches öffentliches Recht und Völkerrecht 197, 207.

[435] See contradicting opinions on the question of the CMS's role in the protection of habitats in Young, *Trading Fish, Saving Fish* (n 31) 62; Matz, 'Chaos or Coherence? – Implementing and Enforcing the Conservation of Migratory Species through Various Legal Instruments' (n 434) 201–202.

[436] 120 parties as of 1 May 2014 ('Parties to the Convention on the Conservation of Migratory Species of Wild Animals' (2014) http://www.cms.int/en/parties-range-states accessed 4 July 2015).

[437] Bowman, Davies and Redgwell (n 57) 536, 544; several major range states such as Russia, China, Japan, Brazil or the USA are not parties to the CMS.

[438] Lyster (n 404) 16; see also Birnie, Boyle and Redgwell (n 64) 684.

[439] Migratory Marine Species 2008, UNEP/CMS/Resolution 9.9 (CMS CoP) §4(b); By-Catch 2008, UNEP/CMS/Resolution 9.18 (CMS CoP) §9.

[440] Caddell, 'International Law and the Protection of Migratory Wildlife: An Appraisal of Twenty-Five Years of the Bonn Convention' (n 412) 117.

non-compliance mechanism to encourage states to abide by their obligations. Furthermore, it is considered that listing in neither Appendix I nor Appendix II has been used to its fullest.[441]

Moreover, no precautionary principle is to be applied in the listing of species.[442] Finally, even if the CMS has an in-house scientific council envisioned in the Convention, the decisions of making additions to a list are taken at the CoP; they are consequently political, even if based on the advice of the Scientific Council.[443]

3. Conclusions

The CMS aims at protecting migratory species by promoting collaboration between range states but also, for Appendix I species, by prohibiting the taking of specimens. It provides 'softer' obligations regarding habitats and other aspects of the ecosystem than some other regimes. Even if it is not used for main commercially-exploited fish species (yet), this regime already applies to several cetaceans and sharks.

III. 1946 International Convention on the Regulation of Whaling

1. Development

The International Convention on the Regulation of Whaling (ICRW)[444] was signed in 1946 by 15 states and entered into force two years later. The IWC was then founded as a resource management institution.

The ICRW has two distinct types of objectives: on the one hand it aims at "safeguarding for future generations the great natural resources represented by the whale stocks" and on the other hand it tries "to provide for the proper conservation

---

[441] Birnie, Boyle and Redgwell (n 64) 684.

[442] Positive evidence that a species is endangered is needed for it to be listed in Appendix I (CMS (n 408) article III(2)).

[443] Ibid. article VIII(5).

[444] The North Atlantic Marine Mammal Commission (NAMMCO) was established in 1992 by pro-whaling states which were frustrated with the IWC (Norway, Iceland, Greenland and the Faroe Islands). It was meant as an alternative to the IWC but has remained, so far, an advisory body only. Also, since states that wanted to resume whaling have done so within the IWC – by entering into reservations or undertaking scientific whaling –, NAMMCO's role is not crucial (S. Andresen, 'The International Whaling Regime: Order at the Turn of the Century' in D. Vidas and W. Ostreng (eds), *Order for the oceans at the turn of the century* (Kluwer Law International 1999) 222–223). On NAMMCO, see D. D. Caron, 'The International Whaling Commission and the North Atlantic Marine Mammal Commission: The Institutional Risks of Coercion in Consensual Structures' (1995) 89 The American Journal of International Law 154, 163–166.

of whale stocks and thus make possible the orderly development of the whaling industry".[445] Hence, this treaty originally was neither a preservationist one, nor promoting unlimited exploitation.[446]

The IWC used to be a club of whaling nations,[447] very much like most RFMOs are clubs of fishing nations—as such this regime could also have found its place in the previous section about management regimes. Its early years' membership "consist[ed] almost entirely of whaling nations [. . .,] unlikely to agree to radical reforms."[448] The results of its decisions were, at first, not satisfactory in terms of conservation, with most whale species getting depleted.[449] Data was not sufficient and scientific advice not followed.

In the 1970s, the IWC became a rather conservation-oriented and then a preservation-oriented regime. Indeed, with many non-whaling states joining, the membership of the IWC changed in the last quarter of the twentieth century, which led to a modification in the approach taken to manage whales.[450] Also, due to the depletion of stocks, whaling was not an activity with clear economic benefits anymore.[451] Finally, pressures from some countries, international institutions and NGOs towards a moratorium were strongly felt.[452] In particular the Stockholm Resolution of 1972 required, at the unanimity of the 53 members present—though in a soft-law manner—a 10-year ban on commercial whaling.[453] It lasted until 1982 for such a moratorium to be agreed upon within the IWC for the 1986 whaling season.[454] While some states opted out in that regard, the USA used strong

[445] 'International Convention for the Regulation of Whaling' (ICRW) 1946, 161 *United Nations Treaty Series* 72 preamble.

[446] Bowman, Davies and Redgwell (n 57) 152–153. On preservationism and sustainable use, see infra Chapter 4 A. II. 4.

[447] S. Freeland and J. Drysdale, 'Co-Operation or Chaos? – Article 65 of United Nations Convention on the Law of the Sea and the Future of the International Whaling Commission' (2005) 2 Macquarie Journal of International and Comparative Environmental Law 1, 3.

[448] W. C. Burns, 'The International Whaling Commission and the Future of Cetaceans: Problems and Prospects' (1997) 8 Colorado Journal of International Environmental Law and Policy 8, 35.

[449] M. Arbour and S. Lavallée, *Droit international de l'environnement* (Bruylant 2006) 406; Bowman, Davies and Redgwell (n 57) 164; Burns (n 448) 35–41; M. Fitzmaurice, 'Divided we stand: the legal issues concerning the international whaling commission' in H. P. Hestermeyer and others (eds), *Coexistence, Cooperation and Solidarity: Liber Amicorum Rüdiger Wolfrum* (Martinus Nijhoff Publishers 2011) 832–833.

[450] Bowman, Davies and Redgwell (n 57) 157: in 1982, of the 25 votes in favor of the moratorium, 15 were from members who had joined since 1979.

[451] Andresen (n 444) 218.

[452] Fitzmaurice, 'Divided we stand: the legal issues concerning the international whaling commission' (n 449) 833, 838.

[453] Andresen (n 444) 219; C. P. Carlarne, 'Saving the Whales in the New Millennium: International Institutions, Recent Developments and the Future of International Whaling Policies' (2005–2006) 24 Virginia Environmental Law Journal 1, 7.

[454] C. Epstein, 'The Making of Global Environmental Norms – Endangered Species Protection' (2006) 6 Global Environmental Politics 32, 45.

diplomatic pressure to 'encourage' them to remove such reservations—successfully in the cases of Peru and Japan, but not for Norway and Russia.[455] Nevertheless, Norway implemented an autonomous ban on whaling until 1993.[456] As for Japan, it was 'encouraged' to withdraw its reservation, with the risk, otherwise, to suffer economic sanctions applied by the USA.[457]

Originally, the moratorium was to be assessed in 1990[458] and a Revised Management Procedure was to be developed before any quotas above zero could be set again. Mainly, this new management is meant to take account, in a more successful manner, of uncertainty as well as of data imperfection.[459] However, it has not yet been adopted because it was later agreed that such management procedure should be part of a more global Revised Management Scheme—of which the inspection and observation section is still under negotiation.[460] Pro-whaling states are under the impression that such adoption is delayed in order to make the resuming of whaling impossible, while some species such as the minke whale could sustain some exploitation.[461] Those states consequently consider that the moratorium is not based on the scientific data available for some species. Other states argue that strong monitoring and enforcement mechanisms must still be put in place if past mistakes are not to be reproduced as soon as the moratorium is lifted.[462]

---

[455] Freeland and Drysdale (n 447) 8–9.

[456] Eldridge (n 399) 553–554; S. Suhre, 'Misguided Morality: The Repercussions of the International Whaling Commission's Shift from a Policy of Regulation to One of Preservation' (1999–2000) 12 Georgetown International Environmental Law Review 305, 313–314. Japan signed the Murazawa-Baldridge Pact in 1987 according to which it was allocated fishing rights in the American EEZ in exchange for withdrawing its reservation against the moratorium (K. Hirata, 'Why Japan Supports Whaling' (2005) 8 Journal of International Wildlife Law & Policy 129, 132).

[457] Hirata (n 456) 132.

[458] The Schedule of the ICRW §10(e) (reference omitted) reads as "Notwithstanding the other provisions of paragraph 10, catch limits for the killing for commercial purposes of whales from all stocks for the 1986 coastal and the 1985/86 pelagic seasons and thereafter shall be zero. This provision will be kept under review, based upon the best scientific advice, and by 1990 at the latest the Commission will undertake a comprehensive assessment of the effects of this decision on whale stocks and consider modification of this provision and the establishment of other catch limits." On this, see Arbour and Lavallée (n 449) 405; Bowman, Davies and Redgwell (n 57) 166; Eldridge (n 399) 554; Suhre (n 456) 310–311.

[459] Bowman, Davies and Redgwell (n 57) 166–167.

[460] On the Revised Management Procedure, see Burns (n 448) 53–73; on the Revised Management Scheme, see Gillespie, *Whaling Diplomacy* (n 401) 357–385 chapter 12.

[461] Bowman, Davies and Redgwell (n 57) 153–154, 167–168; Burke, *The New International Law of Fisheries* (n 55) 108; J. Morishita, 'Multiple analysis of the whaling issue: Understanding the dispute by a matrix' (2006) 30 Marine Policy 802, 805.

[462] P. J. Clapham and others, 'The whaling issue: Conservation, confusion, and casuistry' (2007) 31 Marine Policy 314, 317–318.

All through the years, a number of states voiced their discontent regarding the moratorium and kept whaling, more or less openly. Norway for example put an end to its voluntary moratorium and resumed whaling of minke whales—because the IWC moratorium had not been reviewed as was supposed to be done in 1990.[463]

The most recent evolution of the IWC is the debate about resuming whaling and the (ab)use by Japan of scientific research permits, which led to the case that Australia has brought against Japan in front of the ICJ.[464] The Court recently adjudicated that Japan's JARPA II program does not correspond to scientific whaling as provided for in article VIII of the ICRW and that it is consequently in breach of several of its obligations under the regime.[465] After declaring that it would comply with the judgment, Japan also announced that it would undertake scientific whaling in the Pacific on the basis of a research program following the criteria highlighted by the ICJ.[466] Another development is that many states, allegedly sponsored by Japan, are joining the ICRW and tend to vote in favor of suspending the moratorium.[467]

2. Relevant Provisions

The ICRW's management is based on a Schedule which is to be adopted annually, if not by consensus, at least by a three-quarter majority of parties voting.[468] It sets, inter alia, annual levels of whaling allowed, originally on the basis of a 'blue whale unit' benchmark and, since 1976, under a new management procedure, of stock-by-stock quotas.[469] The amendments to the Schedule are binding upon the states who have not objected to them[470] and they "shall be based on scientific findings".[471] In

[463] Suhre (n 456) 312–313.

[464] ICJ – Press Release, 'Australia institutes proceedings against Japan for alleged breach of international obligations concerning whaling' (2010) http://www.icj-cij.org/docket/files/148/15953.pdf accessed 4 July 2015; D. K. Anton, 'Dispute Concerning Japan's JARPA II Program of "Scientific Whaling" (Australia v. Japan)' (2010) http://www.asil.org/insights/volume/14/issue/20/dispute-concerning-japan%E2%80%99s-jarpa-ii-program-%E2%80%9Cscientific-whaling%E2%80%9D accessed 4 July 2015. Morishita considers that the scientific program is source of useful data and is not 'commercial whaling in disguise' ('Multiple analysis of the whaling issue: Understanding the dispute by a matrix' (n 461) 804).

[465] *Whaling in the Antarctic (Australia v. Japan; New Zealand intervening)* (n 347) 71 §247; for more on this case, see infra Chapter 4 A. IV. 3. c) aa).

[466] 'Japan to launch reduced Pacific whale hunt next week' *BBC news* (18 April 2014) http://www.bbc.com/news/world-asia-27079462 accessed 6 July 2015.

[467] Carlarne (n 453) 35; Epstein (n 454) 47–48; Molenaar, 'Marine Mammals: The Role of Ethics and Ecosystem Considerations' (n 401) 40.

[468] ICRW (n 445) article III(2).

[469] Ibid. article V(1); Bowman, Davies and Redgwell (n 57) 164–165; Burke, *The New International Law of Fisheries* (n 55) 256.

[470] ICRW (n 445) article V(3).

[471] Ibid. article V(2).

1982, the Schedule's quotas were set at zero, implementing a de facto ban on whaling.[472] No-catch zones can also be established; the coordinates of the two existing sanctuaries (in the Indian Ocean and in the Southern Ocean) are to be found in the Schedule.[473]

States which do not want to be bound by the Schedule can enter into a reservation within 90 days after its adoption.[474] This measure has been used all through the history of the ICRW's regime, with states such as the USSR, the USA, Japan and Canada already opting out of a 1954 conservation measure towards blue whales or, later, with similar declarations of Japan, Peru, Norway and Russia (replacing the then-USSR) towards the moratorium.[475] The two latter countries have not withdrawn their reservations and are, legally, allowed to catch whales, activity which Norway is known to carry on with.[476] Another, more dramatic possibility for states is to leave the treaty regime.[477]

For those states that have undertaken to respect the Schedule, there are also exemptions to the moratorium, particularly in the case of aboriginal whaling.[478] This basis for particular authorizations is supported by the USA. Two groups living there were granted such a right to whale—the Makah tribe and the Inuit. However, other countries, mainly Japan, question the legitimacy of such allowance since, until recently, the Makah tribe had not caught a whale for 70 years.[479] Also, aboriginal whaling seems not to include coastal communities, even though some were, prior to the moratorium, highly dependent on such activity which represented an important part of their culture.[480] This differentiation between aboriginal groups

---

[472] Bowman, Davies and Redgwell (n 57) 152.

[473] IWC, 'Whale Sanctuaries – Establishment of the International Whaling Commission's sanctuaries' http://iwc.int/sanctuaries accessed 4 July 2015. They are adopted on the basis of the ICRW (n 445) articles V(1)(c)-VI, as presented in Beer-Gabel and Lestang (n 213) 115–117. The creation of sanctuaries is however not without controversy (J. Braig, 'Whaling' Max Planck Encyclopedia of Public International Law www.mpepil.com §28).

[474] ICRW (n 445) article V(3).

[475] Bowman, Davies and Redgwell (n 57) 163; D. R. Rothwell and T. Stephens, *The International Law of the Sea* (Hart Publishing 2010) 309.

[476] Bowman, Davies and Redgwell (n 57) 163.

[477] This has happened at several occasions, for example with Iceland, which left in 1992 – and rejoined ten years later – or Ecuador which withdrew in 1994 – and rejoined in 2007 (Bowman, Davies and Redgwell (n 57) 151 n 14).

[478] Schedule of the ICRW (n 458) §13. On aboriginal whaling, see for example J. Firestone and J. Lilley, 'Aboriginal Subsistence Whaling and the Right to Practice and Revitalize Cultural Traditions and Customs' (2005) 8 Journal of International Wildlife Law & Policy 177; Fitzmaurice, 'Divided we stand: the legal issues concerning the international whaling commission' (n 449) 845–848; Gillespie, *Whaling Diplomacy* (n 401) 194–246 chapter 8; H. N. Scheiber, 'Historical Memory, Cultural Claims, and Environmental Ethics: The Jurisprudence of Whaling Regulation' in H. N. Scheiber (ed), *Law of the Sea: The Common Heritage and Emerging Challenges* (Martinus Nijhoff Publishers 2000) 142–146.

[479] Bowman, Davies and Redgwell (n 57) 170.

[480] Carlarne (n 453) 11–12.

and small coastal communities is criticized by Japan which calls for an allocation of quotas of minke whales for such communities.[481]

Moreover, scientific whaling under a system of nationally granted permits is allowed.[482] Even though the Scientific Committee can comment on these permits and guidelines have been adopted on this practice,[483] their issuance falls under national sovereign rights.[484] Notoriously, Japan has set high levels of catches for scientific research, against the recommendations of the IWC to stop lethal research in some areas.[485] It is worth noting that the ICJ has considered that the Japanese JARPA II program was not scientific whaling according to the ICRW and thus had to stop.[486]

Compliance control took at first the form of national inspectors placed on board whaling vessels and, after 1971, of international observers.[487] No international enforcement was envisioned in the regime apart from investigating the members' actions and reporting on them.[488] Measures against violations must be taken nationally.[489] However, the USA enacted a number of laws with extra-territorial implications such as the Pelly Amendment to the Fisherman's Protective Act of 1967 or the Packwood-Magnuson Amendment to the Fishery Conservation and Management Act of 1976.[490] According to these pieces of legislation, unilateral

---

[481] Bowman, Davies and Redgwell (n 57) 178; Hirata (n 456) 136; Suhre (n 456) 325–326. Against the validity of this view, see for example Scheiber (n 478) 164–166.

[482] ICRW (n 445) article VIII(1); P. H. Sand, 'Japan's 'Research Whaling' in the Antarctic Southern Ocean and the North Pacific Ocean in the Face of the Endangered Species Convention (CITES)' (2008) 17 RECIEL 56, 57. On scientific whaling, see for example Gillespie, *Whaling Diplomacy* (n 401) 109–147 chapter 5.

[483] Schedule of the ICRW (n 458) §30; IWC, Resolution 1995–9, Resolution on whaling under special permit 1995, (IWC Annual meeting of the Commission); Arbour and Lavallée (n 449) 402–403.

[484] Bowman, Davies and Redgwell (n 57) 175; Braig (n 473) §21–25.

[485] IWC, Resolution 2007-1, Resolution on JARPA 2007; Fitzmaurice, 'Divided we stand: the legal issues concerning the international whaling commission' (n 449) 848–849; Sand, 'Japan's 'Research Whaling' in the Antarctic Southern Ocean and the North Pacific Ocean in the Face of the Endangered Species Convention (CITES)' (n 482) 57–58.

[486] For more on the *Whaling* case, see infra Chapter 4 A. IV. 3. c) aa).

[487] Schedule of the ICRW (n 458) §21(a) and (c); A. Gillespie, 'The Search for a New Compliance Mechanism Within the International Whaling Commission' (2003) 34 Ocean Development and International Law 349, 350.

[488] Suhre (n 456) 316.

[489] CITES (n 294) article VIII.

[490] Respectively 22 USC §1978 and 16 USC §1821. On these, see for example G. S. Martin and J. W. Brennan, 'Enforcing the International Convention for the Regulation of Whaling: The Pelly and Packwood-Magnuson Amendments' (1989) 17 Denver Journal of International Law and Policy 293; D. M. Wilkinson, 'The Use of Domestic Measures to Enforce International Whaling Agreements: A Critical Perspective' (1989) 17 Denver Journal of International Law and Policy 271.

economic sanctions can be used towards other states if they do not respect some environmental standards, including the regulation of whaling.[491] Such sanctions have never been applied but several states (for example Japan, Iceland and Norway) have been threatened thereof.[492]

3. Conclusions

The IWC regime evolved from being a—failed—management one to a conservationist and then a preservationist one; in neither extreme is science the actual basis for decisions.[493] This was a dramatic shift of paradigm, culminating with the debate about the right of whales to life. However, there is no unanimity about the moratorium which was forced through; "[c]oercion may be effective as a means to change behaviour, but not necessarily as a means to change perceptions and mind-sets."[494] A small number of states find ways to continue whaling, either under the treaty regime exemptions or following a reservation to the moratorium. Hence, in legal terms, such states are—at least arguably—acting in compliance with their international obligations.

To assess this regime's efficacy one should first clarify the objectives of the ICRW. In reality, this is made difficult by the ambiguity of the treaty and its quite stunning evolution; "the situation is characterised by deep conflict of interests as well as over values. [... T]he parties are unable to agree over the real purpose of the IWC; to protect whales or to utilise whales".[495]

---

[491] Andresen (n 444) 224. Under the Pelly Amendment, the Secretary of Commerce (as well as the Secretary of the Interior in situations involving endangered or threatened species) can certify to the President that nationals of another country are "conducting fishing operations in a manner [...] which diminish[es] the effectiveness of an international fishery conservation program [...or] any international program for endangered or threatened species" (22 USC §1978(a)(1)(2)). In this case, the President can give the order to the Secretary of the Treasury "to prohibit [...] the importation into the United States of any products from the offending country for any duration as the President determines appropriate" (22 USC §1978(a)(4)). As for the Packwood-Magnuson Amendment, it functions in a similar manner, but without the discretion given to the President to require or not sanctions; in this case, as soon as a country is certified by the Secretary of Commerce as "diminish[ing] the effectiveness of the International Convention for the Regulation of Whaling" (16 USC §1821(e)(2)(A)), the Secretary of State must decrease this country's allocation of fishing rights in American waters (16 USC §1821(e)(2)(B)). The Secretary of Commerce has been granted quite a large discretion in its certification process, since the Supreme Court has decided that violations of IWC quotas would not necessarily trigger certification (Japanese Whaling Association v. American Cetacean Society – 478 US 221, 241 (1986)). On this, see L. Jenkins, 'Trade Sanctions: An Effective Enforcement Tool' (1993) 2 RECIEL 362, 364–365; Suhre (n 456) 317–319.

[492] Bowman, Davies and Redgwell (n 57) 184.

[493] Morishita, 'Multiple analysis of the whaling issue: Understanding the dispute by a matrix' (n 461) 803.

[494] Andresen (n 444) 227.

[495] Ibid. 215.

It is anything but ideal that the states still actively whaling should be out of the regime, without many options to internationally review the sustainability of their catches. A compromise solution towards a more nuanced regime has been recommended.[496] This would probably entail a return to some positive quotas.[497] In any case the particular vulnerabilities of cetaceans need to be taken into account if one is not to fall back into overexploitation, stocks' depletion and potential species' extinction.[498]

Beyond the issues of states not participating or opting out, usual in international law, the lack of international compliance and enforcement mechanisms further weakens this regime. Indeed, the use of unilateral trade sanctions by a powerful anti-whaling champion such as the USA is unlikely to prove effective in the future, or even to be actually used, since such sanctions are presumably breaching the World Trade Organization (WTO) law.

D. World Trade Organization Law: Trade Regime

The WTO aims at the diminution and elimination of restrictions to trade, in order to allow the free international trade of goods and services.[499] With nearly 40 % of fish entering international trade,[500] the WTO rules are very relevant to questions of fisheries. The pillars of the freedom of trade are the rule of the most-favored nation, national treatment, elimination of quantitative restrictions and access to ports.

---

[496] Suhre (n 456) 315.

[497] A. Proelß, 'Marine Mammals' Max Planck Encyclopedia of Public International Law www.mpepil.com §17.

[498] Molenaar, 'Marine Mammals: The Role of Ethics and Ecosystem Considerations' (n 401) 37.

[499] 'Agreement establishing the World Trade Organization' (WTO Agreement) 1994, 1867 *United Nations Treaty Series* 154 preamble; Tyler (n 40) 83.

[500] Statistics valid for 2000 (L. A. Chaves, 'Illegal, Unreported and Unregulated Fishing: WTO-consistent Trade related Measures to Address IUU Fishing: AUS:IUU/2000/16' (Background paper for the Expert Consultation on Illegal, Unreported and Unregulated Fishing Organized by the Government of Australia in Cooperation with FAO, 2000) http://www.fao.org/docrep/005/Y3274E/y3274e0i.htm#bm18 accessed 4 July 2015, Executive Summary).

I. Relevant Provisions

1. Towards Free-Trade[501]

The first relevant obligation, article I of the General Agreement on Tariffs and Trade (GATT), requires that "any advantage, favour, privilege or immunity granted by any contracting party to any product originating in or destined for any other country shall be accorded immediately and unconditionally to the like product originating in or destined for the territories of all other contracting parties."[502] The most-favored nation principle means that all states parties to the WTO trade regime have to be treated in the same way with regard to 'like' products and no preference can be accorded to a trading partner.

Also in order to avoid discrimination, GATT article III states that imports shall be treated in the same way as 'like' domestic products.[503] The term 'like product', referred to in both articles I and III, is central in determining the respect or the violation of those rules. The main controversy about its meaning, the so called product-versus-process debate, is linked to "whether products may be distinguished

---

[501] On the core GATT obligations, see for example Calley (n 131) 210–219; B. J. Condon, *Environmental Sovereignty and the WTO: Trade Sanctions and International Law* (Transnational Publishers 2006) chapter 3.

Other aspects of trade law, such as subsidies, are also relevant to fisheries governance but are not the focus of the present study (on this aspect, see for example E. A. Bilsky, 'Conserving Marine Wildlife Through World Trade Law' (2008–2009) 30 Michigan Journal of International Law 599; C.-J. Chen, *Fisheries Subsidies under International Law* (Springer 2010); C. D. Stone, 'Too Many Fishing Boats, Too Few Fish: Can Trade Laws Trim Subsidies and Restore the Balance in Global Fisheries?' in K. P. Gallagher and J. Werksman (eds), *The Earthscan Reader on International Trade and Sustainable Development* (Earthscan 2002); O. R. Young, 'Fragmentation or interaction: the WTO, fisheries subsidies, and international law' (2009) 8 World Trade Review 477). Also, the Agreement on Technical Barriers to Trade could come into question, since states shall not implement unnecessary barriers to trade. However, protecting the environment is a legitimate goal to adopt technical barriers (on this aspect, see M. Ahmed, 'Market Access and Trade Liberalisation in Fisheries' ICTSD Natural Resources, International Trade and Sustainable Development Series Issue Paper 4 (Geneva 2006) 15–17; Calley (n 131) 220–221; R. G. Tarasofsky, 'Regional Fisheries Organizations and the World Trade Organization: Compatibility or Conflict?' (TRAFFIC International 2003) 9–10). Finally, sanitary and phytosanitary measures can be of relevance (on this aspect, see Ahmed (n 501) 14–15; Young, *Trading Fish, Saving Fish* (n 31) 72–73; T. L. McDorman, 'Fisheries Conservation and Management and International Trade Law' in E. Hey (ed), *Developments in International Fisheries Law* (Kluwer Law International 1999) 508–510).

[502] 'General Agreement on Tariffs and Trade' (to be read together with GATT 1947) (GATT 1994) 1994, 1867 *United Nations Treaty Series* 187 article I(1).

[503] Ibid. article III(4).

based solely on the physical characteristics of the product or whether they may also be distinguished based on the manner in which they were produced."[504]

According to an old Panel report thereabout, in the *Tuna-Dolphin I* case,[505] production processes cannot be used to differentiate products, in particular if they do not influence the end product:[506] this signifies that the importing country "has no say in how the exporting country protects its own environment or the global commons."[507] This opinion has been nuanced in the more recent jurisprudence, in particular in the *Shrimp-Turtle* cases.[508] However, no clear answer is provided by jurisprudence or doctrine yet.[509]

As for article V, it provides for the freedom of transit across the territory of a contracting party;[510] "it has been argued that Article V of the GATT implies a freedom of entry into the ports of states parties, at least for the purpose of transshipment."[511]

---

[504] Condon (n 501) 53. On the debate, see for example N. Bernasconi-Osterwalder and others, *Environment and Trade: A Guide to WTO Jurisprudence* (Earthscan 2006) 203–205; Calley (n 131) 212–213.

[505] The Tuna/Dolphin cases (*Restrictions on Imports of Tuna (Mexico v. United States), GATT Panel Report circulated not adopted,* 3 September 1991, DS21/R – 39S/155, (1991) 30 ILM 1594; *Restrictions on Imports of Tuna (European Economic Community (EEC) and the Netherlands v. United States), GATT Panel Report circulated not adopted,* 16 June 1994, DS29/R, (1994) 33 ILM 839 – respectively *Tuna Dolphin I* and *Tuna Dolphin II*) were related to a ban by the USA on tuna imports in situations where dolphins were not sufficiently protected in the course of the tuna fishing activities.

[506] *Tuna Dolphin I* (n 505) §5.8–5.16; Bernasconi-Osterwalder and others (n 504) 30.

[507] Condon (n 501) 55.

[508] The Shrimp-Turtle cases (*United States – Import Prohibition of Certain Shrimp and Shrimp Products (India, Malaysia, Pakistan and Thailand v United States of America), WTO Dispute Settlement Body – Appellate Body Report Adopted*, 6 November 1998, WT/DS58/AB/R and *United States – Import Prohibitions of Certain Shrimp and Shrimp Products (Recourse to Article 21.5 of the DSU by Malaysia), WTO Dispute Settlement Body – Appellate Body Report Adopted,* 22 October 2001, WT/DS58/AB/RW, which is an examination by the Appellate Body of Malaysia's allegation that the USA did not comply with the ruling – respectively *Shrimp Turtle* case and *Shrimp Turtle 21.5 Appellate Body*) were related to a ban by the USA on shrimp imports coming from countries that did not, in their view, sufficiently protect turtles in the course of the shrimp catching activities. On these cases, see Bernasconi-Osterwalder and others (n 504) 233–235.

[509] The criteria to determine 'likeness' might indeed be changing as hinted by other case-law (*Measures Affecting Asbestos and Asbestos-Containing Products (Canada v. European Communities), WTO Dispute Settlement Body – Appellate Body Report Adopted*, 12 March 2001, WT/DS135/AB/R §102); on this, see for example Condon (n 501) 58–59.

[510] GATT 1994 (n 502) article V.

[511] L. de La Fayette, 'Access to Ports in International Law' (1996) 11 International Journal of Marine and Coastal Law 1, 19.

Finally, GATT article XI provides that "[n]o prohibitions or restrictions other than duties, taxes or other charges, whether made effective through quotas, import or export licenses or other measures" are to be authorized towards imported products of other contracting parties.[512]

2. Exception Clause

The only exceptions to these rules are provided in article XI itself—allowing duties, taxes or other charges—and in article XX. The latter article provides, inter alia, at lit b) and g), that restrictions to trade for 'like' products can be allowed where "necessary to protect human, animal or plant life or health" (paragraph b) or when they relate "to the conservation of exhaustible natural resources if such measures are made effective in conjunction with restrictions on domestic production or consumption" (paragraph g). It has been suggested that paragraph b is rather applicable to domestic matters while transnational or global issues should be approached under paragraph g[513] but also that the latter would be preferred in environmental disputes as 'relating to' is a lower threshold to reach than proving that the measures are 'necessary'.[514]

Some frictions between trade and environmental law can exist, particularly when trade restrictions are used by a state as a means to enforce or encourage compliance by other states with some environmental measures. Such potential conflicts can, under certain circumstances, be avoided through this exception clause.

a) Possible Conflicts Between Trade Restrictions and GATT

Trade measures in fisheries which possibly conflict with the most-favored nation rule are numerous. First, blanket trade bans are at times—or at least can be—implemented against a state because some of its vessels are put on a blacklist or it otherwise fails to respect the rules set by an RFMO.[515] For example, ICCAT recommended and implemented a ban on import of Bluefin Tuna from

[512] GATT 1994 (n 502) article XI(1).

[513] Condon (n 501) 10.

[514] Calley (n 131) 214–215.

[515] Tarasofsky (n 501) 5–6.

non-members Belize, Honduras and Panama in 1996, in order to tackle their continued disregard of CMMs.[516] Moreover, some RFMOs such as NAFO take specific measures against non-contracting parties by reversing the burden of proof for authorizing landing. As to the trade bans in CITES, they are used as a sanction against a state for its continued violation of obligations under the Convention and are susceptible of criticism, particularly when based on unilateral stricter measures.[517]

Then, potentially problematic with regard to the concept of national treatment is that CITES provides for different regulations in cases of domestic and international trade. Indeed, nationally traded products do not need a CITES permit, hence do not actually need to respect CITES requirements of legality and sustainability.

Measures in fisheries that have the potential to conflict with the prohibition of restrictions to trade are trade bans, which can be implemented for example through an ad hoc landing ban because of minimum size violations or the failure to provide a required certification. Such measures can be set by RFMOs.[518] The same can be said of the CITES permitting system, which explicitly aims at restricting trade.[519]

Finally, port state measures are particularly relevant to the question of access to ports, since they can either result in altogether denial of access or refusal of transshipment/landing services. Many RFMOs apply port state measures, which are also promoted by the Port State Measures Agreement.

b) Applicability of Article XX

Trade restrictions as a management measure or "sanction[s] for the violation of conservation measures to protect dwindling fishery resources" can fall under article XX (g) environmental exception.[520] To fulfil the criteria of lit. (g), trade-restricting measures must "relat[e] to the conservation of exhaustible natural resources", not be discriminatory and not be a hidden protectionist policy.[521]

---

[516] Recommendation by ICCAT regarding Belize and Honduras Pursuant to the 1994 Bluefin Tuna Action Plan Resolution, 1996; K. W. Riddle, 'Illegal, Unreported, and Unregulated Fishing: Is International Cooperation Contagious?' (2006) 37 Ocean Development and International Law 265, 285–286; Schmidt (n 241) 7 about Belize; D. Warner-Kramer, 'Control Begins at Home: Tackling Flags of Convenience and IUU Fishing' (2004) 34 Golden Gate University Law Review 497, 514; ICTSD, 'Fishing, International trade and Sustainable Development: Policy Discussion Paper' Natural Resources, International Trade and Sustainable Development Studies (2006) http://www.ictsd.org/themes/environment/research/fisheries-international-trade-and-sustainable-development accessed 4 July 2015, 85. ICCAT delayed the imposition of a similar ban in relation to Panama, providing this country with approximately six months to rectify the situation, since Panama had pledged its willingness to do so – it was unsuccessful in this attempt (Chaves (n 500) §20).

[517] Yeater and Vasquez (n 380) 273.

[518] Tarasofsky (n 501) 2–3.

[519] Reeve, *Policing International Trade in Endangered Species* (n 296) 311–312.

[520] La Fayette, 'Access to Ports in International Law' (n 511) 20.

[521] GATT 1994 (n 502) article XX chapeau and (g).

There is no past jurisprudence which can fully inform on the application of such provision in one of the possible conflicts highlighted here above, as no case directly involving the implementation of an international treaty has yet come to the WTO dispute settlement body.[522] Indeed, while the dispute settlement body of the WTO has clarified in the *Tuna/Dolphin* and the *Shrimp/Turtle* cases the conditions under which restrictions to trade could be allowed, these cases were about the implementation of unilateral measures.[523] Hence, the links with international environmental standards were only indirect.

It is nevertheless useful to examine the WTO jurisprudence in order to clarify the current understanding of article XX's terms and to analyse this provision's applicability to the above-mentioned potential conflicts.[524] In the *Tuna/Dolphin* case II, the panel considered that the fulfilment of three requirements was needed in order for restrictive measures to fit into the article XX (g) exemption: first, the policy in question had to be in "the range of policies to conserve exhaustible natural resources"; second, the measure had to be 'related to' such conservation and be supplemented by domestic measures of a restrictive nature; and third, it had to be conform with the chapeau of article XX.[525] The recent case-law indeed shows a tendency to consider whether a restriction can be justified under one of the reasons listed in article XX, before examining the compliance with the chapeau.[526]

Regarding the first condition, in the case at hand, fish are obviously 'exhaustible natural resources', as many studies prove that fish stocks are declining and hence that they are not infinite. Furthermore, while the meaning of words was not

---

[522] Tarasofsky (n 501) 1.

[523] Reeve, *Policing International Trade in Endangered Species* (n 296) 307.

[524] It is worthwhile noting that article XX is interpreted in an evolutionary way (*Shrimp Turtle* case (n 508) §130).

Of particular interest are the Shrimp Turtle cases and Tuna Dolphin cases. Both the dolphin and turtle species were listed under CITES, but the issue was the unilateral implementation by the USA of rules linked to the by-catch of such species.

[525] *Tuna Dolphin II* (n 505) §5.12.

[526] R. Wolfrum, 'Article XX GATT [Introduction]' in R. Wolfrum, P.-T. Stoll and A. Seibert-Fohr (eds), *WTO – Technical Barriers and SPS Measures* (Martinus Nijhoff Publishers 2007) 64; *Shrimp Turtle* case (n 508) §123, 146–147.

necessarily clear or the same at the time of GATT's drafting, the Appellate Body considered these terms as "by definition, evolutionary".[527] The Appellate Body has even explicitly clarified that "[l]iving resources are just 'as finite' as petroleum, iron ore and other non-living resources."[528] Hence, "endangered marine resources can qualify as 'exhaustible natural resources' under article XX(g)."[529]

The questions of whether a sufficient nexus between the natural resources and the state implementing the restrictions is needed and how to determine it are still not completely settled. In the *Shrimp/Turtle* case, this issue was solved by the fact that turtles go through American waters regularly.[530] A risk consequently remains that it would be difficult to identify a sufficient nexus in the case of a species staying exclusively in the high seas.[531] However, an argument can be developed that the existence of a MEA gives a 'legal connection' to all its parties.[532]

Then, as to the second condition, the wording 'relating to the conservation' traditionally means that the measures have to be 'primarily aimed at'

---

[527] *Shrimp Turtle* case (n 508) §130.

Whether a treaty ought to be interpreted with the critical date being the time of its adoption (*ex tunc* interpretation) or whether later developments should be taken into account (*ex nunc* interpretation) remains to be fully clarified. The former is based on the justification that it actually reflects the legislative intent of the drafters, while the latter recognizes that subsequent practice exists and affects legal relationships. Overall, the decision to choose one interpretative method over the other depends on the circumstances of the case and the text of the instrument itself (ILC, 'Fragmentation of international law: difficulties arising from the diversification and expansion of international law' Report of the Study Group of the International Law Commission (finalized by M. Koskenniemi) 13 April 2006, A/CN.4/L.682, 240–243 §475–478).

This tension was already acknowledged in the doctrine of inter-temporal law, as defined by Judge Huber (*Island of Palmas case (Netherlands/United States of America), Award,* 4 April 1928, Permanent Court of Arbitration II R.I.A.A. 839, 845) since it encompassed both aspects of contemporaneity (with the drafting) and evolution. As for the ICJ, it presented arguments for an *ex tunc* interpretation in *Case concerning rights of nationals of the United States of America in Morocco (France v. United States of America), Judgment,* 27 August 1952, I.C.J. Reports 1952 176, 189, but also found support in the wording of the treaty for an *ex nunc* interpretation in its *Legal consequences for states of the continued presence of South Africa in Namibia (South West Africa) notwithstanding Security Council Resolution 276 (1970), Advisory Opinion,* 21 June 1971, I.C.J. Reports 1971 16, 31 §53. On inter-temporal law and evolutionary interpretation of WTO provisions, see Condon (n 501) 37–41; P.-M. Dupuy, 'Evolutionary Interpretation of Treaties: Between Memory and Prophecy' in E. Cannizzaro (ed), *The Law of Treaties Beyond the Vienna Convention* (Oxford University Press 2011) 128–130.

[528] *Shrimp Turtle* case (n 508) §128, making reference to World Commission on Environment and Development, *Our Common Future,* (usually referred to as the Brundtland Report) (Oxford University Press 1987) 13; on this, see N. Matz-Lück and R. Wolfrum, 'Article XX lit. g GATT' in R. Wolfrum, P.-T Stoll and A. Seibert-Fohr (eds), *WTO – Technical Barriers and SPS Measures* (Martinus Nijhoff Publishers 2007) 145–147.

[529] Tyler (n 40) 88 (reference omitted).

[530] *Shrimp Turtle* case (n 508) §133.

[531] Matz-Lück and Wolfrum (n 528) 149.

[532] Condon (n 501) 143–144, 197–198; on more general questions of standing and judicial dispute settlement, see infra Chapter 4 A. IV. 3. c) ee).

conservation.[533] A slightly more flexible line of reasoning based on 'reasonableness' was followed in *Shrimp/Turtle*: "[i]n determining whether the measures 'related to' conservation, the AB [Appellate Body] considered whether the means were reasonably related to the ends. In ascertaining the importance of the policy goal (the 'ends'), the AB referred to CITES and found that the policy of protecting sea turtles was shared by all disputing parties and third parties, and the 'vast majority of the nations of the world'."[534] CITES and RFMOs measures try to avoid that fish caught respectively illegally and unsustainably, and in violation of CMMs, reaches markets—in order to make IUU fishing less attractive and hence to better conserve fish stocks. They appear to be related to the goal of conservation and consequently fulfill this criterion.

Furthermore, "in conjunction with domestic measures" can indicate that vessels flying a state's flag must be subjected to at least the same measures as the ones implemented by that state for all ships, or, more generally, that there also must be domestic restrictions for conservation purposes.[535]

Finally, to respect the third condition, the measures must not be "applied in a manner which would constitute a means of arbitrary or unjustifiable discrimination between countries where the same conditions prevail, or [be] a disguised restriction on international trade".[536]

The first aspect of the chapeau criterion states that measures must not be arbitrarily or unjustifiably discriminatory between countries: restrictions must be applied equally to all actors who disregard the conservation measures. Obviously, CITES restrictions to trade in the form of import/export permits would respect this condition as well as CMMs of RFMOs—at least with regard to members of these regimes who have agreed to such restrictions.[537] As for CITES sanctions against non-compliant states and RFMOs sanctions measures, the lawfulness of these would depend on whether the measures are applied in an equitable way and negotiations with the non-compliant party have been attempted before using sanctions.[538] It has been suggested that CITES sanctions have not always been applied in an equitable manner.[539]

The second aspect provides that measures must not be a disguised restriction to trade. The adoption and implementation of CITES and (arguably) RFMOs' trade measures are of transparent and multilateral nature (as opposed to unilateral

---

[533] *Canada — Measures Affecting Exports of Unprocessed Herring and Salmon (United States v. Canada), GATT Panel Report adopted,* 22 March 1988, BISD 35S/98 §4.6; Condon (n 501) 98–101; Tarasofsky (n 501) 11.

[534] Young, *Trading Fish, Saving Fish* (n 31) 200 (references omitted); *Shrimp Turtle* case (n 508) §135, 141–142.

[535] La Fayette, 'Access to Ports in International Law' (n 511) 21.

[536] GATT 1994 (n 502) article XX chapeau.

[537] Reeve, *Policing International Trade in Endangered Species* (n 296) 311.

[538] On the need to pursue equitable negotiations, see *Shrimp Turtle* case (n 508) §166, 172; *Shrimp Turtle 21.5 Appellate Body* (n 508) §122.

[539] Reeve, *Policing International Trade in Endangered Species* (n 296) 311–312.

measures). The Panel has shown that it prefers a multilateral approach,[540] but it does necessarily require that a formal agreement be concluded.[541] Also, these measures have usually been developed after other efforts to deal with stocks' depletion were proven insufficient. Then, there is a "global consensus, reflected in several instruments, on the necessity of such measures".[542] Finally, dialogue with the flag states of targeted vessels or exporting states exists prior to the implementation of sanctions—the Panel emphasized the need of good faith negotiations and this conclusion was upheld by the Appellate Body in Shrimp 21.5.[543]

In conclusion, restrictions to trade of endangered fish species for conservation purposes, equally applied to all actors attempting to trade species listed on CITES Appendices or violating the CITES provisions or CMMs, in the case where there are also domestic restrictions, have good chances of not contravening WTO law.[544] Indeed, the restrictions to trade under CITES and some under RFMOs, though *prima facie* contradicting obligations under WTO law, comply with the requirements of GATT article XX (g) and hence fit in this exception.

A strong sign of this conclusion is that "[t]he Appellate Body's interpretation of the article XX environmental exceptions in the *Shrimp-Turtle* dispute endorsed the utilization of trade measures to prevent certain environmental harms. The circumstance of global fisheries, and the multilateral regimes that have developed for their protection, suggest that trade measures used to enforce compliance with RFMOs would fall within the zone of environmental protection that the Appellate Body in *Shrimp-Turtle* held as permissible under the WTO."[545] An additional indicator that fisheries measures can respect trade law is that, recently, "[t]he WTO Secretariat has indicated that ICCAT trade restrictions 'provide examples of appropriate and WTO-consistent (i.e., non-discriminatory) use of trade measures in multilateral environment agreements.'"[546] It similarly seems that the CCAMLR CDS has been developed with the WTO jurisprudence in mind.[547] Finally, the *Shrimp-Turtle 21.5* case shows that even unilateral measures not directly based on the

[540] *United States – Import Prohibitions of Certain Shrimp and Shrimp Products (Recourse to Article 21.5 of the DSU by Malaysia), WTO Dispute Settlement Body – Panel Report*, 15 June 2001, WT/DS58/RW §5.88 (*Shrimp Turtle 21.5 Panel*).

[541] *Shrimp Turtle 21.5 Appellate Body* (n 508) §124.

[542] Tarasofsky (n 501) vi. For RFMOs, it indeed appears that most trade restricting measures reflect the international consensus as found in the UNFSA, PSMA, CCRF, IPOA-IUU, but questions related to the RFMO's inclusiveness and decision-making methods – i.e. adoption rules – would be probably discussed (Tarasofsky (n 501) 22–23; Condon (n 501) 194).

[543] *Shrimp Turtle 21.5 Panel* (n 540) 74 §5.43–81 §5.67; *Shrimp Turtle 21.5 Appellate Body* (n 508) §115–134.

[544] La Fayette, 'Access to Ports in International Law' (n 511) 20–21; Reeve, *Policing International Trade in Endangered Species* (n 296) 311–312.

[545] Tyler (n 40) 82 (reference omitted).

[546] B. K. Sovacool, 'A Game of Cat and Fish: How to Restore the Balance in Sustainable Fisheries Management' (2009) 40 Ocean Development and International Law 97, 117 (reference omitted).

[547] Agnew (n 239) 369–370.

implementation of an MEA obligation can fall under article XX, if enough good faith effort has been put in negotiations with the trading partner.[548] Consequently, it is likely that trade restriction measures that are related to an international treaty could also fulfill article XX's conditions.

c) Limits to the Applicability of Article XX

Nevertheless, some trade restrictions are controversial as they could hide protectionist policies[549] or are otherwise discriminatory. It is most likely that not all fisheries trade restrictions would fall under article XX.

One cannot exclude that some measures taken on the basis of CITES or an RFMO could be considered by the sanctioned state as a violation of WTO rules and brought to dispute settlement. For example, Japan had made clear that it would retaliate by complaining to the WTO if the USA were to apply economic sanctions against it due to their different opinions on whaling.[550]

Indeed, for example, measures against non-parties are highly controversial. RFMOs are known to have sanctioned some non-parties with trade bans and still apply different burden of proofs to contracting parties and non-contracting parties.[551] While these measures can arguably be justified because contracting parties take other actions within the RFMO framework in fulfilment of their obligation to cooperate,[552] they can also be perceived as discriminatory.[553] Indeed, should a scheme only target non-contracting parties, whether in theory or in

---

[548] *Shrimp Turtle 21.5 Appellate Body* (n 508) §134; Reeve, *Policing International Trade in Endangered Species* (n 296) 308–309.

[549] O. S. Stokke, 'Trade Measures and the Combat of IUU Fishing: Institutional Interplay and Effective Governance in the Northeast Atlantic' (2009) 33 Marine Policy 339, 342; on the expected effects of a trade ban against Iceland because of its whaling activities, see T. L. McDorman, 'Iceland, Whaling and the U.S. Pelly Amendment: The International Trade Law Context' (1997) 66 Nordic Journal of International Law 453.

[550] Carlarne (n 453) 40.

[551] On this, see supra Chapter 3 B. II. 3.

[552] Stokke, 'Trade Measures and the Combat of IUU Fishing: Institutional Interplay and Effective Governance in the Northeast Atlantic' (n 549) 347.

[553] WTO, 'The Doha mandate on multilateral environmental agreements (MEAs)' http://www.wto.org/english/tratop_e/envir_e/envir_neg_mea_e.htm accessed 4 July 2015.

practice[554] "[s]uch discrimination may be unjustifiable and thereby inconsistent with international trade law."[555] Moreover, it can be considered an attempt to force a state to change its actions in order to comply with detailed rules it never agreed to be bound by. However, such discrimination might, arguably, be justifiable as a countermeasure[556] based on the fact that the non-party violates the general obligation to cooperate in the conservation of marine resources.

Also, measures decided unilaterally are more likely to reveal problematic: not stemming from a multilateral regime, they can more easily be perceived as disguised protectionist policies or as an attempt to force another state to change its laws. For example, stricter domestic measures under CITES, if decided without the recommendation of the Standing Committee or the Secretariat, are possibly questionable. Indeed, such measures usually target one state and "it is not acceptable, in international trade relations, for one WTO Member to use an economic embargo to require other Members to adopt essentially the same comprehensive regulatory program, to achieve a certain policy goal, as that in force within that Member's territory, without taking into consideration different conditions which may occur in the territories of those Members."[557]

II. Applicability and Appropriateness

The WTO regime is applicable worldwide and is considered very efficient.[558] Its rules are relevant to all types of trade, including that of fish. Hence, some measures restricting trade, adopted under other regimes, can enter into conflict with substantive GATT rules. However, the exceptions under article XX encompass most of the existing measures taken towards better management and conservation of fisheries.

The existence of a dispute settlement mechanism is very useful in order to clarify the rules applicable and, of course, to solve inter-state disagreements. For states

---

[554] For example on the one applied by IOTC, see Rayfuse, *Non-Flag State Enforcement in High-seas Fisheries* (n 267) 191. Also, the shifting of the burden of proof might be problematic (D. Bialek, 'Sink or Swim: Measures Under International Law for the Conservation of the Patagonian Toothfish in the Southern Ocean' (2003) 34 Ocean Development and International Law 105, 124–125). On a general presentation of the issues at stake in applying trade restrictions to non-contracting parties, in terms of international economic law, see D. König, 'Flags of Convenience' Max Planck Encyclopedia of Public International Law www.mpepil.com §22–23.

[555] E. J. Molenaar, 'Port State Jurisdiction: Towards Mandatory and Comprehensive Use' in D. Freestone, R. Barnes and D. Ong (eds), *The Law of the Sea – Progress and Prospects* (Oxford University Press 2006) 207.

[556] ILC, Draft Articles on the Responsibility of States for Internationally Wrongful Acts with commentaries 2001, articles 22, 52.

[557] *Shrimp Turtle* case (n 508) §164 (italics omitted).

[558] Tyler (n 40) 82.

which consider that another party to the WTO violates their rights to the point that the benefits they are entitled to under the regime are nullified or impaired, the possibility exists to go in front of the Dispute Settlement Body (DSB). An ad hoc panel is then composed,[559] whose report becomes binding on the parties to the dispute once it has been adopted by the DSB.[560] Then, an appeal mechanism has been created in the form of the Appellate Body.[561] Finally, the surveillance of the implementation and the possibility of subsequent trade sanctions against non-compliant parties strengthen the likelihood that decisions are respected.[562] These procedures are considered some of the major strengths of the regime.[563] Cases such as the *Tuna/Dolphin* ones or the *Shrimp/Turtle* ones have allowed discussions and/or decisions on what states can or cannot do regarding marine environmental protection.

It is worth noting that, while it is possible to envision measures from RFMOs being challenged in front of the WTO Dispute Settlement Body, it is less likely that restrictions based on the CITES regime will. Indeed, issues linked to this treaty seem to always have, in the last 30 years, been dealt with within the regime and not to have been formulated in terms of violation of WTO law.[564] Nevertheless, the possibility of a non-party to CITES using the WTO as a forum to complain about CITES restrictions to trade cannot be denied.[565]

III. Conclusions

Freedom of trade is a right encompassed in the WTO regime. The rules underlying it are applicable to fisheries as they would be to any other traded good. Many trade restrictions measures taken by RFMOs or under the CITES regime might conflict with the relevant provisions of WTO law.

---

[559] 'Understanding on Rules and Procedures Governing the Settlement of Disputes' 1994, 1869 *United Nations Treaty Series* 401 article 6; P.-T. Stoll and F. Schorkopf, *WTO – World Economic Order, World Trade Law* (Martinus Nijhoff Publishers 2006) 77–78.

[560] 'Understanding on Rules and Procedures Governing the Settlement of Disputes' (n 559) article 16.

[561] Ibid. article 17; Stoll and Schorkopf (n 559) 78–79.

[562] 'Understanding on Rules and Procedures Governing the Settlement of Disputes' (n 559) articles 21–22; Stoll and Schorkopf (n 559) 90–95; Stokke, 'Trade Measures and the Combat of IUU Fishing: Institutional Interplay and Effective Governance in the Northeast Atlantic' (n 549) 342.

[563] M. W. Gehring, 'WTO law and sustainable development' in D. Armstrong (ed), *Routledge Handbook of International Law* (Routledge 2009) 377–378.

[564] Condon (n 501) 45.

[565] Ibid. 195; E. Franckx, 'The Protection of Biodiversity and Fisheries Management: Issues Raised by the Relationship between CITES and LOSC' in D. Freestone, R. Barnes and D. Ong (eds), *The Law of the Sea – Progress and Prospects* (Oxford University Press 2006) 229.

However, it is accepted that "the two bodies of law could be compatible."[566] The preamble of the Agreement Establishing the WTO mentions "the objective of sustainable development".[567] Also, in practice, the article XX exception clause foresees that restrictions to trade are licit in specific cases, such as for the conservation of exhaustible resources. Measures aiming at conservation and stemming from multilateral regimes such as RFMOs or CITES have strong chances of being considered WTO-compatible.[568] Thanks to the compulsory dispute settlement mechanism the applicable rules have been clarified and issues between states can be solved.

E. Conclusion

A general framework to regulate fishing activities is provided by UNCLOS, with rights and duties entrusted to both coastal and flag states. The regime's content varies depending on the jurisdictional area where fishing takes place. While coastal states enjoy sovereign rights over the resources found in their EEZs, they must cooperate with other states when stocks are transboundary, straddling or highly migratory. All flag states fishing in the high seas are under a similar obligation of cooperation in the conservation of marine living resources.

While this framework reflects CIL and is hence applicable to all states, the relevant provisions are vague and do not take into account later developments in the protection of the environment. Dispute settlement is also limited to certain situations.

Hence, in an attempt to clarify, detail and to some extent develop the rules applicable to fishing activities, the UNFSA was adopted in 1995. It states the obligation to apply a precautionary approach, takes on board some aspects of the ecosystem approach and spells out the central role of RFMOs in the cooperation to conserve marine resources.

While the joint UNCLOS-UNFSA regime provides a reasonably clear framework to the rights and obligations of various categories of states in the high seas and in EEZs, the actual relationship between these two treaties and the nature of the latter agreement are uncertain. It remains consequently unclear whether the obligation to join—or at least to cooperate with—the relevant RFMO is a prerequisite to the right to fish in the high seas for all states or only for the parties to UNFSA. This lack of legal certainty is highly problematic in relation to the protection of global commons.

Actors such as the FAO and RFMOs are of outmost importance in the regulation of fisheries. The former gathers and disseminates expertise on fisheries, plays a role

[566] Reeve, *Policing International Trade in Endangered Species* (n 296) 306.

[567] WTO Agreement (n 499) preamble; Gehring (n 563) 381–382.

[568] Chaves (n 500) §32–36.

in promoting actions and coordinating states and other institutions' efforts, as well as adopts instruments of a binding and non-binding nature of major importance in the field. It is sufficient to mention in that respect the Code of Conduct for Responsible Fisheries and the recent Port State Measures Agreement. As for RFMOs, they are the institutions through which states cooperate in the management and conservation of a species or of several species in a particular area.

However, a number of problems plague these organizations' performance, not least being free-riding by non-parties, opting out of measures by parties, non-compliance with rules and geographical fragmentation. Also, no institution is entrusted with the task to review the consistency of measures adopted by RFMOs with their founding documents and international law.

The multi-layered regime of fisheries governance increasingly includes conservation treaties, such as CITES and the CMS—which protects migratory species, including sharks and cetaceans—as well as regimes which evolved from being management-oriented to now being at the forefront of preservationism. Indeed, the IWC experienced a dramatic change of purpose with the adoption and the continuance of a moratorium on whaling. The decision to refrain from any catches—except in the cases of aboriginal and scientific whaling—is not based on unanimity of views in the international community and such measures justified by ideology rather than science are problematic; the regime cannot be fully functional.

The rationale of CITES, which is the focus of this study, is that a decrease in harvesting pressure, caused by reducing the opportunities to trade internationally certain species, will benefit their biological status. It goes without saying that restrictions to trade carry the risk of violating WTO law, but, under certain circumstances, compatibility can be achieved through GATT's article XX (g)'s exception clause. CITES' multilateral nature and its focus on protecting exhaustible natural resources strongly suggest that measures adopted under its auspices are consistent with trade law.

The CITES system can have a role to play in improving fisheries' governance. Indeed, this regime is quasi-universal, functions through issuance of national permits, which are then overviewed by a very active Secretariat, and the existence of sanctions mechanisms is important to fight non-compliance. The added value of using CITES for commercially-exploited fish species is analyzed in Chapter 5 D. In order to provide a benchmark to such assessment, Chapter 4 focuses on the structural and governance issues facing fisheries, first by thematically highlighting the problems of the current regimes, before presenting the characteristics of appropriate governance.

Chapter 4
Structural and Governance Issues

The international governance of fisheries suffers from several structural deficiencies (A), in particular the lack of a clear framework as to the rights and obligations of states fishing on the high seas (I), the traditional problem of fragmentation (II), inadequate conservation and management measures as well as imperfect adoption procedures (III), and finally weak compliance and enforcement mechanisms (IV). The analysis of these issues allows a presentation, *a contrario*, of the characteristics of appropriate governance (B), which will serve, in Chapter 5, to demonstrate the added value of using CITES.[1]

Governance is understood, for the purpose of this research, as the various sets of principles and rules relevant to the field examined—here the conservation and management of marine fisheries which aim at sustainability—the adoption procedures applicable, the actors involved and their interrelations, as well as the mechanisms available to verify compliance and, if needed, hold wrongdoers accountable.[2] Governance issues can be found in each of those aspects and are those problems which impair the achievement of the sustainability goal.

---

[1] See infra Chapter 5 D. II.

[2] This definition is inspired by Institute on Governance, 'Defining Governance' http://iog.ca/about-us/defining-governance/ accessed 5 July 2015; FAO, 'Fisheries governance' http://www.fao.org/fishery/topic/2014/en accessed 5 July 2015; and M. A. Young, *Trading Fish, Saving Fish: The Interaction between Regimes and International Law* (Cambridge Studies in International and Comparative Law, Cambridge University Press 2011) 17. Hence, this part focuses on public institutions and rules but does not examine the individual and private sector, also included in the definition of governance provided for in Commission on Global Governance, *Our global neighbourhood: the report of the Commission on Global Governance* (Oxford University Press 1995) 2; on this, see M. P. Karns and K. A. Mingst, *International Organizations: The Politics and Processes of Global Governance* (Lynne Rienner Publishers 2004) 4.

S. Guggisberg, *The Use of CITES for Commercially-exploited Fish Species*,
Hamburg Studies on Maritime Affairs 35, DOI 10.1007/978-3-319-23702-2_4

A. Analysis of Structural Issues

I. Legal Uncertainty

First, the lack of clarity as to the current principle of governance applicable on the high seas is an overarching issue. Indeed, even though the freedom of fishing is now undoubtedly qualified and hence the tragedy of the commons is likely to be less acute, opinions differ as to whether the oceans beyond 200 nm are, with regard to fisheries, still free for all to access and/or under which conditions they can be accessed. This is highly problematic: states do not have legal certainty as to the extent of their obligations, a situation which can incentivize free-riding.

1. Disappearing Freedoms of the High Seas

a) Restrictions to the Freedoms of the High Seas

On the high seas, traditionally, states have no right to any sovereign claims and enjoy freedom of activity without interference by third states—this is the *mare liberum* described by Grotius.[3] He based his view on the fact that "[t]he sea is common to all, because it is so limitless that it cannot be a possession of one, and because it is adapted for the use of all, whether we consider it from the point of view of navigation or of fisheries."[4] In an area of free access for all with common resources—i.e. which can be accessed by all participants equally[5]—the tragedy of the commons, as described by Hardin, is an issue to be expected and which has become a problematic reality nowadays. Indeed, there, "fishery resources belong to no one. As a result, anyone can harvest such resources without restraint."[6] With the increase in demand for fish products and developments in technology, Grotius'

[3] On the development of the *mare liberum* concept at the time of Grotius, see for example T. Scovazzi, '"Dici Quodammodo Potest Pisces Exhauriri": Fishing in the Mare Liberum' in R. Casado Raigón and G. Cataldi (eds), *L'évolution et l'état actuel du droit international de la mer: Mélanges de droit de la mer offerts à Daniel Vignes* (Bruylant 2009).

[4] H. Grotius, *The Freedom of the Sea or the Right Which Belongs to the Dutch to Take Part in the East Indies Trade: J. Brown Scott (ed) and R. Van Daman Magoffin (translation)* (Elsevier/Oxford University Press 1609/1916), as presented in R. P. Anand, 'Changing Concepts of Freedom of the Seas: A Historical Perspective' in J. M. van Dyke, D. Zaelke and G. Hewison (eds), *Freedom for the Seas in the 21st Century: Ocean Governance and Environmental Harmony* (Island Press 1993) 75.

[5] G. Hafner, 'The Division of the Commons? The Myth of the Commons: Divide or Perish' in H. P. Hestermeyer and others (eds), *Law of the sea in dialogue* (Beiträge zum ausländischen öffentlichen Recht und Völkerrecht vol 221, Springer 2011) 96.

[6] P. A. Nickler, 'A tragedy of the commons in coastal fisheries: contending prescriptions for conservation, and the case of the Atlantic Bluefin Tuna' (1998–1999) 26 Environmental Affairs Law Review 549, 552 (references omitted).

views of unlimited resources are proven wrong and Hardin's predictions have turned out to be correct.[7]

However, this concept of total freedom of the high seas was already qualified by certain obligations in the 1958 Law of the Sea Conventions, for example by the obligation to exercise these freedoms "with reasonable regard to the interests of other States",[8] or, in the field of fisheries, to adopt conservation measures.[9] While the 1958 Convention on High Seas Fisheries did not benefit from a large membership, the ICJ considered, in the *Fisheries Jurisdiction* case, that it reflects CIL.[10]

More recently, UNCLOS and later the UNFSA have added further restrictions to the freedoms of the high seas. The obligation to cooperate in the conservation and management of the living resources on the high seas[11] is an accepted rule of CIL, which is of the utmost importance here. As a norm of CIL, it applies to all states, whether they ratified UNCLOS or not. However, the normative content of this provision is subject to debate. Indeed, it is quite vague—it does not clarify through which mechanisms and to what extent cooperation is required—and hardly enforceable, since no compliance mechanism has been envisioned for non-cooperating or non-complying states[12] beyond the general international law of state responsibility and little used judicial settlement mechanisms.[13]

---

[7] B. H. Thompson (Jr.), 'Tragically difficult: the obstacles to governing the commons' (2000) 30 Environmental Law 241, 247; M. Wijkman, 'Managing the global commons' (1982) 36 International Organization 511, 511.

[8] 'Convention on the High Seas' 1958, 450 *United Nations Treaty Series* 11 article 2.

[9] 'Convention on Fishing and Conservation of the Living Resources of the High Seas' 1958, 559 *United Nations Treaty Series* 285 article 1(2). On earlier debates about needed limitations, such as in the *Arbitration between the United States and the United Kingdom relating to the rights of jurisdiction of United States in the Bering's sea and the preservation of fur seals, Award,* 15 August 1893, 2007 XXVIII Reports of International Arbitral Awards 263, see P. Birnie, A. Boyle and C. Redgwell, *International Law and the Environment* (3rd edn Oxford University Press 2009) 707–708; A. Gros, *La Convention sur la pêche et la conservation des ressources biologiques de la haute mer* (Recueil des Cours vol 97, A. W. Sijthoff 1960) chapter I; P. C. Jessup, *L'exploitation des richesses de la mer* (Recueil des Cours vol 29, Librairie Hachette 1929) chapter V.

[10] *Fisheries Jurisdiction Case (United Kingdom v. Iceland), Judgment,* 25 July 1974, I.C.J. Reports 1974 3, 22 §50. The ICJ also said, at 31 §72, that the rule applicable was not that of 'laissez-faire' anymore; as it "has been replaced by a recognition of a duty to have due regards to the rights of other States and the needs of conservation for the benefit of all" (the same opinions were expressed, in *Fisheries Jurisdiction Case (Germany v. Iceland), Judgment,* 25 July 1974, I.C.J. Reports 1974 174 respectively at 191 §42 and 200 §64).

[11] 'United Nations Convention on the Law of the Sea' (UNCLOS) 1982, 1833 *United Nations Treaty Series* 397 article 118.

[12] T. Henriksen, G. B. Hønneland and A. K. Sydnes, *Law and Politics in Ocean Governance: The UN Fish Stocks Agreement and Regional Fisheries Management Regimes* (Publications on Ocean Development, Martinus Nijhoff Publishers 2006) 3.

[13] On compliance mechanisms, sanctions and the use of the law of state responsibility in the field of fisheries, see infra Chapter 4 A. IV.

b) Decreasing High Seas

In the recent history of the law of the sea's development, following the 1945 Truman Proclamation and subsequent claims by other countries over large portions of the seabed and/or the water column, sovereign rights of states have been officially extended up to 200 nm (or possibly beyond 200 nm for the continental shelf), encroaching what used to be the high seas.[14] In this area, coastal states have now sovereign rights, particularly with regard to natural resources. Such areas amount to 36 % of the whole oceans' surface and to nearly 90 % of the fisheries.[15] It has been said to be "the largest transfer of resources to national jurisdiction in history."[16]

Furthermore, UNCLOS created a new source of sovereign jurisdiction over large portions of the oceans in the form of archipelagic states. Indeed, these states exercise sovereignty not only over the islands that compose them but also over the waters between such land masses.[17]

UNCLOS reflects a delicate balance between coastal states' interests and their tendency to make increasing sovereign claims and third states' interests to conserve the status quo on the high seas. In any case, this treaty clarifies the extent of acceptable sovereign claims in rules which have become CIL. The concept of EEZ has indeed been recognized as having reached this status, for example by the ICJ in the *Gulf of Maine* case[18] or in the *Continental Shelf* case between Libya and Malta.[19] Hence, by putting large areas of the seas under national jurisdiction or by creating functional sovereign rights where previously the high seas regime applied, UNCLOS restricts some freedoms of the high seas—the freedom of fishing being one of them—to significantly smaller portions of the oceans.

---

[14] R. R. Churchill and A. V. Lowe, *The law of the sea* (3rd edn Juris Publishing – Manchester University Press 1999) 144–145; N. Oral, 'Protection of Vulnerable Marine Ecosystems in Areas Beyond National Jurisdiction: Can International Law Meet the Challenge?' in A. Strati, M. Gavouneli and N. Skourtos (eds), *Unresolved Issues and New Challenges to the Law of the Sea: Time Before and Time After* (Martinus Nijhoff Publishers 2006) 90; Y. Tanaka, *The International Law of the Sea* (Cambridge University Press 2012) 132–133.

[15] H. Tuerk, 'The Waning Freedom of the Seas' in R. Casado Raigón and G. Cataldi (eds), *L'évolution et l'état actuel du droit international de la mer: Mélanges de droit de la mer offerts à Daniel Vignes* (Bruylant 2009) 918. Most fish stocks are indeed located in shallow waters closer to the coast (W. T. Burke, *The New International Law of Fisheries* (Clarendon Press 1994) 26).

[16] Tuerk (n 15) 918 (reference omitted).

[17] UNCLOS (n 11) article 49; N. Schrijver, *Sovereignty Over Natural Resources: Balancing Rights and Duties* (Cambridge Studies in International and Comparative Law, Cambridge University Press 1997) 207; Tuerk (n 15) 916.

[18] *Case of the delimitation of the maritime boundary in the Gulf of Maine Area (Canada/United States of America), Judgment,* 12 October 1984, I.C.J. Reports 1984 246, 294 §94.

[19] *Case concerning the Continental Shelf (Libyan Arab Jamahiriya/Malta), Judgment,* 3 June 1985, I.C.J. Reports 1985 13, 33 §34.

2. Another Step Taken with the UNFSA?

a) Enclosure of the High Seas

The UNFSA, as seen above, brings a number of clarifications and further developments to UNCLOS' provisions.[20] A rather crucial one is to be found in article 8, in the form of the obligation for states to participate in the relevant RFMO in order to be allowed to fish in its management area. This provision, without saying it explicitly, marks the end of the free unregulated access to fish on the high seas. Indeed, only members of RFMOs and cooperating non-members can legally fish in the managed area.

Also, if no RFMO exists, states are obliged to found one. Hence, the UNFSA gives RFMOs an exclusive competence over the management of straddling and highly migratory fish species[21] and creates "a kind of RFMO dominion over parts of the high seas, which might be described as a fine-tuned *Mare clausum* in the limited context of fisheries regulation enforcement."[22]

b) Nature of the UNFSA

The extent of the applicability of such rules depends on the nature of the UNFSA. The question is whether it is new law—in which case it only applies to its limited membership—whether it reached CIL status and/or whether it provides the authoritative interpretation of UNCLOS.

As discussed in detail above, this is a debated question.[23] Widespread states' practice, usually taking place through RFMOs, shows a high level of agreement that non-parties to RFMOs should refrain from fishing in the area of management, or at least respect the CMMs. This view is also shared by many scholars and practitioners. Nevertheless, not all commentators agree and no clarification which would provide legal certainty has been presented in the recent years by the UN or an international court. Hence, until this question is authoritatively settled, the exact content of the obligation to cooperate in the conservation of high seas living resources and the extent of required participation in the existing institutions will remain uncertain, as well as, consequently, the nature of the measures taken against free-riders.

---

[20] See supra Chapter 3 A. II. 1.

[21] Henriksen, Hønneland and Sydnes (n 12) 16.

[22] Tuerk (n 15) 932–933 (reference omitted); see also M. Gorina-Ysern, 'World Ocean Public Trust: High Seas Fisheries After Grotius Towards a New Ocean Ethos?' (2004) 34 Golden Gate University Law Review 645, 684.

[23] See supra Chapter 3 A. II. 3. b) and 4.

3. Conclusions

A regime of commons is a major challenge to the effective governance of the high seas:[24] "the open-access regime enshrined in the principles espoused by Grotius means that the high seas are still plagued by legal overfishing, illegal fishing, unregulated and unreported fishing, free-riding, and damaging impacts on marine ecosystems, habitats and species."[25] Also, the regime cannot be efficient or sustainable here since the potential users are all states. It is unlikely that they will all agree to sustainable measures if this does away with the possibility to maximize benefits without any restrictions. Furthermore, no one has an interest in taking strong conservation measures if there is no consensus as to the obligation for all to do so: the non-complying states, fishing in an illegal or unregulated manner, harvest the benefits of other states' sacrifices, a situation which can only encourage free riding.[26] Hence, under the current circumstances, the tragedy of the commons is likely to continue,[27] where "[e]ach man is locked into a system that compels him to increase his herd without limit – in a world that is limited."[28]

The current law of the sea in this regard is in major evolution. However, the enclosure of large high seas areas into EEZs has not solved the problem of high fishing pressure, but has rather moved it further away from the coasts. Also, as highlighted implicitly by the wide range of views on the nature of the UNFSA, the legal regime applicable to fishing on the high seas is not fully clarified.[29] While there is, for lack of control, de facto free access to high seas fishing grounds, it is necessary to clarify whether there is also a de jure free access—and if not, what the legal consequences of a violation as well as the possible redress mechanisms would be. This lack of a clear framework as to the rights and obligations of states fishing on the high seas is problematic especially since other structural governance deficiencies further complicate the picture.

---

[24] R. Rayfuse and R. Warner, 'Securing a Sustainable Future for the Oceans Beyond National Jurisdiction: The Legal Basis for an Integrated Cross-Sectoral Regime for High Seas Governance for the 21st Century' (2008) 23 International Journal of Marine and Coastal Law 399, 407.

[25] K. M. Gjerde, 'High Seas Fisheries Governance: Prospects and Challenges in the 21st Century' in D. Vidas and P. J. Schei (eds), *The World Ocean in Globalisation: Climate Change, Sustainable Fisheries, Biodiversity, Shipping, Regional Issues* (Martinus Nijhoff Publishers 2011) 221.

[26] E. J. Molenaar, 'Participation, Allocation and Unregulated Fishing: The Practice of Regional Fisheries Management Organisations' (2003) 18 International Journal of Marine and Coastal Law 457, 460.

[27] Rayfuse and Warner (n 24) 407–408.

[28] Hardin (n 1) 1244.

[29] F. Orrego Vicuña, 'Le droit de la pêche en haute mer: développement et conservation des ressources' in Mélanges offerts à Laurent Lucchini et Jean-Pierre Quéneudec (ed), *La mer et son droit* (Pedone 2003) 469.

II. Fragmentation

It is well \accepted "that international law is inherently a law of a fragmented world."[30] The law of the sea is disunited in more than one way, starting with its characteristic geographical fragmentation (1). Also, fisheries management is traditionally species specific (2). Then, rules related to management of resources and biodiversity conservation have been developed in a rather piecemeal approach, creating both institutional and substantial fragmentation (3). Moreover, a major reason for the lack of unity is that different values underlie negotiations and various goals are pursued in the development of rules, goals that sometimes are in opposition with each other (4). Finally, it is important to note that not only are the substantive rules fragmented but the dispute settlement mechanisms are also complex and varied (5).

1. Geographical

First, the regime is obviously fragmented because of the geographical approach of law and politics. The law of the sea creates separate areas in the oceans, divisions which are not based on any scientific zoning. Hence, ecosystems do not always fall within a legal zone of either an EEZ or the high seas.[31] Fish stocks, needless to develop on that, do not respect political boundaries.[32] One of the consequences of this zonal management is that the ecosystem-based approach to fisheries cannot be implemented easily.[33]

This causes issues of coherence and implementation, as marine species migrate and ecosystem realities are not related to political structures. Hence, managing fish efficiently requires highly coordinated management policies between jurisdictions. Otherwise, "different states may take disparate actions in a shared sea. Such diverse actions may undercut each other, making national action ineffective."[34] Another negative consequence that can possibly arise is that, even if the national management of a stock is adequate in the EEZ, a distant-water fishing vessel can undermine the coastal state's effort by catching high levels of fish on the adjacent high seas.[35]

---

[30] ILC, Report of the Study Group on Fragmentation of International Law (54th session) 1 August 2002, A/CN.4/L.628 2 §6; A. Mahiou, *Le droit international ou la dialectique de la rigueur et de la flexibilité: Cours général de droit international public* (Recueil des Cours vol 337, Martinus Nijhoff Publishers 2008) 93–95.

[31] Burke, *The New International Law of Fisheries* (n 15) 58–59; Y. Tanaka, 'The Changing Approaches to Conservation of Marine Living Resources in International Law' (2011) 71 Zeitschrift für ausländisches öffentliches Recht und Völkerrecht 291, 295.

[32] D. R. Rothwell and T. Stephens, *The International Law of the Sea* (Hart Publishing 2010) 303.

[33] E. A. Kirk, 'Maritime Zones and the Ecosystem Approach: A Mismatch?' (1999) 8 RECIEL 67; on the ecosystem approach, see infra Chapter 4 B. III. 1. b).

[34] L. Juda and R. H. Burroughs, 'The prospects for comprehensive ocean management' (1990) 14 Marine Policy 23, 32.

[35] R. Bratspies, 'Finessing King Neptune: Fisheries Management and the Limits of International Law' (2001) 25 Harvard Environmental Law Review 213, 218.

In the facts, the tension between coastal states and distant water fishing states with regard to straddling stocks is quite obvious.[36] It is not clear how the cooperation between these two groups of states is to be organized and implemented. Indeed, the UNFSA speaks only generally of the need for compatible measures, not clarifying whether sovereign rights might be encroached upon or whether the high seas quotas have to be in line with the ones decided for the EEZ.[37] The risk of conflicts between these two types of states on issues of conservation and management consequently remains high.[38]

Then, the regional nature of regimes for fisheries seems to disregard the realities of the industry. Fishing is indeed linked to a global market with boats moving from one area to the other and landing their harvest in different ports, which are not necessarily close-by the fishing ground.[39] Global markets and lack of global obligations can encourage the development of ports of convenience.[40] Transshipment at sea is an added issue, since it makes traceability more difficult and/or allows circumventing regional measures.[41] Indeed, with some fishing fleets, such as

[36] G. Vigneron, 'Compliance and International Environmental Agreements: A Case Study of the 1995 United Nations Straddling Fish Stocks Agreement' (1997–1998) 10 Georgetown International Environmental Law Review 581, 585.

[37] 'Agreement for the Implementation of the Provisions of the United Nations Convention on the Law of the Sea of 10 December 1982 Relating to the Conservation and Management of Straddling Fish Stocks and Highly Migratory Fish Stocks' (UNFSA) 1995, 2167 *United Nations Treaty Series* 3 article 7(2); T. L. McDorman, 'Implementing Existing Tools: Turning Words Into Actions – Decision-Making Processes of Regional Fisheries Management Organisations (RFMOs)' (2005) 20 International Journal of Marine and Coastal Law 423, 437–438; P. Oerebech, K. Sigurjonsson and T. L. McDorman, 'The 1995 United Nations Straddling and Highly Migratory Fish Stocks Agreement: Management, Enforcement and Dispute Settlement' (1998) 13 International Journal of Marine and Coastal Law 119, 127–129. As to W. T. Burke ('Compatibility and Precaution in the 1995 Straddling Stock Agreement' in H. N. Scheiber (ed), *Law of the Sea: The Common Heritage and Emerging Challenges* (Martinus Nijhoff Publishers 2000) 114), he considers compatibility as meaning that flag states fishing in the high seas are the ones who should ensure that their measures are compatible with the ones of coastal states.

[38] Or has even increased, according to M. A. Orellana ('The Law on Highly Migratory Fish Stocks: ITLOS Jurisprudence in Context' (2004) 34 Golden Gate University Law Review 459, 460), because of the ambiguities of the UNFSA.

[39] Gjerde, 'High Seas Fisheries Governance: Prospects and Challenges in the 21st Century' (n 25) 228.

[40] E. J. Molenaar, 'Port State Jurisdiction: Towards Comprehensive, Mandatory and Global Coverage' (2007) 38 Ocean Development and International Law 225, 226; J. Swan, 'Port State Measures to Combat IUU Fishing: International and Regional Developments' (2006–2007) 7 Sustainable Development Law and Policy Review 38, 38.

[41] UNGA, Resolution adopted by the General Assembly on 17 November 2004 – Sustainable fisheries, including through the 1995 Agreement for the Implementation of the Provisions of the United Nations Convention on the Law of the Sea of 10 December 1982 relating to the Conservation and Management of Straddling Fish Stocks and Highly Migratory Fish Stocks, and related instruments 17 January 2005, A/RES/59/25 §34; WCPFC Secretariat, 'Transhipment: First meeting of the Technical and Compliance Committee of the WCPFC (5–9 December 2005)' (WCPFC/TCC/17 Rev. 1, 22 November 2005) http://www.wcpfc.int/system/files/WCPFC-TCC1-17.pdf

longline fishing vessels for tuna, staying at sea for months, catches are transferred onto other boats, potentially breaking the existing chain of custody. Measures only adopted at a regional level hence risk causing relocation of IUU or unsustainable practices rather than forcing a decrease in such activities.[42] As for the existing cooperation between RFMOs, it is quite recent—for example through the Kobe process—and limited in any case to particular species or regions.

The example of CCAMLR is quite telling of the problems caused by geographical fragmentation. While it is one of the most efficient management regimes,[43] this RFMO obviously only applies to its area of jurisdiction[44] which does not comprise the whole migratory range of toothfish.[45] Its CDS hence is only relevant for toothfish caught in the managed area and when landed in member or cooperating states' ports. The need for management is unfortunately different.[46] Toothfish stocks straddle in high seas areas beyond CCAMLR's mandate zone and are regularly caught there, where no management regime applies. Hence, catches in these unregulated areas negatively affect the efficacy of management measures. Also, larger proportions of catches are declared to originate from some of these unregulated areas than scientific previsions envision available;[47] declaring fishing in these unregulated areas indeed allows to 'launder' fish caught illegally within the CCAMLR area.[48]

---

accessed 5 July 2015, §6–12; O. S. Stokke, 'Trade Measures and the Combat of IUU Fishing: Institutional Interplay and Effective Governance in the Northeast Atlantic' (2009) 33 Marine Policy 339, 341; Environmental Justice Foundation, 'Transhipment at sea: The need for a Ban in West Africa' (2013) http://ejfoundation.org/sites/default/files/public/ejf_transhipments_at_sea_web_0.pdf accessed 5 July 2015.

[42] M. Lack, 'Catching On?: Trade-related Measures as a Fisheries Management Tool' (TRAFFIC International 2007) v.

[43] R. Warner, *Protecting the Oceans Beyond National Jurisdiction* (Martinus Nijhoff Publishers 2009) 123–124.

[44] R. Herr, 'The International Regulation of Patagonian Toothfish: CCAMLR and High Seas Fisheries Management' in O. S Stokke (ed), *Governing High Seas Fisheries: The Interplay of Global and Regional Regimes* (Oxford University Press 2001) 308; P. Sands and others, *Principles of International Environmental Law* (3rd edn Cambridge University Press 2012) 414–415.

[45] ICTSD, 'Fishing, International trade and Sustainable Development: Policy Discussion Paper' Natural Resources, International Trade and Sustainable Development Studies (2006) http://www.ictsd.org/themes/environment/research/fisheries-international-trade-and-sustainable-development accessed 5 July 2015, 86–87. Applicability to only a part of a stock's area is a major problem (J. Ziemer, *Das gemeinsame Interesse an einer Regelung der Hochseefischerei: Dargestellt am Beispiel des Fish Stocks Agreement* (Veröffentlichungen des Walther-Schücking-Instituts für Internationales Recht an der Universität Kiel vol 128, Duncker & Humblot 2000) 79).

[46] M. Lack, 'Continuing CCAMLR's Fight Against IUU Fishing for Toothfish' (WWF Australia and TRAFFIC International 2008) 1–2.

[47] A. Willock, 'Uncharted Waters/Implementation Issues and Potential Benefits of Listing Toothfish in Appendix II of CITES' (TRAFFIC 2002) 9.

[48] D. Bialek, 'Sink or Swim: Measures Under International Law for the Conservation of the Patagonian Toothfish in the Southern Ocean' (2003) 34 Ocean Development and International Law 105, 126; D. S. Calley, *Market Denial and International Fisheries Regulation: The Targeted*

2. Species Wise

Most fish species are managed under a single-species approach,[49] "based on the presumption that productivity of a given species is only a function of the characteristics of its own population dynamics".[50] In reality, species interact one with another[51] in a given ecosystem and also with the physical environment surrounding them. Indeed, "species do not exist in a vacuum, but form part of an intricate web of life".[52]

The extinction or important decrease in one species has major consequences for the whole ecosystem of which it is part. An example of unexpected consequences in a single-species management—or rather mismanagement here—is that of sea otters in the North East Pacific. These animals, which were nearly hunted to extinction at the end of the nineteenth century, are predators for urchins, which themselves eat kelp. With the sharp decrease in sea otters, urchins flourished with as a result a "large-scale loss of kelp-beds, crucial habitat for fish and invertebrates."[53]

Also, species interact with the physical environment of their ecosystem, both positively and negatively. Some natural phenomena like El Niño but also human-inferred habitat destruction, climate change or pollution have an impact on the living organisms in the ecosystem[54] and, with regard to exploitation, on their development and ability to cope with harvesting pressure.

---

and Effective Use of Trade Measures Against the Flag of Convenience Fishing Industry (Martinus Nijhoff Publishers 2012) 153, 178; K. W. Riddle, 'Illegal, Unreported, and Unregulated Fishing: Is International Cooperation Contagious?' (2006) 37 Ocean Development and International Law 265, 283.

[49] J. Morishita, 'What is the ecosystem approach for fisheries management?' (2008) 32 Marine Policy 19, 21.

[50] E. J. Molenaar, 'Marine Mammals: The Role of Ethics and Ecosystem Considerations' (2003) 6 Journal of International Wildlife Law and Policy 31, 45.

[51] Morishita, 'What is the ecosystem approach for fisheries management?' (n 49) 21.

[52] Orellana, 'The Law on Highly Migratory Fish Stocks: ITLOS Jurisprudence in Context' (n 38) 482.

[53] Molenaar, 'Marine Mammals: The Role of Ethics and Ecosystem Considerations' (n 50) 45; J. C. Kunich, 'Losing Nemo: The Mass Extinction Now Threatening the World's Ocean Hotspots' (2005) 30 Columbia Journal of Environmental Law 1, 31; A. Longhurst, *Mismanagement of Marine Fisheries* (Cambridge University Press 2010) 122–123; World Ocean Review, *Living with the oceans* (Maribus, Future Ocean 2010) 117.

[54] R. T. Carson and others, 'Fisheries Management Under Cyclical Population Dynamics' (2009) 42 Environmental and Resource Economics 379, 380–381; Molenaar, 'Marine Mammals: The Role of Ethics and Ecosystem Considerations' (n 50) 45. El Niño is known to influence enormously the level of anchovetas (P. A. Larkin, 'Concepts and issues in marine ecosystem management' (1996) 6 Reviews of Fish Biology and Fisheries 139, 152).

3. Structural

a) Substantive Piecemeal Approach

Structural fragmentation stems from the obvious fact that relevant rules are found in various conventions with no automatic coherence or hierarchy. This is a general issue in international law[55] which is felt particularly acutely in fisheries matters. Indeed, not only were 293 treaties related to fish recorded at the beginning of the twenty-first century[56] but general conservation or environmental protection treaties, such as CITES or the CMS, are also relevant.

The regulation of fisheries is indeed based on a piecemeal approach. Solutions have been sought under different angles and single-issue conventions have been developed where a more global approach could have been desirable.

Moreover, this proliferation of treaties also means that states are not all bound by the same rules. This problem is not particular to fisheries governance, but more generally linked to the horizontal nature of international law, a system based on sovereignty where states develop the rules and are only bound by them if they give their consent.

b) Institutional Fragmentation Without Coordinator

The law of the sea regime has been said to miss an international organization overseeing its application in a global way. Indeed, no institution was created to deal with the overall implementation of UNCLOS, a treaty which encompasses most aspects of the law of the sea.[57] Since coordination does not exist in the sense of a global marine international organization with binding powers over its members, several institutions of the UN, amongst others the UNGA and the UNFSA Review Conference, analyze the issues at stake, recommend measures and encourage actors to proceed towards more sustainable practices.

The UNGA is somehow filling the gap, but in a soft-law manner. Indeed, it has the potential to "foster a coordinated approach"[58] by overseeing the implementation of UNCLOS and other relevant instruments, encouraging the actors to strengthen the

---

[55] K. Zemanek, *The legal foundations of the international system: general course on public international law* (Recueil des Cours vol 266, Martinus Nijhoff Publishers 1997) 62–65.

[56] M. Giordano, 'The Internationalization of Wildlife and Efforts Towards its Management: A Conceptual Framework and the Historic Record' (2001–2002) 14 Georgetown International Environmental Law Review 607, 616.

[57] On the various but specialized institutions established by UNCLOS, see for example T. Treves, 'The Law of the Sea "System" of Institutions' (1998) 2 Max Planck Yearbook of United Nations Law 325, 325–340.

[58] H. Corell, 'Oversight of the Implementation of the Global Ocean Regime: The Role of the United Nations' in D. Vidas and W. Ostreng (eds), *Order for the oceans at the turn of the century* (Kluwer Law International 1999) 338.

regime, identifying new problems and thus making recommendations to states, international and regional organizations.[59] Nevertheless, however comprehensive the UNGA is in its Resolutions, it relies on exhortation to induce change. Indeed, UNGA Resolutions are political instruments, hence not legally binding on states.[60]

As for the UNFSA Review Conference, while it can be seen as having some coordination powers—it indeed reviews a wide-ranging convention—it is not a global law of the sea or fisheries review mechanism. One can nevertheless note that, by making recommendations on how to reach the goal of sustainable use, it has the potential to go beyond the provisions of the UNFSA—it mentioned for example fishing capacity, deep-sea fisheries and the development of a global record of fishing vessels in its 2010 exercise.[61] Also, even within the framework of the convention, the special role that the UNFSA allocates to RFMOs allows the review conference to look at these entities' balance sheets and to guide them towards best-practices—it did so, for example, by encouraging them to undertake performance reviews or by looking at the decision-making procedures used in these bodies.[62] That the Review Conference is not permanent and only meets every couple of years is however a major issue.

Furthermore, neither the global environmental regime nor the fisheries one is arranged with a centralized organization granted actual coordinating powers. While UNEP and the FAO exist and somehow oversee the Secretariats and governing bodies of several relevant treaties,[63] the recommendations they issue have no binding force.[64] Indeed, the role of the FAO towards RFMOs is limited: it plays a facilitator role for the Regional Fisheries Body Secretariats Network but is not directly involved in managing fisheries[65] and does not have actual power over RFMOs. Hence, no mechanism to formally coordinate RFMOs' work exists, a situation which is at odds with the global nature of some of the problems facing fisheries.[66]

---

[59] Ibid. 338, 340–341.

[60] 'Charter of the United Nations' 1945, 1 *United Nations Treaty Series* xvi articles 10–14 show that UNGA Resolutions are recommendatory in nature; see M. J. Peterson, 'General Assembly' in T. G. Weiss and S. Daws (eds), *The Oxford Handbook on The United Nations* (Oxford University Press 2007) 103.

[61] UNFSA, 'Report of the resumed Review Conference on the Agreement for the Implementation of the Provisions of the United Nations Convention on the Law of the Sea of 10 December 1982 relating to the Conservation and Management of Straddling Fish Stocks and Highly Migratory Fish Stocks' 27 July 2010, A/CONF.210/2010/7 §57, 65, 109.

[62] Ibid. §76–77, 86–87.

[63] Young, *Trading Fish, Saving Fish* (n 2) 67–68; on the supervisory role of the FAO over RFMOs, see supra Chapter 3 B. II. 4.

[64] R. Wolfrum and N. Matz, *Conflicts in international environmental law* (Beiträge zum ausländischen öffentlichen Recht und Völkerrecht, Springer 2003) 182; on UNEP's more general coordinating role, see ibid. 181–191; N. Matz, *Wege zur Koordinierung völkerrechtlicher Verträge: Völkervertragsrechtliche und institutionelle Ansätze* (Springer 2005) 220–229.

[65] J. Harrison, *Making the Law of the Sea: A Study in the Development of International Law* (Cambridge University Press 2011) 228.

[66] Warner (n 43) 126.

c) Consequences of Structural Fragmentation

Because of the piecemeal development of the relevant rules, as well as because of the large number of treaties and lack of coordination between them,[67] one can fear a lack of coherence in the system.[68]

In certain cases, decentralization and partial overlap can lead to doubling of work or tensions and, arguably, even to conflicts between norms stemming from different specialized regimes.[69] Such potential conflicts arise, for example, between rules for the protection of the environment and trade law.[70] Indeed, several sets of rules applicable to fisheries can be seen as "[c]ompeting regulations where different regulation could become applicable to the same situations or facts".[71] A lack of coordination between the fisheries and the environmental regimes is also noticeable at times.[72]

That several institutions are responsible for aspects of marine environmental matters not only causes overlap and risks of competition, but also increases the risks of conflicts of norms. Indeed, institutional fragmentation, where specialized institutions generate their own legal norms, as if in a vacuum, makes conflicts between regimes more likely.[73]

Besides overlaps, protection gaps are also a big problem,[74] such as in the high seas areas where no RFMO exists, for example in the international waters adjacent to the CCAMLR area, where fishing is unregulated. Other major geographical gaps still exist where no management regime has been established, such as in the South West Atlantic.[75] Furthermore, some species are left mostly unregulated. Worth mentioning

---

[67] Birnie, Boyle and Redgwell (n 9) 704; M. Bowman, P. Davies and C. Redgwell, *Lyster's International Wildlife Law* (2nd edn Cambridge University Press 2010) 122–123.

[68] C. McLachlan, 'The Principle of Systemic Integration and Article 31(3)(C) of the Vienna Convention' (2005) 54 International and Comparative Law Quarterly 279, 283.

[69] ILC, 'Fragmentation of international law: difficulties arising from the diversification and expansion of international law' Report of the Study Group of the International Law Commission (finalized by M. Koskenniemi) 13 April 2006, A/CN.4/L.682, 34 §55; G. Hafner, 'Pros and Cons Ensuing from Fragmentation of International Law' (2003–2004) 25 Michigan Journal of International Law 849, 850–851; N. Matz, 'The Interaction between the Convention on Biological Diversity and the UN Convention on the Law of the Sea' in P. Ehlers, E. Mann-Borgese and R. Wolfrum (eds), *Marine Issues: From a Scientific, Political and Legal Perspective* (Kluwer Law International 2002) 203.

[70] Hafner, 'Pros and Cons Ensuing from Fragmentation of International Law' (n 69) 852.

[71] Ibid. 855 (reference omitted).

[72] Rayfuse and Warner (n 24) 402–403.

[73] A. Lindroos, 'Addressing Norm Conflicts in a Fragmented Legal System: The Doctrine of Lex Specialis' (2005) 74 Nordic Journal of International Law 27, 32–33.

[74] K. M. Gjerde, 'High-seas Fisheries Management under the Convention on the Law of the Sea' in D. Freestone, R. Barnes and D. Ong (eds), *The Law of the Sea – Progress and Prospects* (Oxford University Press 2006) 295; E. J. Molenaar, 'Managing Biodiversity in Areas Beyond National Jurisdiction' (2007) 22 International Journal of Marine and Coastal Law 89, 96.

[75] D. Freestone, 'Problems of High Seas Governance' in D. Vidas and P. J. Schei (eds), *The World Ocean in Globalisation: Climate Change, Sustainable Fisheries, Biodiversity, Shipping, Regional Issues* (Martinus Nijhoff Publishers 2011) 103–104; Rayfuse and Warner (n 24) 402. A management regime will enter into force on 19 July 2015 in the North Pacific (Multilateral Meeting on Management of High Seas Fisheries in the North Pacific Ocean, 'North Pacific Fisheries Commission' http://nwpbfo.nomaki.jp/index.html accessed 6 July 2015).

are the discrete fish stocks which do not fall within the framework developed by the UNFSA.[76] Also, the obligations with regard to transboundary (or shared) stocks—i.e. those which migrate only between states' EEZs—are very vague and are not included in the UNFSA.[77] A further gap can be witnessed with regard to deep-sea species which are likely to interact with the exploration and exploitation of the deep seabed[78] but do not belong to the Area's regime. Deep-sea fisheries only came onto the agenda in the last decade[79] and even though progress has been made at the RFMO level[80] some gaps still remain.[81] Lastly, the low level of binding regional or international regulation towards aquaculture practices must be mentioned.[82]

In conclusion, such a complex and fragmented framework is not fully efficient, not least because it lacks clarity and hence puts legal security at risk.[83] It has even been said, in quite dramatic terms, that "[t]he disintegration of the legal order jeopardizes the credibility, reliability, and, consequently, the authority of international law."[84]

---

[76] Rayfuse and Warner (n 24) 403. Nevertheless, at the 2006 UNFSA Review Conference, the participants "encouraged States, as appropriate, to recognize that the general principles of the Agreement should also apply to fisheries for discrete fish stocks on the high seas" (D. A. Balton and H. R. Koehler, 'Reviewing the United Nations Fish Stocks Treaty' (2006–2007) 7 Sustainable Development Law and Policy Review 5, 7). A number of RFMOs have implicitly or explicitly included such stocks in their management (Molenaar, 'Managing Biodiversity in Areas Beyond National Jurisdiction' (n 74) 99–103).

[77] Indeed, the only applicable rule for these species is UNCLOS (n 11) article 63(1); on this see for example R. R. Churchill, 'The Management of Shared Fish Stocks: the Neglected 'other' Paragraph of Article 63 of the UN Convention on the Law of the Sea' in A. Strati, M. Gavouneli and N. Skourtos (eds), *Unresolved Issues and New Challenges to the Law of the Sea: Time Before and Time After* (Martinus Nijhoff Publishers 2006).

[78] Freestone, 'Problems of High Seas Governance' (n 75) 102.

[79] Compare the absence of reference to bottom trawling in UNGA, Resolution adopted by the General Assembly on 12 December 2002 – Agreement for the Implementation of the Provisions of the United Nations Convention on the Law of the Sea of 10 December 1982 relating to the Conservation and Management of Straddling Fish Stocks and Highly Migratory Fish Stocks 26 February 2003, A/RES/57/143 §8 and its insertion in Resolution adopted by the General Assembly on 17 November 2004, A/RES/59/25 (n 41) §66–70.

[80] A number of RFMOs (CCAMLR, GFCM, NAFO, NEAFC, SEAFO) have taken measures of some sort about deep-sea fishing (A. Jackson, 'The Convention on the Conservation and Management of Fishery Resources in the South East Atlantic Ocean, 2001: an introduction' (2002) 17 International Journal of Marine and Coastal Law 33, 38; Sands and others (n 44) 441).

[81] T. Treves, 'Do fisheries organisations and agreements provide an appropriate framework for sustainable management of deep-sea fisheries in the high seas?' (2009) 35 Océanis 71, 84. As many deep-sea species are caught using bottom trawling, it is worth mentioning the continued lack of a global instrument to deal with such destructive practice (Gjerde, 'High-seas Fisheries Management under the Convention on the Law of the Sea' (n 74) 295; Warner (n 43) 16).

[82] On the few existing measures, see supra Chapter 3 B. II. 2. c). The FAO however highlighted "a good overall status of governance in aquaculture" at the national level (FAO, 'The State of the World Fisheries and Aquaculture' (Rome 2014) 89).

[83] ILC, 'Fragmentation of international law' (n 69) 32 §52.

[84] Hafner, 'Pros and Cons Ensuing from Fragmentation of International Law' (n 69) 856.

4. Philosophical

Another type of fragmentation is due to the existence of different goals pursued by states, according to their interests and values. These goals can also vary within one state and different opinions be expressed, depending on the institutional culture as well as the representatives that are sent to the negotiation table.

a) Different Values

There is no globally shared understanding of the reason behind protection of species: it rests on different perceptions of the role of humans in nature, as well as on moral or utilitarian views on wildlife:[85] "[o]n one side are preservationists, who hold that nature and wildlife should be preserved as such and that the wide-spread utilization of (usually select) species should be abandoned in favour of a new environmental ethic. On the other side, conservationists have no problem with utilization, provided it is not excessive and does not threaten a species they want to utilize in the future."[86] As can be expected, putting forward the animals' *per se* right to life or the need to manage them as resources does not generate the same legal regime.

Conservationism, a synonym of sustainable use, can be consumptive or not but always includes "an element of human care and control."[87] A consumptive conservationism relates to harvesting while the non-consumptive one focuses for example on "aesthetic, scientific, cultural [and] recreational" uses—[88] such as ecotourism, parks, etc.[89] Sustainable use, one facet of sustainable development,[90] was already mentioned by the IUCN in 1980 in its World Conservation Strategy as the "[u]se of an organism, ecosystem or other renewable resource at a rate within its capacity for renewal."[91] Views differ as to whether consumptive or

---

[85] Bowman, Davies and Redgwell (n 67) 62–64; on a presentation of the consequences in terms of treaties' development of the change of focus, see Matz, *Wege zur Koordinierung völkerrechtlicher Verträge* (n 64) 77–83.

[86] P. Stoett, 'To trade or not to trade?: The African elephant and CITES' (1996–1997) 52 International Journal of Marine and Coastal Law 567, 569 (italics omitted).

[87] C. L. Krieps, 'Sustainable use of endangered species under CITES: is it a sustainable alternatives' (1996) 17 University of Pennsylvania Journal of International Law 461, 476.

[88] CITES Resolution Conf. 8.3 (Rev. CoP13) – Recognition of the benefits of trade in wildlife 1992 preamble.

[89] D. S. Favre, 'Debate within the CITES Community: What Direction for the Future?' (1993) 33 Natural Resources Journal 875, 914.

[90] World Commission on Environment and Development, *Our Common Future*, (usually referred to as the Brundtland Report) (Oxford University Press 1987) 348 annex 1 Summary of Proposed Legal Principles for Environmental Protection and Sustainable Development Adopted by the WCED Experts Group on Environmental Law §3; Sands and others (n 44) 207.

[91] IUCN (World Conservation Union), UNEP (United Nations Environment Programme) and WWF (World Wide Fund for Nature), *Caring for the earth: A strategy for sustainable living* (1991) 211.

non-consumptive uses can best strengthen sustainability,[92] but the underlying philosophy accepts a balance between human and wildlife interests as long as the activities are sustainable.

As for preservationism, it aims not only at preventing harm to the species, but also at respecting the animals,[93] whatever the costs to human societies. The views on human intervention are varied; for some, preservationism focuses on non-consumptive uses and inherent values given to animals and plants,[94] and hence is somehow similar to non-consumptive conservationism. For others, the goal is rather to "preserv[e] wilderness areas in their natural state for their own intrinsic aesthetic or spiritual value, without much ongoing human intervention."[95] In all cases, nothing, not even the needs of the local population coupled with the proven sustainability of a harvesting activity, can justify the consumptive use of wildlife.[96]

The fragmentation arising from different underlying philosophies towards nature can for example be seen in the lack of overall coherence in rules related to biodiversity, and also, quite strikingly, in the situation within the IWC. Some states view whales as deserving to live as charismatic megafauna[97]—hence defending the moratorium without requiring any scientific data showing the need for the survival of the species—while other states point out that some species could sustainably be harvested again and that, after all, whales are a source of food.[98]

---

[92] J. G. Robinson and K. H. Redford, 'The Use and Conservation of Wildlife' in J. G. Robinson and K. H. Redford (eds), *Neotropical Wildlife Use and Conservation* (The Chicago University Press 1991) 3–4.

[93] Favre, 'Debate within the CITES Community: What Direction for the Future?' (n 89) 880. Sometimes the reasoning goes even further, with the concept of entitlement, of a right to life for certain animals (A. D'Amato and S. K. Chopra, 'Whales: Their Emerging Right to Life' (1991) 85 American Journal of International Law 21).

[94] S. Young, 'Contemporary Issues of the Convention on International Trade in Endangered Species of Wild Fauna and Flora (CITES) and the Debate Over Sustainable Use' (2003) 14 Colorado Journal of International Environmental Law and Policy 167, 182.

[95] Krieps (n 87) 476 (reference omitted).

[96] D'Amato and Chopra (n 93) 45.

[97] On the argument of protectionists in favor of a right to life of whales, see H. N. Scheiber, 'Historical Memory, Cultural Claims, and Environmental Ethics: The Jurisprudence of Whaling Regulation' in H. N. Scheiber (ed), *Law of the Sea: The Common Heritage and Emerging Challenges* (Martinus Nijhoff Publishers 2000) 139–140.

[98] C. P. Carlarne, 'Saving the Whales in the New Millennium: International Institutions, Recent Developments and the Future of International Whaling Policies' (2005–2006) 24 Virginia Environmental Law Journal 1, 36 n 221, quoting John Vidal, 'Weeping and Whaling', Guardian, 7 May 1993 (quoting Gro Brundtland, Prime Minister of Norway); J. Morishita, 'Multiple analysis of the whaling issue: Understanding the dispute by a matrix' (2006) 30 Marine Policy 802, 806; Scheiber (n 97) 139–140.

b) Various Goals

There are different goals pursued by states and other stakeholders in designing and implementing fisheries-related regulations.[99] They can be categorized in several overarching groups: biological objectives such as the maximum yield, ecosystem protection or preservation of vulnerable species; economic objectives such as maximizing the benefits from fishing activities; and social objectives, such as food security, or employment targets.[100]

In theory, such various objectives can co-exist through a compromise solution, such as the MSY. However, it seems that one of the major problems is the priority given to socio-economic goals over the biological reality.[101] Indeed, "[t]he objective of preserving marine ecosystems is broadly in conflict with all other objectives because the more you protect an ecosystem the less resource is available for utilization in the form of yield, economic rents, or jobs."[102] All in all, it has been estimated that all three goals are being failed, with unsustainable fishing practices for the species, as well as economic and jobs losses.[103]

Different groups and institutional structures defend these various objectives. While fishermen associations usually push for higher quotas, environmental lobbies defend the protection of the ecosystems including vulnerable species. Within the state, these various trends are also reflected across the departments: usually, the ministries for economy/agriculture have different agendas than the ministry for the environment. Hence, in their relations with other states, it is likely that the ministerial portfolio of the state's representative will, to some extent, influence the negotiations and their results.[104] Consequently, against the expectation of a state's unity,[105] one must admit that different interests are defended within one state's foreign policy.[106] Also, the institutional particularities of the forum where

---

[99] D. J. Dankel, D. W. Skagen and Ø. Ulltang, 'Fisheries management in practice: review of 13 commercially important fish stocks' (2008) 18 Reviews of Fish Biology and Fisheries 201, 202.

[100] R. Hilborn, 'Defining success in fisheries and conflicts in objectives' (2007) 31 Marine Policy 153, 153.

[101] K. L. Cochrane, 'Reconciling sustainability, economic efficiency and equity in fisheries: the one that got away?' (2000) 1 Fish and Fisheries 3, 7; P. Vincent, *Droit de la mer* (Larcier 2008) 225.

[102] Hilborn, 'Defining success in fisheries and conflicts in objectives' (n 100) 155.

[103] Cochrane, 'Reconciling sustainability, economic efficiency and equity in fisheries: the one that got away?' (n 101) 5–6.

[104] M. A. Young, 'Protecting Endangered Marine Species: Collaboration Between the Food and Agriculture Organization and the CITES Regime' (2010) 11 Melbourne Journal of International Law 1, 38.

[105] S. Beslier, 'Pêche et biodiversité marine: complémentarité ou concurrence au sein de l'ordre juridique international? Le cas du thon rouge' (Paris November 2010) Idées pour le débat n°9, 6.

[106] This was for example noticed during the negotiations of the CITES-FAO MoU (Young, *Trading Fish, Saving Fish* (n 2) 174).

negotiations take place are likely to influence the nature of decisions taken.[107] Consequently, states might be bound by overlapping, or even conflicting obligations, although this situation contradicts the idea of a unitary rational state.

It is nevertheless worth remembering that states are not, as a matter of law, free to choose as they please between the various objectives here above mentioned. Indeed, the law of the sea clearly requires states, in their EEZs to make sure that their living resources are not endangered by overexploitation and all states on the high seas have to conserve living resources. Hence, depletion of stocks goes against UNCLOS.[108] Additionally and more generally, under the principle of sustainability, or of prevention of harm—the extinction of a species definitely constituting a level of harm—, some restraints to the freedom of decision-makers are set.

5. Dispute Settlement

In addition to the fragmented substantive legal system outlined above, secondary rules and mechanisms can also be described as fragmented. Indeed, the existence, in parallel, of several potentially competent dispute settlement bodies allows states to shop for a preferred forum and puts the coherence of the application and interpretation of international law at risk.[109] Conflicts in both jurisdiction and jurisprudence are potential sources of concern.[110]

A first problem is that some disputes are based on more than one legal regime and can be adjudicated in different courts. Since each mechanism will apply the set of rules it is related to, at least in priority, it is likely that states will go to different bodies, depending on where they think their interests will be best defended, engaging in what is called forum shopping.[111] And, if decisions or reasoning of the different dispute settlement mechanisms are not in line, parallel developments or interpretation of general rules risk existing side-by-side, with the logical problems this lack of cohesiveness would cause to legal security.

[107] Whether it qualifies as an epistemic community (defined as "a network of professionals with recognized expertise and competence in a particular domain and an authoritative claim to policy-relevant knowledge within that domain or issue-area" (P. M. Haas, 'Introduction: epistemic communities and international policy coordination' (1992) 46 International Organization 1, 3; see also S. B. Kaye, *International Fisheries Management* (Kluwer Law International 2001) 31–39)) or is simply the consequence of a particular institutional culture is less clear.

[108] FAO, 'State of the World Fisheries and Aquaculture' (Rome 2004) 86.

[109] Some concerns were voiced regarding the danger caused to the cohesion of international law by a proliferation of courts, but these fears were not unanimously shared (see a presentation of views in D. L. Morgan, 'Implications of the Proliferation of International Legal Fora: The Example of the Southern Bluefin Tuna Cases' (2002) 43 Harvard International Law Journal 541, 543–544).

[110] T. Treves, 'Le Tribunal International du Droit de la Mer dans la pléiade des juridictions internationales' in O. Delas and others (eds), *Les juridictions internationales: complémentarité ou concurrence* (Bruylant 2005) 11.

[111] Hafner, 'Pros and Cons Ensuing from Fragmentation of International Law' (n 69) 857.

For example, cases related to restrictions, for conservation measures, to trade in fish products and to access to ports fall under both the WTO and UNCLOS dispute settlement mechanisms. The *Swordfish* case between Chile and the European Union (EU) was presented both in front of ITLOS and the WTO dispute settlement mechanism.[112] Even though an agreement between the parties was found before the proceedings in front of the courts were finalized, this situation exemplified several issues since both fora had jurisdiction and could have clearly reached distinct conclusions.[113] It would have been interesting to see to which extent the WTO dispute settlement body could have considered the law of the sea and ITLOS examined trade law.[114] If a case is adjudicated from two regimes' perspectives, conclusions are likely to differ, putting the legitimacy of dispute settlement mechanisms at risk and possibly endangering the coherence of international law.[115]

Then, overlapping jurisdictions can, in certain cases, weaken the system if the choice is not between two equally binding mechanisms. In the *Southern Bluefin Tuna* cases, ITLOS in the first place and for the purpose of indicating provisional measures decided that the Arbitral Tribunal, constituted under Annex VII, would have *prima facie* jurisdiction.[116] Reaching the opposite conclusion, the Arbitral Tribunal itself considered that the existence of a regional specialized non-compliance mechanism—even though it does not refer to a judicial body with compulsory jurisdiction but relies on measures to be decided by the parties

---

[112] M. A. Orellana, 'The Swordfish Dispute between the EU and Chile at the ITLOS and the WTO' (2002) 71 Nordic Journal of International Law 55; R. Rayfuse, 'The Future of Compulsory Dispute Settlement Under The Law of the Sea Convention' (2005) 36 Victoria University of Wellington Law Review 683, 702–703.

[113] On the basis of GATT, Chile should probably have authorized access to its ports, but on the basis of UNCLOS, as the coastal state, it was entitled to exercise sovereignty over its ports for purposes of conservation (Orellana, 'The Swordfish Dispute between the EU and Chile at the ITLOS and the WTO' (n 112) 65; R. Salama, 'Fragmentation of International Law: Procedural Issues Arising in Law of the Sea Disputes' (2005) 19 Australian and New Zealand Maritime Law Journal 24, 25, 31; Treves, 'Le Tribunal International du Droit de la Mer dans la pléiade des juridictions internationales' (n 110) 22–23).

[114] Salama (n 113) 32; on a comparison of the two dispute settlement mechanisms, see for example R. J. McLaughlin, 'Settling Trade-Related Disputes Over the Protection of Marine Living Resources: UNCLOS or the WTO?' (1997–1998) 10 Georgetown International Environmental Law Review 29, 47–51.

[115] Hafner, 'Pros and Cons Ensuing from Fragmentation of International Law' (n 69) 858. Others highlight the potential for cross-fertilization and its positive impact on the development of international law (T. Treves, 'Cross-fertilization between different international courts and tribunals: the Mangouras case' in H. P. Hestermeyer and others (eds), *Coexistence, Cooperation and Solidarity: Liber Amicorum Rüdiger Wolfrum* (Martinus Nijhoff Publishers 2011) in particular between the European Court of Human Rights and ITLOS (1788–1791)).

[116] UNCLOS (n 11) articles 288(1), 290(5); *Southern Bluefin Tuna cases (New Zealand v. Japan; Australia v. Japan), Provisional Measures,* 27 August 1999, International Tribunal for the Law of the Sea (ITLOS) cases n°3 & 4 §62; B. Kwiatkowska, 'The Southern Bluefin Tuna (New Zealand v Japan; Australia v Japan) Cases' (2000) 15 International Journal of Marine and Coastal Law 1, 18–21.

by consensus[117]—was a reason to reject competence.[118] Hence, instead of a binding compulsory mechanism under UNCLOS, a legally weaker mechanism was used. If this line of reasoning serves as a precedent, then it does exclude a number of treaties which could otherwise fall *ratione materiae* under UNCLOS dispute settlement procedures but have a clause related to some dispute settlement mechanism—usually a weak one—of their own.[119] It must nevertheless be noted that the cases of the *Southern Bluefin Tuna*, where ITLOS and the Arbitral Tribunal reached different conclusions regarding jurisdiction do not necessarily prove that coherence is at risk. Indeed, the threshold to reach is different for provisional measures, where only *prima facie* jurisdiction has to be established, and for the merits, where the courts often have to examine in detail their competence to hear a case, in particular if a party presents objections.

Finally, if a dispute arises about fish stocks which can be found in both the high seas and an EEZ, the compulsory dispute settlement is applicable to the former but not to the latter part[120]—the sovereign rights of a coastal state over fisheries in its EEZ are indeed exempted from compulsory dispute settlement entailing binding decisions, as stated in article 297 (3).[121] This situation would require the use of two different dispute settlement mechanisms, with different legal consequences, and, potentially, different conclusions: while the high seas part of the dispute can be settled through the binding mechanisms developed in UNCLOS Section 2 of Part XV,[122] the EEZ part is limited to being presented to a conciliation commission in certain particular cases.[123] Hence, it appears that, for a straddling stock, a coastal state could bring a dispute in front of ITLOS, the ICJ or arbitration against a distant-water fishing state fishing in the adjacent high seas area but that the distant-water fishing state could not do the same against a coastal state for its management of the resources in its EEZ.[124] This fragmentation and unequal treatment can be said to "seriously diminish the overall integrity of the Convention."[125]

---

[117] CCSBT ('Convention for the Conservation of Southern Bluefin Tuna' (CCSBT Convention) 1993, 1819 *United Nations Treaty Series* 360 article 16.

[118] *Southern Bluefin Tuna (New Zealand v. Japan, Australia v. Japan), Arbitral Award,* 4 August 2000, Arbitral Tribunal, (2006) XXIII Reports of International Arbitral Awards 1 §53–59, 72; Salama (n 113) 36–37.

[119] Morgan (n 109) 550; Salama (n 113) 36–37; see supra Chapter 3 B. II. 3.

[120] A. Boyle, 'Dispute Settlement and the Law of the Sea Convention: Problems of Fragmentation and Jurisdiction' (1997) 46 International and Comparative Law Quarterly 37, 42–44; J. G. Merrills, *International Dispute Settlement* (5th edn Cambridge University Press 2011) 191.

[121] UNCLOS (n 11) article 297(3); see supra Chapter 3 A. I. 2. and infra Chapter 4 A. IV. 3. a).

[122] UNCLOS (n 11) part XV section 2.

[123] Ibid. article 297(3)(b)(c); Tanaka, *The International Law of the Sea* (n 14) 400–402.

[124] R. R. Churchill, 'The Jurisprudence of the International Tribunal for the Law of the Sea Relating to Fisheries: Is There Much in the Net?' (2007) 22 International Journal of Marine and Coastal Law 383, 390.

[125] Boyle, 'Dispute Settlement and the Law of the Sea Convention: Problems of Fragmentation and Jurisdiction' (n 120) 42.

A second issue is that of forum shopping within the same legal regime.[126] The major risk arising from such phenomenon would be that, if jurisprudence was not coherent, states would face different legal consequences depending on the dispute settlement mechanism they chose.[127]

Due to the particularities of the dispute settlement structure developed in UNCLOS, this problem is quite acute in the law of the sea, where states have the choice of various fora to bring a dispute. UNCLOS indeed leaves the choice open to go to the ICJ, ITLOS, or to arbitration. Several judges of the ICJ expressed their opposition to the creation of ITLOS "saying among other things that a new court would jeopardize the unity and consistency of international law."[128] While it is quite unlikely that these judicial institutions will reach clearly conflicting conclusions,[129] they could still apply, interpret or even develop the law into different directions.

However, many commentators disagree with such a grim view.[130] Particularly, Professor Charney, who extensively studied the question of jurisdictional overlap and the risks it causes for the coherence of international law, concluded that, in reality, no such fear is justified, particularly owing to the weight given to the ICJ's existing jurisprudence.[131] While some slight differences can exist, it is unlikely that different courts or tribunals would adjudicate in such an unexpected way that international law's cohesiveness would be at risk. As for forum shopping, it is a reality but might also

[126] Hafner, 'Pros and Cons Ensuing from Fragmentation of International Law' (n 69) 857; Morgan (n 109) 548.

[127] ILC, 'Fragmentation of international law' (n 69) 247 §489; Morgan (n 109) 544.

[128] Morgan (n 109) 543; for example Judge Oda, Judge Schwebel and Judge Guillaume, as presented in Rayfuse, 'The Future of Compulsory Dispute Settlement Under The Law of the Sea Convention' (n 112) 686 or in H. Caminos, 'The International Tribunal for the Law of the Sea and its Alleged Threat to the Unity of International Law' in R. Casado Raigón and G. Cataldi (eds), *L'évolution et l'état actuel du droit international de la mer: Mélanges de droit de la mer offerts à Daniel Vignes* (Bruylant 2009) 75–76.

[129] It has nevertheless happened that international courts interpret general international law differently. On the International Criminal Tribunal for the former Yugoslavia (ICTY)/ICJ interpretations of 'effective control' (*Military and Paramilitary Activities in and against Nicaragua (Nicaragua v. United States of America), Judgment,* 27 June 1986, I.C.J. Reports 1986 14, 64–65 §115; *Prosecutor v. Dusko Tadic, Judgment,* 15 July 1999, ICTY IT-94-1-A 205–207 §585–586), see ILC, 'Fragmentation of international law' (n 69) 31–32 §49–51.

[130] Boyle, 'Dispute Settlement and the Law of the Sea Convention: Problems of Fragmentation and Jurisdiction' (n 120) 41; Rayfuse, 'The Future of Compulsory Dispute Settlement Under The Law of the Sea Convention' (n 112) 710; S. Rosenne, 'Establishing the ITLOS' (1995) 89 American Journal of International Law 806, 814; A. Yankov, 'The International Tribunal for the law of the Sea: Its Place Within the Dispute Settlement System of the UN Law of the Sea Convention' (1997) 37 Indiana Journal of International Law 356, 364.

[131] J. I. Charney, *Is International Law Threatened by Multiple International Tribunals?* (Recueil des Cours vol 271, Martinus Nijhoff Publishers 1998) 101–382 in particular 319–321, also presented in Caminos (n 128) 77.

encourage states to agree to at least one jurisdiction and also to bring cases in front of the institutions best positioned to deal with them, in terms of expertise for example.[132]

The risk of conflicting jurisprudence between ITLOS and the ICJ, in particular, cannot be ruled out since there are overlaps in jurisdiction *ratione materiae*, but should not be overestimated as the approach followed has been one of further development within the existing jurisprudential pool.[133] Hence, while the ICJ has played and will certainly keep playing an important role in the law of the sea field, the fears of "substantial inconsistencies and fragmentation in the area of the law of the sea have not materialized".[134] For example, the first delimitation case presented in front of ITLOS was envisioned as a major compatibility test[135] and could reassure the most skeptical commentators regarding the consideration ITLOS has for the ICJ. Indeed, the Tribunal made numerous references to the ICJ's jurisprudence,[136] followed the current delimitation method[137] and explicitly declared the importance of, and its respect for,

[132] Caminos (n 128) 79; R. Higgins, 'Respecting Sovereign States and Running a Tight Courtroom' (2001) 50 International and Comparative Law Quarterly 121, 122.

[133] In particular the references made in its *The M/V "Saiga" (Saint Vincent and the Grenadines v. Guinea), Judgment,* 1 July 1999, ITLOS case n°2 §120, 133, 170 to the *Case concerning certain German interests in Polish Upper Silesia (Germany v. Poland), Judgment,* 25 May 1926, Series A Publications of the Permanent Court of International Justice 1926 4; to the *Factory at Chorzów (Germany v. Poland), Judgment,* 13 September 1928, Series A Publications of the Permanent Court of International Justice 1928 4; and to the *Gabcikovo-Nagymaros Project (Hungary/Slovakia), Judgment,* 25 September 1997, I.C.J. Reports 1997 7; see Treves, 'Le Tribunal International du Droit de la Mer dans la pléiade des juridictions internationales' (n 110) 27 n 38.

[134] B. Sepúlveda-Amor, 'Contribution of the International Court of Justice to the development of the international law of the sea' (2010) 1 Aegean Review of the Law of the Sea and Maritime Law 5, 6.

[135] Treves, 'Le Tribunal International du Droit de la Mer dans la pléiade des juridictions internationales' (n 110) 28.

[136] In *Dispute concerning delimitation of the maritime boundary between Bangladesh and Myanmar in the Bay of Bengal (Bangladesh/Myanmar), Judgment,* 14 March 2012, ITLOS case n°16, reference was made for example at paragraph 90 to *Maritime Delimitation and Territorial Questions between Qatar and Bahrein (Qatar v Bahrain), Jurisdiction and Admissibility,* 1 July 1994, I.C.J. Reports 1994 112, 120 §23 in relation to the variety of forms an international agreement can take; at paragraph 264 to *Maritime Delimitation in the Black Sea (Romania v. Ukraine), Judgment,* 3 February 2009, I.C.J. Reports 2009 61, 108 §137 in relation to the determination of basepoints; and at paragraphs 294–295 to the *North Sea Continental Shelf Case (Federal Republic of Germany/Denmark; Federal Republic of Germany/Netherlands), Judgment,* 20 February 1969, I.C.J. Reports 1969 3, 17 §8 and 49 §89 in relation to the issues linked to concave/convex coasts.

[137] In *Dispute concerning delimitation of the maritime boundary between Bangladesh and Myanmar in the Bay of Bengal (Bangladesh/Myanmar)* (n 136) §240, the Tribunal decided to apply the three-stage method which was also used, for example in *Maritime Delimitation in the Black Sea (Romania v. Ukraine)* (n 136); see I. Papanicolopulu, 'From the North Sea to the Bay of Bengal: Maritime Delimitation at the International Tribunal for the Law of the Sea' (23 March 2012) http://www.ejiltalk.org/from-the-north-sea-to-the-bay-of-bengal-maritime-delimitation-at-the-international-tribunal-for-the-law-of-the-sea/ accessed 5 July 2015.

existing case-law.[138] In conclusion, while the limited number of cases at ITLOS impairs a clear answer on whether the proliferation of courts will lead to fragmentation of jurisprudence, it nevertheless seems "that ITLOS has endeavoured to secure consistency with the jurisprudence of the ICJ as well as the development of the rules of international law in general. Thus it would be wrong to lay too great an emphasis on the risk of the fragmentation of international law in the ITLOS jurisprudence."[139]

6. Conclusions

In conclusion, the different values and variety of goals existing in the field of marine biodiversity are two reasons to explain the multiplicity of institutions active in the field, which reflect states' numerous interests.[140] The proliferation of fora, rules and dispute settlement mechanisms may trigger serious problems: lack of consistency or even conflicting obligations, forum shopping, inefficiency or treaty congestion, also called 'treaty fatigue' as well as overlaps or protection gaps. A fragmented approach with specialized answers but without the necessary overview is not sufficient nor does it always offer an adequate solution to the issues faced by fisheries.[141]

III. Inadequate and Political Decisions

1. Problems with Scientific Data

a) Lack or Imperfection of Data

Fisheries management relies on data, which is highly complex as well as difficult and costly to gather; even for a single-species' management, information on location and abundance of stocks is needed in particular in relation to recruitment—recruitment shows "the changes in abundance from one generation to the next"[142]—, growth and natural mortality rates.[143] In practice, there is in many cases a lack of adequate data on nearly all of these aspects.[144] Indeed, even if states are required under a number of

---

[138] *Dispute concerning delimitation of the maritime boundary between Bangladesh and Myanmar in the Bay of Bengal (Bangladesh/Myanmar)* (n 136) §184, 226, 240.

[139] Tanaka, *The International Law of the Sea* (n 14) 421.

[140] M. Ivanova and J. Roy, 'The Architecture of Global Environmental Governance: Pros and Cons of Multiplicity' http://www.centerforunreform.org/node/251 accessed 5 July 2015, 3.

[141] Ibid. 3–5.

[142] Larkin (n 54) 142.

[143] C. Frid, 'The role of marine science in participatory fisheries governance' in T. S. Gray (ed), *Participation in Fisheries Governance* (Springer 2005) 233.

[144] Burke, *The New International Law of Fisheries* (n 15) 28–29; Longhurst (n 53) 70. Of the 1500 stocks commercially exploited, comprehensive knowledge about abundance exists only for about

treaties, such as the one founding IATTC or other RFMOs, to collect and share information[145] this does not appear to be sufficient.

Moreover, although obligations to take into account more than the target species are envisioned in several treaties, implementing a multi-species or even an ecosystem approach is very complex. Data not only on the target species but also on dependent species in the ecosystem is needed, as well as information on the physical environment in which the marine species live and on the various interactions at stake. Such scientific data and knowledge are incomplete as well as suffer from a high level of uncertainty in most if not all regions[146] and, consequently, do not permit to develop reliable predictions.[147]

Furthermore, random fluctuations influence stocks, while being by nature unforeseeable.[148] The level of catches is indeed not the only factor affecting their abundance, with stocks showing high levels of natural short term non-expectable variability and a long term reaction to environmental conditions.[149] However, such natural oscillations are usually not taken into account due to their complexity.[150]

All in all, biological uncertainty is high, which is quite typical for marine species:[151] the dynamics of the resources are relatively badly known and it is rather more complicated to observe marine species than land-based ones.[152] Even though scientific bodies such as RFMOs' scientific committees are created to analyze the data available, the understanding of the marine life remains full of knowledge gaps. Indeed, the truth is that little is understood about an environment which has not been fully explored.[153] In some cases, such as the exploitation of the deep-sea

---

500 (World Ocean Review, *The Future of Fish – The Fisheries of the Future* (Maribus, Future Ocean 2013) 43).

[145] G. Lugten, 'The role of international fishery organizations and other bodies in the conservation and management of living aquatic resources' FAO Fisheries and Aquaculture Circular 1054 (Rome 2010) 24–25.

[146] Henriksen, Hønneland and Sydnes (n 12) 28; Orellana, 'The Law on Highly Migratory Fish Stocks: ITLOS Jurisprudence in Context' (n 38) 486–487.

[147] Molenaar, 'Marine Mammals: The Role of Ethics and Ecosystem Considerations' (n 50) 46; on RFMOs' opinions on the question, see Lugten (n 145) 9.

[148] Dankel, Skagen and Ulltang (n 99) 203.

[149] Burke, *The New International Law of Fisheries* (n 15) 26–27; Cochrane, 'Reconciling sustainability, economic efficiency and equity in fisheries: the one that got away?' (n 101) 6.

[150] S. M. Garcia and K. L. Cochrane, 'Ecosystem approach to fisheries: a review of implementation guidelines' (2005) 62 ICES Journal of Marine Science 311, 312.

[151] Cochrane, 'Reconciling sustainability, economic efficiency and equity in fisheries: the one that got away?' (n 101) 6; K. L. Cochrane, 'Complexity in fisheries and limitations in the increasing complexity of fisheries management' (1999) 56 ICES Journal of Marine Science 917, 917.

[152] Cochrane, 'Complexity in fisheries and limitations in the increasing complexity of fisheries management' (n 151) 919; R. I. C. C. Francis and R. Shotton, "Risk' in fisheries management: a review' (1997) 54 Canadian Journal of Fisheries and Aquaculture Science 1699, 1701–1702.

[153] Orellana, 'The Law on Highly Migratory Fish Stocks: ITLOS Jurisprudence in Context' (n 38) 483.

orange roughy, it appears that fish stocks can be depleted even before they are studied sufficiently.[154]

It also happens that the data on stocks' abundance, which is used to make scientific recommendations, is not only imperfect but also already biased. Indeed, data comes from both scientific research and information from the fishing sector.[155] In reality, since not enough independent stocks' assessments can be made, much of the data used derives from reports of landings: stocks' abundance is usually measured on the basis of catch per unit effort,[156] in other words in extrapolating how many fish are in the stock on account of how many were caught per unit of effort to catch them.

However, capture statistics are not always accurate: data stemming from landing declarations is indeed known to be partial, uncertain if not outright false.[157] This comes from the nature of such information gathering, either in port or on the beach, from the incentives not to report correctly, from the data consolidation processes, and from the high likelihood that discards and real levels of fishing are not reported.[158] Also, IUU fishing is, by nature, off the record,[159] making official data incomplete: it is thus a threat to the accuracy of stock assessments, since fishing mortality is underestimated.[160] Moreover, other factors influence the fishing levels and the adequacy of assuming a direct relationship between the level of catches and the size of the stock has been questioned in the light of the lack of data on the evolution of natural versus fishery-caused mortality.[161]

Additionally, information provided by states is at risk of being voluntarily biased by the states themselves.[162] There are also occurrences of lack of transparency and where reports meant to be transmitted to the RFMOs were withdrawn based on a decision to protect political sensitivities.[163]

---

[154] Freestone, 'Problems of High Seas Governance' (n 75) 100.

[155] G. Hubold, 'Fishery and Sustainability' in P. Ehlers, E. Mann-Borgese and R. Wolfrum (eds), *Marine Issues: From a Scientific, Political and Legal Perspective* (Kluwer Law International 2002) 191; World Ocean Review, *The Future of Fish – The Fisheries of the Future* 2013 (n 144) 42–43.

[156] Cochrane, 'Reconciling sustainability, economic efficiency and equity in fisheries: the one that got away?' (n 101) 6.

[157] Larkin (n 54) 145; R. Watson, U. R. Sumaila and D. Zeller, 'How much fish is being extracted from the oceans and what is it worth?' in V. Christensen and J. MacLean (eds), *Ecosystem Approaches to Fisheries: A Global Perspective* (Cambridge University Press 2011) 57–60.

[158] Longhurst (n 53) 161–163, 169.

[159] Gjerde, 'High-seas Fisheries Management under the Convention on the Law of the Sea' (n 74) 290; on the similar problems in estimating the importance of flags of convenience in fishing, see Calley (n 48) 50.

[160] Lack, 'Catching On?' (n 42) 4; Longhurst (n 53) 175.

[161] Longhurst (n 53) 177.

[162] W. T. Burke, 'Unregulated High Seas Fishing and Ocean Governance' in J. M. van Dyke, D. Zaelke and G. Hewison (eds), *Freedom for the Seas in the 21st Century: Ocean Governance and Environmental Harmony* (Island Press 1993) 244.

[163] T. Polacheck, 'Politics and independent scientific advice in RFMO processes: A case study of crossing boundaries' (2012) 36 Marine Policy 132.

b) Unclear Scientific Advice

The availability and quality of the data is a first cornerstone to science-based decision-making and it has been demonstrated above that the situation is not an optimal one. However, a second crucial step presents some further difficulties. Indeed, the data has to be analyzed and transformed into scientific advice with regard to the status of the stock and the level of catches it can sustain. This is usually undertaken by the RFMO's scientific body, or in cases of externalization, by a provider such as ICES. Advice is based on models which try to predict the future of a stock and the related sustainable levels of fishing, using the scientific data available.

Since they are based on an imperfect and simplified knowledge of natural processes, these models themselves are often uncertain.[164] Hence, they leave space for mistakes and for interpretation, which is for example illustrated by the fact that predictions for a particular species seem to differ depending on whether a single-species or multi-species model is used.[165] Moreover, while making predictions for a stock's abundance is already difficult, a multi-species model requires taking into account the interactions between the stocks, and an ecosystem-based model requires the inclusion of the interactions with the physical environment. Currently, this is technically too challenging to be put in practice[166] and any tentative ecosystem-based model is very simplified.[167]

Hence, a strong criticism towards the models used is that they do not take into consideration the physical components which influence fish stocks, such as currents, climate change, pollution as well as other relevant natural causes.[168] The effects of changing weather patterns and physical environment were recognized as of significance quite late in time and many fish stocks are still managed as if they were living in an invariant ocean.[169] However it seems that such influences can have a major role in fish stocks' abundance, recovery or collapse, such as with the North West Atlantic cod.[170] Due to the unpredictability of ecosystem structures, introducing natural causes into models is complex and it has even been claimed that "to stimulate the consequences of different levels of fishing mortality on a population of fish in its natural habitat – in order to take rational management decisions – remains out of reach."[171] It even appears that some processes, for example that of a stock collapse, are not understood well enough for a model to be developed.[172]

---

[164] Cochrane, 'Complexity in fisheries and limitations in the increasing complexity of fisheries management' (n 151) 919.

[165] Frid (n 143) 234; C. Frid, O. Paramor and C. Scott, 'Ecosystem-based fisheries management: progress in the NE Atlantic' (2005) 29 Marine Policy 461, 463.

[166] Frid (n 143) 244.

[167] Longhurst (n 53) 73–74.

[168] Ibid. 10–11.

[169] Ibid. 79–80.

[170] Ibid. 148–149.

[171] Ibid. 75.

[172] Cochrane, 'Complexity in fisheries and limitations in the increasing complexity of fisheries management' (n 151) 921.

A phenomenon known as the shifting of baselines further interferes with the provision of adequate advice.[173] Indeed, it is difficult to determine the baseline—that is the status of the fisheries before human intervention occurred—and, generation after generation, fisheries managers seem to modify it to the level they first came across. What they consider as a 'pristine' population, in terms of size and abundance, is hence subjective. This greatly changes perceptions of stocks' health.[174]

It is also known that scientists have in some cases been pressured into conclusions and recommendations which have little to do with science.[175] With regard to the Canadian cod collapse, discussions have arisen about "assessments being deliberately slanted to favour particular value judgment about appropriate risk."[176] It indeed seems that some scientists were forced to withdraw their reports if conclusions were not suitable to the political will.[177]

Even more frequent is the use of uncertainty as a justification to avoid making clear (unpopular) recommendations.[178] For reasons not necessary to discuss here but which can rank from personal agenda or institutional membership to financing of research projects, some scientists might decide to moderate their conclusions and use uncertainty as an excuse.

[173] D. Pauly, 'Anecdotes and the shifting baseline syndrome of fisheries' (1995) 10 TRENDS in Ecology and Evolution 430; J. Jackson and J. Jacquet, 'The shifting baselines syndrome: perception, deception, and the future of our oceans' in V. Christensen and J. MacLean (eds), *Ecosystem Approaches to Fisheries: A Global Perspective* (Cambridge University Press 2011).

[174] Longhurst (n 53) 124.

[175] For example this has allegedly happened in relation to the TACs recommended for South Africa (Cochrane, 'Reconciling sustainability, economic efficiency and equity in fisheries: the one that got away?' (n 101) 12, on K. L. Cochrane and others, 'Management procedures in a fishery based on highly variable stocks and with conflicting objectives: experiences in the South African pelagic fishery' (1998) 8 Reviews of Fish Biology and Fisheries 177); or in Canada with regard to Atlantic cod (J. A. Hutchings, C. Walters and Haedrich R. L. 'Is scientific enquiry incompatible with government information control?' (1997) 54 Canadian Journal of Fisheries and Aquatic Sciences 1198).

[176] Cochrane, 'Complexity in fisheries and limitations in the increasing complexity of fisheries management' (n 151) 921, quoting C. Walters and J.-J. Maguire, 'Lessons for stock assessment from the northern cod collapse' (1996) 6 Reviews of Fish Biology and Fisheries 125, 135.

[177] Longhurst (n 53) 142, 147.

[178] Cochrane, 'Reconciling sustainability, economic efficiency and equity in fisheries: the one that got away?' (n 101) 12; E. Masood, 'Scientific Caution blunts efforts to conserve fish stocks' (1996) 379 Nature 481, 481.

2. Inadequate Decision-Making

a) Decisions Taken Not Following Scientific Advice

Fisheries management is doubtlessly complex, entailing a high level of uncertainty and thus advice is based on predictions, but this situation is only aggravated by the fact that such problems have been used by decision-makers as an excuse to continue with unsustainable practices.[179] Usually, science is only able to determine that a stock may be endangered, while the fishing industry will be clear that it must have high quotas. In such a situation, the certainty of 'must' usually takes over the possibility of 'may'.[180] Indeed, "commercial 'imperatives' almost inevitably overwhelm uncertain science."[181] For example, in the case of management of the orange roughy, the scientific uncertainty was used to set and keep TACs at high levels, too high, as can be observed by the subsequent decline.[182] In cases of uncertainty, a precautionary approach to fisheries management was late to appear and is indeed still unequally present in practice, partly because fish species have been perceived more as commodities than as wildlife.[183]

Even when data is available and scientific advice is clear decisions are taken, in a number of cases, in partial or full disregard of such guidance.[184] Indeed, since many other interests than the ecological one are at stake, scientific advice is not always followed; in the face of conflicting objectives, sustainability is not necessarily the highest on the list, usually with socio-economic objectives dominating the discussion.[185]

Legally speaking, UNCLOS already makes clear that TACs should be determined on the basis of the best scientific information available—although in the EEZ, the coastal state is only obliged to 'take into account' this information.[186] This

[179] Cochrane, 'Reconciling sustainability, economic efficiency and equity in fisheries: the one that got away?' (n 101) 9.

[180] W. E. Schrank, 'Is there any hope for fisheries management?' (2007) 31 Marine Policy 299, 304.

[181] J. Montgomery and L. Carter, 'Marine Science in the Past 25 Years: Main Findings and Trends' in D. Vidas and P. J. Schei (eds), *The World Ocean in Globalisation: Climate Change, Sustainable Fisheries, Biodiversity, Shipping, Regional Issues* (Martinus Nijhoff Publishers 2011) 47.

[182] Gjerde, 'High-seas Fisheries Management under the Convention on the Law of the Sea' (n 74) 284–285; Montgomery and Carter (n 181) 47–48.

[183] S. Fordham and C. Dolan, 'A Case Study in International Shark Conservation: The Convention on International Trade in Endangered Species and the Spiny Dogfish' (2004) 34 Golden Gate University Law Review 531, 540.

[184] R. R. Churchill, 'Legal uncertainties in international high seas fisheries management' (1998) 37 Fisheries Research 225, 226.

[185] Cochrane, 'Complexity in fisheries and limitations in the increasing complexity of fisheries management' (n 151) 917; Frid (n 143) 238.

[186] Respectively for the EEZ and the high seas UNCLOS (n 11) articles 61(2), 119; see Burke, *The New International Law of Fisheries* (n 15) 56; Freestone, 'Problems of High Seas Governance' (n 75) 123.

is reiterated by the UNFSA in slightly more binding terms: "coastal States and States fishing on the high seas shall [. . .] ensure that such measures are based on the best scientific evidence available".[187]

However, in practice, the relevant legal obligation is not always to base measures on scientific evidence but to take it into account.[188] Moreover, it is often further qualified, making reference only to the 'scientific evidence *available*'.[189] This illustrates states' willingness to retain leeway in their decision-making and is evidently a shortcoming for sustainable management.

Many institutions involved in fisheries management are examples of this problem.[190] For example, the records of ICCAT are unfortunately indisputable on the matter: until recently, decisions taken in setting the TACs for some species such as the Atlantic Bluefin Tuna were not respecting the advice, even of their own

---

[187] UNFSA (n 37) article 5(b).

[188] Some RFMOs have weaker wording than others: GFCM and NAFO refer to the need to 'take into account' respectively the best scientific evidence available and scientific information and advice ('Agreement for the establishment of the General Fisheries Commission for the Mediterranean' (GFCM Agreement) 1949, 126 *United Nations Treaty Series* 237 article III(2); 'Convention on Future Multilateral Cooperation in the Northwest Atlantic Fisheries' (NAFO Convention) 1978, 1135 *United Nations Treaty Series* 369 article XI(2)) while CCAMLR "shall formulate, adopt and revise conservation measures on the basis of the best scientific evidence available" ('Convention on the Conservation of Antarctic Marine Living Resources' (CCAMLR Convention) 1980, 1329 *United Nations Treaty Series* 47 article IX(1)(f)); IATTC shall "adopt measures that are based on the best scientific evidence available" ('Convention for the Strengthening of the Inter-American Tropical Tuna Commission Established by the 1949 Convention Between the United States of America and the Republic of Costa Rica' (Antigua Convention) 2003 article VII(1)(c)); IOTC has "to adopt [. . .] on the basis of scientific evidence, conservation and management measures" ('Agreement for the Establishment of the Indian Ocean Tuna Commission' (IOTC Agreement) 1993, 1927 *United Nations Treaty Series* 329 article V(2)(c)); NEAFC shall "ensure that such recommendations are based on the best scientific advice" ('Convention on the Future Multilateral Cooperation in North-East Atlantic Fisheries' (NEAFC Convention) 1980, 1285 *United Nations Treaty Series* 129 (revised 2006, 'New Convention' http://www.neafc.org/system/files/Text-of-NEAFC-Convention-04.pdf accessed 4 July 2015)) article 4(2)(a)); SEAFO "shall [. . .] adopt measures, based on the best scientific evidence available" ('Convention on the Conservation and Management of Fishery Resources in the South East Atlantic Ocean' (SEAFO Convention) 2001, 2221 *United Nations Treaty Series* 189 article 3(a)); SIOFA's measures "shall be adopted on the basis of the best scientific evidence available to ensure the long-term conservation of fishery resources" ('Southern Indian Ocean Fisheries Agreement' (SIOFA Agreement) 2006 article 4(a)); SPRFMO's "decisions shall be based on the best scientific and technical information available" ('Convention on the Conservation and Management of High Seas Fishery Resources in the South Pacific Ocean' (SPRFMO Convention) 2009 article 3(1)(a)); and WCPFC "shall [. . .] ensure that [. . .] measures are based on the best scientific evidence available" ('Convention on the Conservation and Management of Highly Migratory Fish Stocks in the Western and Central Pacific Ocean' (WCPFC Convention) 2000, 2275 *United Nations Treaty Series* 43 article 5(b)). See also McDorman, 'Implementing Existing Tools: Turning Words Into Actions – Decision-Making Processes of Regional Fisheries Management Organisations (RFMOs)' (n 37) 434–435.

[189] Burke, *The New International Law of Fisheries* (n 15) 56, 115–116.

[190] Bowman, Davies and Redgwell (n 67) 146.

scientific body, and resulted in clear mismanagement.[191] It is also worth noting that mismanagement is as much a problem of fishing in the high seas as in waters under national jurisdiction.[192]

At the other extreme, the decision-making at the IWC also has flaws in its relation to science. It is indeed ideologically biased pro-preservation and scientific advice according to which some whale species (in particular the minke whale) could sustain a limited, but nevertheless above zero level of harvesting, appears to be disregarded.[193]

b) Lack of Rules Binding on All

A major issue is that, even when a CMM has been adopted within an RFMO, the possibility exists for members to opt out.[194] Such an exit door allows free-riders from the inside to evade obligations. This practice has not been explicitly outlawed, even by the UNFSA.[195] It creates some uncertainty as to the measures' actual scope of application.[196] Indeed, if a TAC is adopted but then several fishing nations use their right to opt out, the scientific soundness of the level of fishing agreed upon is unlikely to remain valid. The opt-out right is used extensively in most RFMOs, for example by Iceland and Russia, within NEAFC, for mackerel.[197]

To add to that, none of the regimes presented here is truly universal.[198] That they are not binding on non-parties is a normal characteristic of international law,[199] but this becomes problematic when dealing with endangered global commons outside of national jurisdiction. "The combination of two basic principles of international law – the freedom of the high seas and the principle of consent – means that a State

[191] Anonymous, 'Report of the Independent Performance Review of ICCAT' (Madrid 2009), Executive summary; Monaco, 'CoP15 Inf. 12 – Supplementary information on the proposal to include Atlantic Bluefin Tuna (*Thunnus Thynnus* Linnaeus 1758) on Appendix I of CITES in accordance with article II 1 of the Convention' (13–25 March 2010) 5–6; S. Lieberman, 'Science versus Politics: Tales from CITES' (Elisabeth Mann Borgese Lecture, 2010) http://internationaloceaninstitute.dal.ca/EMBlecturetext_2010.pdf accessed 5 July 2015, 5.

[192] G. van Bohemen, 'High Seas Fisheries Management: Reflections on Experience with Regional Fisheries Management Organisations in the South Pacific' in D. Vidas and P. J. Schei (eds), *The World Ocean in Globalisation: Climate Change, Sustainable Fisheries, Biodiversity, Shipping, Regional Issues* (Martinus Nijhoff Publishers 2011) 234.

[193] S. Ásmundsson, 'Whaling' in M. H. Nordquist and others (eds), *Law, Science & Ocean Management* (Martinus Nijhoff Publishers 2007) 459, 461, 467.

[194] See supra Chapter 3 B. II. 1.

[195] Oerebech, Sigurjonsson and McDorman (n 37) 125–126.

[196] Churchill, 'Legal uncertainties in international high seas fisheries management' (n 184) 227.

[197] NEAFC Convention (n 188) article 12(2); Henriksen, Hønneland and Sydnes (n 12) 105.

[198] Ziemer (n 45) 78–79.

[199] 'Vienna Convention on the Law of Treaties' (VCLT) 1969, 1155 *United Nations Treaty Series* 331 article 34.

can fish on the high seas for stocks regulated by a regional fisheries commission without being obliged to become a member of that commission."[200]

In practice, in certain RFMOs, fishing by non-contracting parties is considered to undermine CMMs' effectiveness, unless proven otherwise.[201] It must be stressed that access to the fisheries is thus denied to non-members,[202] which is questionable as a matter of law, since non-contracting parties are, under the *pacta tertiis* rule, not bound by CMMs[203]—except if the UNFSA's content is binding on all states through its (debated) CIL status.[204]

Under these conditions, free-riding is tempting. Indeed, the free-riders benefit from any measure of restraint taken by the members since conservation of stocks is improved, through the sacrifices of others: this creates a win-win situation for the free-rider and a loose-loose situation for the complying states.[205] Hence, the incentive is strong for states fishing in an area not to join an RFMO, to leave it or to opt out of the CMMs. Also, it brings about uncertainty as to the measures to adopt and, in particular, the quotas to set, since the level of fishing by free-riders is not known or predictable.[206]

However, even for those states who would wish to join an RFMO for example to respect their obligations under the UNFSA, it is not always an easy task. This was a difficulty that Iceland faced when it came to the Barents Sea Loophole and Russia and Norway refused at first to negotiate.[207] To define the 'real interest in the fisheries concerned' which is referred to in the UNFSA as the condition to be allowed membership in the fishery is complex—and it is evident that to reserve membership to states with an interest in the fishery intends to keep non-fishing states, which are possibly more interested in environmentally based decisions, away from the RFMO.[208]

---

[200] Churchill, 'Legal uncertainties in international high seas fisheries management' (n 184) 229 (parentheses omitted); see also A. Gillespie, *Conservation, biodiversity and international law* (Edward Elgar 2011) 445–447.

[201] See supra Chapter 3 B. II. 3.

[202] See an explicit example thereof in the NEAFC Fishing Licence Guide http://neafc.org/fishinglicence accessed 5 July 2015.

[203] Tanaka, *The International Law of the Sea* (n 14) 246–247.

[204] See supra Chapter 3 A. II. 3. b) and 4.

[205] M. Hayashi, 'Regional Fisheries Management Organisations and Non-Members' in T. M. Ndiaye and R. Wolfrum (eds), *Law of the sea, environmental law and settlement of disputes: Liber amicorum Judge Thomas A. Mensah* (Nijhoff 2007) 751–752; Warner (n 43) 126.

[206] Churchill, 'Legal uncertainties in international high seas fisheries management' (n 184) 229–230.

[207] O. S. Stokke, 'Governance of High Seas Fisheries: The Role of Regime Linkages' in D. Vidas and W. Ostreng (eds), *Order for the oceans at the turn of the century* (Kluwer Law International 1999) 165.

[208] UNFSA (n 37) article 11; Henriksen, Hønneland and Sydnes (n 12) 19–20; on conditions of entry into RFMOs, see supra Chapter 3 B. II. 1.

Finding a balance between keeping a stable level of members in order to have effective conservation measures implemented and keeping the RFMO open enough in order to avoid the incentive to free-ride is a major difficulty.[209] NEAFC for example has a system where 75 % of the parties must agree to the entrance of a new member;[210] this semi-closed system can be seen as problematic, particularly since decision-making is not transparent. However, as the status of cooperating non-contracting party exists, with possibilities for them to be attributed fishing quotas, it is argued that even such a system respects the UNFSA.[211]

3. Conclusions

One of the main governance challenges identified in resource management is linked to the "generation of adequate and reasonably consensual *scientific knowledge* to permit informed judgments about whether and how exploitation of resources shall be conducted".[212] Some RFMOs have good records in their scientific advice (such as CCAMLR), while others have experienced rather problematic overestimation of stocks (such as NAFO and the cod disaster).[213]

It seems that, often, the data can be erroneous, or partly or fully missing. In any case, it is then inadequate to serve as the basis for scientifically sound measures, at least regarding decisions taken at the margin, such as the MSY.[214]

Not to have robust independent scientific advice on which a political decision can be based is highly problematic. Indeed, if this happens too often, it puts the validity of scientific recommendations into question. Also, it creates a sort of triple-blind, where input data and advice are already politically motivated, but then tend to be treated as science when policy makers negotiate on the political level—with misconceptions as to the risks appurtenant to several management options.[215]

Another, at least equally serious governance issue is related to the difficulty to adopt "legitimate and appropriate regulatory measures"[216] and the impossibility to

---

[209] Henriksen, Hønneland and Sydnes (n 12) 21–22.

[210] NEAFC Convention (n 188) article 20(4); Molenaar, 'Participation, Allocation and Unregulated Fishing: The Practice of Regional Fisheries Management Organisations' (n 26) 465.

[211] Henriksen, Hønneland and Sydnes (n 12) 103–104.

[212] Stokke, 'Governance of High Seas Fisheries: The Role of Regime Linkages' (n 207) 159 (italics in the original text).

[213] Schrank (n 180) 303; Stokke, 'Governance of High Seas Fisheries: The Role of Regime Linkages' (n 207) 160.

[214] Schrank (n 180) 301. The MSY is traditionally based on TACs decided at the upper limit of estimated sustainability, leaving no margin for unforeseen natural cycles, which are proven to exist – this concept has however evolved, see supra footnote 65 of chapter 3.

[215] Polacheck, 'Politics and independent scientific advice in RFMO processes: A case study of crossing boundaries' (n 163) 139.

[216] Stokke, 'Governance of High Seas Fisheries: The Role of Regime Linkages' (n 207) 159 (italics omitted); see also California Environmental Associates, 'Charting a Course to Sustainable Fisheries' (16 January 2012) http://www.chartingacourse.org/downloads/ accessed 4 July 2015, 24.

have fully inclusive regimes where all states involved in the fisheries are bound by the adopted measures.

IV. Insufficient Implementation and Enforcement

Compliance mechanisms are necessary to verify that substantive rules are implemented successfully by vessels and by states. Implementation refers to measures that states can or should take in order to secure the respect, by their vessels, of national laws and regulations stemming from international obligations. Enforcement refers to mechanisms which can be taken at the international level to hold a state accountable for its non-compliance with international obligations and ultimately to promote compliance.[217] In the context of fisheries, measures taken by states other than the flag state when that state is unwilling or unable to control its vessels include both implementation (e.g. boarding and inspection) and enforcement (e.g. sanctions against states), although they are often, in a generic manner, referred to as 'non-flag state enforcement measures'.

In that field of law, both aspects suffer from a number of major deficiencies which, although they involve different main actors, are of equal gravity in the waters under national jurisdiction and in the high seas.[218] Low levels of compliance of vessels are indeed often coupled with the existence of a number of coastal and flag states who are unwilling or unable to efficiently control the boats, while not being subjected to an international review of performance (1). Moreover, remedies and sanctions against both states and RFMOs are extremely limited (2). Finally, dispute settlement mechanisms suffer from restrictions in relation to fisheries, and recourse to international courts appears not to be simple or adequate for the protection of global commons (3).

1. Low Levels of Compliance and Verification of Implementation Measures

a) States' Difficulties in Controlling their Areas and Vessels

Surveillance, control, boarding and inspection, arrest and judicial sanctions against a vessel or people are measures securing the implementation of CMMs and/or the sanction of their violations.[219] According to the *lex lata*, the coastal state is responsible for ensuring compliance with its CMMs in the territorial sea and in

---

[217] Definition inspired by the 'international enforcement' to which R. Reeve (*Policing International Trade in Endangered Species* (Earthscan 2002) 17) makes reference.

[218] Burke, *The New International Law of Fisheries* (n 15) 308–310, 345–346.

[219] Ibid. 303.

the EEZ, while the flag state is, except if an agreement explicitly provides to the contrary, the only one allowed to board and to inspect a vessel flying its flag on the high seas.

In the EEZ, coastal states are entitled to a wide array of measures in order to ensure compliance, although imprisonment or corporal punishment are prohibited and an arrested vessel and its crew must be released promptly in exchange of a reasonable bond.[220] One must note that ITLOS, in prompt release cases, has not accepted to broaden the factors involved in its assessment of the reasonableness of a bond in a way that would really take into account the level of international concerns about IUU fishing, nor has it decided to include in the bond a payment meant to prevent future violations.[221] In practice, it can be expensive and difficult for the coastal state to control vessels' activities in its EEZ. Indeed, this area is vast and countries' resources are limited.[222] Additionally, some means of control are controversial such as forbidding unlicensed vessels from transiting through states' EEZ or forcing the use of particular sea lanes by certain vessels, since such measures go against the freedom of navigation principle.[223]

The size of the high seas also makes surveillance, this time by the flag state, a complicated matter. Indeed, governmental vessels are limited in numbers and have several other tasks in addition to verifying fishing vessels' compliance with the relevant CMMs. That fishing often is a decentralized practice, with smaller boats than for shipping and with no formal IMO number requirement,[224] only makes the situation worse. It also seems that obligations towards vessels of less than 24 m are unclear: the Compliance Agreement includes this size limitation with regard to the obligation to have vessels on records and to grant authorizations to fish.

In addition to the inability of some states to control their vessels, some states are not willing to do so. Indeed, the largely recognized problem of flags of convenience is important in the fishing sector.[225] Flag state obligations are detailed in the

---

220 UNCLOS (n 11) article 73.

221 *The "Volga" Case (Russian Federation v. Australia), Judgment on prompt release*, 23 December 2002, ITLOS case n°11 §68–69, 80; D. R. Rothwell and T. Stephens, 'Illegal Southern Ocean Fishing and Prompt Release: Balancing Coastal and Flag State Rights and Interests' (2004) 53 International and Comparative Law Quarterly 171, 182–184.

222 Burke, *The New International Law of Fisheries* (n 15) 305–309.

223 Ibid. 322–328.

224 S. Flothmann and others, 'Closing Loopholes: Getting Illegal Fishing Under Control' (2010) 328 Science 1235, 1236.

225 S. Beslier, 'The exercise of jurisdiction over vessels: new developments in the sector of fisheries' in E. Franckx and P. Gautier (eds), *The Exercise of Jurisdiction over Vessels: New Developments in the Fields of Pollution, Fisheries, Crimes at Sea and Trafficking of Weapons of Mass Destruction* (Bruylant 2010) 48–50; Calley (n 48) chapter two (in particular he lists some of the main flags of convenience states at 21); Gillespie, *Conservation, biodiversity and international law* (n 200) 448; H. Slim, 'Les pavillons de complaisance' (Le pavillon – Colloque international, Institut Océanographique de Paris, 2–3 March 2007 (Pedone 2008)) 100–103; J. Swan, 'Fishing Vessels operating under open registers and the exercise of flag State responsibilities: Information and options' FAO Fisheries Circular 980 (Rome 2002) 2–4; Tanaka, *The International Law of the Sea* (n 14) 242;

Compliance Agreement and in other instruments, but flags of non-compliance will most certainly not be party to them[226] and this is one of the reasons for which vessels are flagged there.[227] Such free-riding attitude is not only a threat to the sustainability of any CMMs adopted, but also to their legitimacy.

In addition to systems based on the flag state's competence (vessel monitoring system, log books examination), some non-flag state mechanisms, mainly inspections, IUU lists and port state measures,[228] have been developed to detect non-compliance by vessels and to fight this paradoxical situation where the flag state enjoys primary control, but appears to have no strong obligation. Since port state measures usually include an aspect of sanctions, they will be examined here below.[229] As for non-flag state boarding and inspection schemes on the high seas, they remain based on consent by the flag state.[230] Non-compliant states are unlikely to agree to these schemes, developed in the UNFSA, or in RFMOs such as CCMALR or NAFO.

Finally, the generally low probability to be apprehended and the limited range of sanctions at the national level are not sufficiently deterrent compared to the high potential profits linked to IUU fishing.[231] However, in some cases, the main example being the Lacey Act in the USA, sanctions are envisioned under pieces of legislation, which allow the prosecution of anyone in the country who conducted fishing activities in violation of any fishery-related legislation.[232]

D. Warner-Kramer, 'Control Begins at Home: Tackling Flags of Convenience and IUU Fishing' (2004) 34 Golden Gate University Law Review 497, 497–498; Watson, Sumaila and Zeller (n 157) 64; A. Yankov, 'Reflagging of fishing vessels: a critical assessment of its impact on the enforcement of fishing regulations and the responses thereto' in P. Ehlers, E. Mann-Borgese and R. Wolfrum (eds), *Marine Issues: From a Scientific, Political and Legal Perspective* (Kluwer Law International 2002) 195.

The wordings flag of convenience and flag of non-compliance (the latter having been coined in 'Flags of non-compliance' CCAMLR Resolution 19/XXI, 2002) are differentiated by some commentators in that the latter does not have an open registry but is known to be deficient in controlling its vessels, respecting CMMs and/or fighting IUU fishing (Calley (n 48) 17; T. Lobach, 'Port State control of foreign fishing vessels' FAO Fisheries Circular 987 (Rome 2003) 8; Environmental Justice Foundation, 'Lowering the Flag – Ending the use of Flags of Convenience by Pirate Fishing Vessels' Vessels' (2009) http://ejfoundation.org/sites/default/files/public/Lowering%20the%20flag.pdf accessed 6 July 2015, 11).

[226] Tanaka, *The International Law of the Sea* (n 14) 243–244; Warner-Kramer (n 225) 501.

[227] Calley (n 48) 11. This problem was already identified in the early twentieth century with regard to the regulation of whaling (Jessup (n 9) 502).

[228] Burke, *The New International Law of Fisheries* (n 15) 339.

[229] See infra Chapter 4 A. IV. 2. b).

[230] Calley (n 48) 80–82; D. Guilfoyle, *Shipping Interdiction and the Law of the Sea* (Cambridge Studies in International and Comparative Law, Cambridge University Press 2009) 168–169.

[231] B. Le Gallic and A. Cox, 'An economic analysis of illegal, unreported and unregulated (IUU) fishing: Key drivers and possible solutions' (2006) 30 Marine Policy 689, 690–691; U. R. Sumaila, J. Adler and H. Keith, 'Global scope and economics of illegal fishing' (2006) 30 Marine Policy 696, 697–701.

[232] E. M. McOmber, 'Problems in Enforcement of the Convention on International Trade in Endangered Species' (2001–2002) 27 Brooklyn Journal of International Law 673, 699; S. Patel,

b) Lack of Overview on States' Compliance

Verification of states' compliance is crucial to any functioning regime. In the fisheries field, pure free-riding is a major issue, but non-compliance with international obligations is probably as problematic. Indeed, an important part of IUU fishing is undertaken by RFMO members themselves.[233]

Since coastal and flag states are and remain the main actors and an important share of information is based on their reports, surveillance, monitoring and sanctioning can, in reality, prove difficult. Indeed, if they are not willing to implement the measures of the relevant RFMO, information-gathering to examine their compliance will certainly be insufficient too.

Also, unlike in some other treaty regimes, states' compliance with fisheries measures is not reviewed by a single authority. Even though reporting procedures exist within RFMOs and reports are then usually checked by a committee within the RFMO, the lack of a global control and review mechanism leaves monitoring fragmented.

Holding flag states accountable for their failure to respect their international obligations is necessary. However, this would require a clear catalogue of flag state obligations, which does not exist at present in a binding form;[234] the violation of a CMM by a vessel is indeed not to be attributed to the flag state—the flag state only has to exercise sufficient control over its vessels to implement its international obligations. This was fully clarified by ITLOS who, in the advisory opinion it issued on 2 April 2015, following a request submitted by the Sub-Regional Fisheries Commission (SRFC), commented on the nature of flag state obligations in relation to IUU fishing in third countries' EEZs, while remaining quite in the abstract. In the Tribunal's words, to fulfil its duty, the flag state has "to take necessary measures, including those of enforcement, to ensure compliance by vessels flying its flag with the laws and regulations" of the relevant coastal state in its EEZ. ITLOS characterized those obligations as of 'due diligence' and made perfectly clear that a flag state is not liable for violations of measures by its vessels, but for its "failure to comply with its "due diligence" obligations".[235]

Also, in the fight against flags of convenience, the requirement of the 'genuine link' has not been fully clarified: although the efficiency of the relationship between the flag state and its vessels is key—in that the flag state must be able to implement

---

'The Convention on International Trade in Endangered Species and the Last Unicorn' (1995–1996) 18 Houston Journal of International Law 157, 178–180. Only few countries have similar procedures available (Riddle (n 48) 272).

[233] For GFCM for example, see Lack, 'Catching On?' (n 42) 13–14; Lugten (n 145) 8.

[234] The obligations are indeed included in several instruments to which not all states are parties. They were usefully compiled in Consultative Group on Flag State Implementation, 'Report of the Secretary General', 5 March 2004, A/59/63.

[235] *Request for an Advisory Opinion Submitted by the Sub-Regional Fisheries Commission (SRFC), Advisory Opinion*, 2 April 2015, ITLOS case n°21 60–61. For more on this advisory opinion, see infra Chapter 4 A. IV. 3. c) aa).

its duties towards vessels flying its flag—it has been recognized that no general criteria to determine the genuine link could be authoritatively established.[236]

An overview mechanism, the flag state performance review, has been recently developed. Standards and procedures to this end were discussed in two fora, the FAO and the UN headquarters.[237] The establishment of a review process made slow progress at the FAO,[238] until the adoption of voluntary guidelines in early 2013 and their endorsement in 2014.[239] This is a positive step, but this document is quite general, restates the most accepted obligations of flag states and, most problematic of all, does not provide for a procedure against states that do not want to undertake a performance review.[240] Hence, although flag states should theoretically not allow their vessels to fish on the high seas if they are unable to control them,[241] the adopted

[236] *The M/V "Saiga" (Saint Vincent and the Grenadines v. Guinea)* (n 133) §83; P. Gautier, 'Comments on flag state's responsibility, enforcement measures against IUU fishing activities, and the settlement of disputes' in E. Franckx and P. Gautier (eds), *The Exercise of Jurisdiction over Vessels: New Developments in the Fields of Pollution, Fisheries, Crimes at Sea and Trafficking of Weapons of Mass Destruction* (Bruylant 2010) 83–84; UNGA, 'Report of the Ad Hoc Consultative Meeting of Senior Representatives of International Organizations on the "Genuine Link"' 17 July 2006, A/61/160, 5–6 §10–11; D. König, 'Flags of Convenience' Max Planck Encyclopedia of Public International Law www.mpepil.com §9–10. On the evolution of the requirement of genuine link, see Calley (n 48) 84–101; D. König, 'Flag of Ships' Max Planck Encyclopedia of Public International Law www.mpepil.com §6–15.

[237] Beslier, 'The exercise of jurisdiction over vessels: new developments in the sector of fisheries' (n 225) 52. Such discussion was undertaken at the FAO following the invitation of the UNGA to do so (UNGA, Resolution adopted by the General Assembly on 18 December 2007 – Sustainable fisheries, including through the 1995 Agreement for the Implementation of the Provisions of the United Nations Convention on the Law of the Sea of 10 December 1982 relating to the Conservation and Management of Straddling Fish Stocks and Highly Migratory Fish Stocks, and related instruments 28 February 2008, A/RES/62/177 §41).

[238] First mentioned in Committee on Fisheries, 'Report of the Twenty-seventh Session of the Committee on Fisheries (Rome, 5–9 March 2007)' FAO Fisheries Report 830 (Rome 2007) 11 §71; Freestone, 'Problems of High Seas Governance' (n 75) 117–118.

[239] FAO, 'Adopted Voluntary Guidelines for Flag State Performance', 8 February 2013 ftp://ftp.fao.org/FI/DOCUMENT/tc-fsp/2013/VolGuidelines_adopted.pdf accessed 5 July 2015; Committee on Fisheries, 'The Voluntary Guidelines for Flag State Performance, the 2009 FAO Agreement on Port State Measures and other Instruments Combating IUU Fishing', presented at the Thirty-first Session (Rome, 9–13 June 2014) COFI/2014/4.2/Rev.1; Committee on Fisheries, 'Report of the 31st Session of the Committee on Fisheries (Rome, 9–13 June 2014)' C 2015/23, 9–10 §37.

[240] In terms of comparison, it is worth noting that the IMO has established a Voluntary IMO Member State Audit Scheme which will become compulsory as of 2015 ('Further development of the voluntary IMO member state audit scheme (adopted on 25 November 2009)' 18 January 2010, Resolution A.1018(26) (IMO)). This scheme verifies compliance with a set number of IMO instruments. See IMO, 'Voluntary IMO Member State Audit Scheme' http://www.imo.org/blast/mainframe.asp?topic_id=841 accessed 6 July 2015; L. D. Barchue, 'Making a case for the Voluntary IMO Member State Audit Scheme: Paper delivered at a seminar on 'Auditing Flag States: New Directions for Smaller States' in Malmö, World Maritime University' (2005) http://www.imo.org/OurWork/Safety/Implementation/Documents/Voluntary.pdf accessed 6 July 2015; M. Lemke, *Erfüllungsdefizite des Flaggenstaats* (Nomos 2011).

[241] FAO Agreement on Port State Measures and other Instruments Combating IUU Fishing' (n 239) 10 §30; UNCLOS (n 11) article 94; 'FAO Agreement to Promote Compliance with

procedure does not permit to hold states accountable without their agreement, to actually clarify the content of obligations or to apply sanctions in case of non-compliance.

2. Insufficient Remedies and Piecemeal Sanctions Against Non-compliant States

a) International Responsibility of States

In public international law, it is recognized that states not fulfilling their obligations can see their international responsibility invoked.[242] However, in the present case, determining the actual content of the flag state or coastal state obligations which could be breached and solving questions of attribution of conduct could be difficult, in addition to not necessarily leading to adequate remedies.

aa) Content of Obligation and Attribution

For the international responsibility of a state to arise, it must have breached an international obligation by which it was bound and the conduct in question must be attributable to it.[243] Both aspects are likely to be complex in the case of fisheries.

First, the obligations of coastal states are vague and their implementation is subject to those states' discretion. As to the regulation of the high seas, the only fully universal and uncontroversial obligation of states whose vessels are fishing there is to cooperate in the conservation and management of the living resources in that area. It would however be challenging to prove the violation of such a softly worded duty.[244] As far as the detailed rules are concerned, except for states parties to the UNFSA, only the measures adopted by RFMOs to which states are members can be unquestionably used against them, due to the debated status of the 1995 Implementing Agreement.

Second, questions of attribution would also present some difficulties. Indeed, the actions of a vessel are not, as such, attributable to the flag state.[245] Hence, the mere

International Conservation and Management Measures by Fishing Vessels on the High Seas: Compliance Agreement' (Compliance Agreement) 1993, 2221 *United Nations Treaty Series* 91 article 3(3); FAO, 'Implementation of the International Plan of Action to Prevent, Deter and Eliminate Illegal, Unreported and Unregulated Fishing' FAO Technical Guidelines for Responsible Fisheries 9 (Rome 2002) 20–21.

[242] ILC, Draft Articles on the Responsibility of States for Internationally Wrongful Acts with commentaries 2001, article 1.

[243] Ibid. article 2.

[244] See supra Chapter 3 A. I. 1.-2.

[245] Y. Takei, 'Assessing Flag State Performance in Legal Terms: Clarifications of the Margin of Discretion' (2013) 28 International Journal of Marine and Coastal Law 97, 130. ITLOS made a statement to that effect (*Request for an Advisory Opinion Submitted by the Sub-Regional Fisheries Commission (SRFC), Advisory Opinion* (n 235) 61–62 reply to second question).

violation, by a vessel, of a CMM, would not automatically entail the responsibility of its flag state. However, the lack of actions of a state to control its vessels and prevent IUU fishing as well as its passivity or inadequate reaction in the face of these violations could entail the responsibility of the flag state.[246]

bb) Invocation

Moreover, only certain categories of states might invoke such responsibility. Invocation is to be understood, according to the Commentary to the International Law Commission (ILC) articles on state responsibility as related to "measures of a relatively formal character, for example, the raising or presentation of a claim against another State or the commencement of proceedings before an international court or tribunal."[247]

First of them are the injured states:[248] as the relevant obligations under UNCLOS or RFMOs are not owed to any particular state (except maybe if only two states are fishing a transboundary stock),[249] the state wanting to invoke the wrongdoing state's responsibility would have to fulfil the criteria of lit (b) of article 42 for obligations owed to a group of states. Hence, it has either to be particularly affected or the situation is to be such as that "each party's performance is effectively conditioned upon and requires the performance of each of the others" (in what is also known as absolute obligations).[250] The former case could be that of a coastal state if a straddling stock is overfished in the high seas adjacent to its EEZ. While the latter could be envisioned, as the efficiency of cooperation towards conservation of species depends on the compliance of all states involved, it has not yet been invoked by a state against free-riders, which leaves the adequacy of using such avenue unsettled. Moreover, it is probably inapplicable for detailed RFMOs rules, as 'simply' disrespecting quotas by which a state is bound cannot fit into this latter category: other states could still comply with their own quantified obligations and hence perform their obligations.[251]

---

[246] European Union, 'Written Statement: Request for an Advisory Opinion Submitted by the Sub-Regional Fisheries Commission' (29 November 2013) 18 §56; Takei (n 245) 131–132. ITLOS also clarified that point (*Request for an Advisory Opinion Submitted by the Sub-Regional Fisheries Commission (SRFC), Advisory Opinion* (n 235) 61–62 reply to second question).

[247] Articles on the Responsibility of States (n 242) commentary on article 42 §2.

[248] Ibid. article 42 reads as:

"A State is entitled as an injured State to invoke the responsibility of another State if the obligation breached is owed to: (a) that State individually; or (b) a group of States including that State, or the international community as a whole, and the breach of the obligation: (i) specially affects that State; or (ii) is of such a character as radically to change the position of all the other States to which the obligation is owed with respect to the further performance of the obligation."

[249] A. Serdy, 'Accounting for Catch in Internationally Managed Fisheries: What Role for State Responsibility?' (2010) 15 Ocean and Coastal Law Journal 23, 31.

[250] Articles on the Responsibility of States (n 242) commentary on article 42 §13.

[251] Serdy, 'Accounting for Catch in Internationally Managed Fisheries: What Role for State Responsibility?' (n 249) 32.

The second category of states is that of the non-injured ones.[252] They can invoke a state's responsibility if the obligation is owed to a group of states to which they belong and is established to protect of a collective interest of the group (here for example the sustainable use of resources within an RFMO),[253] or if the obligation is owed to the international community. The obligation to cooperate is arguably one such obligation as it is endowed in an international treaty to which most states are parties, and one of CIL which is directed to all states. It also aims at protecting a collective interest. Under this title of invocation the non-injured state can only require cessation of the wrongful act and reparations for the injured state(s) or the beneficiaries of the obligation.[254]

cc) Remedies

The remedies envisioned under the ILC articles on state responsibility are cessation, reparation and countermeasures. Reparations are due for the injury caused, which "includes any damage, whether material or moral, caused by the internationally wrongful act".[255]

Determining who the injured state or the beneficiaries of the obligation are if a non-injured state invokes the responsibility of another state is a necessary condition of reparation[256] but is problematic for damages inflicted to the global commons.[257] Proving causality between damage caused to fisheries and the wrongful act is also deemed complicated as many factors influence the status of stocks.[258]

Moreover, restitution, which is the preferred form of reparation requires to know how the situation actually was before the wrongful act took place, in order to re-establish this situation, and must be chosen if it "is not materially impossible [and] does not involve a burden out of all proportion".[259] Re-improving the status of stocks after illegal fishing does not appear an easy task and might be out of proportion with the benefit expected. Hence, compensation is the next form of reparation, but assessing monetary value for the loss of fish stocks might prove

---

[252] Articles on the Responsibility of States (n 242) article 48 reads as:

"1. Any State other than an injured State is entitled to invoke the responsibility of another State in accordance with paragraph 2 if: (a) the obligation breached is owed to a group of States including that State, and is established for the protection of a collective interest of the group; or (b) the obligation breached is owed to the international community as a whole."

[253] Serdy, 'Accounting for Catch in Internationally Managed Fisheries: What Role for State Responsibility?' (n 249) 35.

[254] Articles on the Responsibility of States (n 242) article 48(2).

[255] Ibid. article 31.

[256] Ibid. article 48(2)(b); G. Gaja, 'Obligations and Rights Erga Omnes in International Law: Preparatory Work' (2005) 71-I Annuaire de l'Institut de Droit International 117, 137.

[257] Gaja (n 256) 138.

[258] Serdy, 'Accounting for Catch in Internationally Managed Fisheries: What Role for State Responsibility?' (n 249) 38–39.

[259] Articles on the Responsibility of States (n 242) article 35.

difficult too, although generally easier than for other types of environmental damages which might not have a market price attached to them.

dd) Countermeasures

The use of countermeasures in the field of fisheries, which any state can take against non-compliant states according to Rayfuse,[260] is not, legally speaking, an uncontroversial suggestion.[261] Indeed, as useful as such an approach would be, countermeasures appear to be, *de lege lata*, only possible for the injured state.[262] Except in the case of a particularly affected state or if the obligation to cooperate was recognized as absolute, identifying which state is actually injured would be complex. In most circumstances indeed overfishing or illegal fishing induces a decrease to the stocks, but does not cause damage to one state in particular.

Also, it has been said that "[t]he use of countermeasures against a state [... is] often inappropriate – partly because it is difficult to establish a causal link between an environmental damage and the source attributed to a state, and partly because the environmental damages may be irreversible."[263]

However, other opinions exist and some argue in favor of a right for non-injured states to use countermeasures to enforce multilateral obligations owed to the international community.[264]

In conclusion, using the regime of state responsibility and the remedies it includes for issues related to violations of fisheries rules and regulations seems complicated and not necessarily successful to ensure compliance by the states active in a fishery.

---

260 R. Rayfuse, 'Countermeasures and High Seas Fisheries Enforcement' (2004) 51 Netherlands International Law Review 41, 43.

261 Guilfoyle (n 230) 164–168.

262 The use of countermeasures by a non-injured state would go beyond the options explicitly envisioned in the Articles on the Responsibility of States (n 242). The previous draft which included countermeasures by non-injured states was criticized by a large number of states, leading to the deletion of this provision and the insertion of a simple saving clause (Articles on the Responsibility of States (n 242) article 54; J. R. Crawford, *Brownlie's Principles of Public International Law* (8th edn Oxford University Press 2012) 588).

263 Henriksen, Hønneland and Sydnes (n 12) 53 (reference omitted).

264 The Institut de Droit International (Resolution on the Obligations and rights erga omnes in international law 2005, Session de Krakow article 5(c)) reached the conclusion that countermeasures are legal in case of a grave breach of an *erga omnes* obligation; on this, see C. Dominicé, 'A la recherche des droits erga omnes', *Droit du pouvoir, pouvoir du droit: Mélanges offerts à Jean Salmon* (Bruylant 2007) 364–366. See also the detailed examination of the subject in C. J. Tams, *Enforcing Obligations Erga Omnes in International Law* (Cambridge University Press 2005) chapter 6, 249–251 for the conclusions.

b) Piecemeal Sanctions Against States for Non-compliance

In the fisheries-related regimes themselves, the rare sanctions which exist are structurally and geographically piecemeal, rarely used, and some potentially violate WTO law.

Indeed, sanctions as such do not exist in UNCLOS, but some were developed in other regimes, in particular in the RFMOs. For example, sanctions against vessels suspected of IUU fishing are implemented in many RFMOs through port state measures. Several measures can be taken both against parties and non-parties if it is suspected that vessels flying these states' flags engage in activities that undermine the effectiveness of CMMs. Conceivable measures are inter alia port state sanctions and the use of CDS.[265]

However, due to their regional nature and lack of universality, such sanctions based on ports are not fully efficient.[266] Indeed, in the same way that flags of non-compliance exist, ports of non-compliance can put a whole enforcement system in a region at risk; "the effectiveness of a [catch. . .] documentation scheme depends crucially on the number of participating states."[267]

Also, it is not totally clear whether port states can or must implement port state measures.[268] For UNFSA parties, it might be argued that article 23 read together with article 20 (1) provides for an obligation for regional states to use their inspection and enforcement rights in port.[269] This is however debatable since the article on port state measures is somewhat contradictory. Indeed, the first paragraph, referring to the 'duty' of port states, supports an obligation while the following ones, as they use the word 'may' and mention the port state's sovereignty, rather suggest that states retain discretionary powers in the measures they take.[270]

In practice, a recent study shows that only a small proportion of vessels on the IUU lists are checked in ports, that RFMOs do not monitor their member states' implementation of port state measures and that they have no sanction mechanism against states that are non-compliant with such an obligation—when one exists.[271]

---

[265] Hayashi, 'Regional Fisheries Management Organisations and Non-Members' (n 205) 752–759, presenting the measures taken by the tuna RFMOs, CCAMLR, NAFO and NEAFC. On port state measures as a method to fight IUU fishing see for example Swan, 'Port State Measures to Combat IUU Fishing: International and Regional Developments' (n 40). For more on port state measures and CDS, see supra Chapter 3 B. II. 2.-3.

[266] R. Rayfuse, 'Building sustainable high seas fisheries through certification processes: issues and perspectives' (2009) 35 Océanis 93, 98–99.

[267] Stokke, 'Trade Measures and the Combat of IUU Fishing: Institutional Interplay and Effective Governance in the Northeast Atlantic' (n 41) 343.

[268] Molenaar, 'Port State Jurisdiction: Towards Comprehensive, Mandatory and Global Coverage' (n 40) 244–245.

[269] Molenaar, 'Participation, Allocation and Unregulated Fishing: The Practice of Regional Fisheries Management Organisations' (n 26) 473.

[270] UNFSA (n 37) article 23; Henriksen, Hønneland and Sydnes (n 12) 50.

[271] Flothmann and others (n 224) 1236.

Finally, while trade measures against particular vessels such as denial of landing and/or port services are partly efficient, their legality when implemented towards non-parties for example can be questioned and could eventually be the basis for a WTO case.[272] Other economic sanctions against states are even more controversial.

The limited number and scope of measures that can be taken against IUU fishing vessels or flags of non-compliance is problematic since free-riding is such an attractive option. Canada's reaction against the EU in what became the *Estai* case—the arrest of a Spanish vessel beyond its 200 nm—, however illegal under international law, is an interesting illustration of the lack of options open to that coastal state to respond to the Spanish and Portuguese vessels.[273] These European vessels acted in full legality, fished straddling stocks just beyond the Canadian EEZ and enjoyed the freedom of fishing without respecting the quotas adopted at NAFO, the EU having once more objected to them[274]—the objection procedure was used by the EU 48 times between 1985 and 1991.[275]

c) Measures Against RFMOs

RFMOs do not always respect their mandates. Indeed, when they adopt quotas well above the scientific recommendations, they obviously disregard sustainability, which is, in most cases, one of the principles set out in their founding treaties. Not performing in accordance with one's founding treaty is not only a bad governance practice,[276] but also a breach of obligations.

In general, with the increasing tasks entrusted to international organizations, the necessity of being able to hold them accountable for their actions has been recognized and it was tentatively addressed by the ILC in its draft articles of 2011 on the

---

[272] Tanaka, *The International Law of the Sea* (n 14) 248–249; on this see supra Chapter 3 D. I. 2. c).

[273] In 1994 Canada enacted legislation applicable to straddling stocks on the high seas and gave powers to its national authorities to enforce them towards some vessels not flagged in Canada. Hence it appears to have been a fully assumed strategy (Churchill and Lowe, *The law of the sea* (n 14) 306; Longhurst (n 53) 146–147; L. Lucchini, 'La loi canadienne du 12 mai 1994: la logique extrême de la théorie du droit préférentiel de l'Etat côtier en haute mer au titre des stocks chevauchants' (1994) 40 Annuaire français de droit international 864; D. Vanderzwaag, 'Canadian Struggles with Freedoms of Navigation and Fisheries: a Short "Legal Cruise"' in E. Franckx and P. Gautier (eds), *The Exercise of Jurisdiction over Vessels: New Developments in the Fields of Pollution, Fisheries, Crimes at Sea and Trafficking of Weapons of Mass Destruction* (Bruylant 2010) 32). On the ICJ case thereabout (*Fisheries Jurisdiction (Spain v. Canada), Judgment on Jurisdiction,* 4 December 1998, I.C.J. Reports 1998 432), see for example J. Pfeil, 'Fisheries Jurisdiction Case (Spain v Canada)' Max Planck Encyclopedia of Public International Law www.mpepil.com.

[274] Vincent, *Droit de la mer* (n 101) 228.

[275] Vanderzwaag, 'Canadian Struggles with Freedoms of Navigation and Fisheries: a Short "Legal Cruise"' (n 273) 31.

[276] International Law Association (ILA), 'Accountability of International Organisations (Berlin Conference): Final Report' (2004) 12–13.

responsibility of international organizations (ILC Draft articles of 2011),[277] by the International Law Association,[278] as well as by the Institute of International Law.[279]

aa) International Responsibility of International Organizations

As international organizations with a legal personality,[280] RFMOs could potentially be held accountable for their actions if these amount to wrongful acts imputable to them.[281] International organizations are indeed subject to international law and have to abide by their treaty engagements as well as the rules of general international law incumbent upon them.[282] In the case of RFMOs, this could be the obligation to cooperate in the conservation of living resources, which is in part given shape by their creation.

Furthermore, international organizations are bound by their founding treaties.[283] Hence, in addition to the possibility of the international responsibility of the

[277] International Law Commission (ILC), 'Draft Articles on the Responsibility of International Organizations, with Commentaries' 2011.

[278] It had a Committee working on this topic from 1996 to 2004 (ILA, 'Accountability of International Organisations (1996–2004)' http://www.ila-hq.org/en/committees/index.cfm/cid/9 accessed 6 July 2015), culminating with its 2004 Report (ILA, 'Accountability of International Organisations (Berlin Conference)' (n 276)) and attached Resolution (ILA, 'Accountability of International Organisations (Berlin Conference): Resolution No. 1/2004' (2004)). It has currently again a working group addressing it (ILA, 'Study Group on the Responsibility of International Organizations (Sofia Conference): Report' (2012)).

[279] It is worth noting the *travaux préparatoires* (R. Higgins, 'The Legal Consequences for Member States of the Non-Fulfilment by International Obligations of their Obligations towards Third Parties' (1995) 66-I Annuaire de l'Institut de Droit International 251) as well as the Resolution of 1995 (Institut de Droit International, The Legal Consequences for Member States of the Non-fulfilment by International Organizations of their Obligations toward Third Parties 1995, Resolution – Session of Lisbonne).

[280] Henriksen, Hønneland and Sydnes (n 12) 2; see supra Chapter 3 B. II. 1.

[281] Draft Articles on the Responsibility of International Organizations (n 277) articles 3–4; M. Bettati, 'Création et personnalité juridique des organisations internationales' in R.-J. Dupuy (ed), *Manuel sur les organisations internationales – A Handbook on International Organizations* (2nd edn Martinus Nijhoff Publishers 1998) 58; P. Sands, P. Klein and D. W. Bowett, *Bowett's law of international institutions* (6th edn Sweet & Maxwell 2009) 523–524; H. G. Schermers and N. Blokker, *International institutional law: Unity within diversity* (5th edn Nijhoff 2011) §1583.

[282] Schermers and Blokker (n 281) §1139, 1579; Schermers argues that another basis to justify that international organizations are bound by CIL is that all their member states are, and that they cannot transfer competences in order to evade their obligations (H. G. Schermers, 'The legal bases of international organization action' in R.-J. Dupuy (ed), *Manuel sur les organisations internationales – A Handbook on International Organizations* (2nd edn Martinus Nijhoff Publishers 1998) 402–403).

[283] *Interpretation of the Agreement of 25 March 1951 between the WHO and Egypt, Advisory Opinion,* 20 December 1980, I.C.J. Reports 1980 73, 89–90 §37: "International organizations are subjects of international law and, as such, are bound by any obligations incumbent upon them under general rules of international law, under their constitutions or under international agreements to which they are parties." See also Schermers (n 282) 404–405.

international organization being triggered if the act went against its obligations towards the outside, a situation could arise where the RFMO's actions or omissions are considered to go beyond (or against) its own mandate.[284] However, establishing the *ultra vires* nature of actions or omissions is complicated,[285] and finding a forum where such determination could be made remains a challenge to this day, especially if no dispute settlement mechanism is envisioned in the treaty.[286]

The example of Japan and Norway within the IWC is quite telling: they argue that the regime is now acting *ultra vires*,[287] but have not found any method to challenge the legality of the IWC's actions in relation to its mandate. Indeed, mandates can change and no overarching rule states that sustainable use is the goal in the management of marine mammals. This situation also generally reflects the complicated problem of treaties' amendments through the adoption of collective decisions of parties.[288] Indeed, to face the difficult issue of formal amendments, a number of changes are adopted through other means, such as CoP documents, or, in the case of the IWC, modifications in the Schedule. Whether this is actually *intra* or *ultra vires* is debatable, in particular when new obligations are created—in other words questions of a non-technical nature are clarified[289]—and when the decision is not a unanimous one.

Finally, some more general issues are still problematic and/or unsolved in relation to holding international organizations accountable through the invocation of their international responsibility. Indeed, uncertainties remain as to the exact content of rules binding these bodies.[290] Also, no judicial forum is available to invoke international organizations' responsibility. Hence, unsustainable RFMO

[284] On *ultra vires* acts, see for example E. Cannizzaro and P. Palchetti, 'Ultra vires acts of international organizations' in J. Klabbers and A. Wallendahl (eds), *Research Handbook on the Law of International Organizations* (Edward Elgar 2011) 365–397; J. Klabbers, *An introduction to international institutional law* (Cambridge University Press 2009) 53, 56–57, 59. On the powers granted to international organizations, according to their founding treaties and implied powers, see supra Chapter 3 B. II. 1.

[285] There is a presumption of an action being *intra vires* if it "warrants the assertion that it was appropriate for the fulfilment of one of the stated purposes of the" international organization (*Certain Expenses of the United Nations (Article 17, Paragraph 2, of the Charter), Advisory Opinion*, 20 July 1962, I.C.J. Reports 1962 151, 168; see also ILA, 'Accountability of International Organisations (Berlin Conference)' (n 276) 26; Klabbers, *An introduction to international institutional law* (n 284) 215, 218–219). An argument could always be developed that the action reflects the parties' will and/or subsequent practice (Klabbers, *An introduction to international institutional law* (n 284) 254).

[286] P. W. Birnie, 'Are Twentieth-Century Marine Conservation Conventions Adaptable to Twenty-First Century Goals and Principles?: Part I' (1997) 12 International Journal of Marine and Coastal Law 305, 326.

[287] A. Gillespie, 'Forum Shopping in International Environmental Law: The IWC, CITES, and the Management of Cetaceans' (2002) 33 Ocean Development and International Law 17, 37.

[288] Klabbers, *An introduction to international institutional law* (n 284) 82.

[289] Ibid.

[290] Sands, Klein and Bowett (n 281) 524.

practices, such as setting quotas well above the scientific recommendations, cannot as such be reviewed by a court for lack of competence towards these bodies.

Moreover, the detailed framework provided by the ILC Draft articles of 2011 (on the responsibility of international organizations) does not benefit from the same recognition as its counterpart on the responsibility of states. It is indeed characterized as 'progressive development' of international law in opposition to codification thereof.[291] Hence, the detailed rules, very similar to those related to state responsibility, for example on attribution, circumstances precluding wrongfulness, reparation and invocation are not of CIL nature and consequently not binding.

In the face of these difficulties and of some RFMOs' decision-making procedures, it might be advisable and arguably possible to hold member states accountable together with or instead of the international organizations.

bb) International Responsibility of their Member States

While there is no judicial instance that has competence to review the legality of an international organization's actions or omissions, it might still be possible to review RFMOs' doings through the prism of their member states. This might either be done by holding both subjects of international law responsible together or by focusing only on the member states.

In relation to the first scenario, a doctrinal discussion centered itself on whether and how member states of an international organization can or should be held responsible for the international organization's acts.[292] Indeed, on the one hand, while it is generally accepted that a "State cannot by delegation (even if this be genuine) avoid responsibility for breaches of its duties under international law"[293] it appears that when an international organization has legal personality, its responsibility is considered to be the one arising, in opposition to that of the member states for their collective actions within the organization.[294] Membership does not suffice to trigger the member state's responsibility.[295] However, on the other hand, such

[291] Draft Articles on the Responsibility of International Organizations (n 277) general commentary §5.

[292] Klabbers, *An introduction to international institutional law* (n 284) 273, 285–288; Sands, Klein and Bowett (n 281) 518, 526–531.

[293] I. Brownlie, 'State responsibility: the problem of delegation' in K. Ginther and K. Zemanek (eds), *Völkerrecht zwischen normativem Anspruch und politischer Realität: Festschrift für Karl Zemanek zum 65. Geburtstag* (Duncker & Humblot 1994) 300; ILA, 'Accountability of International Organisations (Berlin Conference)' (n 276) 18.

[294] J. d'Aspremont, 'Abuse of the Legal Personality of International Organizations and the Responsibility of Member States' (2007) 4 International Organizations Law Review 91, 95; Klabbers, *An introduction to international institutional law* (n 284) 279; Sands, Klein and Bowett (n 281) 529.

[295] P. Sturma, 'The Responsibility of International Organizations and their Member States' in M. Ragazzi (ed), *Responsibility of international organizations: Essays in Memory of Sir Ian Brownlie* (Martinus Nijhoff Publishers 2013) 318. Another related point is whether member states are obliged to contribute to reparations of an act of the organization by simple virtue of their

member states' responsibility might arise on the basis of additional acts.[296] Hence, under certain circumstances, the responsibility of states can be triggered jointly or in addition to that of the international organization.[297]

As stated in the Commentary to the ILC Draft articles of 2011, "[t]he joint responsibility of an international organization with one or more States is envisaged in articles 14 to 18, which concern the responsibility of an international organization in connection with the act of a State, and in articles 58 to 62, which deal with the responsibility of a State in connection with the internationally wrongful act of an international organization."[298] Hence, the joint responsibility of member states might arise if they assist or control the international organization in committing an internationally wrongful act.[299] Alternatively, and probably most relevant here is the situation of a member state "taking advantage of the fact that the organization has competence in relation to the subject-matter of one of the State's international obligations [to] circumven[t] that obligation by causing the organization to commit an act that, if committed by the State, would have constituted a breach of the obligation."[300]

However, while the ILC Draft articles of 2011 recognize the potential parallel responsibility of the international organization together with that of (some of) its member states,[301] this process of lifting the veil of international organizations' exclusive responsibility is not uncontroversial[302] and these draft articles, as examined in the previous sub-section, are not considered to reflect CIL.

---

membership to it (Higgins, 'The Legal Consequences for Member States of the Non-Fulfilment by International Obligations of their Obligations towards Third Parties' (n 279); Institut de Droit International, Resolution on the Legal Consequences for Member States of the Non-fulfilment by International Organizations of their Obligations toward Third Parties (n 279) are both of the opinion that membership alone does not create such liability. On this Resolution see I. Brownlie, 'The Responsibility of States for the Acts of International Organizations' in M. Ragazzi (ed), *International Responsibility Today: Essays in Memory of Oscar Schachter* (Brill 2005) 355–362. On the relation between international organizations and theirs members in terms of liability, see Schermers and Blokker (n 281) §1585–1588). This question is also addressed in general terms in Draft Articles on the Responsibility of International Organizations (n 277) article 40(2), although the actual consequences of this provision remain unclear.

[296] Sands, Klein and Bowett (n 281) 529–530; S. Yee, "Member Responsibility' and the ILC Articles on the Responsibility of International Organizations: Some Observations' in M. Ragazzi (ed), *Responsibility of international organizations: Essays in Memory of Sir Ian Brownlie* (Martinus Nijhoff Publishers 2013) 331.

[297] On this, see K. Nakatani, 'Responsibility of Member States towards Third Parties for an Internationally Wrongful Act of the Organization' in M. Ragazzi (ed), *Responsibility of international organizations: Essays in Memory of Sir Ian Brownlie* (Martinus Nijhoff Publishers 2013) 294–301.

[298] Draft Articles on the Responsibility of International Organizations (n 277) commentary on article 48 §1.

[299] Ibid. articles 58–59.

[300] Ibid. 61(1).

[301] Ibid. articles 19, 48, 63; see also ILA, 'Accountability of International Organisations (Berlin Conference)' (n 276) 30.

[302] Salmon for example considers that if an international organization has legal personality, then its member states cannot, reasonably, be held responsible for its actions (in Higgins, 'The Legal

The second scenario is that of the exclusive member states responsibility for their actions within the RFMO. This might first happen, as envisioned in the ILC Draft articles of 2011 in article 61, if the state uses the international organization to circumvent its international obligations. In such cases indeed, the action of the international organization does not need to be wrongful for this subject of international law.[303] Hence, while the international organization might or might not be committing a wrongful act (depending on its international obligations), the member state is.

Another possibility is that, in certain cases, the international organization might only be a mere "veil through which the member states are clearly visible"[304] even if it is endowed with formal legal personality. It is generally admitted that distinct responsibility stems from distinct will.[305] Hence, the international responsibility of international organizations is arguably not applicable when the measures in question reflect an aggregated decision of the states[306] and do consequently not represent the will of the international organization.

Defining the degree and nature of distinct will—or reversely of the control by member states over decisions[307]—necessary to trigger this situation is complex.[308] It is also probably a controversial process as it actually questions the adequacy of having granted some organizations with legal personality;[309] this might however be necessary to avoid states giving legal personality to an international organization with the only goal to shield themselves from responsibility for actions they will continue to control.

Despite the difficulty related to the determination of the actual role of the international organization in a decision, one could nevertheless use the following questions in such an attempt: did the representatives behave as representatives of the member states only, acting as if at an ad hoc diplomatic conference, or were they somehow influenced

Consequences for Member States of the Non-Fulfilment by International Obligations of their Obligations towards Third Parties' (n 279) 336, as presented in Bettati (n 281) 580).

[303] Draft Articles on the Responsibility of International Organizations (n 277) article 61(2) and commentary on part V at 90 §3.

[304] N. Blokker, 'International Organizations and Their Members: International Organizations Belong to All Members and to None' (2004) 1 International Organizations Law Review 139, 144.

[305] Klabbers, *An introduction to international institutional law* (n 284) 274; Schermers and Blokker (n 281) §44; N. D. White, *The law of international organisations* (2nd edn Juris Publishing – Manchester University Press 2005) 30–32; d'Aspremont (n 294) 91.

[306] Schermers and Blokker (n 281) §44A.

[307] D'Aspremont argues in favour of a joint responsibility of the member states using excessive control over decision-making and of the international organization but rather envisions the problem of a few member states controlling the decision-making process rather than the decision-making process in itself being questioned in its capacity to produce decisions attributable to the international organization (d'Aspremont (n 294) 92, 101, 103, 109–111, 116–117).

[308] Klabbers, *An introduction to international institutional law* (n 284) 262.

[309] Schermers and Blokker (n 281) §1584.

by the membership in the international body and/or restricted by the existence of a formal permanent structure serving as a forum for their meetings?[310]

On this basis, an RFMO, when setting quotas for example can in certain circumstances lack a distinct will and consequently the legal personality/responsibility for its actions. Indeed, should the representatives of member states meet in plenary, adopt overall quotas at unanimity, and negotiate their shares with only the interests of their states in mind, the decisions resulting from the meeting of the RFMO are in fact the aggregated decisions of states.[311] However, as is increasingly the case now, states are somehow restricted in their actions by their membership in the said RFMO,[312] for example by the advice of the scientific body, by the corpus of past decisions, by rules of procedure setting majority vote and/or by the founding agreement.[313] Hence, it is not totally clear-cut whether such decision-making process "remains the creature of the states members who are thus liable for its acts"[314] or whether the stated—but not always convincing—legal personality of the international organization comes into play. More generally, as Klabbers put it: "it is not always clear where the organization begins and its member states end";[315] well beyond the particular questions at stake here, this remains one of the main hurdles of the framework related to the responsibility of international organizations.

Since using member states' responsibility to hold international organizations accountable is not a straight-forward avenue and appears not fully consensual, it is worth looking at other review processes for RFMOs.

cc) Performance Reviews

As already mentioned, many RFMOs have undergone some review process in the last years.[316] Some of these performance reviews are undertaken internally, with all the pitfalls this creates in terms of transparency, and none of them includes a sanction mechanism beyond naming and shaming. Such latter technique has been used, for example, against ICCAT, with both positive and negative consequences. Indeed, on the one hand it increases the visibility of the issues, but on the other hand, it has the potential to strengthen a defensive approach as well as a polarization of views, rendering further negotiations more difficult.

---

[310] Blokker (n 304) 149; I. Österdahl, 'International organizations – institutions and organs' in J. Klabbers and A. Wallendahl (eds), *Research Handbook on the Law of International Organizations* (Edward Elgar 2011) 178; Schermers and Blokker (n 281) §44.

[311] White, *The law of international organisations* (n 305) 31.

[312] Blokker (n 304) 144.

[313] Ibid. 149.

[314] Higgins, 'The Legal Consequences for Member States of the Non-Fulfilment by International Obligations of their Obligations towards Third Parties' (n 279) 254.

[315] Klabbers, *An introduction to international institutional law* (n 284) 274.

[316] See supra Chapter 3 B. II. 5.

Regarding the legal overview that the FAO can have in relation to RFMOs, it is quite an indirect one, if any. Indeed, while it is true that many instruments which refer to RFMOs' role in fisheries, guidelines, codes of action and best-practices are developed under or put together by the FAO, RFMOs are under no obligation to comply with them. For the RFMOs created under the auspices of the FAO, references to the principles developed in FAO instruments have usually found their way into the founding text when it was adopted after the development of the relevant instruments. Also, the FAO can request the review of these particular bodies, and it has done so.[317] However, it has no mandate to make such demands towards the many more RFMOs that were created outside its ambit. It has been said that "[u]ltimately, the FAO has little direct influence over these other institutions with an interest in fisheries".[318]

One can conclude that RFMOs have been granted, by the UNFSA, a special role in the management of fisheries and, within a good governance system, such prerogative should also include some accountability, "that is the duty to account for its exercise."[319] However, responsibilities and accountability did not seem to come together with these rights. Indeed, it is clear that RFMOs themselves, as non-state actors, are not bound by the UNFSA (except in its potential CIL form),[320] so it is at best an indirect responsibility, through member states. Moreover, the complexities linked to invoking the international responsibility of international organizations and the lack of a formal mechanism to overview their work makes it difficult, if not impossible to hold them accountable for their potential failure to respect the applicable law of the sea.[321] This is far from acceptable as "[n]o situation should arise where an [international organization] would not be accountable to some authority for an act that might be deemed illegal."[322]

3. Insufficient Dispute Settlement Mechanisms

UNCLOS compulsory dispute settlement is quasi-unique in international law and provides a choice of various procedures which reflect the variety of desires of states at the negotiation table.[323] Also, according to UNCLOS article 288, the dispute settlement mechanism of this treaty can be used for other instruments on related

[317] Committee on Fisheries, 'Strengthening FAO Regional Fishery Bodies', presented at the Twenty-second Session of the Committee on Fisheries (Rome, 17–20 March 1997) COFI/97/4 (Rome 1997); J. Swan, 'Regional Fishery Bodies and Governance: Issues, Actions and Future Directions' FAO Fisheries and Aquaculture Circular 959 (Rome 2000) 5–7.

[318] Harrison (n 65) 236.

[319] ILA, 'Accountability of International Organisations (Berlin Conference)' (n 276) 5.

[320] Henriksen, Hønneland and Sydnes (n 12) 12–13.

[321] Gjerde, 'High-seas Fisheries Management under the Convention on the Law of the Sea' (n 74) 298.

[322] ILA, 'Accountability of International Organisations (Berlin Conference)' (n 276) 26.

[323] Merrills (n 120) 169–170.

matters. The UNFSA for example links its dispute settlement mechanism to the one of UNCLOS, even for states that are only parties to the former instrument.[324] Additionally, it also widens the reach of Part XV by declaring its applicability to issues related to regional or other fisheries treaties:[325] "[t]he effect of this provision seems to be that as between Parties to the Fish Stocks Agreement, the dispute settlement machinery of the LOSC is incorporated into existing treaties with regard to straddling or highly migratory fish stocks."[326] Finally, some RFMOs directly refer to UNCLOS compulsory dispute settlement mechanism.[327] However, even within the law of the sea, a field with compulsory binding dispute settlement, exceptions exist where states' actions are not reviewable.

a) Jurisdiction Against States in Waters Under National Jurisdiction

UNCLOS dispute settlement is prevented to work in its entirety for fisheries matters since the coastal state enjoys quasi full discretion in its decisions. Issues on the coastal states' sovereign rights regarding living resources in the EEZ are indeed listed in the exceptions to the compulsory dispute settlement principle found in UNCLOS.[328] In certain cases, conciliation under Annex V section (2) is possible if a party decides to use this procedure, but its results are non-binding[329] and the coastal states' discretion is not to be put into question.[330] Also, this mechanism is not even applicable to all disputes: article 297 (3) (b) explicitly provides that conciliation can be used to assess issues linked to the coastal state's manifest failure to keep living resources from serious danger, to arbitrary refusals to determine TACs and to arbitrary refusals to allocate surpluses. The use of words such as 'manifest' and 'arbitrary' shows that such procedure would be limited to extreme situations.[331]

As for cases triggered by the lack of agreement on compatible measures, as provided for in the UNFSA,[332] UNCLOS dispute settlement mechanisms apply.[333] However, it is unclear how exactly this relates to UNCLOS article 297 (3) which

[324] UNFSA (n 37) article 30(1); Tanaka, *The International Law of the Sea* (n 14) 392.

[325] UNFSA (n 37) article 30(2); Merrills (n 120) 192.

[326] Tanaka, *The International Law of the Sea* (n 14) 392 (reference omitted).

[327] SEAFO (SEAFO Convention (n 188) article 24(4)); SIOFA (SIOFA Agreement (n 188) article 20(1)); SPRFMO (SPRFMO Convention (n 188) article 34(2)); WCPFC (WCPFC Convention (n 188) article 31). For more on RFMOs' dispute settlement mechanisms, see supra Chapter 3 B. II. 3.

[328] UNCLOS (n 11) article 297(3)(a); Tanaka, *The International Law of the Sea* (n 14) 401.

[329] UNCLOS (n 11) article 297(3)(b)(c) and annex V articles 7(2) and (14).

[330] Merrills (n 120) 172.

[331] Churchill, 'The Jurisprudence of the International Tribunal for the Law of the Sea Relating to Fisheries: Is There Much in the Net?' (n 124) 389.

[332] UNFSA (n 37) article 7(4) and Part VIII.

[333] Ibid. article 7(5).

exempts coastal states management in their EEZ from review and which is mentioned explicitly in the UNFSA.

Finally, the practice of international courts regarding fisheries matters in waters under national jurisdiction is so far limited to prompt release procedures. Also, ITLOS might seem, due to its conservative interpretation of the Convention, nearly protective of fishermen who were involved in IUU fishing. The *Volga* case, for example, was related to Australia's arrest of a Russian fishing vessel and the subsequent imposition of a bond which included a guarantee of non-repetition to be verified through vessel monitoring system, in other words a 'good behavior bond'. ITLOS rejected this guarantee on future activities and, more generally, the imposition of non-financial bonds such as the installation of a vessel monitoring system.[334] Judge Anderson and Judge ad hoc Shearer, as they mentioned in their Dissenting Opinions, would have preferred a rather more progressive application of UNCLOS. Indeed, one could have wished for some penalties related to the breach of the obligation to cooperate in the conservation of marine living resources, but environmental considerations were left nearly unaddressed.[335]

However, it must be noted that some commentators consider the recognition by ITLOS of the concerns linked to IUU fishing to be a positive sign.[336] Also and mainly, prompt release proceedings are only related to the release of vessels[337] and not to the legality of their activities; ITLOS already moves into uncertain grounds when somehow considering the severity of the violations as a factor to determine a reasonable bond.[338]

b) Jurisdiction Against States in Waters Beyond National Jurisdiction

For problems arising on the high seas, Part XV is potentially relevant. However, according to article 281, if the parties to a dispute have decided to settle the issue by

---

[334] *The "Volga" Case (Russian Federation v. Australia)* (n 221) §75–80; Churchill, 'The Jurisprudence of the International Tribunal for the Law of the Sea Relating to Fisheries: Is There Much in the Net?' (n 124) 409–410; D. R. Rothwell, 'The Contribution of ITLOS to Oceans Governance through Marine Environmental Dispute Resolution' in T. M. Ndiaye and R. Wolfrum (eds), *Law of the sea, environmental law and settlement of disputes: Liber amicorum Judge Thomas A. Mensah* (Nijhoff 2007) 1016–1019.

[335] H. Tudela, 'La protection du milieu marin et la pratique de la pêche devant le Tribunal international du droit de la mer' in O. Lecucq and S. Maljean-Dubois (eds), *Le rôle du juge dans le développement du droit de l'environnement* (Bruylant 2008) 362–364.

[336] J. M. van Dyke, 'Giving teeth to the environmental obligations in the LOS Convention' in A. G. Oude Elferink and D. R. Rothwell (eds), *Oceans Management in the 21st Century: Institutional Frameworks and Responses* (Martinus Nijhoff Publishers 2004) 171–172; Rothwell, 'The Contribution of ITLOS to Oceans Governance through Marine Environmental Dispute Resolution' (n 334) 1023.

[337] UNCLOS (n 11) article 292(3).

[338] Churchill, 'The Jurisprudence of the International Tribunal for the Law of the Sea Relating to Fisheries: Is There Much in the Net?' (n 124) 404.

other peaceful means, UNCLOS procedures are only applicable if the procedure chosen has not led to a settlement and if the agreement between the parties does not exclude the posterior use of Part XV.

The very wide interpretation to this second requirement given by the Arbitral Tribunal in the *Southern Bluefin Tuna* case is a source of concern regarding the applicability of Part XV to fisheries issues. Indeed, while the treaty binding Australia, New Zealand and Japan did not explicitly reject the referral of a dispute to ITLOS, the ICJ or an arbitral tribunal, the Arbitral Tribunal reached, on this basis, a negative conclusion as to its competence.[339] This decision has been widely criticized,[340] and rightly so. It shows—or creates—a weak system, where non-binding regional dispute settlement mechanisms can easily take priority over UNCLOS mechanisms.[341] And it "may have eroded the pivotal role of compulsory dispute settlement in developing the global public order of the high seas established by the UNCLOS constitutional regime".[342]

c) Limited Use of Bilateral Dispute Settlement for Fisheries Issues

aa) In Practice

As for the practice of international courts and tribunals to settle fisheries disputes, it has been limited so far. Special chambers have been established by ITLOS[343] and the ICJ to deal with issues related to marine living resources' conservation, to fisheries and to environmental matters, but none has not been used so far.[344] Except

---

[339] *Southern Bluefin Tuna Arbitral Award (New Zealand v. Japan, Australia v. Japan)* (n 118) 1390 §57–59, based mostly on the reasoning that, since the dispute settlement under CCSBT article 16 requires states' consent for further procedure, this regime is removed from UNCLOS Part XV compulsory dispute settlement; see Tanaka, *The International Law of the Sea* (n 14) 394–396.

[340] A. Boyle, 'The Southern Bluefin Tuna Arbitration' (2001) 50 International and Comparative Law Quarterly 447, 451; J. Peel, 'A Paper Tiger which Dissolves in the Rain? The Future for Resolving Fisheries Disputes under UNCLOS in the Aftermath of the Southern Bluefin Tuna Arbitration' (2002) 3 Melbourne Journal of International Law 53; C. Romano, 'The Southern Bluefin Tuna Dispute: Hints of a World to Come...Like It or Not' (2001) 32 Ocean Development and International Law 313, 331. For a more neutral analysis of this award's implications, see for example B. H. Oxman, 'Complementary Agreements and Compulsory Jurisdiction' (2001) 95 American Journal of International Law 277.

[341] On this case, see Morgan (n 109) 543–548; Tudela (n 335) 358–359.

[342] Orellana, 'The Law on Highly Migratory Fish Stocks: ITLOS Jurisprudence in Context' (n 38) 464. Only ICCAT and NAFO have no dispute settlement provision at all (McDorman, 'Implementing Existing Tools: Turning Words Into Actions – Decision-Making Processes of Regional Fisheries Management Organisations (RFMOs)' (n 37) 441); see supra Chapter 3 B. II. 3.

[343] ITLOS, 'Chambers' http://www.itlos.org/index.php?id=19 accessed 6 July 2015.

[344] S. Rosenne, 'Reflections on Fisheries Management Disputes' in R. Casado Raigón and G. Cataldi (eds), *L'évolution et l'état actuel du droit international de la mer: Mélanges de droit de la mer offerts à Daniel Vignes* (Bruylant 2009) 835; Sands and others (n 44) 172; Tudela (n 335) 359. See ICJ, 'List of Cases referred to the Court since 1946 by date of introduction' http://www.

for provisional measures or prompt release of fishing vessels or vessels supporting fishing operations,[345] ITLOS and the ICJ have not adjudicated on the merits of a case related purely to fisheries.[346] While some recent cases are to be noted and hence the tendency of not using dispute settlement mechanisms for marine resources disputes might be changing, a number of issues remain.

The recent case-law to consider starts first with the award given by a Review Panel, formed in order to examine Russia's objection to a decision of the SPRFMO. The founding treaty of this RFMO indeed provides strict conditions to allow reservations—an objecting state must present the reasons of its objection, which are only acceptable if the "the decision unjustifiably discriminates in form or in fact against the member", and "adopt alternative measures that are equivalent in effect to the decision to which it has objected".[347] A Review Panel is to be automatically established in order to determine whether the reasons given for the objection are justifiable, as well as whether the alternative measures are actually equivalent.[348]

In the case at hand, the first one to implement this provision, Russia objected to its exclusion from quotas allocated for Chilean jack mackerel as an unjustifiable discrimination.[349] That decision had been taken on the basis of the relative share of fishing undertaken by member states in the benchmark year 2010.[350] Russia reported in 2010 some catches of Chilean jack mackerel, but these operations had been criticized as unlikely to have taken place, since the Russian vessel mentioned to be fishing appeared to be, at best, a processing vessel.[351] Hence, the data was not taken into account to allocate quotas. The Review Panel found this to be discriminatory, but did consider that the measures adopted by Russia—a self-allocation of the percentage it could have received according to its report, with the undertaking to

icj-cij.org/docket/index.php?p1=3&p2=2 accessed 6 July 2015; ITLOS, 'List of Cases' http://www.itlos.org/index.php?id=35 accessed 6 July 2015.

345 Amongst others: *The M/V "Saiga" (Saint Vincent and the Grenadines v. Guinea), Judgment on prompt release*, 4 December 1997, ITLOS case n°1; *Southern Bluefin Tuna cases (New Zealand v. Japan; Australia v. Japan), Provisional Measures* (n 116); *The "Volga" Case (Russian Federation v. Australia)* (n 221).

346 In *Southern Bluefin Tuna Arbitral Award (New Zealand v. Japan, Australia v. Japan)* (n 118), the Arbitral Tribunal decided that it did not have jurisdiction; the ICJ reached the same conclusion in the *Fisheries Jurisdiction (Spain v. Canada)* (n 273); and the Swordfish case was suspended since the parties found a provisional arrangement in 2001 and removed from the list of cases in 2009 as a more structured framework for cooperation was agreed upon (*Case concerning the Conservation and Sustainable Exploitation of Swordfish Stocks in the South-Eastern Pacific Ocean (Chile/European Union), Order*, 16 December 2009, ITLOS case n°7).

347 SPRFMO Convention (n 188) article 17(2).

348 Ibid. article 17(5).

349 *The Objection by the Russian Federation to a Decision of the Commission of the South Pacific Regional Fisheries Management Organisation, Findings and Recommendations*, 5 July 2013, Review Panel – Permanent Court of Arbitration, 4 §3, 14–15 §53.

350 Ibid. 12 §50.

351 Ibid. 8–10 §25–45.

stop fishing as soon as the global TAC was reached[352]—were not equivalent as they would affect the allocations already made to other states and which created justified expectations.[353]

Although a finding of the Review Panel is not meant to carry the same weight as a judgment from an international court or tribunal, it is worth noting that such a procedure allows disputes from within an RFMO to be settled, in a manner that is not just bilateral. Indeed, since all the member states of the RFMO as well as the Commission can intervene it is actually the collective interest which is examined. Moreover, while the SPRFMO restrictions to opting out are already the strictest ones in all RFMOs, the Review Panel considered the equivalence of measures in a non-lenient manner and actually made practical recommendations.

Second, the ICJ judged, in the *Whaling* case, that the JARPA II program was not scientific whaling as understood in accordance with the ICRW and that, consequently, whaling undertaken under this program contravened the moratorium on commercial whaling, the factory ship moratorium as well as the Southern Ocean sanctuary.[354] The remedies ordered were to revoke authorizations already granted and to "refrain from granting any further permits under Article VIII, paragraph 1, of the Convention, in pursuance of that [JARPA II] programme."[355]

While not *stricto sensu* a fisheries case, it nevertheless is related to the protection of marine biodiversity and scientific research. However, the Court avoided to look in details into the concept of scientific research,[356] but rather examined "whether the killing, taking and treating of whales in pursuance of JARPA II is *for purposes of* scientific research".[357] In that respect, the Court considered several points which led to its negative conclusion. In particular, different research-periods for minke, fin and humpback whales are difficult to reconcile with the stated objectives of ecosystem and multispecies research,[358] as well as is the fact that very few fin whales and no humpback whales were actually taken in the program.[359] The gap between the sample size and the actual number of specimens caught is best explained by quotas being set too high in comparison to the research needs.[360] Moreover, the program is open-ended, has produced limited scientific output and cooperation with other scientific institutions is nearly inexistent.[361]

---

[352] Ibid. 21 §83–86.

[353] Ibid. 23–24 §98–100.

[354] *Whaling in the Antarctic (Australia v. Japan; New Zealand intervening), Judgment,* 31 March 2014, ICJ 66–68 §228–233, 71–72 §247.

[355] Ibid. 70 §245.

[356] Ibid. 33 §86, 41 §127.

[357] Ibid. 41–42 §127.

[358] Ibid. 53–54 §181, 57 §194.

[359] Ibid. 59 §201, 62 §210.

[360] Ibid. 59 §201–202, 61 §209.

[361] Ibid. 63 §216, §219, 64 §222.

The case could have become the opportunity for larger discussions about whaling or marine biodiversity issues. Indeed, Australia's institution of proceedings stated that Japan, by its implementation of JARPA II was in violation of its obligations under the ICRW "as well as its other international obligations for the preservation of marine mammals and the marine environment".[362] However, the scope of the case remained limited, not least because of the clear decision of the Court to stay away from examining issues related to whaling policy: it "is aware that members of the international community hold divergent views about the appropriate policy towards whales and whaling, but it is not for the Court to settle these differences."[363]

Third, the case-law has been further enriched by an advisory opinion, which was requested by the SRFC to ITLOS on questions related to IUU fishing, coastal state obligations and flag state obligations as well as the potential liability of the latter in cases of IUU fishing. ITLOS, after examining its competence and making a positive finding in that respect,[364] provided the requested answers to the following four questions:

> "1. What are the obligations of the flag State in cases where illegal, unreported and unregulated (IUU) fishing activities are conducted within the Exclusive Economic Zone of third party States?
> 2. To what extent shall the flag State be held liable for IUU fishing activities conducted by vessels sailing under its flag?
> 3. Where a fishing license is issued to a vessel within the framework of an international agreement with the flag State or with an international agency, shall the State or international agency be held liable for the violation of the fisheries legislation of the coastal State by the vessel in question?
> 4. What are the rights and obligations of the coastal State in ensuring the sustainable management of shared stocks and stocks of common interest, especially the small pelagic species and tuna?"

First, it declared that "[t]he flag State has the obligation to take necessary measures, including those of enforcement, to ensure compliance by vessels flying its flag with the laws and regulations enacted by the SRFC Member States concerning marine living resources within their exclusive economic zones for

---

[362] Ibid. 9 §1.

[363] Ibid. 29 §69.

[364] ITLOS Statute does not provide for an explicit competence to give an advisory opinion in general matters (in opposition to the ones related to the deep-seabed, UNCLOS (n 11) article 191), but its Rules do. The Tribunal had to address some of the issues of jurisdiction brought up by intervening states (See arguments inter alia from China (People's Republic of China, 'Written Statement: Request for an Advisory Opinion Submitted by the Sub-Regional Fisheries Commission' (26 November 2013) 24–24 §53–55); Australia (Australia, 'Written Statement: Request for an Advisory Opinion Submitted by the Sub-Regional Fisheries Commission' (28 November 2013) 6–14 §11–39); the United Kingdom (United Kingdom, 'Written Statement: Request for an Advisory Opinion Submitted by the Sub-Regional Fisheries Commission' (28 November 2013) 4–20 §6–35); the USA (United States of America, 'Written Statement: Request for an Advisory Opinion Submitted by the Sub-Regional Fisheries Commission' (27 November 2013) 2–9 §8–27)).

purposes of conservation and management of these resources". It further clarified that such obligations are of a due diligence nature. Second, it analyzed the liability of flag states and concluded that it did not arose as a consequence of the violation by a vessel flying that flag of relevant conservation and management measures. However, it would be triggered if the flag state failed to comply with its due diligence obligations. Third, it examined the sharing of competences between member states and the European Union and considered that "[i]n cases where an international organization, in the exercise of its exclusive competence in fisheries matters, concludes a fisheries access agreement with an SRFC Member State, which provides for access by vessels flying the flag of its member States to fish in the exclusive economic zone of that State, the obligations of the flag State become the obligations of the international organization." Finally, in relation to the obligation of coastal states to ensure the sustainable management of shared stocks and stocks of common interest, ITLOS clarified that the obligation "to seek to agree", as well as the duty to cooperate (as found respectively in UNCLOS articles 63 (1) and 64 (1)) are also due diligence in nature, hence prescribing a type of behavior—good faith and meaningful consultation—but not a result.[365] This advisory opinion might open the door to a future similar practice by groups of states desiring to have some aspects of the law of the sea clarified authoritatively. It is worth noting that ITLOS, in several places, made sure to explicitly limit the scope of its answers to the EEZs of SRFC member states,[366] but its views could easily be transposed to other areas, in particular when it was analyzing UNCLOS rather than the relevant regional treaty.

Nevertheless, recourse to bilateral dispute settlement for issues related to marine living resources remains rare. The potential role of judicial dispute settlement mechanisms for such matters is not subject to unanimous comments, which range from gloomy realism[367] to more positive opinions, for example praising ITLOS judges' "willingness to go beyond a narrow assessment of the law and take into account environmental principles".[368] In practice, several factors can explain the limited use of dispute settlement for questions related to fisheries.

---

[365] *Request for an Advisory Opinion Submitted by the Sub-Regional Fisheries Commission (SRFC), Advisory Opinion* (n 235) 60–67.

[366] Ibid. 25 §87, 50 §179, 54 §200, 55 §204.

[367] M. Fitzmaurice notes that environmental law cases are still the 'poor relative' ('The International Court of Justice and International Environmental Law' in C. J. Tams and J. Sloan (eds), *The Development of International Law by the International Court of Justice* (Oxford University Press 2013) 353); see also Sands and others (n 44) 151.

[368] Rothwell, 'The Contribution of ITLOS to Oceans Governance through Marine Environmental Dispute Resolution' (n 334) 1024; for a nuanced assessment of the ICJ's contribution to environmental law, see for example J. E. Viñuales, 'The Contribution of the International Court of Justice to the Development of International Environmental Law: A Contemporary Assessment' (2008) 32 Fordham International Law Journal 232, 232–258.

bb) Issues of Jurisdiction, Clarity and Fragmentation

To start with, the jurisdictional issues examined above[369] largely restrict access to a judicial forum for a compulsory dispute settlement of fisheries disputes, both in relation to the EEZs and the high seas.

Moreover, "[s]tates may be more reluctant to litigate where the rules of customary law are themselves unsettled or an underlying consensus on what they should be is not yet fully established."[370] Here, the legal status of the obligation for all to cooperate and apply RFMOs measures has not been clarified yet. Hence, in the face of uncertainty, negotiated solutions are sometimes preferred by states which do not want to take the risk of an unwelcome precedent.[371]

Finally, as pointed out previously,[372] a risk exists that one single dispute would have to be solved through two different mechanisms[373] since the use of UNCLOS Part XV is possible—even though difficult—for fishing in waters beyond national jurisdiction, but not for EEZ fishing. Such fragmentation of foras and of decisions is unpractical and might contribute to keeping states away from courts.[374]

However, it seems that even when treaties provide states with compulsory dispute settlement, they tend not to use it for environmental matters[375] for reasons to be seen here below.

cc) Inadequacy of Bilateral Dispute Settlement

The adequacy of a judicial dispute settlement mechanism in relation to public interest issues can be questioned.[376] Dispute settlement in front of international courts is indeed typically aiming at solving bilateral disputes, while fisheries matters—same as most global environmental issues—might rather need to be considered multilaterally.[377] Usually, all parties to MEAs and fisheries-related treaties are affected by non-compliance and this particularity is not easily taken into account by judicial dispute settlement, as it is fundamentally bilateral.[378]

369 See supra Chapter 4 A. IV. 3. a)-b).

370 Birnie, Boyle and Redgwell (n 9) 252.

371 Ibid.

372 See supra Chapter 4 A. II. 5.

373 Rosenne, 'Reflections on Fisheries Management Disputes' (n 344) 843–844.

374 Birnie, Boyle and Redgwell (n 9) 252.

375 M. Bothe, 'Compliance' Max Planck Encyclopedia of Public International Law www.mpepil.com §45.

376 Birnie, Boyle and Redgwell (n 9) 212; Bothe (n 375) §45; P. Okowa, 'Environmental Dispute Settlement: Some Reflections on Recent Developments' in M. D Evans (ed), *Remedies in International Law: The Institutional Dilemma* (Hart Publishing 1998) 166–167.

377 R. R. Churchill and G. Ulfstein, 'Autonomous Institutional Arrangements in Multilateral Environmental Agreements: A Little-Noticed Phenomenon in International Law' (2000) 94 American Journal of International Law 623, 644–645.

378 Ibid.

In addition to questions of standing which will be addressed separately here below, dealing with non-bilateral issues would call for a right of intervention by third states. While it does not amount to becoming a party to a case, it is allowed to parties to a multilateral treaty,[379] but otherwise only under restrictive circumstances for disputes related to CIL.[380] This "may deny third parties the opportunity to intervene in cases where the respondent state is violating the rights of all states, or of the international community as a whole in the case of obligations *erga omnes*."[381]

Moreover, having recourse to a court is usually examined in comparison to using a treaty-built compliance mechanism; the advantages of the latter—and hence weaknesses of the former—are its non-adversarial approach and the fact that it can encourage compliance rather than just punish its absence.[382] The goal of a compliance procedure is indeed the respect of the rules by all states rather than the determination of responsibilities.[383]

dd) Evidentiary Issues

States' traditional reluctance to go to court over marine biodiversity issues[384] might also be linked to expected evidentiary issues: one must note the difficulty, if not the potential impossibility, to prove non-compliance with UNCLOS obligation to cooperate in the conservation of marine species.

Indeed, while a stock's collapse can be identified, other factors than fishing influence fish stocks and the scientific data on which (un)sustainable behavior can be determined is never totally consensual or is sometimes actually inconclusive.[385] Furthermore, except in particular cases, it is difficult to prove that one state's actions were the cause of depletion.

---

[379] 'Annex to the United Nations Charter: Statute of the International Court of Justice' (Statute of the ICJ) 1945, 1 *United Nations Treaty Series* xvi article 63; 'Annex VI of UNCLOS – Statute of the International Tribunal for the Law of the Sea' (Statute of ITLOS) 1982 article 32.

[380] Statute of the ICJ (n 379) article 62; Statute of ITLOS (n 379) article 31; Birnie, Boyle and Redgwell (n 9) 252–253.

[381] Birnie, Boyle and Redgwell (n 9) 253.

[382] Churchill and Ulfstein (n 377) 644–645; E. Louka, *International Environmental Law: Fairness, Effectiveness, and World Order* (Cambridge University Press 2006) 128.

[383] Louka (n 382) 128.

[384] U. Beyerlin and J. Grote Stoutenburg, 'Environment, International Protection' Max Planck Encyclopedia of Public International Law www.mpepil.com §86; Churchill, 'The Jurisprudence of the International Tribunal for the Law of the Sea Relating to Fisheries: Is There Much in the Net?' (n 124) 418. On the general reasons explaining this reluctance, such as the inadequacy of international litigation for multilateral treaty obligations, see A. Boyle, 'Environmental Dispute Settlement' Max Planck Encyclopedia of Public International Law www.mpepil.com §3–5; Bothe (n 375) §45; Churchill and Ulfstein (n 377) 629, 644.

[385] Orellana, 'The Law on Highly Migratory Fish Stocks: ITLOS Jurisprudence in Context' (n 38) 480.

ee) Problem of Standing

A major practical problem is that of the determination of the entity competent to bring a case to court. Except for advisory opinions[386] and arguably in circumstances where the parties consensually decide otherwise in front of ITLOS,[387] only those states whose legal interests are infringed have standing to initiate proceedings in front of an international court.[388] It is indeed generally accepted that a state "should be able to establish a legal interest in respect of the claim brought before an international tribunal"[389] and that "a mere interest" is insufficient.[390] Traditionally, those are states defending their rights or legal interests, in opposition to those of the international community.[391]

In issues involving multilateral obligations and global commons, such as high seas fisheries or the protection of biodiversity in waters beyond national jurisdiction, no particular state's rights are usually at risk.[392] A right of standing for non-injured states is needed for anyone to be able to invoke those obligations against a wrongdoing state.[393] *De lege lata*, the existence of a right for a state to act for the protection of the international community's rights and on behalf of it, in what is called an *actio popularis*, is however still controversial and unsettled even in relation to obligations owed to the international community.[394]

---

[386] International organizations can request advisory opinions. The list is however very limited in the case of the ICJ ('Charter of the United Nations' (n 60) article 96; International Court of Justice, 'Organs and Agencies of the United Nations Authorized to Request Advisory Opinions' http://www.icj-cij.org/jurisdiction/index.php?p1=5&p2=2&p3=1 accessed 6 July 2015) but appears less strict in ITLOS, especially since the request of the SRFC for an advisory opinion was granted. On this, see supra Chapter 4 A. IV. 3. c) aa).

[387] Birnie, Boyle and Redgwell (n 9) 251–252 which states, in relation to on the Statute of ITLOS (n 379) article 20(2), that "[i]n consensual proceedings brought before the ITLOS (but *not* in compulsory jurisdiction cases) the range of potential parties may include not only international organizations, NGOs, and private parties but also entities of uncertain status, such as Taiwan."

[388] Birnie, Boyle and Redgwell (n 9) 252; I. Brownlie, *International law at the fiftieth anniversary of the United Nations: general course on public international law* (Recueil des Cours vol 255, Martinus Nijhoff Publishers 1995) 103.

[389] P. Okowa, 'Issues of Admissibility and the Law on International Responsibility' in M. D Evans (ed), *International Law* (3rd edn Oxford University Press 2010) 472.

[390] *Case concerning the Barcelona Traction, Light and Power Company, Limited (Belgium v. Spain) – second phase, Judgment*, 5 February 1970, I.C.J. Reports 1970 3, 36 §46; Tams (n 264) 29–31.

[391] Boyle, 'Environmental Dispute Settlement' (n 384) §6; Fitzmaurice, 'The International Court of Justice and International Environmental Law' (n 367) 354; A. Kiss and D. Shelton, *Guide to International Environmental Law* (Martinus Nijhoff Publishers 2007) 15.

[392] Birnie, Boyle and Redgwell (n 9) 233; M. Fitzmaurice, *International Protection of the Environment* (Recueil des Cours vol 293, Kluwer Law International 2002) 166.

[393] Birnie, Boyle and Redgwell (n 9) 233; K. Zemanek, 'New Trends in the Enforcement of erga omnes Obligations' (2000) 4 Max Planck Yearbook of United Nations Law 1, 10.

[394] Birnie, Boyle and Redgwell (n 9) 253; R. Wolfrum, 'Enforcing Community Interests through International Dispute Settlement: Reality or Utopia' in U. Fastenrath and others (eds), *From*

That particular type of obligations, so-called *erga omnes*, is related to rights that all state have a legal interest in seeing protected,[395] because "they articulate basic interests and needs as well as fundamental values of the international community as a whole".[396]

A distinction is to be made here between obligations *erga omnes partes* and obligations *erga omnes*. While they share many characteristics and are not always differentiated in the same manner,[397] the former rely exclusively on a multilateral treaty. Of interest for the present discussion is that, in some cases specific jurisdictional clauses in these treaties provide for a basis for standing and "[n]o special interest needs to be shown".[398] Hence, states wishing to react to a breach of these conventional frameworks do not have to demonstrate an injury or that the obligations in question are *erga omnes*.[399] A relevant example is article 286 of UNCLOS which provides that "any dispute concerning the interpretation or application of this Convention shall [...] be submitted *at the request of any party* to the dispute", in which case the dispute settlement procedures envisioned in Part XV apply.

As many rules of interest in fisheries law are based on CIL and/or do not benefit from such a specific clause, issues are likely to arise on the ability to bring a claim related to obligations *erga omnes* under the general rules of standing—in other words in the absence of a particular conventional link connecting the Court, the claimant and the potential respondent, and providing for special rights of enforcement.[400] These obligations of relevance hence mostly include CIL but can also encompass treaty obligations when they are governed by the general rules of standing only.

It is argued that the protection of the marine environment as well as that of biodiversity beyond national jurisdiction or biodiversity in general qualify as *erga omnes*. Those obligations are indeed directed to as well as owed by all states, are related to values of importance for the international community and their violation

---

Bilateralism to Community Interest: Essays in Honour of Bruno Simma (Oxford University Press 2011) 1132.

[395] Dominicé (n 264) 359.

[396] J. Delbrück, '"Laws in the Public Interest" – Some Observations on the Foundations and Identification of erga omnes Norms in International Law' in V. Götz, P. Selmer and R. Wolfrum (eds), *Liber amicorum, Günther Jaenicke – Zum 85. Geburtstag* (Springer 1998) 18; see also Dominicé (n 264) 360–361 on the *Barcelona Traction* dictum; Gaja (n 256) 120–122.

[397] The ILC does not differentiate them on the basis of the source (Articles on the Responsibility of States (n 242) commentary on article 48 §6, 8); nor does the Institut de Droit International, Resolution on the Obligations and rights erga omnes in international law 2005 (n 264) article 1; see Tams (n 264) 120–123, but also Beyerlin and Grote Stoutenburg (n 384) §78.

[398] Gaja (n 256) 123; Tams (n 264) 36–37. In *Questions relating to the Obligation to Prosecute or Extradite*, the Court addressed the question of obligations *erga omnes partes* according to article IX of the Genocide Convention: "any State party to the Convention may invoke the responsibility of another State party with a view to ascertaining the alleged failure to comply with its obligations *erga omnes partes*" (*Questions relating to the obligation to prosecute or extradite (Belgium v. Senegal), Judgment*, 20 July 2012, I.C.J. Reports 2012 422, 450 §69).

[399] Tams (n 264) 125.

[400] Ibid. 125–127.

injures no state in particular. Nevertheless, while the doctrine appears to widely agree on these obligations fulfilling the requirements of *erga omnes* ones,[401] no statement by a court has been made yet in the matter, the identification of such norms is still difficult,[402] and, more generally, the question of who is to authoritatively decide what falls into the category of a public/common interest remains unsettled.[403]

As to the particular questions of standing, they are not automatically clarified by the recognition of the existence of community interests.[404] The justiciability of such obligations is indeed still subject to discussion,[405] with arguments in favor of it[406] such as the insertion of article 48 in the ILC articles on state responsibility[407] but also with no clear answer having been given by international courts yet, in particular the ICJ which is examined here below.[408]

---

[401] Articles on the Responsibility of States (n 242) commentary on article 48 §10; Institut de Droit International, Resolution on the Obligations and rights erga omnes in international law 2005 (n 264) preamble.

Amongst those who explicitly mention the protection of areas beyond national jurisdiction (such as the high seas) as well as of the (marine) environment and/or the conservation of biodiversity as being community interests, hence with *erga omnes* effects, are U. Beyerlin and T. Marauhn, *International Environmental Law* (Hart Publishing; Verlag CH Beck 2011) 287–288; Delbrück (n 396) 27–28; Fitzmaurice, 'The International Court of Justice and International Environmental Law' (n 367) 357–358; Kiss and Shelton (n 391) 15; E. M. Kornicker Uhlmann, 'State Community Interests, Jus Cogens and Protection of the Global Environment: Developing Criteria for Peremptory Norms' (1998) 11 Georgetown International Environmental Law Review 101, 102; M. Ragazzi, *The concept of international obligations erga omnes* (Oxford University Press 2000) 155–162; Sands and others (n 44) 149; Y. Shigeta, 'Obligation to Protect the Environment in the IC's Practice: To What Extent Erga Omnes?' (2012) 55 Japanese Yearbook of International Law 176, 201–202; Wolfrum, 'Enforcing Community Interests through International Dispute Settlement: Reality or Utopia' (n 394) 1333–1335.

[402] J. Brunnée, 'International Law and Collective Concerns: Reflections on the Responsibility to Protect' in T. M. Ndiaye and R. Wolfrum (eds), *Law of the sea, environmental law and settlement of disputes: Liber amicorum Judge Thomas A. Mensah* (Nijhoff 2007) 38; Tams (n 264) chapter 4.

[403] Delbrück (n 396) 31–36.

[404] Fitzmaurice, *International Protection of the Environment* (n 392) 168; Wolfrum, 'Enforcing Community Interests through International Dispute Settlement: Reality or Utopia' (n 394) 1137–1139.

[405] Tams ((n 264) chapter 5, in particular 197) even if he argues in favour of a right for all states to have standing in cases of violations of *erga omnes* obligations, also recognizes that there is, so far, no definite answer.

[406] It is worth noting that the Institute of International Law, in its Resolution on the subject, declared that a non-injured state can claim cessation of the violation as well as performance of the obligation (Institut de Droit International, Resolution on the Obligations and rights erga omnes in international law 2005 (n 264) article 2) as well as has standing to bring a claim "[i]n the event of there being a jurisdictional link between" the two states concerned (ibid. article 3; on which Dominicé (n 264) 366). This additional condition simply clarifies that the competence of the Court does depend on such a link, not on the *erga omnes* nature of an obligation.

[407] J. R. Crawford, 'State Responsibility' Max Planck Encyclopedia of Public International Law www.mpepil.com §43–46.

[408] Sands and others (n 44) 148–149; E. B. Weiss, 'Invoking State Responsibility in the Twenty-First Century' (2002) 96 American Journal of International Law 798, 801–808. The GATT dispute settlement panel appeared to point against the possibility of an *actio popularis* when it rejected the

The ILC, going beyond the traditional bilateral nature of inter-states relations,[409] recognized in its articles on state responsibility first that a state might respond as an injured state even if the obligation is multilateral, as long as certain conditions are met.[410] Second, it also acknowledged that a non-injured state can invoke a wrongdoing state's responsibility in order to protect collective interests, hence pointing out to the possibility for an *actio popularis*.[411] Article 48 indeed provides that "[a]ny State other than an injured State is entitled to invoke the responsibility of another State [. . .if. . .] (b) the obligation breached is owed to the international community as a whole." Since invocation is understood by the ILC as including recourse to judicial dispute settlement,[412] the articles on state responsibility actually provide for the justiciability of *erga omnes* obligations. It is also worth noting in that respect that Rapporteur Ago considered, already in 1976, that "[e]very State even if it is not immediately and directly affected by the breach, should be considered justified in invoking the responsibility of the State committing the internationally wrongful act."[413]

However, while the articles are otherwise generally considered a reflection of CIL,[414] the consequences of the ability for a non-injured state to invoke responsibility were acknowledged in the Commentary to be a progressive development of international law.[415] More generally, the rules included in article 48 remain special: it has been declared by Special Rapporteur Crawford that they "were not part of international law before they were formulated in that article"[416] and were subject to controversy during the drafting.[417]

USA's claim of extra-jurisdictional protection of dolphins on the high seas (*Restrictions on Imports of Tuna (Mexico v. United States), GATT Panel Report circulated not adopted,* 3 September 1991, DS21/R – 39S/155, (1991) 30 ILM 1594 §5.32. as presented in Sands and others (n 44) 150). The findings were slightly more open in the Shrimp-Turtle case (*United States – Import Prohibition of Certain Shrimp and Shrimp Products (India, Malaysia, Pakistan and Thailand v United States of America), WTO Dispute Settlement Body – Appellate Body Report Adopted,* 6 November 1998, WT/DS58/AB/R §133, but avoided clarifying the point by claiming that the turtles were of a specific interest for the USA since they were regularly to be found in USA waters.

409 Brunnée, 'International Law and Collective Concerns: Reflections on the Responsibility to Protect' (n 402) 39.

410 Articles on the Responsibility of States (n 242) article 42(b).

411 Ibid. article 48(1)(b); on see this Beyerlin and Grote Stoutenburg (n 384) §78; Birnie, Boyle and Redgwell (n 9) 233–234; Crawford, 'State Responsibility' (n 407) §46; Institut de Droit International, Resolution on the Obligations and rights erga omnes in international law 2005 (n 264).

412 Articles on the Responsibility of States (n 242) commentary on article 42 §2.

413 Special Rapporteur Ago, 'Report on State Responsibility' (1976) II Yearbook of the International Law Commission 3, 29; Ragazzi (n 401).

414 The ILC articles, as they are regularly referred to by courts, seem to reflect CIL (Crawford, 'State Responsibility' (n 407) §65).

415 Articles on the Responsibility of States (n 242) commentary on article 48 §12.

416 J. R. Crawford, 'The International Court of Justice and the Law of State Responsibility' in C. J. Tams and J. Sloan (eds), *The Development of International Law by the International Court of Justice* (Oxford University Press 2013) 74.

417 Weiss (n 408) 805.

As for the ICJ, although it was instrumental in launching the *erga omnes* concept and its jurisprudence evolved on the subject over the decades, it did not make its position as clear as the ILC on the question of standing by a non-injured state to defend the international community's interests.

It originally stated that Liberia and Ethiopia had no standing in the *South West Africa* cases because *actio popularis* does not exist in public international law[418]—in what turned out to be a much criticized judgment.[419] Later, in the *Barcelona Traction* case, the ICJ made a distinction "between the obligations of a State towards the international community as a whole, and those arising vis-à-vis another State in the field of diplomatic protection. By their very nature the former are the concern of all States. In view of the importance of the rights involved, all States can be held to have a legal interest in their protection; they are obligations *erga omnes*."[420]

This *obiter dictum*, always quoted in relation to the topic examined here, is not interpreted in an unanimous fashion: some consider that it implies that all states would have standing if *erga omnes* obligations are in play,[421] while others have pointed out that it does not unequivocally give states "the right of recourse to international courts to protect those interests."[422] Moreover, in the 1974 *Nuclear Tests* cases while the majority of judges avoided discussing the legality of an *actio popularis* on the basis that the case was moot,[423] seven Judges examined the concept of obligations *erga omnes* in their separate or dissenting opinions, showing a whole range of views on the question.[424]

On top of these older cases which predate the ILC articles on state responsibility and have influenced them,[425] other cases have arisen in which related matters have been addressed. However, there again, the ICJ has not clarified the question of

418 "[T]he equivalent of an '*actio popularis*', or right resident in any member of a community to take legal action in vindication of a public interest [...] is not known to international law as it stands at present" (*South West Africa cases (Ethiopia v South Africa; Liberia v. South Africa) – second phase. Judgment,* 18 July 1966, I.C.J. Reports 1966 6, 47 §88).

419 J. R. Crawford, 'The General Assembly, the International Court and self-determination' in V. Lowe and M. Fitzmaurice (eds), *Fifty years of the International Court of Justice: Essays in honour of Sir Robert Jennings* (Cambridge University Press 1996) 587–588.

420 *Case concerning the Barcelona Traction, Light and Power Company, Limited (Belgium v. Spain) – second phase* (n 390) 32 §33.

421 Birnie, Boyle and Redgwell (n 9) 233.

422 H. Fox, 'Jurisdiction and immunities' in V. Lowe and M. Fitzmaurice (eds), *Fifty years of the International Court of Justice: Essays in honour of Sir Robert Jennings* (Cambridge University Press 1996) 234.

423 Tams (n 264) 112–114, 180, on the *Nuclear Tests Case (Australia v. France), Judgment,* 20 December 1974, I.C.J. Reports 1974 253 and *Nuclear Tests Case (New Zealand v. France), Judgment,* 20 December 1974, I.C.J. Reports 1974 457.

424 Tams (264) 180–182, on Judge Petrén's separate opinion, as well as Judges Castro, Onyeama, Dillard, Jiménez de Aréchaga, Sir Humphrey Waldock, and Judge ad hoc Barwick's dissenting opinions.

425 Articles on the Responsibility of States (n 242) commentary on article 48.

standing.[426] Indeed, in the advisory opinion on the *Legal Consequences of the Wall*, the Court 'only' addressed the legal consequences of violations of *erga omnes* obligations.[427] Also, in the case of *Armed activities* opposing the Democratic Republic of the Congo to Rwanda, it made some observations on the fact that consent to jurisdiction is not to be presumed because *erga omnes* obligations are at play,[428] but it did not approach the question of standing.

In conclusion, although from the number of references to *erga omnes* obligations in judgments and the existence of article 48 of the ILC articles, an *actio popularis* might well be an option open to states to defend common interests, a case still remains to be brought to court where the ICJ or ITLOS clarifies beyond doubt the question of standing.

4. Conclusions

In conclusion, the implementation and enforcement of the rules related to fisheries are strongly deficient. States are not always able or willing to control their vessels and areas, which is only made worse by the limited options for non-flag state enforcement and by the lack of a mechanism to overview flag states' compliance. Indeed, the development of a compulsory flag state performance review is not faring well and there is in any case no universal benchmark to determine compliance.

While the international responsibility of a flag or coastal state could arise for an international wrongful act attributable to it, this regime of secondary rules seems ill-suited to questions related to fisheries. Indeed, the determination of which state can invoke such responsibility would be complex and the remedies envisioned unlikely to be adequate. The existing sanctions which can be used against wrong-doing states are limited to port state measures, arguably not compulsory, and to trade restrictions that could be, depending on the circumstances, in violation of trade law. As to the measures that can be taken against RFMOs if they breach their obligations or do not fulfil their mandates, their extent remains debatable. Indeed, the regime of the responsibility of international organizations could be applicable, if it was not falling so clearly in the 'progressive development of international law' category, and with no forum to invoke it. The performance reviews undertaken by several RFMOs are positive, but neither compulsory nor based on the same criteria.

Finally, recourse to judicial dispute settlement mechanisms is restrained by jurisdictional limitations in the field of fisheries, both for waters under and beyond

[426] Zemanek, 'New Trends in the Enforcement of erga omnes Obligations' (n 393) 11–12.

[427] *Legal consequences of the construction of a wall in the occupied Palestinian territory, Advisory Opinion*, 9 July 2004, I.C.J. Reports 2004 136, 199–200 §154–160.

[428] *Case concerning armed activities on the territory of the Congo (new application: 2002) (Democratic Republic of the Congo v. Rwanda), Judgment on jurisdiction and admissibility*, 3 February 2006, I.C.J. Reports 2006 6, 32 §64, 52 §125.

national jurisdiction. The little amount of actual practice bears witness not only of the problems expected, such as the question of standing or evidentiary issues, but also of the probable inadequacy of a bilateral and adversarial system for environmental issues of common interest.

V. Conclusions

In conclusion, governance of fisheries suffers from numerous flaws, such as lack of legal clarity, fragmentation, inadequate and politically-motivated decisions as well as insufficient implementation and enforcement measures.

Indeed, first, fishing on the high seas was traditionally considered an open-access regime, with the associated issue of the tragedy of the commons. Although the size of the high seas has decreased and limitations to the freedom of fishing have been set through the 1958 Conventions and UNCLOS, the actual paradigm shift which would signify a closure of the high seas is to be found in UNFSA and the scope of application of this treaty is unclear. Not knowing whether states can still access fishing grounds without joining or really cooperating with the relevant RFMO and not being sure on which legal basis to treat the free-riders is a thorn on the side of legal security.

Moreover, the regime regulating fisheries suffers from fragmentation. It is indeed subdivided geographically, tends to consider species separately from each other and/or from the physical environment in which they live, is structurally piecemeal due to the high number of instruments and rules applicable, is based on different values as well as pursues several goals. Finally, disputes in that field can be brought to various dispute settlement fora, sometimes simultaneously. Such a situation might cause lack of consistency, conflicting obligations, forum shopping, treaty fatigue and protection gaps.

Additionally, the CMMs applicable to fisheries are all too often not based on scientific advice, which is also plagued by insufficient knowledge and uncertainties. Furthermore, the measures adopted are rarely binding on all states and not even always on all states parties to the relevant RFMO, due to the existence of opt-out clauses.

Finally, in addition to a limited implementation of the rules applicable by the coastal and flag states, no formal overview mechanism exists to control their compliance with their international obligations. Also, neither general international law nor fisheries specific regimes provide for sufficient remedies and/or sanctions against states and/or RFMOs breaching their obligations. Finally, the dispute settlement mechanisms developed by UNCLOS are little used, not least because of the limitations to compulsory jurisdiction for fisheries disputes both in EEZs and the high seas.

Under such an imperfect framework, with little international supervision and nearly no enforcement mechanism, attaining compliance with CMMs or national quotas is difficult. This is particularly serious, since TACs are often high, sometimes set well above the scientific recommendations and any additional non-compliance or IUU fishing with them further threatens sustainability.[429]

B. Characteristics of Appropriate Governance

I. Legal Clarity and Political Unity

1. Legal Clarity

It would be ideal if a strong political consensus pro-conservation/sustainable use, as already reflected in existing treaties and general principles, were to be achieved in practice. However, states' interests and values vary, as revealed by the tension between preservation, conservation and management regimes; this reality ought to be acknowledged.

Nevertheless, the compromise between socio-economic and biological objectives should at least seriously take the latter objectives into account. Ways to attain sustainable development ought to be pursued, with enough flexibility to broker compromises between the diverging goals followed by states. Overall a development that "meets the needs of the present without compromising the ability of future generations to meet their own needs"[430] corresponds well with an equilibrium objective in fisheries.

For such a solution to be found, the different dependency on the resources of the various stakeholders shall be recognized and acted upon if needed.[431] The goal of development for poorer countries—in the sense of improving the people's living conditions[432]—cannot be put aside. In particular, when food security is at stake, it is crucial that environmental obligations of states be nuanced and measures taken to balance the negative impacts on livelihoods.

Moreover, ideology-based goals, such as the preservation of all whales, are a rather dangerous path. Indeed, beyond the whaling controversy, the real issue with that type of approach is that it polarizes the debate in such a manner that no compromise can ever be reached. If scientific advice is to lead the way towards a sustainable use of resources, quasi faith-based decisions disregarding scientific data shall be avoided, whether in one extreme or in the other.

---

[429] Lack, 'Catching On?' (n 42) 4; Stokke, 'Trade Measures and the Combat of IUU Fishing: Institutional Interplay and Effective Governance in the Northeast Atlantic' (n 41) 341–342.

[430] As defined in the World Commission on Environment and Development, *Our Common Future* (n 90) 8.

[431] Cochrane, 'Reconciling sustainability, economic efficiency and equity in fisheries: the one that got away?' (n 101) 10.

[432] M.-C. Cordonier Segger, 'Sustainable development in international law' in D. Armstrong (ed), *Routledge Handbook of International Law* (Routledge 2009) 356.

Also, legally speaking, some points would benefit from clarification. First, the sustainability-oriented objectives for fisheries should be spelt out in more binding terms in global treaties, in order to strengthen the legal framework guiding states in their negotiations. Then, the debated possibility to review states' actions within their EEZs, through the UNFSA compatible measures' clause, should be scrutinized, since it is not easy to see how this provision interacts with the quasi-absolute discretion granted to those states under UNCLOS.

Furthermore, with regard to straddling and highly migratory fish stocks, the status of the UNFSA, which encompasses most of the necessary principles for the sustainable use of fisheries, needs to be explicitly stated. If it is part of CIL or the authoritative interpretation of UNCLOS, then the precautionary and ecosystem approaches as well as the compulsory participation in RFMOs, three crucial aspects of fisheries governance, are legally binding on all states.

The rules applicable to high seas fisheries would also benefit from being further clarified. A new implementation agreement has been under discussion for some time, which would apply to the marine biological diversity in areas beyond national jurisdiction, hence further modifying the legal framework; the relevant Ad Hoc Open-ended Informal Working Group has now fulfilled its mandate and recommended to the UNGA the development of a new agreement in the coming years.[433] The desirable content and scope of such a new instrument as well as its potential to strengthen cooperation and global applicability of regional rules regarding fisheries[434] are however beyond the scope of this study.

[433] UNGA, Letter dated 13 February 2015 from the Co-Chairs of the Ad Hoc Open-ended Informal Working Group to the President of the General Assembly, A/69/780 Annex – Outcome of the Ad Hoc Open-ended Informal Working Group to study issues relating to the conservation and sustainable use of marine biological diversity beyond national jurisdiction and Co-Chairs' summary of discussion §1(e). On the past discussions in the Ad Hoc Open-ended Informal Working Group ('Co-Chairs' summary of discussions, advance and unedited, 1 April 2014 http://www.un.org/depts/los/biodiversityworkinggroup/documents/Advance%20and%20Unedited%20BBNJ-AHWG-7.pdf accessed 6 July 2015) in relation to the commitment made in *The Future We Want* – endorsed in UNGA Resolution A/RES/66/288 27 July 2012 (Document adopted at Rio + 20) §162. In that document, states had indeed agreed to decide before the 69th session of UNGA (September 2014) whether and how to move forward on the question of a new Implementing Agreement. See E. Druel, R. Billé and J. Rochette, 'Getting to yes? Discussions towards an Implementing Agreement to UNCLOS on biodiversity in ABNJ' (2013) 10 Biodiversity http://www.iddri.org/Publications/Collections/Syntheses/PB1013_ABNJ%20IA_druel%20bille%20rochette.pdf accessed 6 July 2015; D. Freestone, 'Protecting Our Oceans: New Challenges, New Solutions An Overview of This Issue' (2006–2007) 7 Sustainable Development Law and Policy Review 2, 4; S. Hart, 'Elements of a Possible Implementation Agreement to UNCLOS for the Conservation and Sustainable Use of Marine Biodiversity in Areas beyond National Jurisdiction' (Gland 2008) IUCN Marine Series 4 https://portals.iucn.org/library/efiles/edocs/EPLP-MS-4.pdf accessed 6 July 2015.

[434] Rayfuse and Warner (n 24) 419–420.

2. Cooperation

A coherent and global legal framework would be ideal, in order to avoid the problem of overlapping regimes and regulatory gaps. Nonetheless, more realistically, in answer to the fragmented nature of the regime, the duty to cooperate has been recognized as part of CIL.[435] Both cooperation between states and coordination between regimes are necessary in the current piecemeal system dealing with global commons located between or outside areas under national jurisdiction.[436]

a) Between States

In the case of fisheries, cooperation between states is essential, because of the transboundary, straddling and migratory nature of fish stocks. The most obvious way to cooperate is for states to conclude agreements bilaterally, regionally or globally, that create similar obligations for them, both in seas under their jurisdiction and in international waters. Cooperation in this geographically and structurally fragmented regime is necessary and exists, but is not sufficient for an effective protection of global commons.

Indeed, universality and measures to fight free-riding as well as non-compliance are crucial since the system can only work if all actors participate and respect their engagements. To keep actual catches at the appropriate level which was set, there is a need not only for members to respect their obligations, but also for enclosure vis-à-vis non-members. RFMOs also ought to be inclusive: new entrants to the fishery should have the opportunity to get involved, at the risk otherwise to see them free-ride.

It is nevertheless true that finding the balance between the necessary openness and the need to sufficiently defend the interests of the existing members to avoid their departure from the regime is a highly complex, if not impossible goal.[437] Indeed, openness only comes at the detriment of those who are already members: with new participants, the total catches will be divided into more shares of lower quotas. Allocation of fishing rights is guided, in the UNFSA, by several non-exhaustive considerations such as the existing level of fishing, fishing patterns and practices, the contribution of states to conservation, data and research, and the needs of specific communities, dependent on fishing and/or developing states. It does however not provide a magic formula as to how to divide the same (or even

---

[435] Birnie, Boyle and Redgwell (n 9) 662.

[436] N. Matz, 'Chaos or Coherence? – Implementing and Enforcing the Conservation of Migratory Species through Various Legal Instruments' (2005) 65 Zeitschrift für ausländisches öffentliches Recht und Völkerrecht 197, 210–212.

[437] On this issue and the tentative to solve it through the 'fully subscribed' quota argument see A. Serdy, 'Postmodern International Fisheries Law, or We Are All Coastal States Now' (2011) 60 International and Comparative Law Quarterly 387, 390.

lower due to the state of resources) total allowable catch into more parts without upsetting members.

Then, collaboration for international implementation is also necessary when compliance is too expensive for developing states or requires specific technology or expert knowledge that is not available to all.

Finally, it is important to have non-fragmented measures, in other words coordination between measures which apply to one and the same ecosystem. In that regard, the compatibility principle developed in the UNFSA and which found its way in some RFMOs is an interesting attempt to make the system of CMMs coherent and coordinated even if adopted by decision-makers of different jurisdictions.[438] This obligation which challenges the usual discretionary powers of the coastal states is rather strong since lack of agreement between the coastal states and the relevant RFMOs triggers the dispute settlement mechanisms.

b) Between International Institutions

Some fragmentation can have positive aspects, since specialized sets of rules possibly better respond to the particularities of specific issues and a plethora of institutions does not necessarily imply competition or tensions.[439] States are more likely to reach agreements within regional bodies if there is some regional specificity to the issue at stake. Different values and goals can also be more adequately respected and represented.[440]

However, it is necessary that cooperation be strengthened in order to avoid duplication of processes and, as much as possible, conflicts.[441] Indeed, the relevant regimes have been developed in an ad hoc manner and there is no sufficient framework where every treaty would fall into place. Moreover, the conventions examined—and so many others—are regulating closely related activities and they do sometimes partly overlap, whether substantially and/or geographically. Hence, coordination of activities undertaken within different regimes is necessary, not only

[438] For example IATTC (Antigua Convention (n 188) article V); SEAFO (SEAFO Convention (n 188) article 19); SPRFMO (SPRFMO Convention (n 188) article 4); WCPFC (WCPFC Convention (n 188) article 8); in general, see McDorman, 'Implementing Existing Tools: Turning Words Into Actions – Decision-Making Processes of Regional Fisheries Management Organisations (RFMOs)' (n 37) 437.

[439] M. Prost and P. K. Clark, 'Unity, Diversity and the Fragmentation of International Law: How Much Does the Multiplication of International Organizations Really Matter' (2006) 5 Chinese Journal of International Law 341, 343–344.

[440] Hafner, 'Pros and Cons Ensuing from Fragmentation of International Law' (n 69) 859.

[441] J. Pauwelyn, *Conflict of Norms in Public International Law: How WTO Law Relates to other Rules of International Law* (Cambridge Studies in International and Comparative Law, Cambridge University Press 2003) 238. For a general framework of cooperation between regimes, see O. S. Stokke, 'The Interplay of International Regimes: Putting Effectiveness Theory to Work' FNI Report 14 (Lysaker 2001) http://www.fni.no/doc&pdf/FNI-R1401.pdf accessed 6 July 2015.

to deal with treaty congestion, a serious problem of environmental law,[442] but also to enhance coherence and cooperation between existing bodies.[443]

Cooperation takes for example the form of the adoption of MoUs or the collaboration through the Kobe process—which actually proved beneficial for example in diminishing the flag of convenience problem in tuna longliners.[444] As for the FAO and the UNGA, they play important roles respectively by coordinating institutions and by pointing states and other actors into the common general direction. Globally, a need is nevertheless felt for stronger links between institutions, at the regional and global level.[445]

Theoretical discussions exist about the added value of a new law of the sea international organization, which would be acting as a global focal point and as a safety net when/if regional organizations fail, or alternatively as a centralized manager.[446] As much as this could solve some problems of fragmentation, it is uncertain whether states would be ready to follow such a path, particularly in this era of treaty fatigue. And it goes without saying that the universality of such an organization would be crucial, since "the success of international cooperation depends largely on the ability to deter free-riding."[447] Finally, it is worth noting that the type of structure in charge is not all that matters and major progress can be made within a regional or specialized institution, in collaboration with others, if the necessary political will exists.[448]

---

[442] D. K. Anton, "Treaty congestion' in contemporary international environmental law' in S. Alam and others (eds), *Routledge Handbook of International Environmental Law* (Routledge 2013) 651–666; R. Caddell, 'International Law and the Protection of Migratory Wildlife: An Appraisal of Twenty-Five Years of the Bonn Convention' (2005) 16 Colorado Journal of International Environmental Law and Policy 113, 148; B. L. Hicks, 'Treaty Congestion in International Environmental Law: the Need for Greater International Coordination' (1998–1999) 32 University of Richmond law review 1643, 1643–1674.

[443] Stokke, 'The Interplay of International Regimes: Putting Effectiveness Theory to Work' (n 441) 13–16.

[444] N. Yagi, 'High Seas Fisheries Today: Challenges and Remedies Under the Global Economy' in D. Vidas and P. J. Schei (eds), *The World Ocean in Globalisation: Climate Change, Sustainable Fisheries, Biodiversity, Shipping, Regional Issues* (Martinus Nijhoff Publishers 2011) 213–216.

[445] Rayfuse and Warner (n 24) 413.

[446] Ibid. 420; on the parallel idea of a world environmental organization, see F. Biermann, 'The Emerging Debate on the Need for a World Environmental Organization' (2001) 1 Global Environmental Politics 45, 45–55; F. Biermann, 'Reforming Global Environmental Governance: The Case for a United Nations Environment Organisation (UNEO)' (2011) http://www.stakeholderforum.org/fileadmin/files/WEO%20Biermann%20FINAL.pdf accessed 6 July 2015; S. Charnovitz, 'A World Environment Organization' (2002) 27 Columbia Journal of Environmental Law 323, 323–362; A. Mahiou, 'De quelques incertitudes institutionnelles et normatives en matière d'environnement' (Le droit international face aux enjeux environnementaux, Aix-en-Provence, 4–6 June 2009) 88–91; A. Najam, 'The Case Against a New International Environmental Organization' (2003) 9 Global Governance 367, 367–384.

[447] M. W. Lodge and others, 'Recommended Best Practices for Regional Fisheries Management Organizations: Report of an independent panel to develop a model for improved governance by Regional Fisheries Management Organizations' (Chatham House 2007) x.

[448] Najam (n 446) 367–369.

3. Legal Coherence

Whereas a coherent international law system developed by a supra-national authority with a global overview could arguably be desirable, it is unconceivable at present. Nevertheless, other methods can be used and further promoted in order to increase consistency between sets of rules, avoid conflicts or, if needed, try solving them.

a) Interpretation

The general methods of interpretation are found in the Vienna Convention on the Law of Treaties (VCLT) at articles 31 to 33: these provisions, after stating that "[a] treaty shall be interpreted in good faith in accordance with the ordinary meaning to be given to the terms of the treaty in their context and in the light of its object and purpose", define what is to be understood as 'the context' and list other instruments which can or should be used for the purpose of interpretation. While interpretation is traditionally used to clarify terms, it can also serve to examine whether two instruments are actually compatible or whether a conflict of norms exists.

Indeed, interpretation helps to determine whether there is a conflict of norms at all[449] before attempting to apply a rule to solve it. Systemic interpretation (also known as reconciliatory interpretation) and mutual supportiveness can or have to be used to avoid *prima facie* conflicts:[450] they allow, in certain cases, two sets of norms to be approached in a manner that makes them compatible with each other and hence applicable simultaneously.[451]

Trying to interpret norms in a harmonious way is linked to the presumption against conflict, according to which states, acting rationally, coherently and in accordance with the existing law, are expected not to undertake conflicting obligations.[452] This principle of interpretation was declared by the ICJ in the *Right of Passage* case.[453] This is also consistent with the law of treaties. In the VCLT

---

[449] McLachlan (n 68) 286. On the difference between apparent and genuine conflicts, see Pauwelyn (n 441) 272. The former may sometimes be solved through interpretation while the latter cannot.

[450] L. Boisson de Chazournes and M. M. Mbengue highlight a difference between mutual supportiveness as a conflict avoidance goal based on the principle of "normative cohesion" and harmonization as a method "linked to the 'presumption against normative conflict'": the former, as it is based on an objective appreciation, cannot be put aside by states, while the latter, as it is subjective, can be put aside if states so wish ('A 'footnote as a principle' Mutual supportiveness and its relevance in an era of fragmentation' in H. P. Hestermeyer and others (eds), *Coexistence, Cooperation and Solidarity: Liber Amicorum Rüdiger Wolfrum* (Martinus Nijhoff Publishers 2011) 1617, 1622–1623).

[451] McLachlan (n 68) 286; on this subject see ILC, 'Fragmentation of international law' (n 69) 206–544 §410–480.

[452] ILC, 'Fragmentation of international law' (n 69) 25–27 §37–40; C. W. Jenks, 'Conflict of Law-Making Treaties' (1953) 30 British Yearbook of International Law 401, 427–429.

[453] Pauwelyn (n 441) 240–241 about the *Case concerning the right of passage over Indian territory (Portugal v. India), Preliminary objections*, 26 November 1957, I.C.J. Reports 1957 125, 142.

indeed, interpreters are encouraged to take into account "[a]ny relevant rules of international law applicable in the relations between the parties" when interpreting a treaty provision.[454] This principle was furthermore already recognized in 1953 as "having a place among the principles of statutory interpretation".[455]

In relation to the different rules applicable to fisheries, systemic interpretation has allowed conflicts to be avoided and even regimes to reinforce each other.[456] For example, GATT article XX contains key terms that were not perfectly clear prior to interpretation, such as 'exhaustible natural resources'. In the *Shrimp/Turtle* case, 'exhaustible' was defined with reference to the risk of extinction that resources, which could otherwise be considered as renewable, might be confronted with.[457] This gives a role to other treaty bodies that serve to determine the biological risks that species are facing. Hence, the regimes of CITES, CMS and the Convention on Biological Diversity (CBD) were mentioned by the Appellate Body:[458] for example, it "referred to the fact that all of the seven recognised species of sea-turtles were listed in Appendix 1 of [...] CITES".[459] As for 'natural resources', opinions differed as to whether only non-living resources qualified as natural resources or whether living resources were covered too. There, the Appellate Body noted that UNCLOS for example included both living and non-living resources in the term natural resources and that it should follow suit.[460]

The fact that the Appellate Body made reference to the law of the sea and international environmental law when interpreting the terms of a GATT article enabled the different regimes not to conflict with each other and the relevant international environmental law rules to fit into article XX exemption. An evolutionary interpretation which considers concepts in their meanings *ex nunc* enables

[454] VCLT (n 199) article 31(3)(c); J. Klabbers, 'Beyond the Vienna Convention: Conflicting Treaty Provisions' in E. Cannizzaro (ed), *The Law of Treaties Beyond the Vienna Convention* (Oxford University Press 2011) 203.

[455] Jenks (n 452) 427–429. This author refers to a statement in Oppenheim's treatise: "[i]t is taken for granted that the contracting parties intend something reasonable, something adequate to the purpose of the treaty, and something not inconsistent with generally recognised principles of International Law, nor with previous treaty obligations towards third States. If, therefore, the meaning of a stipulation is ambiguous, the reasonable meaning is to be preferred to the unreasonable, the more reasonable to the less reasonable, the adequate meaning to the meaning not adequate for the purpose of the treaty, the consistent meaning to the meaning inconsistent with generally recognised principles of International Law and with previous treaty obligations towards third states" (L. Oppenheim, *International Law – A Treatise: Vol. I – Peace*, edited by Lauterpacht (7th edn H. Longmans, Green and Co. 1948) 858–859 (references omitted).

[456] Boisson de Chazournes and Mbengue (n 450) 1619.

[457] *Shrimp Turtle* case (n 408) §128; ILC, 'Fragmentation of international law' (n 69) 223–224 §443; Young, *Trading Fish, Saving Fish* (n 2) 199.

[458] *Shrimp Turtle* case (n 408) §130, 132; Pauwelyn (n 441) 245.

[459] McLachlan (n 68) 303 about *Shrimp Turtle* case (n 408) §132.

[460] *Shrimp Turtle* case (n 408) §130; Young, *Trading Fish, Saving Fish* (n 2) 199.

to keep up with evolving circumstances, without altering or inventing a legislative intent in what is criticized as judicial activism.[461]

However, it must be noted that using systemic interpretation is not the solution to all incoherencies, as it has limitations: some conflicts indeed cannot be avoided, since the provisions are too clearly in opposition. Indeed, reconciliatory interpretation is only possible if the provisions that are *prima facie* conflicting are vague enough to require an interpretation and to permit harmonization.[462]

b) Rules of Reference

Rules of reference are used in a number of treaties to link an obligation to some standards or to a set of benchmarks developed in another instrument, or by another institution. The advantages of such mechanism are twofold. First it might allow the dynamic development of applicable standards without the need to modify the core treaty, permitting timely changes and avoiding the risk of states opting-out.[463] A second positive aspect is the possibility to use instruments adopted in a more specialized and hence expert circle.

Whether the incorporated standards become binding or are of a soft law nature[464] is usually determined by the wording of the obligation. The duty to 'take into account' is for example a weak one, with no obligation of result.[465] This is different when states, in taking certain measures, are "required to conform to generally accepted international regulations, procedures and practices".[466]

It remains complicated to fully clarify which regulations, procedures and practices are referred to, or how one can determine that they are 'generally accepted'.[467] In that respect, it is worth noting that the International Law Association tackled this

[461] P.-M. Dupuy, 'Evolutionary Interpretation of Treaties: Between Memory and Prophecy' in E. Cannizzaro (ed), *The Law of Treaties Beyond the Vienna Convention* (Oxford University Press 2011).

[462] R. Wolfrum and N. Matz, 'The Interplay of the United Nations Convention on the Law of the Sea and the Convention on Biological Diversity' (2000) 4 Max Planck Yearbook of United Nations Law 445, 474. One cannot go beyond the "clear meaning of the words" (C. J. Borgen, 'Resolving Treaty Conflicts' (2005) 37 George Washington International Law Review 573, 639, n 219; see also Pauwelyn (n 441) 245–246).

[463] B. H. Oxman, 'The duty to respect generally accepted international standards' (1991–1992) 24 New York University Journal of International Law and Politics 109, 113–114, 141–143; J. Sommer, 'Environmental Law-Making by International Organizations' (1996) 56 Zeitschrift für ausländisches öffentliches Recht und Völkerrecht 628, 654–656.

[464] W. van Reenen, 'Rules of Reference in the New Convention on the Law of the Sea, in particular in Connection with the Pollution of the Sea by Oil from Tankers' (1981) 3 Netherlands Yearbook of International Law 3, 13–16.

[465] Harrison (n 65) 166.

[466] Such as in UNCLOS (n 11) article 94(5).

[467] Harrison (n 65) 171–179; Oxman, 'The duty to respect generally accepted international standards' (n 463) 148–158; van Reenen (n 464) 8–12.

issue in its Committee on Coastal Jurisdiction over Marine Pollution, highlighted that such rules had to be respected by most states, and hence concluded that "[g]enerally accepted international rules and standards [...] are primarily based on state practice, attaching only secondary importance to the nature and status of the instrument containing the respective rule or standard."[468]

The best known rules of reference, and also the ones with the strongest binding nature in the law of the sea, are these linked to pollution and shipping standards.[469] In contrast to them, the rules of reference regarding fisheries, in UNCLOS, are weakly worded. In developing CMMs, the relevant states must indeed take into account "any generally recommended international minimum standards, whether subregional, regional or global".[470] According to this terminology, states are under no obligation of result, since they do not actually have to obey the said standards.[471] This is further made clear by this treaty's negotiating history: a stronger commitment to standards was proposed but rejected.[472] Mention of 'recommended' standards is arguably to be understood as referring to both binding and non-binding measures.[473] As for the content of global standards, the principles adopted by the FAO, a global institution, rank quite high. For regional standards, they are probably the ones agreed upon by RFMOs. However, to determine the 'generally recommended' condition can be in some cases difficult, particularly for regional standards. In any case, the general rules for RFMOs developed by the UNFSA could serve as a useful benchmark.

The situation is different under the UNFSA, according to which states have to "adopt and apply generally recommended international minimum standards for the responsible conduct of fisheries operations".[474] This means that "the Code [of Conduct] and other relevant FAO instruments [are made] binding on States Parties to the Fish Stocks Agreement."[475] Such incorporation is further drawn

---

[468] Conclusion n°2 as presented in E. Franckx (ed), *Vessel-source Pollution and Coastal State Jurisdiction: The Work of the ILA Committee on Coastal State Jurisdiction Relating to Marine Pollution (1991–2000)* (Kluwer Law International 2001) 107. This topic has indeed been addressed by the Committee on coastal State jurisdiction relating to marine pollution of the ILA and presented in three reports (First Report, Helsinki, 1996; Second Report, Taipei, 1998; Final Report, London, 2000), as presented in Franckx (ed), *Vessel-source Pollution and Coastal State Jurisdiction* (n 468) respectively at 11–31, 45–52, 105–132; see also D. König, 'Marine Environment, International Protection' Max Planck Encyclopedia of Public International Law www.mpepil.com §14.

[469] A. Boyle and C. Chinkin, *The Making of International Law* (Oxford University Press 2007) 219; Harrison (n 65) 166; R. Wolfrum, 'IMO Interface with the Law of the Sea Convention' in M. H. Nordquist and J. N. Moore (eds), *Current Maritime Issues and the International Maritime Organization* (Kluwer Law International 1999).

[470] UNCLOS (n 11) articles 61(3), 119(1)(a).

[471] Harrison (n 65) 225.

[472] Proposal by six Eastern European States, see ibid. 224–225.

[473] Ibid. 225.

[474] UNFSA (n 37) article 10(c).

[475] Harrison (n 65) 226 (reference omitted).

attention to by the reference to “generally accepted standards for the conservation and management of living marine resources” in the dispute settlement provision; these standards are indeed also to be applied by a court.[476]

Other types of rules of reference can be identified. First, most instruments mention the need to take into account the best scientific evidence available. Hence decisions are to be made on the basis of scientific data often developed elsewhere. However, the wording chosen in fisheries as well as in conservation regimes is one of an advisory only role for science, where it serves to inform but does not trigger automatic decisions.

Secondly, some regional measures refer to each other. Indeed, the measures taken in the various RFMOs for sharks are, to a certain extent, linked to each other. Particularly, GFCM has adopted the ICCAT measures applicable in the Mediterranean. As for CCSBT, the RFMO that has not adopted any binding measures for sharks, it encourages its members to comply with measures adopted by IOTC, WCPFC and ICCAT. This mechanism of referral recognizes the interrelatedness, if not overlap, of some management regimes. Nevertheless, it is rather hortatory than binding.

II. Appropriate Decisions and Decision-Making

1. Independent and Sufficient Scientific Advice

First of all, there is a strong need to improve the scientific understanding of the biological processes at stake. For an efficient management regime to be established, as recognized by many RFMOs, sufficient and independent data is indeed required on various aspects of the fisheries,[477] such as the stocks’ abundance, recruitment structure, geographical distribution and age distribution.[478] Information is also needed on the physical factors that impact the stocks and the recruitment as well as the various relationships within the ecosystem. Indeed, even though some structural factors remain stable, the marine habitat is generally of a non-permanent nature.[479] As for an ecosystem-based management to fisheries, it also requires “adequate data on bycatch, species belonging to the same ecosystem, or associated with or dependent upon the target species, and indicators of ecosystem changes”.[480]

Ideally, a sufficient quantity of stocks’ assessments and other research trips would be undertaken in a totally independent manner, at sea and without the

[476] UNFSA (n 37) article 30(5).

[477] Lugten (n 145) 12–14.

[478] Longhurst (n 53) 12.

[479] Ibid. 64–65.

[480] Morishita, ‘What is the ecosystem approach for fisheries management?’ (n 49) 25.

influence of the industry or of political powers. However, the reality is one of limited resources and not enough of such studies can be financed. Instead, information on stocks' statuses is, in large parts, extracted from catch data. These need at least to be as detailed and unbiased as possible.[481]

While models will never be perfect and are based on a simplification of reality, they should nevertheless include rather more than less input data. Also, scientists' role should be trying "to get as close to absolute certainty as possible, and to specify the level of the uncertainty that is inherent in any prediction of future states of the natural world, including that of marine fish populations."[482] Advice, to be legitimate and hence have potentially more influence, should be developed without the involvement of any lobby and outside of political agendas.[483] However, it must be accepted that scientists are humans and as such work on the basis of assumptions; whether acknowledged or not their conclusions will to some extent reflect this situation.[484]

It is also true that, on the scientists' side, there is a clear recognition of the fact that the science on which fisheries advice is based is imprecise, with many uncertain variables.[485] This should nevertheless not serve as a justification for decision-makers not to seriously take it into account or not to adopt measures. Indeed, according to the precautionary principle, the lack of data should not serve as a reason not to act. It is also only when uncertainty decreases that the levels of fishing should be allowed to increase, by proportionally reducing the precautionary buffer zone. As for the argument that the precautionary principle is less relevant to fisheries than to other species because of their high resiliency to catches,[486] it has been proven wrong by the collapse of several fish stocks.

2. Precautionary Approach

The complexity of the marine natural environment and dynamics means that perfectly reliable predictions are unlikely to be made.[487] Within this reality where uncertainty has been described as 'endemic',[488] any advice should take into account the precautionary principle. Indeed, the existence and extent of missing data and

---

[481] Burke, *The New International Law of Fisheries* (n 15) 92.

[482] Longhurst (n 53) 161.

[483] P. M. Haas, 'When Does Power Listen to Truth? A Constructivist Approach to the Policy Process' (2004) 11 Journal of European Public Policy 569, 576.

[484] D. Butterworth, 'Science and Fisheries Management Entering the New Millennium' in M. H Nordquist and G. Moore (eds), *Current Fisheries Issues and the Food and Agriculture Organization of the United Nations* (Kluwer Law International 2000) 45.

[485] Cochrane, 'Complexity in fisheries and limitations in the increasing complexity of fisheries management' (n 151) 921.

[486] Burke, *The New International Law of Fisheries* (n 15) 29.

[487] Schrank (n 180) 300.

[488] Butterworth (n 484) 39.

inherent scientific uncertainty in the fisheries field require such an approach in order to take safe decisions.[489]

The precautionary approach has become increasingly important in environmental law, especially in the last decades[490] and, according to Agenda 21, it applies to the protection of the marine environment.[491] Based on the idea that scientific uncertainty is not a justification not to take measures to preserve the environment,[492] it can be used as a gap-filler when lack of information is temporary and/or as a parameter in decision-making when it is acknowledged that "scientific uncertainty [is. . .] inherently part of a particular issue."[493] Those two aspects are of interest in the case of fisheries' conservation and management, as data is lacking and ecosystem phenomena are too complex to be fully understood, at least for now.

Both the terms 'precautionary approach' and 'precautionary principle' are used in practice. While for many authors, these are not clear cut[494] some commentators differentiate the two concepts, along the following lines: the approach refers to a soft version of precaution while the principle relates to a hard line one.[495] According to the first, as found in the Rio Declaration, scientific uncertainty should not prevent action. The second version provides that in the face of scientific uncertainty, action shall be taken. A strong form of this second version which actually reverses the burden of proof provides that the actor planning to undertake an activity shall prove that it is not harmful for the environment in order to be allowed to proceed.[496]

---

[489] Cochrane, 'Reconciling sustainability, economic efficiency and equity in fisheries: the one that got away?' (n 101) 7.

[490] A. Gillespie, 'The Precautionary Principle in the Twenty-First Century: A Case Study of Noise Pollution in the Ocean' (2007) 22 International Journal of Marine and Coastal Law 61, 66–67.

[491] United Nations, Agenda 21 – The United Nations Programme of Action from Rio (1992) chapter 17 §17.22.

[492] United Nations, Rio Declaration on Environment and Development, 1992 Conference on Environment and Development, principle 15.

[493] A. Wiersema, 'Adversaries or Partners? Science and the Precautionary Principle in International Wildlife Treaty Regimes' (2008) 11 Journal of International Wildlife Law & Policy 211, 218.

[494] M. Fitzmaurice, *Contemporary issues in international environmental law* (Edward Elgar 2009) 8; Sands and others (n 44) 218. Or they simply note the preference of North America for the former term, as in *EC Measures Concerning Meat and Meat Products (Hormones) (European Communities v United States of America), WTO Dispute Settlement Body – Appellate Body Report Adopted,* 13 February 1998 WT/DS26/AB/R and WT/DS48/AB/R §43, 60, as mentioned in D. Vanderzwaag, 'The Precautionary Principle and Marine Environmental Protection: Slippery Shores, Rough Seas, and Rising Normative Tides' (2002) 33 Ocean Development and International Law 165, 166; P. Sands, *Principles of international environmental law* (2nd edn Cambridge University Press 2003) 218.

[495] S. M. Garcia, 'The Precautionary Principle: its Implications in Capture Fisheries Management' (1994) 22 Ocean and Coastal Management 99, 103–104; S. Marr, *The Precautionary Principle in the Law of the Sea: Modern Decision Making in International Law* (Martinus Nijhoff Publishers 2003) 17–18; Vanderzwaag, 'The Precautionary Principle and Marine Environmental Protection: Slippery Shores, Rough Seas, and Rising Normative Tides' (n 494) 166.

[496] J. B. Wiener, 'Precaution' in D. Bodansky, J. Brunnée and E. Hey (eds), *Oxford Handbook of International Environmental Law* (Oxford University Press 2007) 604–606. For a separation in

While its legal status in general international law is still discussed[497] and its application by international courts is timid[498] as well as not uniform,[499] the many

two categories, see Fitzmaurice, *Contemporary issues in international environmental law* (n 494) 8–9.

[497] Wiener (n 496) 601–602, 607; Orellana, 'The Law on Highly Migratory Fish Stocks: ITLOS Jurisprudence in Context' (n 38) 487. According to P.-M. Dupuy, it has not yet reached the sufficient precision and consistency in general international law to show the necessary *opinio juris* ('Le principe de précaution et le droit international de la mer' in Mélanges offerts à Laurent Lucchini et Jean-Pierre Quéneudec (ed), *La mer et son droit* (Pedone 2003) 205–206). Of another opinion, Sands and others consider that the precautionary principle "reflects a principle of customary international law" (Sands and others (n 44) 228) and J. Cameron and J. Abouchar declared, already in 1991, that it "is emerging as a principle of customary international law" ('The Precautionary Principle: A Fundamental Principle of Law and Policy for the Protection of the Global Environment' (1991) 14 Boston College International & Comparative Law Review 1, 21). On the currently unsettled legal nature of the precautionary principle, see for example Fitzmaurice, *Contemporary issues in international environmental law* (n 494) 1, 4, 19, 27–31.

No international court has clarified its legal nature or content yet (Dupuy, 'Le principe de précaution et le droit international de la mer' (n 497) 219–220). The parties' arguments referring to the precautionary principle have not been addressed by any court, such as in *Request for an examination of the situation in accordance with paragraph 63 of the Court's judgment of 20 December 1974 in the Nuclear Tests (New Zealand v. France) case (New Zealand v. France), Order,* 22 September 1995, I.C.J. Reports 1995 288; *Southern Bluefin Tuna cases (New Zealand v. Japan; Australia v. Japan), Provisional Measures* (n 116). In *Gabcikovo-Nagymaros Project (Hungary/Slovakia)* (n 133) also, Hungary made reference to this principle but the Court, in its majority, did not mention it specifically, only addressing at 78 §141 the need for "vigilance and prevention" in protecting the environment. See Dupuy, 'Le principe de précaution et le droit international de la mer' (n 497) 216–217; Fitzmaurice, *Contemporary issues in international environmental law* (n 494) 10–22; Sands and others (n 44) 223–226. The WTO jurisprudence is not more illuminating (see for example the *EC Measures Concerning Meat and Meat Products (Hormones) (European Communities v United States of America)* (n 494) §120–125; Fitzmaurice, *Contemporary issues in international environmental law* (n 494) 22–25; Sands and others (n 44) 226–227).

[498] L. Boisson de Chazournes, 'Precaution in International Law: Reflection on its Composite Nature' in T. M. Ndiaye and R. Wolfrum (eds), *Law of the sea, environmental law and settlement of disputes: Liber amicorum Judge Thomas A. Mensah* (Nijhoff 2007) 28; N. de Sadeleer, 'Le rôle ambivalent des principes dans la formation du droit de l'environnement: l'exemple du principe de précaution' (Le droit international face aux enjeux environnementaux, Aix-en-Provence, 4–6 June 2009) 64–66.

[499] In the *Southern Bluefin Tuna cases (New Zealand v. Japan; Australia v. Japan), Provisional Measures* (n 116), ITLOS spoke of the need of 'prudence and caution' (§77) and of the existence of scientific uncertainty (§80) but refrained from using the term precautionary principle – at the exception of Judge Treves who did so in his Separate Opinion (§8–9). On this, see Fitzmaurice, *Contemporary issues in international environmental law* (n 494) 10–14; Kwiatkowska (n 116) 23; Vanderzwaag, 'The Precautionary Principle and Marine Environmental Protection: Slippery Shores, Rough Seas, and Rising Normative Tides' (n 494) 169–170. This Order was about provisional measures, which are, by nature, precautionary, as Judge Treves declared by stating that "[t]he precautionary approach can be seen as a logical consequence of the need to ensure that, when the arbitral tribunal decides on the merits, the factual situation has not changed" (Separate Opinion Judge Treves (§9); Marr (n 495) 147–148). As for Sands and others, they consider that this case reflects a practical application of the principle (Sands and others (n 44) 421). In *The MOX Plant Case (Ireland v. United Kingdom), Provisional measures,* 3 December 2001, ITLOS case

references made to it in fisheries-related treaties render the precautionary approach quite central in the field.[500] It is indeed generally recognized as applicable to all fisheries.[501]

While one can try to read a precautionary approach into UNCLOS articles on the conservation and management of living resources,[502] it is commonly accepted that this approach was not reflected in the 1982 Convention's provisions, and that it only became an explicitly part of the law of the sea's treaty history with the adoption of the UNFSA.[503] A precautionary approach is indeed required according to the 1995 instrument, with a system of reference points that establishes a buffer zone and directs to increased caution when information is lacking.[504] Many RFMOs also make reference to it and it is recognized in all post-UNFSA RFMOs except IOTC.[505] Moreover, those regional bodies that had not updated their instruments to include the precautionary and ecosystem-based approaches were reminded by the 2006 UNFSA Review Conference of the need for them to do so.[506] These two

n° 10 §75, ITLOS found the precautionary principle inapplicable to the facts of the case. In the *Case concerning Pulp Mills on the River Uruguay (Argentina v. Uruguay), Judgment,* 20 April 2010, I.C.J. Reports 2010 14, 71 §164, the ICJ pointed out that "while a precautionary approach may be relevant in the interpretation and application of the provisions of the [1975 Uruguay River] Statute, it does not follow that it operates as a reversal of the burden of proof". Finally, in *Responsibilities and obligations of States sponsoring persons and entities with respect to activities in the Area (Request for Advisory Opinion submitted to the Seabed Disputes Chamber), Advisory Opinion,* 1 February 2011, ITLOS case n° 17 §122, 125–135, ITLOS noted "the obligation to apply a precautionary approach" and pointed out that the precautionary approach is becoming CIL. On a general overview of the ICJ and ITLOS case-law related to the precautionary approach, see S. Boutillon, 'The Precautionary Principle: Development of an International Standard' (2001–2002) 23 Michigan Journal of International Law 429, 452–456; J. Zander, *The Application of the Precautionary Principle in Practice* (Cambridge University Press 2010) 37–39.

[500] Freestone, 'Problems of High Seas Governance' (n 75) 124.

[501] UNGA, Resolution adopted by the General Assembly on 8 December 2006 – Sustainable fisheries, including through the 1995 Agreement for the Implementation of the Provisions of the United Nations Convention on the Law of the Sea of 10 December 1982 relating to the Conservation and Management of Straddling Fish Stocks and Highly Migratory Fish Stocks, and related instruments, 6 March 2007, A/RES/61/105 §5; Code of Conduct for Responsible Fisheries (FAO 1995) articles 6.1, 6.2, 7.2.3; D. Freestone, 'Principles Applicable to Modern Oceans Governance' (2008) 23 International Journal of Marine and Coastal Law 385, 389.

[502] Marr (n 495) 135–136.

[503] UNFSA (n 37) article 6 and annex II; Burke, *The New International Law of Fisheries* (n 15) 109; D. Freestone, 'International Fisheries Law since Rio: the continued Rise of the precautionary Principle' in A. Boyle and D. Freestone (eds), *International Law and Sustainable Development – Past Achievements and Future Challenges* (Oxford University Press 1999) 149.

[504] Orellana, 'The Law on Highly Migratory Fish Stocks: ITLOS Jurisprudence in Context' (n 38) 489.

[505] McDorman, 'Implementing Existing Tools: Turning Words Into Actions – Decision-Making Processes of Regional Fisheries Management Organisations (RFMOs)' (n 37) 435; Sands and others (n 44) 415. For more on the precautionary approach in RFMOs, see supra Chapter 3 B. II. 1.

[506] UNGA, Resolution adopted by the General Assembly on 8 December 2006, A/RES/61/105 (n 501) §70; Freestone, 'Problems of High Seas Governance' (n 75) 113.

approaches are related, since science is imperfect and ecosystems are extremely complex,[507] too complex to be fully understood for the time being.

That the precautionary approach is applicable to fisheries does however not mean that it is successfully used by decision-makers.[508] A precautionary approach to fisheries is very difficult to implement because of the need to find a balance between the environment and economic development.[509] Hence, it is most often only paid lip-service to. Also, it is not clearly defined, even in the UNFSA: the treaty requires the states to be risk averse, to apply a precautionary approach when data is lacking, but neither this instrument nor any other does clarify which risk is (in)acceptable,[510] which level of scientific knowledge is deemed sufficient[511] or which measures have to be adopted after a possible threat has been identified.[512] This, at the end, is a political decision based on risk-assessments, cost-benefits calculations and trade-off analyses.[513] Hence, the approach is a guiding tool to deal with 'uncertain risks'[514] and not a rule requiring a predetermined result.[515]

---

[507] A. Trouwborst, 'The Precautionary Principle and the Ecosystem Approach in International Law: Differences, Similarities and Linkages' (2009) 18 RECIEL 26, 34; Vanderzwaag, 'The Precautionary Principle and Marine Environmental Protection: Slippery Shores, Rough Seas, and Rising Normative Tides' (n 494) 175.

[508] D. A. Russell and D. Vanderzwaag, 'Ecosystem and Precautionary Approaches to International Fisheries Governance: Beacons of Hope, Seas of Confusion and Illusion' in D. A. Russell and D. Vanderzwaag (eds), *Recasting Transboundary Fisheries Management Arrangements in Light of Sustainability Principles: Canadian and International Perspectives* (Legal Aspects of Sustainable Development vol 8, Martinus Nijhoff Publishers 2010) 61–68.

[509] Dupuy, 'Le principe de précaution et le droit international de la mer' (n 497) 218.

[510] E. Franckx and K. van den Bosche, 'The Influence of Environmental Law on the Development of the Law of the Sea: CITES and the International Law of Fisheries' (2011) 54 Japanese Yearbook of International Law 218, 230.

[511] Henriksen, Hønneland and Sydnes (n 12) 24.

[512] R. Kearney and others, 'Questionable interpretation of the Precautionary Principle in Australia's implementation of 'no-take' marine protected areas' (2012) 36 Marine Policy 592, 593; Tanaka, 'The Changing Approaches to Conservation of Marine Living Resources in International Law' (n 31) 314.

[513] McDorman, 'Implementing Existing Tools: Turning Words Into Actions – Decision-Making Processes of Regional Fisheries Management Organisations (RFMOs)' (n 37) 436; Wiener (n 496) 607–608; Zander (n 499) 17–26.

[514] N. de Sadeleer points to the three types of risks which can be identified: definite ones where the causal links between an activity and a negative consequence are proven, uncertain ones where such links are possible, and hypothetical ones which are not based on any scientific foundation (*Environmental Principles – From Political Slogans to Legal Rules* (Oxford University Press 2002) 91–226 in particular 155–162). Uncertain risks are the ones which require the precautionary approach while the first type of risks is to be dealt with through the preventative approach and the latter is not to be faced by a precautionary approach because a minimum scientific knowledge of the possibility of a risk is needed, as presented in Boisson de Chazournes (n 498) 22–23.

[515] B. Dickson, 'Precaution at the Heart of CITES?' in J. Hutton and B. Dickson (eds), *Endangered Species-Threatened Convention: The Past, Present and Future of CITES* (Earthscan 2000) 40; Marr (n 495) 220.

However, "[t]he greater the uncertainty, the more conservative should be the approach".[516]

The application of this approach to fisheries has created a wave of opposition, predominantly based on the fear that a hardline version would be used, de facto implementing moratoria in most fishing areas. Indeed, if the industry is to prove the harmlessness of fishing on the stocks and on the environment before being allowed to fish, few, if any, fishing grounds would remain open. Such a situation would be highly problematic since this activity is linked to food security—although it is not altogether clear to what extent—[517] and more generally to important socio-economic interests. Reversing the burden of proof would put development at risk[518] and could bring overregulation or too high a number of false-positives.[519] However, this fear appears unfounded since, except in the cases of the driftnet ban—which was criticized by some as too extreme[520] but considered more positively by others[521]—and of the IWC,[522] no such approach has been envisioned. Moreover, neither the UNFSA nor the Code of Conduct requires a hardline precautionary principle.[523]

---

[516] Cochrane, 'Reconciling sustainability, economic efficiency and equity in fisheries: the one that got away?' (n 101) 10.

[517] J. Jacquet, 'Beyond food: fish in the twenty-first century' in V. Christensen and J. MacLean (eds), *Ecosystem Approaches to Fisheries: A Global Perspective* (Cambridge University Press 2011) 121–122.

[518] Garcia, 'The Precautionary Principle: its Implications in Capture Fisheries Management' (n 495) 99, 121–122. Said in other words, the precautionary principle should apply to the environmental sector, as well as to the socio-economic sector (F. González-Laxe, 'The precautionary principle in fisheries management' (2005) 29 Marine Policy 495, 497).

[519] Wiener (n 496) 606, 609.

[520] Burke, *The New International Law of Fisheries* (n 15) 346. For a presentation of the main criticisms, see A. Rengifo, 'Protection of Marine Biodiversity: A New Generation of Fisheries Agreements' (1997) 6 RECIEL 313, 315–316.

[521] Kaye (n 107) 194; Louka (n 382) 359; F. Orrego Vicuña, *The Changing International Law of High Seas Fisheries* (Cambridge University Press 1999) 87–88.

[522] Birnie, Boyle and Redgwell (n 9) 158–159.

[523] The precautionary approach in the UNFSA is reactive and not one that would risk causing a ban on fishing (D. Freestone, 'Implementing Precaution Cautiously: the Precautionary Approach in the Straddling and Highly Migratory Fish Stocks Agreement' in E. Hey (ed), *Developments in International Fisheries Law* (Kluwer Law International 1999) 293; Henriksen, Hønneland and Sydnes (n 12) 24–25; Vanderzwaag, 'The Precautionary Principle and Marine Environmental Protection: Slippery Shores, Rough Seas, and Rising Normative Tides' (n 494) 173). The Code of Conduct for Responsible Fisheries ((n 501) articles 6.5, 7.5) also takes a rather soft approach in the call for a precautionary approach. The FAO Technical guidelines on the precautionary approach to capture fisheries further show the rejection, by fishing nations, of a reversed burden of proof: "although the precautionary approach to fisheries may require cessation of fishing activities that have potentially serious adverse impacts, it does not imply that no fishing can take place until all potential impacts have been assessed and found to be negligible" (FAO, 'Precautionary approach to capture fisheries and species introduction' FAO Technical Guidelines for Responsible Fisheries 2 (Rome 1996) 7 §7(b)).

At the very minimum, the precautionary approach should be implemented in the management models by recognizing a level of uncertainty in the calculation of authorized harvesting levels and the setting of cautious quotas. Such a soft version which acknowledges the limitations of scientific knowledge and takes measures thereabout, by setting precautionary reference points as required by the UNFSA or measures to be taken if these limits are exceeded,[524] is indeed necessary. In CCAMLR, for example, the management plans shall be based on science, but as inter-species relationships are very complex and data is deficient, a precautionary approach is to be applied in the management plans, setting conservative catch limits, in order to "minimise the risk of long-term adverse effects rather than delaying decisions until all necessary data are available."[525]

3. Robust Management

For development to be sustainable[526] management decisions need to be science-based, timely and acceptable to most states involved.[527]

It is true that, except if governments agree to put fisheries-related decisions in the hands of an independent scientific body—an unlikely scenario—, diplomatic negotiations based on national interests will continue to determine the measures applicable to fisheries. Once provided with scientific advice, decision-makers, who follow agendas in which science is only one aspect, have a leeway. This does however not signify a total legal freedom to choose unsustainable measures.[528] Indeed, states individually, but also collectively, shall base their management decisions on science, not only for questions of legitimacy,[529] but also as a matter of legality.

Science is there to objectively inform decision-makers and allow them to have an understanding of the consequences of different possible paths, as well as of the level of uncertainty in the predictions.[530] To increase chances that decisions are taken in accordance with scientific advice, it is crucial to make available objective

[524] Code of Conduct for Responsible Fisheries (n 501) article 7.5.

[525] CCAMLR, 'CCAMLR's Management of the Antarctic' (Hobart, Australia 2001) http://archive.ccamlr.org/pu/E/e_pubs/am/man-ant/e-management.pdf accessed 4 July 2015, 6.

[526] It is not clear what the general legal status of sustainable development is (Birnie, Boyle and Redgwell (n 9) 125; Cordonier Segger (n 432) 371–372) but, in the fisheries field, the MSY comprises the dual goals of development and sustainability.

[527] McDorman, 'Implementing Existing Tools: Turning Words Into Actions – Decision-Making Processes of Regional Fisheries Management Organisations (RFMOs)' (n 37) 430.

[528] See supra Chapter 4 A. II. 4. b).

[529] D. D. Caron, 'The International Whaling Commission and the North Atlantic Marine Mammal Commission: The Institutional Risks of Coercion in Consensual Structures' (1995) 89 The American Journal of International Law 154, 159–163.

[530] Cochrane, 'Reconciling sustainability, economic efficiency and equity in fisheries: the one that got away?' (n 101) 12; Polacheck, 'Politics and independent scientific advice in RFMO processes: A case study of crossing boundaries' (n 163) 140.

and credible data on the scientific, legal, social and economic aspects involved.[531] In order to avoid politically motivated bias at too many stages, it is also important that both processes—advising and deciding—be well demarcated.[532] The use of a separate advisory body is therefore very positive.[533]

Transparency is considered a major factor.[534] It should, for example, be acknowledged if a scientific committee does not reach an agreement, and all views should be reflected in the conclusions.[535] Also, the report should automatically be made public, in order to increase the accountability of states taking decisions on the basis of such recommendations.

Additionally, the decision-making processes should be timely, i.e. be able to act and react rapidly, in order to take into account new information and be capable to deal with uncertainty.[536] In a rapidly changing environment, the importance of flexibility, expeditiousness and efficiency of the procedures to adopt CMMs can indeed not be overemphasized.[537] As mentioned in the UNFSA, procedures to adopt CMMs shall "facilitate the adoption of conservation and management measures in a timely and effective manner."[538]

However, this treaty does not clarify how this is best to be achieved[539] and one can wonder whether majority voting would be an interesting option, since it avoids the pitfall of having to keep on negotiating, sometimes *ad eternam vitam*, until a compromise acceptable to all is found. Nonetheless, majority voting risks leaving some key players out of the picture, which is highly inefficient.[540] Also, even with this decision-making procedure, it happens rather regularly that no agreement on measures can be found by the relevant RFMO, hence resulting in a de facto unregulated fishery. For example, in NEAFC, there were years such as 2005

---

[531] Burke, 'Unregulated High Seas Fishing and Ocean Governance' (n 162) 241.

[532] Polacheck, 'Politics and independent scientific advice in RFMO processes: A case study of crossing boundaries' (n 163) 139–140.

[533] Burke, 'Unregulated High Seas Fishing and Ocean Governance' (n 162) 241.

[534] Polacheck, 'Politics and independent scientific advice in RFMO processes: A case study of crossing boundaries' (n 163) 140.

[535] Stokke, 'Governance of High Seas Fisheries: The Role of Regime Linkages' (n 207) 161. This is for example the procedure in CCAMLR (CCAMLR Rules of Procedure of the Scientific Committee, rule 3); or NAFO (NAFO Convention (n 188) article X(1)). CCBST's practice, *a contrario*, should be mentioned in that its Scientific Committee shall make recommendations by consensus (CCSBT Convention (n 117) article 9(2)(d)).

[536] Gjerde, 'High Seas Fisheries Governance: Prospects and Challenges in the 21st Century' (n 25) 227.

[537] Henriksen, Hønneland and Sydnes (n 12) 39.

[538] UNFSA (n 37) article 10(j).

[539] McDorman, 'Implementing Existing Tools: Turning Words Into Actions – Decision-Making Processes of Regional Fisheries Management Organisations (RFMOs)' (n 37) 427.

[540] Ibid. 429–430. It is interesting to note that some RFMOs, such as SEAFO, couple the double weakness of consensus and opting out (SEAFO Convention (n 188) articles 17, 23). For more on voting procedures and opting-out possibilities in RFMOs, see Chapter 3 B. II. 1.

when only one of the major fish stocks was subjected to measures because the required majority could not be reached for the other stocks.[541]

As to unanimity or consensus, it means respectively that all states agree or that no state is profoundly opposed to the measure proposed.[542] These adoption methods are very inclusive but give a veto right to any member and are usually neither quick nor efficient processes, since the minimum common denominator is chosen.[543]

4. Binding Decisions

As difficult as it is to combine with the sovereignty of states and the freedom of fishing, especially in a time of conservation crisis when the quotas must be decreased, an efficient management regime needs measures binding all actors involved.[544] Ideally, "[a] fishery body should be permitted to prescribe regulations that are effective for all participants in a fishery."[545] More realistically, some solutions must exist to fight the free-riding problem, whether from within the RFMO—with the use of the opt-out clauses—or from outside—with unregulated fishing.

First of all, abolishing the opt-out provisions or at least making their use more difficult by "setting procedural or substantive restraints on its exercise"[546] would be positive. For UNFSA parties two arguments have been developed to defend the existence of a ban on the use of opt-outs. Indeed, these states have to comply with CMMs[547] even if they are not member of the relevant RFMO; it would hence be illogical that membership allowed them to disregard the measures.[548] It has also been suggested that article 18 (1) of UNFSA can be interpreted as obliging states to comply with all measures adopted by RFMOs or at least to act in a way which does not undermine the effectiveness of the CMMs.[549] Finally, under UNCLOS, one could consider that opting-out is a breach of the obligation to cooperate in the conservation of living resources.[550]

---

[541] Henriksen, Hønneland and Sydnes (n 12) 105.

[542] Klabbers, *An introduction to international institutional law* (n 284) 206–208; Schermers and Blokker (n 281) §771–772.

[543] Garcia, 'The Precautionary Principle: its Implications in Capture Fisheries Management' (n 495) 114; Schermers and Blokker (n 281) §785; Warner (n 43) 126.

[544] Burke, *The New International Law of Fisheries* (n 15) 92.

[545] Burke, 'Unregulated High Seas Fishing and Ocean Governance' (n 162) 241.

[546] Churchill, 'Legal uncertainties in international high seas fisheries management' (n 184) 228.

[547] UNFSA (n 37) article 10(a).

[548] Henriksen, Hønneland and Sydnes (n 12) 37–39.

[549] Churchill, 'Legal uncertainties in international high seas fisheries management' (n 184) 228–229.

[550] Ibid. 229.

Some of the most recent RFMOs, such as SEAFO, have restricted the possibility to opt out[551] or totally left it aside (CCSBT, IATTC, SIOFA).[552] In WCPFC, while the adoption procedure is based on consensus, a vote might be taken if there is a stalemate; in that case only, a state who has voted against a measure can seek a review of the measure, but still not opt out from it.[553] As for SPRFMO, it has an innovative procedure, according to which decisions binding on all members can be taken at a three-quarter majority, and any objections to measures must be counterbalanced by measures of a similar nature. Moreover, any objection is to be reviewed by a panel the establishment of which is automatic.[554] Such robust majority decision-making procedures would be of great benefit to many RFMOs.

Second, in the fight against free-riding by non-members, diplomatic pressure, trade measures, such as differential treatment resulting in automatic inspection in port or being considered as undertaking IUU fishing until proven otherwise, have been and are extensively used. They are effective, but they could violate WTO rules.[555]

The counterpart of the fight against free-riding is the need for inclusiveness. As required in the UNFSA, states with a real interest shall be allowed to join an RFMO.[556] However, the extent of rights which will be granted depends on a number of factors, amongst which are the status of stocks, the existing fishing practices and the involvement of the state in conservation and data sharing.[557] It must nevertheless be noted that there is no simple solution to quota allocation[558] since resources are limited and compliance can only be expected if states perceive their share of quotas as sufficient.[559]

---

[551] SEAFO Convention (n 188) article 23(1)(c)-(g), clarifying that the reasons for non-acceptance have to be provided and alternative CMMs prescribed.

[552] CCSBT does however have a unanimity rule (CCSBT Convention (n 117) articles 7–8); IATTC has a consensus rule (Antigua Convention (n 188) article IX); and so does SIOFA (SIOFA Agreement (n 188) article 8).

[553] WCPFC Convention (n 188) article 20.

[554] SPRFMO Convention (n 188) articles 16–17, annex II. Such procedure took place in *The Objection by the Russian Federation to a Decision of the Commission of the South Pacific Regional Fisheries Management Organisation* (n 349); for more on this, see supra Chapter 4 A. IV. 3. c) aa).

[555] Churchill, 'Legal uncertainties in international high seas fisheries management' (n 184) 230–231.

[556] UNFSA (n 37) article 8(3).

[557] Ibid. article 11.

[558] McDorman, 'Implementing Existing Tools: Turning Words Into Actions – Decision-Making Processes of Regional Fisheries Management Organisations (RFMOs)' (n 37) 425–426; Molenaar, 'Participation, Allocation and Unregulated Fishing: The Practice of Regional Fisheries Management Organisations' (n 26) 469–472; see supra Chapter 4 B. I. 2. a).

[559] T. Bjorndal, 'Overview, roles, and performance of the North East Atlantic fisheries commission (NEAFC)' (2009) 33 Marine Policy 685, 687; Molenaar, 'Participation, Allocation and Unregulated Fishing: The Practice of Regional Fisheries Management Organisations' (n 26) 459.

III. Efficient Regional Management and Global Enforcement

An efficient regime is needed to verify compliance and if necessary to sanction non-compliance. This is crucial for both the effectiveness and the credibility of the system.[560] Hence, while the particular CMMs should rather be taken at the regional level, with a view to the ecosystem unity, measures should nevertheless be made globally enforceable in order to close loopholes. Moreover, procedures to review compliance of vessels and of states, to assess RFMOs' performance as well as to facilitate judicial dispute settlement are essential.

1. Scope of Management Measures

a) Regional Focus

It is recognized that CMMs are usually best adopted at the regional level, by the actors who know the specificities of the resources.[561] Such a 'subsidiarity principle' is envisioned in UNCLOS and given full acknowledgment in the UNFSA: both instruments indeed make numerous references to the role of regional and sub-regional institutions in the management of fisheries.[562]

The present situation is quite satisfactory in that regard. Nevertheless, some problems remain such as the lack of binding review of the actions of RFMOs, the non-universality of rules towards the relevant states in an area, and the insufficient coordination between the regional bodies. Also, it is necessary that EEZ waters be included in the regional management, since fish straddles and migrates. Hence, clear procedures to adopt compatible or identical measures should become the norm. While this is already the case in some RFMOs managing highly migratory species and will be so, in the near future, for an RFMO taking care of straddling stocks,[563] the compatibility rule should be further promoted.

---

[560] Burke, 'Unregulated High Seas Fishing and Ocean Governance' (n 162) 241.

[561] Chairman of the United Nations Conference on Straddling Fish Stocks and Highly Migratory Fish Stocks, 'Statement made by the Chairman of the Conference at the Conclusion of the General Debate on 15 July 1993' 21 July 1993, A/CONF.164/12 2(g); Orrego Vicuña, *The Changing International Law of High Seas Fisheries* (n 521) 202.

[562] The various references to the establishment of sub-regional or regional fisheries organizations are found in UNCLOS (n 11) articles 61, 63 and 118 as well as in the UNFSA (n 37) in particular articles 8–9.

[563] Respectively WCPFC (WCPFC Convention (n 188) article 8) and SPRFMO (SPRFMO Convention (n 188) article 4); see van Bohemen (n 192) 243–244; van Dyke (n 336) 175–176.

b) Ecosystem Approach

Since target fish stocks do not exist in a vacuum and their abundance is importantly influenced by various ecological and physical factors, it seems only reasonable to take, as much as possible, the relevant elements into account.[564]

The so-called ecosystem approach has never been defined in a globally accepted manner in relation to fisheries.[565] Two versions of the concept have appeared in marine matters: the ecosystem-based management intends to take into consideration "all human uses of the (marine) environment and must also resolve conflict between uses."[566] Hence, management should integrate questions such as pollution, shipping, together with fishing.[567] As presented by the FAO, it aims at balancing "diverse societal objectives, by taking into account the knowledge and uncertainties about biotic, abiotic, and human components of ecosystems and their interactions and applying an integrated approach to fisheries within ecologically meaningful boundaries."[568]

In comparison, the ecosystem-based fishery management (or also called ecosystem-based approach to fisheries) 'only' focuses on fisheries and what impacts them. It is however a more holistic approach than a multi-species management in that it considers all ecosystem components, not just the species to be fished.[569] In the current practice, this approach involves non-exhaustively the avoidance of by-catch—but the level of mitigation required is a policy decision—, the protection of sensitive ecosystems, multi-species management,[570] and the recognition of prey–predator relationships.[571] Its global implementation would be a major step towards sustainability.

Most RFMOs, starting with CCAMLR, have adopted, formally or in their working methods, an ecosystem-based approach to fisheries.[572] The 2006 UNFSA Review Conference highlighted the need for RFMOs to update their instruments in that regard.[573] This approach is somewhat recognized in the UNFSA,[574] in the FAO Code of Conduct[575] and in fisheries around the world. A global commitment to

---

[564] On this, see for example World Ocean Review, *The Future of Fish – The Fisheries of the Future* 2013 (n 144) 12–23, 105–107.

[565] Tanaka, 'The Changing Approaches to Conservation of Marine Living Resources in International Law' (n 31) 305.

[566] Molenaar, 'Marine Mammals: The Role of Ethics and Ecosystem Considerations' (n 50) 46.

[567] T. J. Pitcher and others, 'An evaluation of progress in implementing ecosystem-based management of fisheries in 33 countries' (2009) 33 Marine Policy 223, 223.

[568] FAO, 'Fisheries Management 2. The ecosystem approach to fisheries' FAO Technical Guidelines for Responsible Fisheries 4 Suppl. 2 (Rome 2003) 6.

[569] Molenaar, 'Marine Mammals: The Role of Ethics and Ecosystem Considerations' (n 50) 46.

[570] Morishita, 'What is the ecosystem approach for fisheries management?' (n 49) 19.

[571] Frid, Paramor and Scott (n 165) 463.

[572] See supra Chapter 3 B. II. 1.

[573] Freestone, 'Problems of High Seas Governance' (n 75) 113.

[574] UNFSA (n 37) article 5(d)-(h).

[575] Code of Conduct for Responsible Fisheries (n 501) articles 6.1, 6.2, 7.2.3.

apply it was moreover taken in 2001, in the Reykjavik Declaration at the Conference on Responsible Fisheries in the Marine Ecosystem.

However, while this approach is already recognized as applicable to all fisheries,[576] its normativity remains debated.[577] Also, it is not clearly defined, difficult to measure and its implementation is insufficient, even by coastal states in the management their EEZs[578]—which leaves little hope of better results in the management of high seas fisheries.[579] Another of the main issues is the lack of sufficient data,[580] a situation which calls for the application of the precautionary principle.[581] Also not to be underestimated is the political resistance against the application of an ecosystem approach because of some misconceptions, and the abuse of the term for controversial measures.[582] Indeed, if, in the name of the ecosystem approach, a zero tolerance to by-catch is expected, many fisheries should be closed. In reality, compromises between species will always be obligatory, since it is impossible to reach MSY jointly for all species in an ecosystem.[583]

In summary, this approach is overall positive for the enhancement of marine species conservation because it requires states to take into account the interrelations in the ecosystem. Indeed, ecosystem management focuses on geographical areas within which there are meaningful interactions between species and with their environment, rather than on a politically defined zone or on the management of one species or one particular use of the seas.[584] However, this complex policy tool has to be based on sufficient scientific knowledge and, more than anything, needs to be actually implemented by decision-makers. On a more positive note, this is the case in some regions even though "the perception is that it remains largely a promise unfulfilled."[585]

[576] UNGA, Resolution adopted by the General Assembly on 8 December 2006, A/RES/61/105 (n 501) §5; Freestone, 'Principles Applicable to Modern Oceans Governance' (n 501) 389.

[577] Tanaka, 'The Changing Approaches to Conservation of Marine Living Resources in International Law' (n 31) 305–306.

[578] Pitcher and others (n 567) 224.

[579] Russell and Vanderzwaag (n 508) 42–56.

[580] Burke, *The New International Law of Fisheries* (n 15) 59; Garcia and Cochrane (n 150) 316; Tanaka, 'The Changing Approaches to Conservation of Marine Living Resources in International Law' (n 31) 306–307.

[581] FAO, 'Fisheries Management 2. The ecosystem approach to fisheries' (n 568) 22 §1.4.7.

[582] Morishita, 'What is the ecosystem approach for fisheries management?' (n 49) 19.

[583] Cochrane, 'Reconciling sustainability, economic efficiency and equity in fisheries: the one that got away?' (n 101) 10; Churchill and Lowe (n 14) 282.

[584] Juda and Burroughs (n 34) 26.

[585] S. A. Murawski, 'Ten myths concerning ecosystem approaches to marine resource management' (2007) 31 Marine Policy 681, 681. In some cases progress can be noted, such as the North East Atlantic, where an ecosystem approach is to be implemented in the next decade or so. However, whatever improved scientific advice this means, no real improvement will be noted if decision-makers continue not to take seriously into account the scientific previsions and advice provided to them (Frid, Paramor and Scott (n 165) 467–468).

2. Global Applicability of Measures and Strong Compliance and Enforcement

For any regime to be efficient actors must be under the expectation that non-compliance is likely to be detected, identified, and that measures can be taken to punish it.[586] Hence, surveillance, monitoring and sanctioning are needed on a global scale. Indeed, technological developments such as processing on board facilities or freezer ships as well as the globalization in trade[587] mean that even regional measures have to be applied globally in order to make unsustainable and illegal practices economically less profitable. Otherwise, free-riding and IUU fishing will continue to take advantage of loopholes in the regulatory and enforcement nets. With respect to global applicability of regional measures, it must nevertheless be noted that "enforcing fishing regulations in respect of non-Parties remains contentious."[588]

a) Verification of Compliance by Vessels

It is crucial that vessels' behavior be controlled and their potential violations of CMMs sanctioned in a sufficiently serious manner that it can serve as a deterrent. Indeed, "[t]he probability of apprehension is vital for inducing compliance. [...] Accordingly, for fisheries management to produce the social and economic benefits sought, the level and effectiveness of enforcement, including surveillance, capacity to apprehend, actual arrest, and penalty, are critical."[589]

In order to promote compliance, it is, before anything else, necessary to gather information to determine problematic behaviors: surveillance and monitoring of vessels are methods to this end.[590] No matter whether it is the flag state only or also the RFMO members who are responsible to verify compliance by vessels, various techniques are used and should be further encouraged to avoid that violations of

[586] Bothe (n 375) §129–139; Churchill, 'Legal uncertainties in international high seas fisheries management' (n 184) 227; R. Rayfuse, 'To Our Children's Children's Children: From Promoting to Achieving Compliance in High-seas Fisheries' (2005) 20 International Journal of Marine and Coastal Law 509, 510.

[587] M. L. McConnell, 'Observations on Compliance and Enforcement and Regional Fisheries Institutions: Overcoming the Limitations of the Law of the Sea' in D. A. Russell and D. Vanderzwaag (eds), *Recasting Transboundary Fisheries Management Arrangements in Light of Sustainability Principles: Canadian and International Perspectives* (Legal Aspects of Sustainable Development vol 8, Martinus Nijhoff Publishers 2010) 74; W. W. Taylor, M. G. Schechter and L. G. Wolfson, *Globalization: Effects on Fisheries Resources* (Cambridge University Press 2007) 25–26.

[588] A. J. Bederman, 'CCAMLR in crisis: a case study of marine management in the southern oceans' in H. N. Scheiber (ed), *Law of the Sea: The Common Heritage and Emerging Challenges* (Martinus Nijhoff Publishers 2000) 190.

[589] Burke, *The New International Law of Fisheries* (n 15) 308 (reference omitted).

[590] Ibid. 311.

CMMs remain undetected. For example, monitoring, control and surveillance include vessel monitoring systems, black and white vessel lists—a system of closed-access where only registered vessels from complying countries can fish—[591] TDS/CDS, licensing of fishing vessels, at sea boarding schemes, inspection and (regional) observers programs at sea and in port.[592]

However, these programs should be made compulsory. In the current situation indeed, where some of these are voluntary measures to be adopted within an RFMO, it is often perceived that they are costly and therefore very often only reluctantly implemented.[593] Similarly, the global record of fishing vessels which is under development at the FAO needs to be obligatory and applicable to all vessels if it is to become an adequate tool.[594]

Furthermore, in order to be able to gather sufficient scientific data, to verify compliance and to implement sanctions, information must be gathered and distributed globally: there is a need for "a global approach to the collection and analysis of catch and effort statistics, monitoring of the size and movements of fishing fleets".[595] Similarly, while IUU lists are shared between some RFMOs, a central focal point should be organized or, if not feasible, cooperation between RFMOs has to be improved.

b) Clarification of States' Obligations and Monitoring of Compliance

Strong enforcement would also require a further clarification of coastal and flag states' obligations in the field of fisheries, although the effort of ITLOS in its advisory opinion should be commended in that regard. For the former, the rules applicable are limited to UNCLOS and the requirement of compatibility in the UNFSA; these are vaguely worded and ought to be developed. As for the flag states obligations, while they are highlighted in several instruments such as the Compliance Agreement or the UNFSA,[596] their scope of applicability is not beyond debate. Hence, as it stands, states have no clear binding list of obligations to which they can be held accountable in their attitude towards fishing vessels flying their flag.

[591] Calley (n 48) 114–121; Gjerde, 'High-seas Fisheries Management under the Convention on the Law of the Sea' (n 74) 301–303.

[592] Bowman, Davies and Redgwell (n 67) 147; Gjerde, 'High-seas Fisheries Management under the Convention on the Law of the Sea' (n 74) 302.

[593] Churchill, 'Legal uncertainties in international high seas fisheries management' (n 184) 233–234.

[594] As opposed to how it is now (Freestone, 'Problems of High Seas Governance' (n 75) 116–117). See Committee on Fisheries, 'The Global Record of Fishing Vessels Refrigerated Transport Vessels and Supply Vessels (Global Record)', presented at the Thirty-first Session (Rome, 9–13 June 2014) COFI/2014/Inf. 12/Rev.1.

[595] Lodge and others (n 447) xi.

[596] Compliance Agreement (n 241) article III; UNFSA (n 37) articles 18–19.

The principles agreed upon for the FAO Flag State Performance could serve as "clear criteria and standards against which the conduct of beneficiaries, including flag states, can be judged and the nature of the response determined".[597] Ideally, a third party would be entitled to assess such compliance even if the flag state is not willing and/or able to ensure the respect of such standards by vessels in case of non-compliance.

c) Deterrent and Globally Applicable Sanctions

Making IUU fishing less profitable, through the inaccessibility of markets, through added costs or through higher—monetary and non-monetary—sanctions against the actors and the vessels engaged in the activity, is a way to decrease this practice.[598] In particular, economic sanctions, forfeiture of the vessel and of the catch have the potential to be "of sufficient severity to effectively prevent, deter and eliminate IUU fishing and to deprive offenders of the benefits accruing for such fishing."[599] Making sure that states have jurisdiction to sanction their nationals when they are involved in IUU fishing, even on board a boat flagged elsewhere is also useful.

As suggested by the IPOA-IUU, trade sanctions can serve as a pressure mechanism since closing markets is a strong motivation to stop IUU fishing.[600] They are indeed meant "[t]o reduce the opportunities and incentives for IUU fishing".[601] For example, the ICCAT ban on import of Atlantic Bluefin Tuna from Belize and Honduras[602] led to an improvement in these states' behavior and ultimately in both countries joining the Commission.[603]

However, trade sanctions cannot reach the desired result, if they are applied in a patchwork manner.[604] Sanctions need to be globally applicable, in order to avoid that the use of flags and ports of convenience undermines their efficiency.[605] Hence, the mechanisms on which the sanctions are based, whether IUU lists or CDS, must be applicable to all states and not only to the members of a particular RFMO, at the risk otherwise that fishing vessels will just go to the next port for landing or change

---

597 Rayfuse and Warner (n 24) 416–417.

598 Calley (n 48) 147–150; Le Gallic and Cox (n 231) 692–694.

599 FAO, International Plan of Action to prevent, deter and eliminate illegal, unreported and unregulated fishing (IPOA-IUU) 2001 §21.

600 Ibid. §66.

601 Lack, 'Catching On?' (n 42) 1.

602 Recommendation by ICCAT regarding Belize and Honduras Pursuant to the 1994 Bluefin Tuna Action Plan Resolution 1996.

603 Calley (n 48) 30–35; Riddle (n 48) 285–286. For another example, see the measures in NEAFC (Stokke, 'Trade Measures and the Combat of IUU Fishing: Institutional Interplay and Effective Governance in the Northeast Atlantic' (n 41)) and in general, see Lack, 'Catching On?' (n 42) viii.

604 Flothmann and others (n 224) 1236.

605 Calley (n 48) 141–147.

regions if spotted as an IUU vessel.[606] Additionally, port states should not only have the right to, but also be under the obligation to use port state measures.[607]

CDS are one of the methods for efficient port state measures. For example, the CCAMLR CDS would be adequate, if it was applicable to all toothfish caught and applied in all ports. Indeed, this scheme has the advantage of being a catch and trade scheme—while many other RFMO schemes only apply to trade—hence making the fish traceable from the fishing point onward.[608] It also significantly decreases the value of toothfish caught in IUU activities, since landing it in a member or cooperating state port is forbidden. However, the complete coverage of the market has not been reached in practice[609] and "too many states are unwilling to scrutinize the documents and many vessels are still able to pass false documents to port officials."[610] Overall the establishment of a general certification mechanism to allow entry into markets would be a very beneficial move.[611]

d) Review of RFMOs Work

Many RFMOs have not reached their sustainability objectives.[612] Hence, performance reviews need to be further undertaken, with independent experts and against some agreed standards. Not only are the CMMs adopted to be generally reviewed, but so do the founding documents, in particular if they do not include the precautionary and ecosystem approaches and other principles examined here above.

It is necessary to hold organizations accountable[613] for both their actions and omissions, but, in the absence of an open judicial forum to invoke their international responsibility, one needs to rely on other procedures. Hence, establishing a regular independent mechanism through which an RFMO's balance sheet would be reviewed against a set of indicators, including their adoption and implementation

---

[606] Flothmann and others (n 224) 1236.

[607] Molenaar, 'Port State Jurisdiction: Towards Comprehensive, Mandatory and Global Coverage' (n 40) 245.

[608] Lack, 'Catching On?' (n 42) 8–9; Rayfuse, 'Building sustainable high seas fisheries through certification processes: issues and perspectives' (n 266) 98–99.

[609] Riddle (n 48) 282–283.

[610] Ibid. 284.

[611] L. A. Chaves, 'Illegal, Unreported and Unregulated Fishing: WTO-consistent Trade related Measures to Address IUU Fishing: AUS:IUU/2000/16' (Background paper for the Expert Consultation on Illegal, Unreported and Unregulated Fishing Organized by the Government of Australia in Cooperation with FAO, 2000) http://www.fao.org/docrep/005/Y3274E/y3274e0i.htm#bm18 accessed 6 July 2015, Executive Summary.

[612] Yagi (n 444) 212.

[613] N. D. White, 'Separate but Connected: Inter-Governmental Organizations and International Law' (2008) 5 International Organizations Law Review 175, 194.

of the UNFSA requirements,[614] would allow the international community to keep track and give advice to ameliorate performance.[615]

e) Dispute Settlement Mechanisms

Ideally, the dispute settlement mechanism applicable to fisheries matters would be compulsory and without exception, inclusive—in other words looking at measures both within and beyond national jurisdiction—and also explicitly open to states acting on behalf of the international community for obligations *erga omnes*. On a positive note, it must be mentioned that dispute settlement procedures for the protection of the marine environment which do not suffer from any limitation of competence are arguably applicable to fisheries.[616]

Moreover, one can hope that the Arbitral Tribunal's decision in the *Southern Bluefin Tuna* case will be 'reversed', bringing disputes between RFMOs members back within the ambit of the UNCLOS Part XV mechanisms. Also, specifications of the principles applicable to fisheries management would be useful, as was provided by ITLOS in its recent advisory opinion. More generally, dispute settlement could help clarify which behavior is expected from a state within an RFMO in order to respect UNCLOS. As for the provision on compatibility in UNFSA, it could possibly allow the review of national measures taken by the coastal state; such an option should be clarified.

Aspects of the governance problems highlighted, such as lack of respect of measures adopted, free-riding, regular opt-out of collective measures, or non-control over vessels flying certain states' flag could benefit from a binding third-party review. States should maybe start trying to hold others accountable in front of international courts for unsustainable fishing practices. Indeed, fishing in violation of RFMOs measures or not cooperating in the conservation of marine living resources breaches international law.

A coastal state whose fisheries is particularly affected by IUU or unsustainable fishing on the high seas might be considered injured and hence able to invoke the responsibility of the state involved in such practice.[617] Other states might also attempt to invoke a wrongdoing state's international responsibility as non-injured states, as provided by article 48 of the ILC articles on state responsibility. The possibility for them to bring a claim on the basis of an *erga omnes* obligation would benefit from a clarification. Questions of reparation, in particular the reasonableness of restitution or the quantification of 'financially assessable damage' for

---

[614] Gjerde, 'High-seas Fisheries Management under the Convention on the Law of the Sea' (n 74) 298.

[615] Ibid. 299.

[616] UNCLOS (n 11) article 297(1)(c); Tudela (n 335) 357–358.

[617] Articles on the Responsibility of States (n 242) commentary on article 42 §12.

compensation would also gain from being examined in more detail in the field of environmental and fisheries law.

Finally, advisory opinions could be increasingly used to clarify points of law—as is the one issued by ITLOS at the request of the SRFC—or to act in the interest of the global commons.[618] It is true that the risk always exists that advisory proceedings are in fact related to a contentious case. However, courts decide whether they are competent and should be trusted in that regard. Advisory opinions are based on a multilateral approach, with all states and many international organizations—even NGOs in the case of ITLOS—invited to make a representation and then comment on each other's' submissions. This type of procedure has been said to "represent perhaps the best method for litigating breaches of *erga omnes* rules."[619]

C. Conclusion

In conclusion, to avoid a tragedy of the commons, the sacrosanct principle of the freedom of fishing on the high seas has to be clearly curbed. Also, while detailed rules which take into account the particularities of a region are crucial and have been adopted in most areas, their applicability would benefit from binding states beyond the region and ideally on a universal scope. Indeed, measures taken by states are costly and complying actors do not want free-riders to be the beneficiaries of their sacrifices. Also, in a globalized economy, the formation of loopholes is a risk to keep in mind. Moreover, sustainable unilateral (or more generally non-universal) measures are likely to fail, as "even if an individual [...] extractor takes measures to diminish his/her impact on a resource, others would continue to behave as profit maximizers, leading sooner or later to the collapse of the resource."[620]

Furthermore, sustainability-oriented principles such as the ecosystem approach, the precautionary principle, decisions based on the best science available and robust management's procedures, such as majority voting and binding decisions on all members of RFMOs, are necessary.

Finally, as many fish stocks bear witness of, even sound decisions are not sufficient if they are not complied with and enforced; these aspects also require improvement.[621] Regional monitoring and surveillance, backed-up by global

[618] Birnie, Boyle and Redgwell (n 9) 253–254; M. Fitzmaurice, 'Environmental protection and the International Court of Justice' in V. Lowe and M. Fitzmaurice (eds), *Fifty years of the International Court of Justice: Essays in honour of Sir Robert Jennings* (Cambridge University Press 1996) 314.

[619] Birnie, Boyle and Redgwell (n 9) 254.

[620] Louka (n 382) 76.

[621] J. L. Garrison, 'The Convention on International Trade in Endangered Species of Wild Fauna and Flora (CITES) and the Debate Over Sustainable Use' (1994) 12 Pace Environmental Law Review 301, 308, 374–379; A. H. Hoel, 'Political uncertainty in international fisheries

implementation to avoid loopholes, enforcement by non-flag states to combat flags of convenience, and widened options for standing in front of a dispute settlement body to protect the global commons are some of the key aspects of effective implementation and enforcement of fisheries regulations.

This chapter highlighted a number of governance problems which seem difficult to solve solely through existing fisheries agreements. Indeed, while RFMOs have been entrusted with the main management powers in terms of fisheries, they cannot apply beyond their limited mandate and, in some cases, have not lived up to the hopes of the international community. Hence, it is interesting to look beyond pure resource management regimes to other mechanisms, which focus rather on biodiversity conservation, to determine whether they can close some loopholes and/or address issues more efficiently.

In that regard, CITES is increasingly discussed as a potential complementary tool[622] and is of particular interest for two main reasons. First, it focuses on species at a risk of extinction. As presented in Chapter 2, the fishing crisis is serious to the point of threatening some species' survival, notwithstanding the particularities of fishes, which are usually highly fecund. Second, CITES created mechanisms to reduce the pressure caused on wild species by international trade. As mentioned in the Introduction, the proportion and value of fishery products entering into international trade is particularly high compared to other natural resources. Since "CITES was created to address over-exploitation driven by international trade, the very pressure that hugely affects many marine species",[623] the use of this regime in relation to commercially-exploited fish species appears to be very relevant and to have some potential to improve the governance of those marine resources.

management' (1998) 37 Fisheries Research 239, 241, presenting the three tasks in the management of living resources, that are the need for a knowledge base, the task of regulating fisheries by the adoption of CMMs and the enforcement of the adopted regulations.

[622] A. C. J. Vincent and others, 'The role of CITES in the conservation of marine fishes subject to international trade' (2014) 15 Fish and Fisheries 563.

[623] Ibid. 585 (references omitted).

Part II
The Use of CITES for Commercially-exploited Fish Species

Chapter 5
Protection of Commercially-exploited Fish Species Under CITES

In this chapter, after discussing the philosophies underlying CITES (A), the legality of listing fish species under CITES will be clarified, (B) the adequacy of the regime for that purpose analyzed (C), and the effects as well as the potential contribution of using CITES towards effective governance of marine fisheries assessed (D).

Contemplating CITES for commercially-exploited fish species is a rather new development.[1] Hence, it is important to examine how the underlying philosophies of the regime could influence the likelihood of a listing. The divide between sustainable use and preservation is indeed at the heart of the debates within the CITES' membership.

Then, the second section of this chapter will help clarify that utilizing CITES for marine species conforms to the mandate given to this regime and hence is legal. It serves to answer criticisms made by some parties. Japan in particular highlighted on several occasions its opinion that CITES is not the proper forum to manage fisheries and that other institutions exist to do so. It declared for example in 2003 that the use of CITES for two shark species was an "inappropriate application of the Convention",[2] and China questioned "whether CITES was intended to deal with such species".[3] In order to handle alleged

---

[1] E. Franckx, 'The Protection of Biodiversity and Fisheries Management: Issues Raised by the Relationship between CITES and LOSC' in D. Freestone, R. Barnes and D. Ong (eds), *The Law of the Sea – Progress and Prospects* (Oxford University Press 2006) 212.

[2] Japan, 'Kesennuma Declaration regarding Sustainable Utilization of Shark Resources and Tuna Longline Fishing' (11 July 2003), as presented in Japan, 'AC19 Doc. 18.3 – Biological and trade status of sharks (Resolution Conf. 12.6 and Decision 12.47) – Progress made by Japan in developing and implementing the IPOA-Sharks' (18 August 2003) annex 2.

[3] P. Thomas, 'Mushrooms and the Future of CITES' (2005) 30 Endangered Species Bulletin http://www.fws.gov/endangered/bulletin/2005/ESB09-05.pdf accessed 6 July 2015, 24, 25; CITES, 'CoP8 Doc. 8.46 (Rev) – Consideration of proposals for amendment of Appendices I and II' (2–13 March 1992) annex 4, 41; China, 'CoP14 Inf. 45 – Shark Issues' (3–15 June 2007) 2. On this, see for example M. A. Young, *Trading Fish, Saving Fish: The Interaction between Regimes and*

S. Guggisberg, *The Use of CITES for Commercially-exploited Fish Species*, Hamburg Studies on Maritime Affairs 35, DOI 10.1007/978-3-319-23702-2_5

conflicts[4] some states have indeed tried to interpret the Convention in an exclusive manner towards some commercially-exploited aquatic species, especially the ones already managed by another regime.[5]

The third section discusses the adequacy of the regime. Some parties' arguments to oppose a listing have indeed focused on the lack of experience and expertise of CITES regarding fish species,[6] have declared that CITES applies an over-precautionary approach and that it suffers from a protectionist bias.[7] The validity of such statements and arguments needs to be addressed by considering the detailed application of CITES in the case studied as well as its ability to evolve.

Finally, the use of CITES for commercially-exploited aquatic species should only be promoted if it is expected to improve the governance of these species. Hence, an examination of the practical effects of such listing, as well as of the added value towards governance will conclude this chapter.

A. CITES' Philosophies

Before clarifying the legality of using CITES for commercially-exploited fish species, it is worth considering the philosophies underlying CITES as they are, explicitly or not, contributing to states' views on whether CITES should be utilized for certain species. Consequently, they influence the debates at the CoPs on the adequacy of using this regime.

Throughout the history of CITES, the relative importance given to preserving wildlife, in opposition to using it sustainably, was debated[8] and two interconnected tensions have been felt. The first relates to the *raison d'être* of the treaty, in other

International Law (Cambridge Studies in International and Comparative Law, Cambridge University Press 2011) 141–142.

[4] FAO, 'Report of the Expert Consultation on Legal Issues Associated to CITES and Commercially-exploited Aquatic Species (22–25 June 2004)' FAO Fisheries and Aquaculture Report 746 (Rome 2004) 2–3 §13 and 10 §47.

[5] Japan would have desired a clear statement, within CITES, that the FAO and RFMOs are the appropriate bodies to deal with fisheries (Japan, 'CoP12 Doc. 16.2.1 – Synergy and Cooperation between CITES and FAO' (3–15 November 2002) annex).

[6] China, 'CoP12 Inf. 30 – "Sharks" and COP12 – A Case for Caution' (3–15 November 2002) 2 §1; A. C. J. Vincent and others, 'The role of CITES in the conservation of marine fishes subject to international trade' (2014) 15 Fish and Fisheries 563, 563.

[7] FAO, 'Report of the Expert Consultation on Legal Issues' (n 4) 6 §28, 7 §31–33; M. A. Young, 'Protecting Endangered Marine Species: Collaboration Between the Food and Agriculture Organization and the CITES Regime' (2010) 11 Melbourne Journal of International Law 1, 13–14.

[8] J. E. Carey, 'Improving the Efficacy of CITES by Providing the Proper Incentives to Protect Endangered Species' (1999) 77 Washington University Law Quarterly 1291, 1292; R. Reeve, *Policing International Trade in Endangered Species* (Earthscan 2002) 14; P. Stoett, 'To trade or not to trade?: The African elephant and CITES' (1996–1997) 52 International Journal of Marine and Coastal Law 567, 569.

words whether species should be protected from extinction for preservationist or for conservationist reasons (I). As for the second, it concerns the means considered most efficient to reach the goal of the treaty, which is to avoid extinction in species of wild fauna and flora (II).

I. Underlying Tensions

1. Tension in the Treaty

The treaty does not clearly state one unique reason for which wildlife should be protected, but it does refrain from setting limits to human actions, except for international trade. It also explicitly envisions some level of acceptable trade for species. Hence, while it has been said that both conservationist and preservationist philosophies "underlie CITES"[9] or even that the "treaty was written with preservation in mind",[10] in reality, the Convention itself rather reflects consumptive and non-consumptive sustainable use views than it does the preservationist one[11]—if that is to be understood as avoiding any human interference with nature.

The Preamble tends to support a non-consumptive view when it declares that "wild flora and fauna in their many beautiful and varied forms are an irreplaceable part of the natural systems of the earth which must be protected for this and the generations to come." Indeed, while it refers to values that are inherent to wildlife and hence could be related to preservationism, this preambular nevertheless finishes by stating that that protection is needed for humans of today and tomorrow to be able to take advantage of nature.

However, other parts of the Preamble by recognizing "the ever-growing value of wild fauna and flora from aesthetic, scientific, cultural, recreational and economic points of view" and the need to protect fauna and flora "against over-exploitation through international trade" acknowledge its consumptive value and the normality of trade.

As to the permit system created by the Convention, it can be argued, on the one hand, that the existence of quasi-trade bans through listing in Appendix I reflects a rather preservationist view, where trade in wildlife is totally curtailed. However, on the other hand, these restrictions are only aimed at "commercial trade in species that are already endangered".[12] As for Appendix II listing, it aims at avoiding utilization incompatible with the survival of the listed species, a goal that closely reflects the

[9] C. L. Krieps, 'Sustainable use of endangered species under CITES: is it a sustainable alternatives' (1996) 17 University of Pennsylvania Journal of International Law 461, 498.

[10] M. J. Hickey, 'Acceptance of sustainable use within the CITES community' (1998–1999) 23 Vermont Law Review 861, 873 (citation omitted).

[11] M. Bowman, 'A Tale of Two CITES: Divergent Perspectives upon the Effectiveness of the Wildlife Trade Convention' (2013) 22 RECIEL 228, 236.

[12] Ibid.

definition of sustainable use. Moreover, the exception for captive-breeding, envisioned in the Convention, shows that the sustainable use of nature, i.e. a use that does not put a species at danger of extinction, is accepted in the CITES framework.[13]

In conclusion, it is not possible to clearly determine which sole philosophy underlies the regime from the treaty itself as it shows some—little—preservationist tendency, while nevertheless mostly reflecting views related to both consumptive and non-consumptive sustainable use.

2. Tension in the CoP Resolutions

In fact, the actual tension was shown by the debates about the criteria to list species, which highlighted the different values and goals favoring rather preservation or sustainable use.[14] Under the earlier Bern criteria, it was difficult to down-/de-list: the level of proof required to delist from Appendix I was higher than that to list in the first place. At the time, this reversed burden of proof was explained by the potential consequences of making a management mistake: if a species was listed that did not really need it, this would result in an administrative burden, while if a species was delisted while needing protection, this would lead to "the permanent loss of the resource. If it errs it should be therefore toward protection of the resource".[15]

The Bern criteria were criticized for being inflexible once a species was listed and by "favor[ing] the complete cessation of wildlife trade rather than the regulation of such trade."[16] The group of states in favor of the economic utilization of the resources by the range states, the 'Consumptive Use Block', constituted mainly of developing countries,[17] made one attempt to update the criteria in Kyoto in 1992 before being successful at the following CoP.

The Fort Lauderdale criteria are oriented towards numerical assessment and, while keeping the precautionary principle on board, envision more clearly down-/de-listing processes. The range states' particular role in the listing procedure is also

[13] Resolution Conf. 12.10 (Rev. CoP15) – Registration of operations that breed Appendix-I animal species in captivity for commercial purposes 2002; D. S. Favre, *International Trade in Endangered Species: A Guide to CITES* (Martinus Nijhoff Publishers 1989) 187; J. L. Garrison, 'The Convention on International Trade in Endangered Species of Wild Fauna and Flora (CITES) and the Debate Over Sustainable Use' (1994) 12 Pace Environmental Law Review 301, 308, 327–329.

[14] S. Young, 'Contemporary Issues of the Convention on International Trade in Endangered Species of Wild Fauna and Flora (CITES) and the Debate Over Sustainable Use' (2003) 14 Colorado Journal of International Environmental Law and Policy 167, 171–172.

[15] Resolution Conf. 1.2 – Criteria for the Deletion of Species and Other Taxa from Appendices I and II (Bern criteria) 1976 (repealed).

[16] Young, 'Contemporary Issues' (n 14) 172 (reference omitted).

[17] D. S. Favre, 'Debate within the CITES Community: What Direction for the Future?' (1993) 33 Natural Resources Journal 875, 883.

affirmed.[18] As for split-listing, it "enables a party with a non-endangered, well-managed population of a species that is endangered in other parts of its range to include its own population in Appendix II and thus allow a limited commercial trade".[19] The adoption of the new listing criteria hence demonstrated an evolution "from an inflexible standard based on preservation, to one that embraces the concept of sustainable use."[20] The down-listing of elephants, criticized by some as the end a strong protection of the environment[21] was seen by many as the practical sign of the "shift of the underlying philosophy of CITES from preservation to sustainable use and conservation."[22] However, the Fort Lauderdale criteria still have very restrictive down-/de-listing criteria with safeguards to be implemented, in addition to the biological status needing to have sufficiently improved.[23]

Furthermore, the regime created the possibility to ranch the populations of Appendix I species which are not endangered anymore.[24] This evolution supports the argument that consumptive use is the main philosophy behind CITES nowadays. Indeed, the creation of the concept of ranching "enabled countries meeting the criteria so established to make legitimate use (including transnational shipments) of Appendix I species at agreed sustainable rates."[25]

Finally, while acknowledging the importance of non-consumptive uses, the CoP "recognize[d] that commercial trade may be beneficial to the conservation of species and ecosystems, and to the development of local people when carried out at levels that are not detrimental to the survival of the species in question".[26] It further noted that the Addis Ababa Principles and Guidelines for the Sustainable Use of Biodiversity "can be utilized in the implementation by CITES Parties" of the relevant provisions, in particular those related to trade in Appendix II species.[27]

---

[18] P. H. Sand, 'Whither CITES? The Evolution of a Treaty Regime in the Borderland of Trade and Environment' (1997) 8 European Journal of International Law 29, 46.

[19] M. Bowman, P. Davies and C. Redgwell, *Lyster's International Wildlife Law* (2nd edn Cambridge University Press 2010) 492.

[20] Hickey (n 10) 874; K. Eldridge, 'Whale For Sale?: New Developments in the Convention on International Trade in Endangered Species of Wild Fauna and Flora' (1994–1995) 24 Georgia Journal of International and Comparative Law 549, 563.

[21] Eldridge (n 20) 560–561.

[22] Hickey (n 10) 862 (references omitted).

[23] Resolution Conf. 9.24 (Rev. CoP16) – Criteria for amendment of Appendices I and II (Resolution Conf. 9.24 (Rev. CoP16) – Lauderdale criteria) 1994 annex 4; on the listing criteria's adequacy for fish species, see infra Chapter 5 C. I.

[24] Resolution Conf. 3.15 – Ranching 1981 (repealed); currently Resolution Conf. 11.16 (Rev. CoP15) – Ranching and trade in ranched specimens of species transferred from Appendix I to Appendix II 2000.

[25] Sand, 'Whither CITES? The Evolution of a Treaty Regime in the Borderland of Trade and Environment' (n 18) 41–42.

[26] CITES Resolution Conf. 8.3 (Rev. CoP13) – Recognition of the benefits of trade in wildlife 1992 preamble and operative part.

[27] Resolution Conf. 13.2 (Rev. CoP14) – Sustainable use of biodiversity: Addis Ababa Principles and Guidelines 2004.

In conclusion, CITES might have started as a treaty implemented in a somehow preservationist manner, but, nowadays, the framework reflects pro-sustainable use values. This does however not mean that all decisions are taken in accordance with the criteria and science rather than with ideology, as states are the ultimate decision-makers. Also, one must admit the existence of uncertainty, which gives room to states to apply a more or less strict version of the precautionary principle, hence defending respectively their more preservationist or more conservationist goals.

II. Arguments in Terms of Efficiency of Means

In addition to the debate related to the main underlying philosophy of the treaty regime, the various actors involved in CITES, such as developed and developing states, range and non-range states, NGOs and lobbies, have different opinions on whether conservationist or preservationist philosophies are to be followed within CITES for it to best fulfil its goal of avoiding extinction in species of wild fauna and flora.[28]

1. Pro-Sustainable Use

The proponents of sustainable use point out the fact that the benefits derived from protective measures can be used to conserve species.[29] Indeed, while the pressure coming from trade was viewed for a long time as necessarily negative for species, it is now recognized that trade "has the potential to be beneficial when conducted in a sustainable manner and when monies derived from international trade are allocated for conservation and/or enforcement purposes".[30]

Through sustainable use, it is also possible to take into account the role of wildlife for local communities as well as to recognize that trade can empower them.[31] This in turn can ease the tension existing between the North, material-rich but species-poor countries, who pushes for strict conservation, and the South, material-poor but species-rich countries.[32] As most range states are developing countries which feel that "they must bear a disproportionate share of the costs for protecting species because no official mechanism for cost sharing exists"[33]

---

[28] Bowman, Davies and Redgwell (n 19) 497–498; Garrison (n 13) 305.

[29] Krieps (n 9) 481–482.

[30] Garrison (n 13) 325.

[31] Hickey (n 10) 872. However it has been commented that the main benefits are in fact made by the middlemen, not the actual hunters (Reeve, *Policing International Trade in Endangered Species* (n 8) 13).

[32] Krieps (n 9) 484; T. Swanson, 'Developing CITES: Making the Convention Work for all the Parties' in J. Hutton and B. Dickson (eds), *Endangered Species-Threatened Convention: The Past, Present and Future of CITES* (Earthscan 2000) 150.

[33] Krieps (n 9) 477.

sustainable use is indeed a way to make the protection of wildlife economically viable[34] if not even to create some return on the resources invested in conservation.[35]

This view also acknowledges a balance between humans and animals.[36] Indeed, it recognizes that wildlife specimens represent important resources (i.e. foreign currency) in certain countries[37] and gives value to the fauna, particularly when wildlife habitats and some forms of economic use are competing for the same land.[38] Elephants for example, while they are perceived as charismatic animals in far-away developed countries, are rather considered as destructive, challenging new forms of land use, as well as economically valuable in the range states.[39]

Finally, a sustainable use philosophy encourages and facilitates the gathering of data for assessing the status of species. There is indeed an economic incentive for the range states not only to take care of their species and fight poaching, but also to develop sound management plans based on science, if they know it is a requirement to access international trade. Thus, under this frame of thought, the range of species protected has the potential to extend to all those which might be traded, and not only to a few charismatic fauna. This particular argument criticizes preservationists' narrow focus on certain specific animals; for example the line of reasoning linked to a 'right to life' of charismatic animals[40] such as whales is always reserved for

---

[34] Swanson (n 32) 150.

[35] R. Cooney and M. Abensperg-Traun, 'Raising Local Community Voices: CITES, Livelihoods and Sustainable Use' (2013) 22 RECIEL 301, 302; Krieps (n 9) 477; Swanson (n 32) 141; Young, 'Contemporary Issues' (n 14) 184.

[36] Hickey (n 10) 871–872; Krieps (n 9) 483–484.

[37] Bowman, Davies and Redgwell (n 19) 483; Sand, 'Whither CITES? The Evolution of a Treaty Regime in the Borderland of Trade and Environment' (n 18) 30.

[38] Stoett (n 8) 571–572; J. G. Robinson and K. H. Redford, 'The Use and Conservation of Wildlife' in J. G. Robinson and K. H. Redford (eds), *Neotropical Wildlife Use and Conservation* (The Chicago University Press 1991) 3–4; Resolution Conf. 8.3 (Rev. CoP13) – Recognition of the benefits of trade in wildlife (n 26) preamble: "RECOGNIZING that the sustainable use of wild fauna and flora, whether consumptive or non-consumptive, provides an economically competitive land-use option; BEING AWARE that, unless conservation programmes take into account the needs of local people and provide incentives for sustainable use of wild fauna and flora, conversion to alternative forms of land use may occur".

[39] P. Birnie, A. Boyle and C. Redgwell, *International Law and the Environment* (3rd edn Oxford University Press 2009) 689; S. Dansky, 'The CITES "Objective" Listing Criteria: Are They Objective Enough to Protect the African Elephant?' (1999) 73 Tulane Law Review 961, 970–971; Young, 'Contemporary Issues' (n 14) 176. To mark their point, some African countries submitted in 1992 a proposal to list red-herring, a species of great commercial importance for some northern countries. The proposal was later withdrawn (Vincent and others (n 6) 569).

[40] C. P. Carlarne, 'Saving the Whales in the New Millennium: International Institutions, Recent Developments and the Future of International Whaling Policies' (2005–2006) 24 Virginia Environmental Law Journal 1, 42; E. J. Molenaar, 'Marine Mammals: The Role of Ethics and Ecosystem Considerations' (2003) 6 Journal of International Wildlife Law and Policy 31, 33–34. See for example A. D'Amato and S. K. Chopra, 'Whales: Their Emerging Right to Life' (1991) 85 American Journal of International Law 21.

megafauna and has consequently never been put forward for animals such as fishes, which are consequently at risk of remaining unprotected.

2. Pro-Preservationism

The proponents of preservationism criticize sustainable use as being based on a totally anthropocentric view of the world.[41] Also, a major issue identified is linked to the actual limitations of science. It can indeed be feared that, in sustainable use programs, states willing to trade under CITES will do so, "without establishing the necessary scientific support structure as a preliminary step",[42] or that management plans will not actually be sustainable, not least for lack of data. The problem caused by insufficient scientific information is that in the absence of another value than a consumptive one given to animals, economic incentives will usually prevail over precaution.[43] This is a serious setback for the precautionary principle[44] and consequently a strong risk for those species in relation to which biological uncertainties remain.

Then, the viability of sustainable use is put into question, in particular with regard to enforcement. Indeed, resources are necessary to verify whether a use is sustainable,[45] and these are scarce. It is also evidently easier to determine the illegality of a particular shipment if no trade at all is allowed[46] and "dual streams involving both legal and illegal products can increase the burden on enforcement."[47] Hence, preservationists argue that cutting all legal entry into the markets would stop illegal harvesting and poaching, or at least reduce it importantly[48] and that, reversely, a level of legal trade offers too many open avenues for illegal trade to take place.[49] However, in many cases, illegal trade in Appendix I species seems to continue, not giving much credit to this assertion.[50]

[41] Hickey (n 10) 869.

[42] Favre, 'Debate within the CITES Community: What Direction for the Future?' (n 17) 890.

[43] Ibid. 912–913; Young, 'Contemporary Issues' (n 14) 185.

[44] Young, 'Contemporary Issues' (n 14) 185.

[45] A. Wiersema, 'Uncertainty and Markets for Endangered Species under CITES' (2013) 22 RECIEL 239, 246.

[46] M. J. Glennon, 'Has International Law Failed the Elephant?' (1990) 84 The American Journal of International Law 1, 27.

[47] Wiersema, 'Uncertainty and Markets for Endangered Species under CITES' (n 45) 246.

[48] Stoett (n 8) 571.

[49] S. J. Keller, 'Is the International Ban on the Importation of Ivory Saving the African Elephant?' (1992) 3 Colorado Journal of International Environmental Law and Policy 381, 383, 391–392; T. Mulliken, 'The Role of CITES in Controlling the International Trade in Forest Products: Implications for Sustainable Forest Management' Non-Wood Forest Products Working Documents 7 (Rome 2009) 30.

[50] For orchids for example, see Mulliken (n 49) 24–26.

Finally, other arguments made against sustainable use are that, if conservation only targets species because they are used by humans, market fluctuations (i.e. the changing value in the eyes of the consumers) mean that protection will equally vary.[51] Also, focus (and hence protection) will be lacking for species without a commercial value attached to them and planning for land use in their area of habitat will not take them into account.

III. Conclusions

Various philosophies intersect in the text of CITES. The evolution of the regime shows that it nowadays follows a sustainable use philosophy rather than a preservationist one. However, the framework does not determine the decisions taken by the CoP, which reflect the whole spectrum of values that parties might wish to defend.

Hence, in practice, the tension between the underlying philosophies has not been fully settled, as states parties' interests and values continue to differ. Even with the evolution of the regime, it is not possible to rule out that a majority of states would vote with a strong precautionary approach in mind. In particular, the very strict down-/de-listing procedures create the fear of CITES being a preservationist regime that cannot be easily dismissed. Indeed, the importance given to precaution in decision-making might ban the resumption of trade even when the status of a species has improved.

In the case at hand, the existence of a strong sustainable use side to CITES is crucial as fisheries are usually managed towards that goal rather than towards the one of preservation. However, the leeway left to states in their decision-making might appear to be a problem if it resulted in a continued listing in Appendix I, even after the commercially-exploited fish species recovered. Fishes are indeed important sources of revenues to many communities and states, as well as serve as food for a non-negligible percentage of the world's population. Any preservationist tendency is source of potential anxieties for range states and likely to impair a listing decision in the first place.[52]

Nevertheless, from the arguments presented by proponents of both philosophies, one can hope that the proposals related to fish species will not fall into the trap of 'once listed never delisted'. Indeed, some characteristics of fisheries make amendments of Appendices less likely to be discussed in preservationist terms.[53] In

[51] Krieps (n 9) 486–488.

[52] Committee on Fisheries – Sub-Committee on Fish Trade, 'Report of the Eighth Session of the Sub-Committee on Fish Trade (Bremen, 12–16 February 2002)' FAO Fisheries Report 673 (Rome 2002) 7 §16; FAO, 'Report of the Expert Consultation on Legal Issues' (n 4) 7 §31.

[53] Bowman, 'A Tale of Two CITES: Divergent Perspectives upon the Effectiveness of the Wildlife Trade Convention' (n 11) 237.

particular and as will be discussed in more details for the special situation of developing countries, fish species are not considered charismatic and are acknowledged as a source of food and revenue to be used sustainably rather than preserved for their inherent value.[54] Furthermore, the debate is likely to be less divided along a North/South line, as some fish species straddle or migrate globally. Hence, the line demarcating groups of states arguing in favor of a listing or against it, and later in favor of down-/de-listing or against it will not necessarily be the usual one, hopefully also avoiding the historical stalemates.

In conclusion, the CITES regime remains somehow at the cross-road between conservation and preservation, not necessarily in the framework itself which is now rather pro-sustainable use, but more so in the decisions taken by states parties in relation to down-/de-listing. This might be a factor influencing some states in declining to use the regime for fish species and some of the arguments presented above do enter debates at CoPs. However, this is not a problem unique to fisheries, which are actually less likely to suffer from preservationist tendencies, as they are non-charismatic, widely distributed and acknowledged as important traded goods.

B. Legality of Listing a Commercially-exploited Marine Species Under CITES

Now that the philosophies underlying CITES have been presented and their relevance for the debate related to listing fish species discussed, it is important to examine the legality of using CITES for commercially-exploited marine species.

I. In the Realm of the Convention

1. Terrestrial and Aquatic Species

The purpose of CITES is to regulate trade in endangered species of wild fauna and flora, species being defined in the Convention as "*any* species".[55] The use of 'any' is clearly inclusive and this definition does not differentiate terrestrial from marine species.[56] As stated in the 2008–2020 Strategic Vision, CITES' vision statement is to "conserve biodiversity and contribute to its sustainable use by ensuring that *no species of wild fauna or flora* becomes or remains subject to unsustainable

---

[54] See infra Chapter 5 C. II.

[55] 'Convention on International Trade in Endangered Species of Wild Fauna and Flora' (CITES) 1973, 993 *United Nations Treaty Series* 243 article I(a) (italics added).

[56] Vincent and others (n 6) 567; C. Wold and E. Thorson, 'The Application of CITES to Atlantic Bluefin Tuna and Other Marine Species' (International Environmental Law Project 2010) 2.

exploitation through international trade".[57] The focus is put here again on the inclusiveness of the Convention, which should consequently not be read in a manner which excludes certain species from its realm, except if explicitly stated in the Convention.

Also, CITES has adopted an evolutionary approach when not sure whether a taxa falls under its jurisdiction. For example mushrooms were species that were not straight-forward to categorize in that regard. Protection under CITES is directed at species of wild fauna and flora. However, since the 1960s already—10 years before the drafting of the Convention—, mushrooms are classified as belonging to a different kingdom than plants. A debate arose in 2002, at CoP12, to decide whether mushrooms could be listed. The adopted recommendation regarding nomenclature, with which Japan expressed its disagreement, clarifies that CITES can also apply to species of the fungi kingdom.[58]

2. Aquatic Species Including Marine Ones

References to marine species are themselves included in the Convention since trade comprises 'introduction from the sea' (IFS). This type of trade is to be understood as the "transportation into a State of specimens of *any* species which were *taken in the marine environment* not under the jurisdiction of any State".[59] In addition to the further use of 'any', species that can be caught in the mentioned waters obviously belong to marine aquatic species—as well as do sea birds which are caught incidentally on the seas—and many are fish species.

It is true that, during the Convention's drafting phase, opinions differed as to whether marine species should be included. These debates can be followed in the travaux préparatoires and opening statements;[60] Japan desired to remove all references to IFS from the working paper[61] and the outcome of the issue was unclear until a late stage of negotiations.[62] However, a compromise solution was found by

---

[57] Resolution Conf. 16.3 – CITES Strategic Vision: 2008–2020 2013 (italics added).

[58] Thomas (n 3) 24; International Institute for Sustainable Development, 'Earth Negotiations Bulletin – Summary of the Twelfth Conference of the Parties to the Convention on International Trade in Endangered Species of Wild Fauna and Flora' (18 November 2002) http://www.iisd.ca/vol21/enb2130e.html accessed 5 July 2015.

[59] CITES (n 55) article I(e) (italics added).

[60] As presented in E. Franckx, 'The Relationship between CITES, FAO and Related Agreements: Legal Issues' FAO Fisheries and Aquaculture Circular 1062 (Rome 2011) 21–22.

[61] CITES' *travaux préparatoires*, Proposed Amendments to the Working Paper (Doc. 3) (Submitted by the delegation of Japan) (Doc PA/Gen/1, 17 February 1973) 1, as presented in E. Franckx, 'The Exercise of Jurisdiction over Vessels: Legal Issues Raised by the Relationship Between CITES, FAO and the UN Convention on the Law of the Sea' in E. Franckx and P. Gautier (eds), *The Exercise of Jurisdiction over Vessels: New Developments in the Fields of Pollution, Fisheries, Crimes at Sea and Trafficking of Weapons of Mass Destruction* (Bruylant 2010) 67 n 61.

[62] E. Franckx and K. van den Bosche, 'The Influence of Environmental Law on the Development of the Law of the Sea: CITES and the International Law of Fisheries' (2011) 54 Japanese Yearbook of International Law 218, 241.

including marine species in the Convention but also introducing paragraphs 4 and 5 into article XIV.[63] These provisions detail the relationships between CITES and other treaty regimes offering protection to marine species, giving priority, for implementation, to the ones already existing at the time of CITES coming into force.[64] The purpose of these provisions was to "mak[e] sure that other international agreements relating to marine species that ensure the survival of the species concerned would not be affected."[65]

Then, mentions of marine species are to be found in several CoP Resolutions,[66] IFS is referred to in such instruments and a clarification of this concept has been adopted.[67] These are strong indications that the listing of marine species is a legal possibility under CITES. Some Resolutions also address particular issues for species such as sharks or toothfish; their adoption further suggests that CITES parties feel entitled to conserve marine fish species under this regime.[68] As examined earlier,[69] such documents are not officially binding on parties, but reflect the parties' common views or understanding of the Convention.

Additionally, in 1997, at CoP10, a Working Group for Marine Species was created, in order to address the increasing likelihood that some fish species might qualify for listing in the Appendices.

Finally, and although conceding that implementation might be very problematic, the Secretariat explicitly agreed that "[t]he inclusion of fish species subject to commercial fisheries in the CITES appendices is perfectly compatible with the Convention."[70] The Secretariat neither has independent decision-making power nor is entrusted to authoritatively interpret the Convention. It indeed works at the service of the deciding body, the CoP. However, it makes recommendations with

---

[63] Franckx, 'The Relationship between CITES, FAO and Related Agreements: Legal Issues' (n 60) 22, 48–49.

[64] For more on article XIV, see infra Chapter 5 B. III.

[65] Franckx, 'The Exercise of Jurisdiction over Vessels: Legal Issues Raised by the Relationship Between CITES, FAO and the UN Convention on the Law of the Sea' (n 61) 68 (reference omitted), presenting the travaux préparatoires.

[66] Resolution Conf. 9.24 (Rev. CoP16) – Lauderdale criteria (n 23) tenth preambular; Resolution Conf. 14.6 (Rev. CoP16) – Introduction from the Sea 2007 ninth preambular.

[67] Respectively Resolution Conf. 11.4 (Rev. CoP12) – Conservation of cetaceans, trade in cetacean specimens and the relationship with the International Whaling Commission 2000; Resolution Conf. 12.4 – Cooperation between CITES and the Commission for the Conservation of Antarctic Marine Living Resources regarding trade in toothfish 2002; and Resolution Conf. 14.6 (Rev. CoP16) – Introduction from the Sea (n 66).

[68] Resolution Conf. 12.6 (Rev. CoP16) – Conservation and management of sharks (Class Chondrichthyes) 2002; Resolution Conf. 12.4 – Cooperation between CITES and the Commission for the Conservation of Antarctic Marine Living Resources regarding trade in toothfish (n 67).

[69] See supra Chapter 3 C. I. 2. f).

[70] CITES, 'CoP8 Doc. 8.46 (Rev) – Consideration of proposals for amendment of Appendices I and II' (n 3) annex 3, 30.

regard to listings, and provides comments on draft Resolutions.[71] As such, and on the basis of the expertise it has developed on the regime it serves, the views of the Secretariat have the potential to be influential on states parties.

3. Commercial Species

The commercial nature of many fisheries is not, legally speaking, a hindrance to their listing in a CITES Appendix. Indeed, it is not stated anywhere in the Convention that only non-commercial species would fall under the scope of this regime.

Rather to the contrary, CITES is a trade instrument and trade is, intrinsically, an activity of a commercial nature. The Preamble of the Convention acknowledges that wildlife can be of economic value.[72] It also hints "that international co-operation is essential for the protection of certain species [. . .] against over-exploitation",[73] thus acknowledging the existence of exploitative activities targeting wildlife species. In practice, having recognized that species of commercial value fall under the Convention's realm, parties have listed several of these species, such as sturgeon, mahogany, European eel or tigers.

4. Conclusion

In conclusion, CITES benefits from a comprehensive jurisdiction towards both terrestrial and aquatic species, and its competence is not curtailed by the commercial nature of the use made of certain species. Hence, this regime has jurisdiction in relation to commercially-exploited aquatic species, except if provided otherwise by the treaty or the intent of states parties. The examples here below provide several practical illustrations of such competence.

---

[71] The functions of the Secretariat include "to invite the attention of the Parties to any matter pertaining to the aims of the present Convention; [. . .] to make recommendations for the implementation of the aims and provisions of the present Convention" (CITES (n 55) article XII(2) (e) and (h)) and to gather advice from various actors, transmit relevant information and make its own recommendations regarding amendments of the Appendices (CITES (n 55) article XV(1) (a) and (2)(b)–(e)).

[72] CITES (n 55) second preambular; Wold and Thorson (n 56) 3.

[73] CITES (n 55) fourth preambular.

II. History of Proposals

1. Species Listed

Due to controversies during the negotiations,[74] it is no surprise that commercially-exploited marine species were not the first listed under CITES (apart from whales). Indeed, "[i]nitially, just one marine fish species, the coelacanth was included in the original Convention, along with four anadromous sturgeons. Since then, Parties have presented listing proposals for an additional 33 marine fish taxa at eleven CoPs over the 34 years from 1976 to 2010, with no proposals at CoP3, CoP5, CoP7 or CoP9."[75]

An increasing (but still limited) number of commercially-exploited fish species have been listed recently.[76] Nowadays, some aquatic species such as the queen conch, hard corals, humphead wrasse, sea horses, paddlefish, whales, several sharks and many dolphins are granted protection under the CITES regime. For most of them, the question of whether they could be protected under CITES was not addressed before the listing, for the simple reason that it did not arise.

a) Marine Mammals

The protection of cetaceans under CITES goes back to CoP1 in 1976, with the listing of the fin whale and the sei whale.[77] At the next meeting in 1979, the whole *Cetacea* order was added to Appendix I and II,[78] and, at CoP4, CITES harmonized its Appendices with the IWC's measures by listing in Appendix I all species protected by the IWC.[79]

Almost all species of whales, as well as several species of dolphins and porpoises are now listed in Appendix I. Only the population of West Greenland of the common minke whale or northern minke whale is listed in Appendix II. The

---

[74] See infra Chapter 5 B. III. 1. a).

[75] Vincent and others (n 6) 568 (brackets omitted).

[76] CITES, *Bluefin tuna main course of CITES world conference: 175 Governments will also discuss urgent measures to tackle illegal wildlife trade and protect the livelihoods of the rural poor* (5 February 2010) http://www.cites.org/eng/news/pr/2010/20100205_cop15.shtml accessed 6 July 2015; D. S. Calley, *Market Denial and International Fisheries Regulation: The Targeted and Effective Use of Trade Measures Against the Flag of Convenience Fishing Industry* (Martinus Nijhoff Publishers 2012) 173.

[77] Amendments to Appendices I and II of the Convention adopted by the Conference of the Parties at its first meeting 1976.

[78] Amendments to Appendices I and II of the Convention adopted by the Conference of the Parties at its second meeting 1979.

[79] Amendments to Appendices I and II of the Convention adopted by the Conference of the Parties at its fourth meeting 1983; M. Arbour and S. Lavallée, *Droit international de l'environnement* (Bruylant 2006) 410. On the relation with the IWC see infra Chapter 6 A. II. 2. b) cc), C. IV. 1, D. IV. 1.

remaining species of the cetacean family are listed in Appendix II. Several states, the main ones being Norway, Iceland and Japan, have entered into reservations with respect to a number of cetacean species under CITES.[80] At CoP9, 10 and 11 Norway and Japan have, without success, proposed to down-list the minke whale.[81]

b) Sturgeons

The first species of sturgeons has been listed in 1976 in Appendix I and, at the following CoP, more species were added to Appendix I and II. In 1998, due to concerns about the overall (un)sustainability of sturgeons' harvesting and the scale of illegal trade, Germany and the USA made a successful proposal to put all remaining species in Appendix II.[82]

In addition to the CITES trade permits requirements and Significant Trade Review,[83] ad hoc mechanisms were developed to tackle the particular issues linked to sturgeons. Hence, exports quotas decided collectively by all (or all but one of) the range states of one water basin are required for international trade to be allowed[84] and exports must take place within the year of quotas having been set. Legal exporters must be licensed and national lists of these licenses sent to the CITES Secretariat. Processing and repackaging plants have to be registered, and national lists of these plants must be sent to the Secretariat.[85] Finally, a universal labeling system for caviar has been developed and must be used—parties are urged not to accept imports of caviar not respecting this rule.[86]

c) Sharks and Rays

Most sawfishes are listed in Appendix I since CoP14. The basking shark, the whale shark, the great white shark and the remaining species of sawfishes are in Appendix II

---

[80] CITES, 'Reservations entered by Parties updated on 5 June 2015' (2013) http://www.cites.org/eng/app/reserve.php accessed 6 July 2015; S. Andresen, 'The International Whaling Regime: Order at the Turn of the Century' in D. Vidas and W. Ostreng (eds), *Order for the oceans at the turn of the century* (Kluwer Law International 1999) 220.

[81] Bowman, Davies and Redgwell (n 19) 187; A. Gillespie, *Whaling Diplomacy: Defining Issues in International Environmental Law* (New Horizons in Environmental Law, Edward Elgar 2005) 340.

[82] CITES Secretariat, 'A brief history of sturgeons & CITES' *CITES World – Official Newsletter of the Parties* (December 2001) 12.

[83] Resolution Conf. 12.8 (Rev. CoP13) – Review of Significant Trade in specimens of Appendix-II species 2002 operative part.

[84] Resolution Conf. 12.7 (Rev. CoP16) – Conservation of and trade in sturgeons and paddlefish 2002.

[85] Ibid.

[86] Ibid.

respectively since CoP12, CoP13 and CoP14. In 2013, at CoP16, on the basis of proposals from several countries, the oceanic whitetip shark, the porbeagle, the scalloped hammerhead, the great hammerhead shark, the smooth hammerhead shark, and manta rays were listed in Appendix II. These species are subjected to the CITES protection since 14 September 2014. An 18-month delay had indeed been agreed upon to allow practical issues of implementation to be solved.[87]

These listings have been praised by the Secretary General of CITES as "mark[ing] a milestone in the involvement of CITES in marine species, which will enable the Convention to serve as a complementary instrument to fisheries agreements and bodies".[88]

Beyond the practical effects of a listing, CITES has taken a strong involvement in sharks and has put a Shark Working Group in place at the Animals Committee following CoP12. Also, already in 1994, a call was made within the CITES community for cooperation on trade and biological data gathering.[89]

In Resolution Conf. 12.6 on *Conservation and management of sharks*, the CoP took a global approach to sharks' conservation, inter alia by urging states which had not yet done so to adopt a national plan of action. It also urged COFI and RFMOs to further undertake relevant research, requested its Secretariat to consider, together with the World Custom Organization, options to improve trade reporting and directed the Animals Committee to report on activities related to sharks and rays at the CoP meetings.

2. Proposals Rejected or Withdrawn

a) Atlantic Bluefin Tuna

In 1992, at CoP8, Sweden made a proposal to list Atlantic Bluefin Tuna in Appendix I. According to ICCAT and other scientists, this species did not meet the CITES criteria and, hence, the Secretariat's advice was to reject the proposal.[90]

This proposal was withdrawn against promises (and pressures) of improving the management under ICCAT—Atlantic Bluefin Tuna is under the management of

---

[87] Amendments to Appendices I and II of the Convention adopted by the Conference of the Parties at its 16th meeting, Bangkok (Thailand), 3–14 March 2013; CITES, *CITES getting ready for sharks and rays* (2013) http://www.cites.org/eng/news/pr/2013/20130914_shark_ray.php accessed 5 July 2015; J. E. Scanlon, 'CITES at Its Best: CoP16 as a 'Watershed Moment' for the World's Wildlife' (2013) 22 RECIEL 222, 222.

[88] Scanlon, 'CITES at Its Best: CoP16 as a 'Watershed Moment' for the World's Wildlife' (n 87) 223.

[89] Resolution Conf. 9.17 – Status of International Trade in Shark Species 1994 (repealed).

[90] Franckx, 'The Exercise of Jurisdiction over Vessels: Legal Issues Raised by the Relationship Between CITES, FAO and the UN Convention on the Law of the Sea' (n 61) 69; CITES, 'CoP8 Doc. 8.46 (Rev) – Consideration of proposals for amendment of Appendices I and II' (n 3) annex 3, 30.

ICCAT, which had had, for a long time, a record of poor results in conservation and management. A short term decrease in TACs followed, but this amelioration did not last more than two years and stocks were not rebuilt. The unsustainability of the regional body's decisions kept on rising afterwards, with the setting of increasing quotas, adopted in complete disregard of the scientific advice provided by its own body.[91]

Nearly two decades later, at CoP15, Monaco proposed to list Atlantic Bluefin Tuna in Appendix I. This proposal was defeated, with 20 votes in favor, 68 against and 30 abstentions.[92] The EU amendment proposal was also unsuccessful.

The arguments in favor of a listing were that the decline of Atlantic Bluefin Tuna, particularly of the populations in the East Atlantic and Mediterranean, was a marked one, hence meeting the CITES criteria for listing—as agreed upon by the SCRS of ICCAT and the FAO in separate studies.[93] Also, in addition to the actual state of stocks, the continued lack of sustainability in the TACs decided by ICCAT, the high IUU fishing rates and the overall inadequate work of ICCAT which did not follow the advice of its own Scientific Committee were underlying this proposal.[94]

As for the parties' arguments against such a listing, they focused on two main points. One is that CITES is not the institution to deal with marine species such as the Atlantic Bluefin Tuna[95] and that it could be overprotective in the long run. The second aspect was that ICCAT had changed and taken more sustainable measures; it had a relatively new CDS, the implementation of which must be given some time.[96] Furthermore, the implementation of a ban on commercial international trade would cause difficulties in further monitoring the stocks' status and trends.[97]

---

[91] S. Lieberman, 'Science versus Politics: Tales from CITES' (Elisabeth Mann Borgese Lecture, 2010) http://internationaloceaninstitute.dal.ca/EMBlecturetext_2010.pdf accessed 5 July 2015, 5.

[92] "Those who spoke in support of the proposal were the EU (although they introduced a confusing compromise proposal), US, Norway, and Kenya. Those who spoke in opposition were Canada, Indonesia, UAE, Venezuela, Chile, Japan, South Korea, Grenada, Senegal, Namibia, Turkey, Iceland, Morocco, Tunisia, and Libya" (ibid. 7).

[93] Monaco, 'CoP15 Prop. 19 – Proposal to include Atlantic Bluefin Tuna (*Thunnus thynnus* (Linnaeus, 1758)) on Appendix I of CITES in accordance with Article II 1 of the Convention' (13–25 March 2010) summary 1–2 §5–9; Monaco, 'CoP15 Inf. 12 – Supplementary information on the proposal to include Atlantic Bluefin Tuna (*Thunnus Thynnus* Linnaeus 1758) on Appendix I of CITES in accordance with article II 1 of the Convention' (13–25 March 2010). It is true that the FAO Expert Panel could not reach consensus on the need for a listing but this was due to the opposition of only one member.

[94] Monaco, 'CoP15 Prop. 19 – Proposal to include Atlantic Bluefin Tuna' (n 93).

[95] CITES, 'CoP15 Com. I Rec. 8 (Rev. 1) – Summary record of the eighth session of Committee I' (13–25 March 2010) 2–3 in particular Canada, Japan, Venezuela, Chile, Grenada, Indonesia and Tunisia.

[96] New measures as presented by F. Hazin, (Chairman of ICCAT), 'CoP15 Inf. 63 – ICCAT Statement Bluefin Tuna' (18 March 2010) 2; CITES, 'CoP15 Com. I Rec. 8 (Rev. 1) – Summary record of the eighth session of Committee I' (n 95) 2–3 in particular Spain for the EU, Canada, Republic of Korea and Senegal.

[97] FAO, 'CoP15 Inf. 26 – Statement from the FAO Fisheries and Aquaculture Department on CoP15 Proposal 19 to List the Atlantic Bluefin Tuna in Appendix I' (13–25 March 2010) annex §11.

In conclusion, the debate moved from the science to the adequacy of using CITES to list Atlantic Bluefin Tuna, highlighting on the one hand the failure of ICCAT and on the other hand the fact that CITES was not developed for commercially-exploited fish species and that using it to manage a species could actually be counter-productive.

b) Patagonian Toothfish

In 2002, at CoP12, the Australian proposal to list Patagonian toothfish in Appendix II was strongly rejected by most of the relevant RFMO's (CCAMLR) members. While the proposal did envisage the continued use of the regional body's measures, most CCAMLR members felt that such a listing would put the RFMOs' competence, credibility and hence efficiency at risk.

Instead of adding toothfish to Appendix II, Resolution Conf. 12.4 on *Cooperation between CITES and CCAMLR regarding trade in toothfish* was adopted, requesting CITES parties to adopt the CCAMLR CDS. This has not been complied with and no solution has been found to deal with the unregulated fishing which takes place outside of the CCAMLR Convention area.

c) Sharks

In 2007, the German proposals (on behalf of the EU) to add the porbeagle and spiny dogfish in Appendix II were rejected at CoP14. The same proposals by Sweden (on behalf of the EU) and by Palau met a similar fate three years later at CoP15. The proposals by the USA and Palau to add Oceanic whitetip shark and several hammerhead sharks to Appendix II were also rejected in 2010.

However, in 2013, at CoP16, all of these species were listed in Appendix II, at the exception of the spiny dogfish which was not included in the proposal that year.

III. Fish Under the Management of Another Regime

From the above mentioned examples, one can note that the opposition to the listing of a commercially-exploited fish species under CITES seems often to stem from issues of expected inadequacy and of inter-institutional overlap.[98]

Indeed, the debate, as exemplified by the position held by Japan and the arguments of some countries at CoP15, that existing RFMOs, rather than CITES,

[98] D. G. Webster, 'The irony and the exclusivity of Atlantic bluefin tuna management' (2011) 35 Marine Policy 249, 249.

are the proper institutions to manage fisheries,[99] contemplates the situation where another regime relevant for the species is applicable. While one could wonder whether CITES has to refrain from protecting species in those circumstances, it is argued here that neither the intention of the parties at the time of the drafting nor the later practice provides convincing legal arguments to require such limitation to the competence of the CITES regime.

As a clause in CITES clarifies its relationship with other treaties related to marine species on the basis of them pre-dating its entry into force, two situations are to be examined: the one of treaties older than CITES and the one of post-1975 instruments.

1. Relationship with Already Existing Regimes

a) Article XIV (4)–(5)

According to article XIV, CITES shows some deference towards measures taken by already existing regimes dealing with marine species. This provision reads as:

> 4. [a] State party to the present Convention, which is also a party to any other treaty, convention or international agreement which is in force at the time of the coming into force of the present Convention and under the provisions of which protection is afforded to marine species included in Appendix II, shall be relieved of the obligations imposed on it under the provisions of the present Convention with respect to trade in specimens of species included in Appendix II that are taken by ships registered in that State and in accordance with the provisions of such other treaty, convention or international agreement.
> 5. Notwithstanding the provisions of Articles III, IV and V, any export of a specimen taken in accordance with paragraph 4 of this Article shall only require a certificate from a Management Authority of the State of introduction to the effect that the specimen was taken in accordance with the provisions of the other treaty, convention or international agreement in question.

Hence, for states which were parties on 1 July 1975 to an RFMO affording protection to the relevant marine species,[100] this provision clearly stipulates that the ships flying the flags of such states are relieved of the obligations under CITES for

---

[99] Japan, 'CoP12 Doc. 16.2.1 – Synergy and Cooperation between CITES and FAO' (n 5) annex; D. Jolly and J. M. Border, 'U.N. Rejects Export Ban on Atlantic Bluefin Tuna' *New York Times* (18 March 2010) http://www.nytimes.com/2010/03/19/science/earth/19species.html?_r=1& accessed 6 July 2015; CITES, 'CoP15 Com. I Rec. 8 (Rev. 1) – Summary record of the eighth session of Committee I' (n 95) 2–3.

[100] A difficulty in the application of this rule is the choice of dating method for treaties. Indeed, some regimes' founding documents have been modified largely, particularly after the generalization of EEZs (NEAFC for example, see T. Henriksen, G. B. Hønneland and A. K. Sydnes, *Law and Politics in Ocean Governance: The UN Fish Stocks Agreement and Regional Fisheries Management Regimes* (Publications on Ocean Development, Martinus Nijhoff Publishers 2006) 99; P. W. Birnie, 'Are Twentieth-Century Marine Conservation Conventions Adaptable to Twenty-First Century Goals and Principles?: Part I' (1997) 12 International Journal of Marine and Coastal

Appendix II species. In that case the state of introduction only has to issue the required certificate.[101]

This exemption was part of the compromise found during the drafting of the Convention in order to integrate marine species into the realm of the regime. Indeed, already at the negotiations' stage, the history of CITES was marked by the existence of the IWC and the refusal by Japan of CITES' involvement in whaling matters. It is worth noting that at some drafting stages, the relevant institutions were explicitly listed, with the IWC as a center-piece but also the International Convention for the Northwest Atlantic Fisheries—replaced in 1979 by NAFO—being mentioned by name.[102]

b) Limited Applicability of the Grandfather Clause

This provision is however not of general applicability and is in no case a legal hindrance to protecting marine species under CITES: article XIV (4)–(5) establishes a special regime only for certain actions under CITES and is restricted to specific cases. Such limitations are clear from the text of the relevant paragraphs read in accordance with the ordinary meaning of the terms.

Indeed, this provision creates a simplified mechanism for the implementation of the protection system, but has no impact on the possibility for the CoP to list a fish species under CITES, even if another treaty existing in 1975 accords protection to the relevant species. Other obligations under CITES such as reporting continue to be applicable.[103]

Also, while it is clear that this provision only applies to pre-CITES conventions, some of these early instruments have been modified in depth since 1975 and one could discuss whether article XIV (4)–(5) remains relevant in such cases.[104] Indeed, the status of treaties "whose founding document has been re-negotiated since the entry into force of CITES may be in question."[105]

Law 305, 335–336) and one can wonder whether the founding treaty's date or its major update's one should be used.

[101] CITES (n 55) article XIV(5). Gillespie considers that the ICRW was the only "convention of relevance that was already in working order at the time of the entry into force of CITES" (A. Gillespie, *Conservation, biodiversity and international law* (Edward Elgar 2011) 459).

[102] On this see examination of the *travaux préparatoires* in Franckx, 'The Protection of Biodiversity and Fisheries Management: Issues Raised by the Relationship between CITES and LOSC' (n 1) 224 n 86–89, in particular the Statement by Mr. T. Yamazaki, delegate of Japan (Doc PR/11, 21 February 1973).

[103] CITES Secretariat, 'CoP15 Inf. 45 – Article XIV, paragraphs 4 and 5, of the Convention' (13–25 March 2010) 3 §19.

[104] See Annex 1 for information on RFMOs' founding documents' entry into force and revision of mandate, when any.

[105] FAO, 'Report of the Expert Consultation on Legal Issues' (n 4) 10 §50. According to the CITES Secretariat this could have practical implications; for example, once amendments to NAFO's treaty enter into force, it is to be considered as not any longer in force at the time of CITES entry

Furthermore, the grandfather clause is restricted to Appendix II species, even though paragraph (5) mentions the exemption, as related to Appendix I, II and III. This is a mistake in the Convention text, already identified in 1976;[106] the correct version is on the list of proposals for amendments of the text.[107]

Then, arguably, only species which are afforded 'protection' are to be treated differently;[108] not all management regimes can be described as protective of species. The threshold is indeed logically at least as high as that of sustainable conservation and management measures. These have been said to be "characterize[d ...] by reference to factual and scientific criteria",[109] hence eliminating from the group measures which blatantly disregard science and are unsustainable.[110] Nevertheless, the practice, particularly with the IWC, shows that when CITES decides to give priority to another regime, the relation of that other regime with science does not matter.[111]

In conclusion, the grandfather clause here examined shows the intent of the parties to CITES, with regard to the relationship of that instrument with treaties already in force. It envisions explicitly a limited exemption related to the implementation of the regime. However, it does not remove from the scope of CITES the possibility to list marine species under the management of another regime which already existed in 1975.

---

into force (CITES Secretariat, 'CoP15 Inf. 45 – Article XIV, paragraphs 4 and 5, of the Convention' (n 103) 2 §10–12).

[106] Favre, *International Trade in Endangered Species* (n 13) 306; W. Wijnstekers, *The Evolution of CITES* (9th edn CIC – International Council for Game and Wildlife Conservation 2011) 454.

[107] Resolution Conf. 4.6 (Rev. CoP16) – Submission of draft resolutions and other documents for meetings of the Conference of the Parties 1983. "A rectification protocol noting editorial mistakes in the authentic treaty texts (inter alia, Arts III, XIV, and XVI CITES) was drawn up in the form of a procès-verbal by the Swiss Government as depositary, communicated to States Parties and signatories on 19 March 1976, and endorsed by the Conference of the Parties ('COP'), the supreme decision-making body of CITES, at its first meeting" (P. H. Sand, 'Endangered Species, International Protection' Max Planck Encyclopedia of Public International Law www.mpepil.com §9) (italics and inter-textual references omitted).

[108] CITES Secretariat, 'CoP15 Inf. 45 – Article XIV, paragraphs 4 and 5, of the Convention' (n 103) 2–3 §15–16. This is indeed a point reported by the Working Group on IFS (CITES Secretariat, 'CoP15 Doc. 27 – Interpretation and implementation of the Convention – Trade control and marking – Introduction from the Sea' (13–25 March 2010) 10 annex 3 §37).

[109] *Fisheries Jurisdiction (Spain v. Canada), Judgment on Jurisdiction,* 4 December 1998, I.C.J. Reports 1998 432, 461 §70; on this see S. Rosenne, 'Reflections on Fisheries Management Disputes' in R. Casado Raigón and G. Cataldi (eds), *L'évolution et l'état actuel du droit international de la mer: Mélanges de droit de la mer offerts à Daniel Vignes* (Bruylant 2009) 833.

[110] CITES Secretariat, 'CoP15 Inf. 45 – Article XIV, paragraphs 4 and 5, of the Convention' (n 103) 3 §16.

[111] See infra Chapter 6 C. IV. 1., D. IV. 1.

c) Conflict Clause Without Conflict

It is interesting to note that, while provisions such as article XIV (4)–(5) are commonly referred to as conflict clauses, they are in fact not necessarily related to any conflict.[112] Usually, it is assumed that such articles, included in many treaties to clarify the relation between that treaty and other past, present or future instruments, which touch upon related matters,[113] exist to "avoid that treaties covering similar or at least partially overlapping subject matters contradict each other."[114] However, two aspects of this statement can be discussed.

First, conflict clauses might be inserted into an instrument for other reasons than to avoid a contradiction. In some cases indeed, the inclusion of such clauses stems from a rather political decision, maybe with the view to avoid duplication of work or the fragmented application of a regime, or simply to increase coherence. Hence, in absence of such a clause, there would be other options to reach the same goal, such as through ad hoc cooperation.

Second, instruments referred to in conflict clauses are not always on similar subject matters.[115] Indeed, while some aspects might overlap, making a conflict clause relevant, treaties are often approaching an issue from different points of view, applying different solutions to similar problems, or just dealing with activities in the same area. Moreover, states, when negotiating a treaty, are free to give precedence to an instrument from another field of law. Hence, that a conflict clause is included does not automatically mean that it refers to a treaty on the same subject matter.

In consequence, in the case at hand, the inclusion of article XIV (4)–(5) in the Convention is not necessarily the sign that CITES and treaties affording protection to marine species conflict or that they are on the same subject matter.

---

[112] J. B. Mus, 'Conflicts Between Treaties in International Law' (1998) 45 Netherlands International Law Review 208, 211.

[113] Report of the International Law Commission on the Work of its 18th Session 4 May 1966, UN Doc A/CN. 4/191 214 commentary on article 26(2); A. Aust, *Modern Treaty Law and Practice* (Cambridge University Press 2000) 174; C. J. Borgen, 'Resolving Treaty Conflicts' (2005) 37 George Washington International Law Review 573, 584.

[114] R. Wolfrum and N. Matz, *Conflicts in international environmental law* (Beiträge zum ausländischen öffentlichen Recht und Völkerrecht, Springer 2003) 121; Borgen thinks that efficient conflict clauses should take into account "different but overlapping areas of concern" (Borgen (n 113) 636–637).

[115] For an interesting discussion on how to determine whether treaties are on the same subject matter, see N. Matz, *Wege zur Koordinierung völkerrechtlicher Verträge: Völkervertragsrechtliche und institutionelle Ansätze* (Springer 2005) 318–324 and her example of how CITES and the WTO do conflict but are obviously not on the same subject matter.

2. Relationship with Later Treaties

With regard to RFMOs instituted after 1975, it is useful to consider how the Convention's drafters intended the relevant instruments to interact, whether states' opinions have changed in that respect and how an alleged conflict could be dealt with.

a) Original Intent

aa) Article XIV (2)–(3)

As examined supra, the drafters of CITES envisioned the relationship with already existing treaties related to the protection of marine species, but stayed silent as to the relationship with future treaties. In order to identify the intent of the states parties, it is important to point to the paragraphs (2) and (3) of article XIV, which deal with the effects of CITES on several international treaties and provide:

> 2. The provisions of the present Convention shall in no way affect the provisions of any domestic measures or the obligations of Parties deriving from any treaty, convention, or international agreement relating to other aspects of trade, taking, possession or transport of specimens which is in force or subsequently may enter into force for any Party including any measure pertaining to the Customs, public health, veterinary or plant quarantine fields.
> 3. The provisions of the present Convention shall in no way affect the provisions of, or the obligations deriving from, any treaty, convention or international agreement concluded or which may be concluded between States creating a union or regional trade agreement establishing or maintaining a common external Customs control and removing Customs control between the parties thereto insofar as they relate to trade among the States members of that union or agreement.

While paragraphs (4) and (5) on marine species focus on treaties already in force, paragraphs (2) and (3) include a pre-emption for future treaties. Indeed, in relation to conventions about "other aspects of trade, taking, possession or transport of specimens" and for treaties "creating a union or regional trade agreement", the Convention refers to any treaty in force/concluded or which may enter into force/ may be concluded.[116]

It can consequently be inferred that the anticipation was that CITES would apply in full in relation to marine species protected under a later regime. Indeed, had the intent of the drafters been different, a wider exemption could have been incorporated on a model similar to paragraphs (2) and (3). As the structure of paragraphs (2) and (3) has not been copied in the following paragraphs, a logical conclusion is that CITES was not meant to give priority to RFMOs created after 1975.

Also, it is worth noting that, according to paragraphs (2) and (3), CITES shall not affect in any way such treaties, which is a broader but also vaguer wording than the

[116] CITES (n 55) article XIV(2)(3).

creation of a limited exemption as provided for in paragraphs (4) and (5). While Wijnstekers, in his book on CITES, mentions article XIV (2) also with regard to marine species,[117] one can question whether this is appropriate, since the following paragraphs explicitly focus on them.[118]

Furthermore, the list of measures mentioned in paragraph 2, while not exhaustive, makes reference to those "pertaining to the Customs, public health, veterinary or plant quarantine fields" which do not resemble or relate to management measures such as adopted by RFMOs.

Finally, even if paragraph (2) or (3) was applicable to marine species, as some treaties related to them can also touch upon 'trade' or 'taking' of listed species, the words "shall in no way affect the provisions of any domestic measures or the obligations of Parties deriving from any treaty" do not prevent the parallel application of rules.[119] In such a case indeed, the obligations regarding the trade or taking would not be affected, but CITES would be equally applicable since it concentrates on regulating international trade. Moreover, such obligations stemming from RFMOs would most likely be measures governing taking through the provision of maximum quotas or trade through restrictions, neither of which would conflict with CITES.[120]

bb) Article XV

It is also worth noting that the Convention itself makes references to cooperation with other bodies involved with marine species. In particular, in the process of listing marine species, consultation with "inter-governmental bodies having a function in relation to those species" is required, in order to gather necessary scientific data and "to ensur[e] co-ordination with any conservation measures enforced by such bodies."[121]

This provision refers to the need for coordination, making it difficult to imagine that CITES' measures are actually conflicting with and/or exclusive towards RFMOs' measures. Also, the specific mention made here of "inter-governmental bodies having a function in relation to" marine species differs from the treaties

---

[117] Wijnstekers (n 106) 433–438, 441–445; see also FAO, 'Report of the Expert Consultation on Legal Issues' (n 4) 10 §49.

[118] Other authors do not explicitly envision the applicability of this paragraph to marine species. For example Franckx, 'The Relationship between CITES, FAO and Related Agreements: Legal Issues' (n 60) 47–48; Gillespie, *Whaling Diplomacy* (n 81) 334–335. For Favre, paragraph 2 refers to laws that "prohibit the importation of certain plants and animals because of concerns over human or animal health, or because of protection against plant pests [and clarifies that] [t]he existence of a CITES permit would not override these laws" (Favre, *International Trade in Endangered Species* (n 13) 305).

[119] Matz, *Wege zur Koordinierung völkerrechtlicher Verträge* (n 115) 264.

[120] For more on the parallel applicability of CITES and RFMOs, see infra Chapter 5 B. III. 2. c) aa).

[121] CITES (n 55) article XV(2)(b).

referred to in article XIV, which are to afford protection to marine species. The former formulation is more general and hence implies that cooperation is possible with a wider range of bodies than the ones to which some deference must be given.

In conclusion, when it comes to the use of CITES for marine species, the original intent of the parties in relation to later treaties does not appear to be one that dismisses the Convention's applicability. Hence, the only two scenarios according to which using CITES for marine species under the management of another later regime would be curtailed would be if the intention of all or some the states parties changed in that regard, or if there was a clear conflict between regimes and procedural rules were applied to the disadvantage of CITES.

b) No Change of Intent

The collective intent of states parties can best be assessed through later practice within the CITES regime, while a development *inter partes* for some states could also take place in a regional forum, such as within RFMOs.

aa) Within CITES

During the negotiations of the MoU on the cooperation between CITES and the FAO the issue of general deference to RFMOs and to the FAO was addressed. Indeed, different views coexisted as to whether the use of CITES for fisheries should or not be conditional to the absence of a management regime, or to the permission of an existing one.[122]

In quite a clear illustration of these opinions, at CoP12, Japan made a proposal for a new Resolution "affirm[ing] that FAO and the mandated regional fisheries management organizations (RFMOs) are appropriate inter-governmental bodies responsible for fisheries and fisheries management; [as well as] agree[ing] that, in cases where there is no responsible fisheries management organization and where trade is having a significant negative impact on conservation, the listing of commercially-exploited fish species in the Appendices may temporarily serve a useful conservation purpose".[123]

Neither this Resolution nor the similarly phrased deference-based draft version of the MoU was adopted, evidencing the compromise solution that states agreed to:

---

[122] Committee on Fisheries – Sub-Committee on Fish Trade, 'Report of the Eighth Session of the Sub-Committee on Fish Trade' (n 52) 7 §18; see also the FAO COFI-led draft where the "primary role of sovereign states, FAO and regional fisheries management organizations in fisheries conservation and management" was recognized and which still included the additional condition of prior agreement by the relevant RFMO (Committee on Fisheries, 'Report of the Twenty-fifth Session of the Committee on Fisheries (Rome, 24–28 February 2003)' FAO Fisheries Report 702 (Rome 2003) Appendix G); see Young, *Trading Fish, Saving Fish* (n 3) 159–162. On the MoU, see infra Chapter 6 A. I. 2.

[123] Japan, 'CoP12 Doc. 16.2.1 – Synergy and Cooperation between CITES and FAO' (n 5) annex.

they desired to keep future relations with RFMOs on an equal footing. Also, the chosen wording would have been a major modification of CITES, since it would have added criteria to the CITES listing process, and such an amendment is unlikely to be addressed in an MoU.[124] The formal amendment procedure would indeed not have been complied with and the MoU, as an instrument developed by the Secretariat, is not the reflection of a formal decision based on the consensus of states parties, as CoP Resolutions can be.

Finally, the CITES CoP has adopted a resolution where it calls for cooperation with RFMOs as part of the procedure to issue IFS certificates.[125] If CITES was to be deprived of competence when fish species are under the management of another regime, this paragraph would have no *raison d'être*.

bb) Outside of the Regime

No treaty establishing an RFMO suggests that the intent of some states has changed in relation to CITES and that the latter regime should not be applicable to species managed by the former. A clear declaration (explicit or not) of such modification would for example take the form of a conflict clause stating the precedence taken by the new treaty over older instruments protecting species from overexploitation driven by international trade, or would include a preambular to such end. Such a clause is absent from all treaties founding RFMOs.

c) No Conflict

The second situation which could cause a later RFMO to supersede CITES would be if there was a clear conflict between the two regimes and procedural rules were applied to the disadvantage of CITES. It is argued here that such conflict is unlikely to be able to arise, in particular as the regimes examined are focused on different subject matters. Moreover, should states nevertheless declare the existence of such a conflict, the existing procedural rules would not be easily applicable, if at all.

---

[124] International Environmental Law Project, 'CITES and Marine Species: An Analysis of the Draft Resolutions on Sharks, Sea Turtles, Toothfish and Whales' (2002) http://www.lclark.edu/live/files/154 accessed 6 July 2015, 2. Young considers that this could have consisted in an *ultra vires* act since it would have de facto amended the Convention, without the normal procedure guarantying the respect of states' sovereignty (Young, *Trading Fish, Saving Fish* (n 3) 169–172). Indeed, "political organs may adopt in their decision-making processes constitutional interpretations which are highly controversial and regarded as *ultra vires* the organs concerned by an outvoted minority of the member states" (Birnie (n 100) 326 (reference omitted)).

[125] Resolution Conf. 14.6 (Rev. CoP16) – Introduction from the Sea (n 66) preamble.

aa) Parallel Applicability but Potential Lack of Coherence

Conflict of norms as a legal term is generally describing a situation in which states cannot simultaneously apply rules from two or more treaties under which they have obligations.[126] In the case at hand, the parallel use of CITES and RFMOs measures would not achieve this result, for two main reasons.

First, while CITES focuses on the regulation of international trade, RFMOs' mandate is the management of specific species or species occurring in a given geographic area: the former can allow or forbid international trade and the latter regulates takings. These activities impact on each other but are mostly distinct.

Second, while it is true that the regulation of IFS by CITES can have a certain influence on takings (in that landing in any country, even the flag state, requires a permit), neither the quotas set by RFMOs nor the NDF undertaken by the national authorities to allow trade provides for the obligation to catch the maximum quotas allowed. These are indeed permissive rules, allowing states to catch *up to* a certain amount of fish. If one figure is lower than the other, they can still be applied simultaneously. Hence, compatibility is an option that remains open to states parties to both CITES and an RFMO.

Beyond incompatible obligations (i.e. genuine conflicts) there might also be conflicting approaches to a similar issue, as well as conflicting objectives or methods.[127] Although these conflicts of treaties do not always result in the impossibility of applying two norms simultaneously, they might "establish divergences or inconsistencies".[128] This in turn has an impact, usually negative, on the effectiveness and coherence of both systems of norms. This situation can best be improved by institutional relations, even though, legally speaking, it does not necessarily need to.

In the case of CITES, it could happen that the parallel application of CITES and RFMOs rules would not lead to the most logical situation. If CITES indeed forbade international trade of a species, which is still harvestable under RFMO measures, a discontinuity would be created. This would then constitute a conflict of logic—to be allowed to catch but not to trade—, but not of norms—being allowed to catch does not automatically entail the obligation to do so, or the right to trade—except if the quotas or a CDS/TDS were expressly to say so.[129] In any event the freedom to fish under RFMOs' mandates is not absolute: species must not be driven to extinction. Should such an unlikely conflict nevertheless arise, one could wonder whether it

[126] Borgen (n 113) 575; Matz, *Wege zur Koordinierung völkerrechtlicher Verträge* (n 115) 11; Mus (n 112) 210.

[127] Wolfrum and Matz, *Conflicts in international environmental law* (n 114) 6–13; Matz, *Wege zur Koordinierung völkerrechtlicher Verträge* (n 115) 8–17.

[128] Wolfrum and Matz, *Conflicts in international environmental law* (n 114) 6.

[129] For a parallel illustration of a narrow understanding of 'conflict' in relation to CITES and another biodiversity treaty (the Protocol concerning Specially Protected Areas and Wildlife), see Borgen (n 113) 612.

would not be the RFMO, by allowing catches at unsustainable levels, which would be acting in violation of its mandate.

Hence, even if the application of one treaty can frustrate the other, or even if the use of CITES and RFMOs measures without coordination between them results in a lack of coherence, management measures and trade permits can in practice coexist. Indeed, the measures used, which are on the one hand trade permits aiming at avoiding species depletion and on the other hand CMMs intending to maximize sustainable catches, are not conflicting. The two sets of obligations can be implemented in a harmonious manner, which would respect the presumption against conflict.[130] In conclusion, the type of relationship between CITES and RFMOs is not one of conflict of norms.

bb) Impracticability of Successive Treaty Rules and of Lex Specialis

In any case, if a dispute between states was to arise and if the parties involved argued that a conflict of rules existed, resorting to the procedural rules dealing with treaties on the same subject matter might be attempted to clarify the intent of the states,[131] and consequently to determine which regime should take precedence over the other.[132]

However, conflicts between rules on different subject matters fall outside the scope of the successive treaty and *lex specialis* rules. It has been demonstrated supra that the existence of a conflict clause in a treaty does not imply that the instruments it refers to are on the same subject matter.[133] Although it is not perfectly clear how 'subject matters' are to be circumscribed,[134] the present competition would

[130] On this see supra Chapter 4 B. I. 3. a).

[131] Franckx, 'The Relationship between CITES, FAO and Related Agreements: Legal Issues' (n 60) 41.

[132] It is Franckx's assumption that the use of CITES for commercially-exploited aquatic species could, should a dispute arise, be followed by a determination of which treaty should take precedence (Franckx, 'The Protection of Biodiversity and Fisheries Management: Issues Raised by the Relationship between CITES and LOSC' (n 1) 216 and following). However this conclusion is debatable and questioned by Young (*Trading Fish, Saving Fish* (n 3) 151–152).

[133] See supra Chapter 5 B. III. 1. c).

[134] M. Fitzmaurice, 'Treaties' Max Planck Encyclopedia of Public International Law www.mpepil.com §110. Also, the test of 'sameness' is not without controversy. On the one hand indeed, Aust (n 113) 183; J. Pauwelyn, *Conflict of Norms in Public International Law: How WTO Law Relates to other Rules of International Law* (Cambridge Studies in International and Comparative Law, Cambridge University Press 2003) 364, presenting Sir Ian Sinclair's views; Wolfrum and Matz argue in favor of a strict interpretation of 'relating to the same subject matter' as it would otherwise be a catch-all phrase with, for example the priority between a human rights and trade treaties to be given through a *lex posterior* rule (Wolfrum and Matz, *Conflicts in international environmental law* (n 114) 149–151).

On the other hand, Pauwelyn considers that there cannot be a conflict of norms if the treaties are not on the same subject matter (Pauwelyn (n 134) 364–367). It has also been said that "[i]f an attempted simultaneous application of two rules to one set of facts or actions leads to incompatible results it can safely be assumed that the test of sameness is satisfied" (ILC, 'Fragmentation of

arguably not be between treaties on the same subject matter, at least when a strict interpretation of 'same' is preferred.

Indeed, while management of fisheries and conservation of endangered species through trade restrictions are partly overlapping themes, the former consists of a set of rules intending to ensure that resources under its management produce the MSY and the latter provides a framework to control international trade in endangered species. In that regard, it has for example been said that "[w]hile ICCAT is responsible for assigning quotas, only CITES can regulate international trade and only CITES has the authority and ability to suspend international commercial trade until the species recovers."[135]

Also, in practice, article 30 VCLT on the successive treaty rule is too vague to be applicable, particularly with regard to the act chosen for dating treaties and to the problem of different memberships.[136] Indeed, since article 30 gives precedence to the later treaty over the earlier one, putting treaties in a chronological order is obviously central. But the provision does not clarify which act (adoption, signature, ratification, entry into force, etc.) determines the point in time to take into account for comparison.[137] Additionally, another difficulty is that, in most cases, the memberships to two treaties are going to differ, which has the potential to create future conflicts if a party to the former treaty expects its legal rights under that treaty to be respected and that this is not compatible with the application of the new treaty by the states which are parties to that later treaty.[138] Hence, the *lex posterior* rule[139] would not be of much practical use.

Similarly, the *lex specialis,* which "[f]ocuses on the scope and precision of the treaties, giving effect to the more narrowly gauged treaty"[140] is of little help. Indeed, this procedural rule is difficult to use *in abstracto* and its mechanical

international law: difficulties arising from the diversification and expansion of international law' Report of the Study Group of the International Law Commission (finalized by M. Koskenniemi) 13 April 2006, A/CN.4/L.682, 18 §22, quoting E. W. Vierdag, 'The Time of the 'Conclusion' of a Multilateral Treaty: Article 30 of the Vienna Convention on the Law of Treaties and Related Provisions' (1988) 59 British Yearbook of International Law 75, 100).

[135] Lieberman (n 91) 6.

[136] Matz, *Wege zur Koordinierung völkerrechtlicher Verträge* (n 115) 332–333.

[137] Mus (n 112) 220. On various opinions on the question, see for example Borgen (n 113) 611–612; Pauwelyn (n 134) 378–379; I. Sinclair, *The Vienna Convention on the Law of Treaties* (2nd edn Manchester University Press 1984) 98; Vierdag (n 134).

[138] J. Klabbers, 'Beyond the Vienna Convention: Conflicting Treaty Provisions' in E. Cannizzaro (ed), *The Law of Treaties Beyond the Vienna Convention* (Oxford University Press 2011) 198.

[139] 'Vienna Convention on the Law of Treaties' (VCLT) 1969, 1155 *United Nations Treaty Series* 331 article 30(3). It seems now well accepted that this article reflects CIL (Aust (n 113) 181; Mus (n 112) 213).

[140] Borgen (n 113) 589. This rule is recognized as applicable in case-law (for example in *Gabcikovo-Nagymaros Project (Hungary/Slovakia), Judgment,* 25 September 1997, I.C.J. Reports 1997 7, 76 §132; Franckx, 'The Relationship between CITES, FAO and Related Agreements: Legal Issues' (n 60) 43).

application can sometimes be considered to reach inappropriate solutions.[141] A problem is indeed, in many cases, to determine which rule is more specialized.[142] One can wonder on which basis specificity is to be decided: scope and precision of the regime, or geographical membership and application.[143] The CITES regime, being global and aiming at all endangered species, could be seen as more general as opposed to regional regimes targeting specific species. However, CITES could also, as an endangered-species convention, be considered as a specialized regime in contrast to management regimes which are concerned with both management and conservation of a species. Determining which set of rules is more specialized is more a value judgment than an exact science.[144] Additionally, it is arguably not useful when rules from different regimes are conflicting[145] and the present situation is one of treaties from different normative areas.

Consequently, such a hypothetical conflict, if it arose between states, would not be resolved by the existing procedural rules. A case-by-case approach would be needed. Most likely, the overlap between regimes which could cause a conflict in the broader understanding of the term would be managed through cooperation of institutions[146] in order to reach a compromise solution.[147]

IV. Conclusions

Being indirectly mentioned in the Convention text and directly in Conference Resolutions, the listing of commercially-exploited marine species falls under the realm of CITES, especially as both aquatic species and commercially-valuable species have already found their way into the Appendices.

---

[141] A. Lindroos, 'Addressing Norm Conflicts in a Fragmented Legal System: The Doctrine of Lex Specialis' (2005) 74 Nordic Journal of International Law 27, 35, 42–43. The ICJ for example refused to disqualify the applicability of human rights in a situation of armed conflict through a blanket application of the *lex specialis* rule, even when humanitarian law came into play. It indeed stated that the International Covenant on Civil and Political Rights continued to apply, although interpretation of terms such as 'arbitrarily deprivation of life' ought to be interpreted in accordance with the *lex specialis*, which is here humanitarian law (*Legality of the Threat or Use of Nuclear Weapons, Advisory Opinion*, 8 July 1996, I.C.J. Reports 1996 226, 240 §25).

[142] ILC, 'Fragmentation of international law' (n 134) 35–36 §58; Klabbers, 'Beyond the Vienna Convention: Conflicting Treaty Provisions' (n 138) 199–200; Lindroos (n 141) 43–44, 48; Matz, *Wege zur Koordinierung völkerrechtlicher Verträge* (n 115) 335; Wolfrum and Matz, *Conflicts in international environmental law* (n 114) 156.

[143] ILC, 'Fragmentation of international law' (n 134) 60–61 §111–112, 62–63 §116–117; Borgen (n 113) 589; Pauwelyn (n 134) 389–391.

[144] Borgen (n 113) 612–613.

[145] Lindroos (n 141) 35, 42.

[146] Young, *Trading Fish, Saving Fish* (n 3) 151–152.

[147] N. Matz-Lück, 'Treaties, Conflicts between' Max Planck Encyclopedia of Public International Law www.mpepil.com §19.

As nothing provides to the contrary, it is possible to state that the CITES framework can undeniably apply to commercially-exploited fish species that are not under the management of another regime.

While it is true that no commercially-exploited fish species which is the focus of a regional fishery management institution is listed, this is not due to a lack of competence to do so. Indeed, article XIV governs the relationships with other treaties and does not impair the ability for the CoP to list such a species. Hence, it seems correct to assume that CITES can be applied to those species, also when they are under the management of another institution. This might simply have practical consequences in the application of the permitting system for species which are afforded protection under pre-1975 regimes.

Since obligations under both RFMOs and CITES can be implemented in parallel, there is no conflict of rules to resolve. However, in theory, it is not crystal clear which institutional regime would have to give deference, if any; should a dispute on the matter arise, it might have to be settled on a case-by-case basis.

Generally, problems stemming from inter-institutional competition are more likely to be solved by cooperation than by a third-party determination of exclusive competence.[148] This approach also fits perfectly with one aspect of the Strategic Vision of CITES which is to "[c]ontribute to significantly reducing the rate of biodiversity loss [...] by ensuring that CITES and other multilateral instruments and processes are coherent and mutually supportive".[149] Hence, overlapping obligations can be stream-lined through cooperation in order to avoid incoherent situations, and, ideally, in order also to increase efficiency.

C. Adequacy of CITES Provisions for Commercially-exploited Fish Species

Since commercially-exploited fish species are different from other aquatic and terrestrial species, it is important to evaluate whether the CITES framework can deal with their particularities, both in terms of biological and trade-related characteristics (I) and in relation to their importance for livelihoods in developing countries (II). The lack of expertise and the alleged bias of this regime towards strong conservation, if not preservation, have indeed been used as arguments against a listing.[150] Hence, inquiring whether CITES provisions are clear enough to be applied to fish species and adequate for their biological particularities, for the quantities traded as well as in their role for developing countries is central to assessing the adequacy of CITES.

[148] Young, *Trading Fish, Saving Fish* (n 3) 292.

[149] Resolution Conf. 16.3 – CITES Strategic Vision: 2008–2020 (n 57) goal 3.

[150] Young, *Trading Fish, Saving Fish* (n 3) 142–144.

Another line of argument used against CITES listing of marine species is linked to the capacity of states to implement such listing. However, as the implementation requirements in terms of NDF are based on the same information needed to manage sustainably a fishery and/or control a fleet, this set of objections is not examined in detail. Indeed, if a country cannot implement CITES, it most probably will be have deeper governance problems, beyond the scope of this research. Also, "[u]nder the terms of the Convention, anxieties about capacity should not be decisive factors in deciding whether to list a species."[151]

I. Adaptability of CITES Provisions

1. Listing Criteria and Procedures

The CITES listing criteria are a crucial aspect of the treaty regime, since it is on this basis that species are proposed to be listed, the decision itself depending on the CoP.

The criteria used to insert a species in CITES Appendices are not included precisely in the Convention text and have evolved over the years. The Convention only states at its article II that:

1. Appendix I shall include all species threatened with extinction which are or may be affected by trade. Trade in specimens of these species must be subject to particularly strict regulation in order not to endanger further their survival and must only be authorized in exceptional circumstances.
2. Appendix II shall include:
(a) all species which although not necessarily now threatened with extinction may become so unless trade in specimens of such species is subject to strict regulation in order to avoid utilization incompatible with their survival; and
(b) other species which must be subject to regulation in order that trade in specimens of certain species referred to in sub-paragraph (a) of this paragraph may be brought under effective control.

Thus, considering the brevity and non-specificity of these provisions, the Parties have given clarification and then adapted the criteria through Resolutions. The Bern criteria, adopted in 1976 in Resolution Conf. 1.1, were the first set of criteria to be developed as a basis for listing species. Biological and trade statuses were to be taken into account when deciding a listing: for Appendix I, if data strongly showed that a species' biological status was "currently threatened with extinction" "only a probability of trade" was needed. Conversely, if trade was "known to occur, information on the biological status [did not need to] be as complete."[152] As for Appendix II, to be listed, species had to be traded or likely to be, in volumes

[151] Vincent and others (n 6) 582, see also 583–585.

[152] Resolution Conf. 1.1 – Criteria for the Addition of Species and Other Taxa to Appendices I and II and for the Transfer of Species and Other Taxa from Appendix II to Appendix I 1976 (repealed).

constituting a potential threat to the species' survival and biological evidence had to show that the species "might become" threatened with extinction.

Criticized for being vague and hence allowing political decisions,[153] the Bern criteria were abandoned in 1994 for the new criteria, to be found in Resolution Conf. 9.24, later amended at several CoPs.[154] The so-called Fort Lauderdale criteria differ from the previous set of criteria mainly in that specific quantitative guidelines are provided,[155] biological status is more important than trade status and split-listing is allowed.

a) Content

aa) Criteria for Inclusion in Appendix I

Appendix I is meant to include species which are or may be affected by trade and the wild population of which is either small and affected by decline, or has very small subpopulations, or is geographically very strongly concentrated at one or more periods of their lives, or shows large short-term fluctuations in population size, or is very vulnerable to intrinsic or extrinsic factors. A second alternative criterion is that the species has a restricted area of distribution and is fragmented, fluctuates largely in the area of distribution, is very vulnerable to intrinsic or extrinsic factors, or that a decline in the area of distribution, its area of habitat, the number of subpopulations, the number of individuals, the quality of habitat or recruitment is observed, inferred or projected. Finally, the last option would be a marked decline in the species' population in the wild.[156]

Terms such as affected by trade, area of distribution, small wild population or decline are clarified in the definitions and, when relevant, numerical benchmarks are provided.[157]

bb) Criteria for Inclusion in Appendix II

Appendix II species, although not currently threatened with extinction, are supposed to be those that could fulfill Appendix I criteria in the near future if trade was not regulated.[158] Also, Appendix II listing is possible if the species is reduced to a

---

[153] Anonymous, 'The CITES Fort Lauderdale Criteria: The Uses and Limits of Science in International Conservation Decisionmaking' (2001) 114 Harvard Law Review 1769, 1776.

[154] Resolution Conf. 9.24 (Rev. CoP16) – Lauderdale criteria (n 23).

[155] Anonymous, 'The CITES Fort Lauderdale Criteria: The Uses and Limits of Science in International Conservation Decisionmaking' (n 153) 1779.

[156] Resolution Conf. 9.24 (Rev. CoP16) – Lauderdale criteria (n 23) annex 1.

[157] Ibid. annex 5.

[158] Anonymous, 'The CITES Fort Lauderdale Criteria: The Uses and Limits of Science in International Conservation Decisionmaking' (n 153) 1780.

level where other influences can threaten its survival.[159] A final category of species listed are those which are not themselves under threat but must be listed for the effective control of an endangered species—usually because they closely resemble an endangered species—or if such a listing is necessary for other "compelling reasons" to effectively control trade.[160]

cc) Precautionary Approach

The precautionary approach takes several forms in the CITES regime and is made use of even if the terminology is not employed in the original treaty.[161] First, mentioning that species shall be listed when they 'may' be threatened seems to imply that uncertainty shall trigger action.[162] However, in practice, this is not implemented in such a radical manner. This provision is rather understood in the sense that "in case of uncertainty either as regards the status of a species or the impact of trade on the conservation of a species, [Parties shall] act in the best interest of the conservation of the species concerned and adopt measures that are proportionate to the anticipated risks to the species."[163]

Also, the current Fort Lauderdale criteria consider the precautionary approach as guiding the decision making processes.[164] In practice, it mainly has implications for down- or de-listing species from the Appendices. Indeed, Appendix I species cannot be delisted directly and have to spend at least two intervals between CoPs in Appendix II. Also, for such down-listing to Appendix II, a safeguard such as lack of demand for international trade, proven sustainable management, the setting of export quotas, or the existence of a ranching proposal must be implemented. Furthermore, an Appendix II species should not be removed from the Appendix if this "would be likely to result in it qualifying for inclusion in the Appendices in the near future."[165]

The application of the precautionary approach to the criteria, where the parties are required to "act in the best interest of the conservation of the species concerned and adopt measures that are proportionate to the anticipated risks to the species",[166] does not mean that trade restrictions are to be automatically implemented in the case of uncertainty. The reference to 'proportionate to the anticipated risks' rather

[159] Resolution Conf. 9.24 (Rev. CoP16) – Lauderdale criteria (n 23) annex 2 a B.

[160] Ibid. annex 2 b A-B.

[161] J. B. Wiener, 'Precaution' in D. Bodansky, J. Brunnée and E. Hey (eds), *Oxford Handbook of International Environmental Law* (Oxford University Press 2007) 601.

[162] S. Marr, *The Precautionary Principle in the Law of the Sea: Modern Decision Making in International Law* (Martinus Nijhoff Publishers 2003) 107.

[163] Resolution Conf. 9.24 (Rev. CoP16) – Lauderdale criteria (n 23) annex 4.

[164] B. Dickson, 'Precaution at the Heart of CITES?' in J. Hutton and B. Dickson (eds), *Endangered Species-Threatened Convention: The Past, Present and Future of CITES* (Earthscan 2000) 42.

[165] Resolution Conf. 9.24 (Rev. CoP16) – Lauderdale criteria (n 23) annex 4.

[166] Ibid.

leaves the door open to conducting risks assessments and evaluating proportionality. For fish species which can usually bounce back even after large reductions in numbers, it will not be needed to err totally on the side of caution, unlike in the case of elephants, rhinos or tigers for whom "the stakes are simply too high"[167] as there are so few specimens left.

Finally, it has been clarified recently that NDF should be based on some precaution, in a manner relative to the risks faced by the species. Indeed, the data required to show that international trade will not be detrimental should be proportionate to the vulnerability of the species.[168] This too is an implementation of the precautionary approach where the level of risk at hand determines the manner in which the availability and adequacy of data will be considered.

dd) Procedures

Proposals to amend the Appendices I and II must be submitted to the Secretariat at least 150 days before the following CoP. The Secretariat then shares them with all parties.[169]

Originally, voting with a two-third majority of parties present and voting was the Convention's default procedure.[170] However, according to the current rules of procedure, decisions on whether to amend the Appendices should be, as far as possible, taken by consensus.[171] Only when this fails is a vote to be suggested;[172] this proposal to vote then necessitates a simple majority in favor of it to be adopted.[173] Postal vote for urgent amendments of Appendices in the intervals between two CoP meetings is also provided for in the Convention.[174]

In both normal and postal procedures, the Secretariat is expected, for marine species, to "communicate [. . . the proposal] to the Parties. It shall also consult intergovernmental bodies having a function in relation to those species especially with a view to obtaining scientific data these bodies may be able to provide and to ensuring co-ordination with any conservation measures enforced by such bodies."[175] Then the Secretariat shall provide these views to the parties as well as make its own recommendations.[176]

---

[167] Wiersema, 'Uncertainty and Markets for Endangered Species under CITES' (n 45) 249.

[168] Resolution Conf. 16.7 – Non-detriment findings 2013.

[169] CITES Rules of Procedure of the Conference of the Parties, as amended at the 16th meeting, Bangkok, 2013 rule 22.1.

[170] CITES (n 55) article XV(1)(b).

[171] Rules of Procedure of the Conference of the Parties, as amended at the 16th meeting, Bangkok, 2013 (n 169) rule 23.1.

[172] Ibid. rule 23.2.

[173] Ibid. rule 26.1.

[174] CITES (n 55) article XV(2).

[175] Ibid. article XV(1) and (2)(b).

[176] Ibid.

Reservations to an amendment are possible within the 90 days following the decision. A state willing to enter into such a reservation has to notify the Depositary Government in writing. In this case, such state is considered as a non-party to the Convention in relation to the species in question.[177]

b) Particular Issues Linked to Fish Species

aa) Different Productivity

The listing criteria have not originally been designed for commercially-exploited fish species. However, their suitability for this category of potentially endangered species has been reviewed on the basis of the advice provided in two FAO Technical Consultations.[178] Hence, the numerical benchmarks to determine decline in such species have been adapted to the different reality of quantity, productivity and vulnerability of fish species.

In general, a marked decline is defined as the situation when 5–30 % of the original population (determined by a baseline) is left. In comparison, the dangerous levels for marine and large freshwater bodies species are set in the range between 5 and 20 %—5–10 % for high productivity species; 10–15 % for medium productivity species; and 15–20 % for low productivity species.[179] This is important because, in general, fish reproduce at a higher rate than terrestrial species and mammals. Recognizing the different reproduction rates between marine species is also necessary since no single numerical criterion could fit for example both the orange roughy and small forage fish.

Then, the fact that a major decrease in abundance is to be expected at the beginning of a fish stock's exploitation is recognized[180] by the statement that "[a] decline that is the result of [... a] harvesting programme that reduces the population to a planned level, not detrimental to the survival of the species, would not normally be covered by the term decline."[181] Hence, the modification of Resolution 9.24 to take into account the FAO advice can be described as an important achievement to make the listing criteria adequate for fish species.

---

177 Ibid. article XV(3).

178 FAO, 'Report of the Technical Consultation on the Suitability of the CITES Criteria for Listing Commercially-exploited Aquatic Species (Rome, 28–30 June 2000)' FAO Fisheries Report 629 (Rome 2000); FAO, 'Report of the Second Technical Consultation on the Suitability of the CITES Criteria for Listing Commercially-exploited Aquatic Species (Windhoek, Namibia 22 October 2001)' FAO Fisheries Report 667 (Rome 2002).

179 Resolution Conf. 9.24 (Rev. CoP16) – Lauderdale criteria (n 23) annex 5 footnote.

180 As opposed to the fear expressed by China that such a reduction would not be taken into account properly by non-fisheries managers (China, 'CoP12 Inf. 30 – "Sharks" and COP12 – A Case for Caution' (n 6) 3 §7); Vincent and others (n 6) 580.

181 Resolution Conf. 9.24 (Rev. CoP16) – Lauderdale criteria (n 23) annex 5.

Additionally, some of the criticisms expressed against these criteria are not justified. For example, it has been said that if the risks faced by a species are estimated on the basis of the number of individuals, this is likely to be problematic for fish species: larvae and juveniles are more numerous than adult individuals and could give too optimistic an opinion on the health of the stocks.[182] However, Resolution 9.24 states that when information on the population size is provided, it should make clear whether it refers to "the total number of individuals or to the effective population size (i.e. individuals capable of reproduction, excluding individuals that are environmentally and behaviourally or otherwise reproductively suppressed in the wild) or to another appropriate measure, index or component of the population." Hence, the regime fully envisions and accommodates the need to use only the relevant individuals in estimating the size of a population.

Furthermore, there was quite a vivid debate about defining whether biological or commercial extinction is the focus of a listing. Indeed, CITES has the declared and clear goal of avoiding biological extinction in species.[183] For commercially-exploited aquatic species, the usual exploitation model stops when a species is not commercially profitable—and this should happen long before biological extinction.[184] Under this point of view, an argument against the appropriateness of using CITES is that such a regime, if properly implemented, would never come into play, in particular once the high fecundity of fish species is taken into account.[185]

However, this opinion can be and has been challenged:[186] it seems indeed that for some high value species, such as the Bluefin Tuna, the exploitation is likely to continue until quasi-biological extinction, since demand is stable (or rises because of rarity and hence exclusivity) even with increasing prices.[187] Similarly, species

[182] D. S. Butterworth, 'Possible interpretation problems for the current CITES listing criteria in the context of marine fish species under commercial harvest' (2000) 42 Population Ecology 29, 31.

[183] "[E]xtinction is usually defined as the total disappearance of a species from the face of the earth and an absence of sightings for a period of at least five decades" (M. A. Du Plessis, 'CITES and the Causes of Extinction' in J. Hutton and B. Dickson (eds), *Endangered Species-Threatened Convention: The Past, Present and Future of CITES* (Earthscan 2000) 14). For example, China reiterated, in the case of porbeagle shark and spiny dogfish, that "[t]he *survival* of neither species is threatened by harvest or trade, despite excessive fishing and reported population declines. *Commercial extinction* is the problem, not *biological extinction*. CITES focus should be to prevent international trade threatening survival, not optimising management to sustain commercial fisheries and profits from trade" (China, 'CoP14 Inf. 45 – Shark Issues' (n 3) 2 §1(a) (italics in the original text)).

[184] World Ocean Review, *The Future of Fish – The Fisheries of the Future* (Maribus, Future Ocean 2013) 24.

[185] China, 'CoP12 Inf. 30 – "Sharks" and COP12 – A Case for Caution' (n 6) 3 §7.

[186] P. A. Larkin, 'Concepts and issues in marine ecosystem management' (1996) 6 Reviews of Fish Biology and Fisheries 139, 148.

[187] N. K. Dulvy, Y. Sadovy and J. D. Reynolds, 'Extinction vulnerability in marine populations' (2003) 4 Fish and Fisheries 25, 44–45; Du Plessis (n 183) 15; C. Huxley, 'CITES: The Vision' in J. Hutton and B. Dickson (eds), *Endangered Species-Threatened Convention: The Past, Present and Future of CITES* (Earthscan 2000) 4–5; World Ocean Review, *The Future of Fish – The Fisheries of the Future* 2013 (n 184) 24–25.

which are incidentally caught can risk biological extinction as they are not the targeted species[188] and consequently do not benefit from the protection afforded by the laws of economic profitability. Furthermore, "fisheries often (and increasingly) proceed well past any bioeconomic equilibrium because of subsidies, lack of opportunity costs, increasing value with increasing rarity, open access and the multispecies nature of many fisheries."[189] Also, it must be kept into account that a minimum size is needed for a stock to replenish[190] and that other factors of vulnerability are likely to be interfering with a stock's survival ability—this becomes problematic if commercial exploitation has brought the stock to a dangerous level where its ability to cope with other external factors, such as the introduction of alien species,[191] is diminished.

bb) Split-Listing

According to split-listing, distinct populations of a species can be listed in different Appendices[192] while otherwise a whole species is listed in one single Appendix regardless of the different biological statuses of some populations.

The existence of this method is important if commercially-exploited fish species are listed.[193] Indeed, the more flexible a listing can be, the more unlikely it is to have healthy stocks unnecessarily subjected to trade restrictions under CITES.

Applying different regulations to separate populations is a common method in fisheries management, since species' statuses vary, inter alia, according to their geographical distribution.[194] While the approach in CITES is usually based on species or sub-species, split-listing is to be implemented on the basis of populations. Hence, the evolution of the regime provides added flexibility, which is more in line with the approach followed in fisheries management. Also noteworthy is that, for sturgeons, the CoP agreed that "[t]he term 'stock' is regarded [. . .] to be synonymous with 'population'".[195] Hence, it appears that the particularities of fisheries management can be taken into account.

[188] Dulvy, Sadovy and Reynolds (n 187) 45.

[189] Vincent and others (n 6) 579 (references omitted).

[190] As is the case for all species (Sand, 'Endangered Species, International Protection' (n 107) §2).

[191] World Ocean Review, *The Future of Fish – The Fisheries of the Future* 2013 (n 184) 55.

[192] Resolution Conf. 9.24 (Rev. CoP16) – Lauderdale criteria (n 23) annex 3.

[193] This has been acknowledged early on, see for example: FAO, 'Report of the Expert Consultation on Implementation Issues Associated with Listing Commercially-exploited Aquatic Species on CITES Appendices (Rome, 25–28 May 2004)' FAO Fisheries and Aquaculture Report 741 (Rome 2004) 11 §59–61; FAO, 'Report of the Expert Consultation on Legal Issues' (n 4) 9–10 §44–46.

[194] FAO, 'Report of the Expert Consultation on Implementation Issues' (n 193) 11 §59.

[195] Resolution Conf. 12.7 (Rev. CoP16) – Conservation of and trade in sturgeons and paddlefish (n 84).

However, split-listing causes enforcement difficulties since it opens the door to 'laundering' the endangered and accordingly highly protected stock of a species under the pretense that it originates from another rather healthier population.[196]

cc) Look-Alike

Under CITES, in order to efficiently control trade in Appendix I and II species, some species not themselves meeting the 'endangered' criterion can be listed,[197] particularly the ones that look so similar to endangered species listed under CITES—or proposed to be added to an Appendix—that it is not possible to differentiate one from the other.

The listing of look-alike species in order to facilitate enforcement is deemed to be problematic for many commercially-exploited aquatic species, which, traded in a processed form, are not easily, if at all, distinguishable.[198] Indeed, listing all species that look alike would have very important socio-economic and administrative impacts[199] as well as go against the principle of optimum utilization of fish stocks, which is central to the law of the sea framework.[200]

At CoP13, on the FAO Expert Consultation's advice, a modification of Resolution Conf. 9.24 (Rev.CoP12) meaning to "increase[e] flexibility from a legal perspective"[201] was accepted. The chapeau paragraph of Annex 2b, which addresses look-alike listing, was consequently changed from "species should be included" to "species may be included". This more discretionary basis for look-alike listings is positive, as it underlines that a case-by-case analysis of the situation is required.

However, listing less look-alike species does in turn increase enforcement issues, a problem which has not, yet, been successfully addressed.[202]

dd) By-Catch

The second paragraph of Annex 2b refers to listing of species for which "[t]here are compelling reasons other than [... look-alike] to ensure that effective control of trade in currently listed species is achieved." This formulation could be understood

[196] FAO, 'Report of the Expert Consultation on Implementation Issues' (n 193) 11 §61; M. 't Sas-Rolfes, 'Assessing CITES: Four Case Studies' in J. Hutton and B. Dickson (eds), *Endangered Species-Threatened Convention: The Past, Present and Future of CITES* (Earthscan 2000) 86.

[197] CITES (n 55) article (II)(2)(b).

[198] FAO, 'Report of the Expert Consultation on Implementation Issues' (n 193) 8 §43. In the case of sturgeons or seahorses for example, the listing of a number of endangered species has led to the listing of all species.

[199] Ibid. 9 §49.

[200] Calley (n 76) 188.

[201] FAO, 'Report of the Expert Consultation on Legal Issues' (n 4) 9 §43.

[202] See infra Chapter 6 B. II. 2.

as a basis to list species which are targeted by fishing activities, during which protected species are also caught incidentally.[203]

Such an interpretation would broaden the potential applicability of CITES so much—nearly no fishing operation only catches one species at a time—that trade regulations would be impracticable and the FAO advised against it.[204] However, the suggested changes in paragraph B were not adopted.[205] Nevertheless, it is worth noting that, since, according to the chapeau paragraph of Annex 2b, species now "may be included" rather than "should be included", the application of paragraph B is not obligatory.

ee) Vulnerability

According to Resolution 9.24, other factors which make the species more vulnerable to extinction can be taken into account in determining whether a species fits in the Appendices' criteria. The first two listing grounds for Appendix I (small size of the wild population or a restricted area of distribution) have to be coupled with another criterion which can be a high vulnerability to intrinsic or extrinsic factors. As for the third listing ground, the marked decline in the population size, it can be inferred or projected on the basis of such a high vulnerability.[206]

Relevant intrinsic and extrinsic factors mentioned in the Fort Lauderdale criteria include, inter alia, the life history, threats from alien species, habitat degradation or loss, or rapid environmental changes.[207] For marine species that are vulnerable because of their low productivity or long life, or due to external factors such as pollution, climate change or habitat destruction, it is important that such aspects can be taken into account, as is the case in the CITES regime.

c) Remaining Inadequacy and Lack of Clarity

aa) Remaining Inadequacy

First, even if, for split-listing, the division is made at the population level, the overall listing criteria apply to species. This is not ideal since fisheries, in general, are better managed at the stock level.[208]

---

[203] FAO, 'An Appraisal of the Suitability of the CITES Criteria for Listing Commercially-exploited Aquatic Species' FAO Fisheries Circular 954 (Rome 2000) 33–35 §4.2; FAO, 'Report of the Expert Consultation on Legal Issues' (n 4) 9 §45.

[204] FAO, 'An Appraisal of the Suitability of the CITES Criteria for Listing Commercially-exploited Aquatic Species' (n 203) 33–35 §4.2.

[205] FAO, 'Report of the Expert Consultation on Legal Issues' (n 4) 9 §45. See infra Chapter 6 B. II. 3.

[206] Resolution Conf. 9.24 (Rev. CoP16) – Lauderdale criteria (n 23) annex 1.

[207] Ibid. annex 5.

[208] Butterworth (n 182) 30–31; CCAMLR, 'CoP12 Inf. 20 – Proposal from Australia (CoP12 Prop. 39)' (3–15 November 2002) 5 §10.11 (Norway).

Second, split-listing is in reality not encouraged under CITES.[209] Also, it is, for the time being, impossible for a species to be totally split-listed in the sense that some populations are listed and others not at all.[210] That signifies that a separate population, totally healthy, could be subjected to CITES restrictions to trade, with important socio-economic implications but for no direct conservation purpose.

Furthermore, even if split-listing was encouraged for commercially-exploited aquatic species, implementation mechanisms would need to be developed in order to counterbalance the increased difficulties in enforcement.[211]

Third, the implementation of the precautionary principle, as found in Annex 4, is potentially inadequate. Indeed, it is so difficult to fulfill the requirements for delisting that, instead of preventing unsustainable delisting, it might currently prevent listing altogether. For species of wildlife that are crucially important for food security and for the economy, and that fluctuate fast and quite widely, there is a need for a more flexible listing and de-listing procedure, as well as for the certainty that the precautionary principle will not be used to block a down-/de-listing—as it has happened in the case of some whale species.

During the debates related to the Atlantic Bluefin Tuna, Monaco suggested the adoption of a Resolution intending to face the difficulties to downlist species and implicitly acknowledging the fears caused by that characteristic of CITES.[212] This Resolution would have requested the Animals Committee, together with ICCAT, to review stocks' status and, if necessary, to mandate the depositary government to submit a down-/de-listing proposal.[213] Very correctly, the FAO commented that although this draft Resolution is an interesting attempt to face the issue of down-/de-listing, the actual decision would still, at the end, be taken by the CoP. "It is therefore not clear that the annotation would significantly improve the likelihood of the species being timeously down-listed or de-listed in response to improved status."[214]

The difficulties existing in the delisting procedures are particularly problematic for fish species, due to their high variability in abundance. That species may stay in Appendix I while their stocks have recovered and reached levels where they do not meet the listing criteria anymore was a concern for the FAO, whose mandate is to promote sustainable exploitation as well as poverty alleviation and food security.[215]

---

[209] Resolution Conf. 9.24 (Rev. CoP16) – Lauderdale criteria (n 23) annex 3.

[210] Ibid.

[211] FAO, 'Report of the Expert Consultation on Legal Issues' (n 4) 9–10 §46.

[212] Monaco, 'CoP15 Doc. 52 (Rev. 1) – Species Trade and Conservation, Atlantic Bluefin Tuna, Draft Resolution of the Conference of the Parties' (13–25 March 2010) preamble.

[213] Monaco, 'CoP15 Prop. 19 – Proposal to include Atlantic Bluefin Tuna' (n 93) summary 3 §17; Monaco, 'CoP15 Doc. 52 (Rev. 1) – Species Trade and Conservation, Atlantic Bluefin Tuna, Draft Resolution of the Conference of the Parties' (n 212) operative part.

[214] FAO, 'CoP15 Inf. 26 – Statement from the FAO Fisheries and Aquaculture Department on CoP15 Proposal 19 to List the Atlantic Bluefin Tuna in Appendix I' (n 97) annex §11.

[215] 'Constitution of the Food and Agriculture Organization of the United Nations' 1945 preamble; FAO, 'CoP15 Inf. 26 – Statement from the FAO Fisheries and Aquaculture Department on CoP15 Proposal 19 to List the Atlantic Bluefin Tuna in Appendix I' (n 97) annex §10.

It also worried representatives of RFMOs.[216] Nevertheless, not all marine species share this characteristic of high variability[217] and it is worth noting that the advice given by the FAO Expert Advisory Panels (Expert Panels) takes this factor into consideration.[218]

Fourth, the normal decision-making schedule at CITES is too slow for fisheries matters. Indeed, the CoP meets every two to three years while fisheries management is usually based on an annual if not bi-annual rhythm. As for the postal voting method, it has not been used extensively.[219]

bb) Remaining Lack of Clarity

First, in the wording of the footnote which provides benchmarks for commercially-exploited aquatic species, it is unclear what refers to Appendix I or to Appendix II. The values of the numerical benchmark are also not unanimously praised as being able to provide an efficient method to determine whether a species is endangered[220]—this is potentially even more difficult for aquatic species than for others due to the high level of uncertainty in the estimations of stocks. Also, further concerns about the adequacy of CITES criteria for fish species are related to the absence of case studies showing the relationship between fish abundance or distribution and the threat of extinction, as well as to the differences between the benchmarks used in fisheries and the CITES criteria.[221]

Second, although specific provisions have been developed for such species and are applicable to them, no definition of 'commercially-exploited aquatic species' is provided. A discussion at the CITES Animals Committee meeting in March 2012 showed a gentleman's (and ladies', as was then rightly pointed out) agreement on the species encompassed by such a concept. These are fish and invertebrate species from the marine environment and from large freshwater bodies, hence excluding mammals, birds and reptiles. However, this does not constitute an authoritative interpretation.

Third, the view could still be defended that Annex 2bB permits listing of species which are not endangered but during the fishing operations of which some endangered species become a victim of by-catch.

Fourth, recently, a debate regarding the interpretation of Annex 2a (listing in Appendix II) has shown some discrepancy in the understanding of terms by various actors. The Annex which causes disagreement reads as:

---

[216] FAO, 'Report of the Second Meeting of FAO and non-FAO regional fishery bodies or arrangements (Rome, 20–21 February 2001)' FAO Fisheries Report 645 (Rome 2001) 9 §46.

[217] Vincent and others (n 6) 579.

[218] See infra Chapter 6 C. I. 1. b).

[219] Favre, *International Trade in Endangered Species* (n 13) 313.

[220] Anonymous, 'The CITES Fort Lauderdale Criteria: The Uses and Limits of Science in International Conservation Decisionmaking' (n 153) 1783–1784.

[221] Butterworth (n 182) 30.

The following criteria must be read in conjunction with the definitions, explanations and guidelines listed in Annex 5, including the footnote with respect to application of the definition of 'decline' for commercially exploited aquatic species.

A species should be included in Appendix II when, on the basis of available trade data and information on the status and trends of the wild population(s), **at least one** of the following criteria is met:

A. It is known, or can be inferred or projected, that the regulation of trade in the species is necessary to avoid it becoming eligible for inclusion in Appendix I in the near future; or

B. It is known, or can be inferred or projected, that regulation of trade in the species is required to ensure that the harvest of specimens from the wild is not reducing the wild population to a level at which its survival might be threatened by continued harvesting or other influences.

On the one hand, the FAO holds that the definition of decline—mentioned in the chapeau paragraph—applies to both paragraphs A. and B. This is the interpretation followed by the FAO Expert Advisory Panels in their assessments. On the other hand, the CITES Secretariat considers that 'decline' only refers to paragraph A. while 'reducing' in paragraph B. can be demonstrated by some decline, which does not need to meet the Annex 5 footnote criteria.[222]

2. Introduction from the Sea

Many marine species do not stay in waters under national jurisdiction but rather migrate through the high seas. Species caught there also (or arguably even to a greater extent) need protection when they are endangered, but the fact that these waters belong to no state is a complicating factor for conservation and management.

This difficulty has been envisioned by CITES with its IFS provisions: as stated in article I (e), international trade indeed includes "transportation into a State of specimens of any species which were taken in the marine environment not under the jurisdiction of any State".

However, while this is a crucial feature of the CITES system with regard to marine species, two aspects of the IFS definition used to suffer ambiguities, which was a main argument against the listing of toothfish in 2002.[223] First, one had to specify what the term 'marine environment not under the jurisdiction of any State' refers to. Second, it was fundamental to identify whether it is the port or the flag state that is the state of introduction. Hence, this provision had to be fully clarified to become applicable.

[222] FAO, 'Report of the FAO Workshop to Review the Application of CITES Criterion Annex 2 a B to Commercially-exploited Aquatic Species (Rome, 19–21 April 2011)' FAO Fisheries and Aquaculture Report 976 (Rome 2011) 7–8 §24–25.

[223] CCAMLR, 'CoP12 Inf. 20 – Proposal from Australia (CoP12 Prop. 39)' (n 208) 6–7 §10.22 (Chile).

a) Waters Not Under the Jurisdiction of Any State

In order to have a system of permits working efficiently at the international level, it is necessary that states are aware of the type of documents that they have to issue. Whether no documentation, an export permit or an IFS certificate is needed depends on where the species are caught as well as how the geographical scope of the relevant provisions is interpreted.

aa) Current International Understanding

The geographic component of "introduction from the sea" has been settled in a Resolution in 2007. Accordingly, "'the marine environment not under the jurisdiction of any State' means those marine areas beyond the areas subject to the sovereignty or sovereign rights of a State",[224] to be understood according to UNCLOS.[225]

In practice then, in line with international law as found in UNCLOS, the areas subject to sovereignty of a state extend to the territorial sea, which is 12 nm wide.[226] The coastal state also benefits from sovereign rights in the EEZ which is 200 nm maximum in breadth,[227] "for the purpose of [...] conserving and managing the natural resources, whether living or non-living, of the waters superjacent to the seabed and of the seabed and its subsoil".[228] Finally, the coastal state has sovereign rights over the living sedentary species[229] of the continental shelf, which extends "throughout the natural prolongation of its land territory to the outer edge of the continental margin",[230] possibly further than 200 nm.

bb) The Added Value of the 2007 Resolution

The adoption of this Resolution is an important step towards a unified application of terms and ensures that CITES evolves with international law. Indeed, if the geographic aspect of IFS had not been settled, the risk theoretically would have

[224] Resolution Conf. 14.6 (Rev. CoP16) – Introduction from the Sea (n 66).

[225] For a summary of the relevant national claims to maritime jurisdiction, see the reference table prepared by the United Nations (as at 15 July 2011) http://www.un.org/Depts/los/LEGISLATIONANDTREATIES/PDFFILES/table_summary_of_claims.pdf accessed 10 July 2015.

[226] 'United Nations Convention on the Law of the Sea' (UNCLOS) 1982, 1833 *United Nations Treaty Series* 397 articles 2–3.

[227] Ibid. article 57.

[228] Ibid. article 56(1)(a).

[229] This means "organisms which, at the harvestable stage, either are immobile on or under the seabed or are unable to move except in constant physical contact with the seabed or the subsoil" (UNCLOS (n 226) article 77(4)).

[230] Ibid. article 76(1).

existed that the coastal state and the state of introduction both consider themselves responsible for permit issuance, or, worse still, that none does so.[231]

Also, had this Resolution not been adopted, an inadequate inter-temporal approach to interpretation might have been used. Accordingly, for the implementation of CITES, waters not under the jurisdiction of any state would have started at 12 nm, as was understood at the time of CITES adoption, in other words *ex tunc*.[232] This would have created a disconnection between the permit system and the fisheries regime under the current international law of the sea, which provides that coastal states' laws and management plans can apply up to 200 nm.

cc) Remaining Potential Problems

While the adoption of a Resolution assists the consistent application of CITES, such a development should, to be fully efficient, be coupled with a settlement of maritime delimitations.

First, to allow a clear application of CITES for marine species, member states that have not done so, should contemplate declaring an EEZ or an exclusive fishing zone.[233] This zone's outer limits must be publicized and charts deposited with the Secretary General of the United Nations.[234]

---

[231] A. Willock, 'Uncharted Waters/Implementation Issues and Potential Benefits of Listing Toothfish in Appendix II of CITES' (TRAFFIC 2002) 20.

[232] As opposed to *ex nunc* (Matz, *Wege zur Koordinierung völkerrechtlicher Verträge* (n 115) 299; E. Franckx, 'Applications of the term "Introduction from the sea"' http://www.cites.org/eng/news/meetings/ifs-05/term_IFS.pdf accessed 6 July 2015, 14, 26; on inter-temporal law and interpretation, see supra Chapter 3 D. I. 2. b).

It was discussed whether the conflict clauses in CITES stating that "[n]othing in the present Convention shall prejudice the codification and development of the law of the sea [. . .] nor the present or future claims and legal views of any State concerning the law of the sea and the nature and extent of coastal and flag State jurisdiction" (CITES (n 55) article XIV(6)) provided any guidance as to the preferred interpretation rules. A careful reading and the context of the time point to the fact that the provision was meant to have no real effect beyond the political one of acknowledging the current negotiations of UNCLOS (Franckx, 'The Relationship between CITES, FAO and Related Agreements: Legal Issues' (n 60) 29–35) and was copied from the 1972 London Dumping Convention (CITES Secretariat, 'CoP15 Inf. 45 – Article XIV, paragraphs 4 and 5, of the Convention' (n 103) 1 §4).

[233] Wold and Thorson (n 56) 13. Declaring an exclusive fishery zone is similar to an EEZ but for the management of living resources only (UNCLOS (n 226) part V; G. Andreone and G. Cataldi, 'Regards sur les évolutions du droit de la mer en Méditerranée' (2010) 66 Annuaire français de droit international 1, 8; J. B. Ingerowski, *Exclusive Fishery Zones (EFZ): Legal Concept & State Practice* (MV Wissenschaft 2009) 183–186). Indeed, according to Nandan, Rosenne and Grandy, "the rights and duties of States that enacted exclusive fishing zone legislation correspond to the applicable rights and duties set out in Part V [of UNCLOS] with regard to exploring and exploiting, and conserving and managing, living natural resources of the waters in question" (*The United Nations Convention on the Law of the Sea: A Commentary: Volume II* (Martinus Nijhoff Publishers 1993) 510 §33) (reference omitted).

[234] UNCLOS (n 226) article 75. In case of opposite or adjacent coasts, the process described in article 74 must be followed.

In the Mediterranean, for example, most waters are high seas because only a few states—but their number is increasing—have declared an EEZ. It is thus 'by accident' that the high seas regime is applicable.[235] By a simple declaration, the relevant states could modify the status quo.[236] The quickly changing situation is confusing and causes legal insecurity, in particular due to the tendency of Mediterranean states to not simply declare EEZs, but rather exclusive fishery or ecological zones.[237] Hence, clarification and publication of this type of information is needed, for fisheries management in the first place and also potentially for the CITES permitting system.

Second and while of no direct interest for fish species themselves,[238] delimiting the outer continental shelf would be a source of legal clarity with regard to sedentary species (such as sponges, sea cucumbers, corals or mollusks), which could also, in the future, be listed. Indeed, species that live on the sea-bed floor of the outer continental shelf are under the sovereign rights of the coastal state and would not need an IFS certificate if harvested by the coastal state for the domestic market. The efficient implementation of the CITES treaty regime for these species would require that relevant continental shelves' outer limits be final and binding, hence set following the recommendations of the Commission on the Limits of the Continental Shelf.[239]

The clarification of maritime boundaries and delimitations would allow the international community to have a clear map of the waters where states exercise sovereignty or sovereign rights. This is important for many activities and, in the case of CITES, would provide some help in determining which type of permits is to be issued, as well as which laws are to be used when verifying legality.

---

[235] C. Chevalier, 'Governance in the Mediterranean Sea – Legal Regime and Prospectives', IUCN Centre for Mediterranean Cooperation http://cmsdata.iucn.org/downloads/legalspects_en_1.pdf accessed 6 July 2015, 14; Andreone and Cataldi (n 233) 4–6.

[236] IUCN, 'Governance of the Mediterranean Sea' http://www.iucn.org/about/union/secretariat/offices/iucnmed/iucn_med_programme/marine_programme/governance/ accessed 6 July 2015.

[237] Andreone and Cataldi (n 233) 10–11.

[238] Even though the definition of sedentary species is not without controversy, particularly with regard to the extent to which the species must be in contact with the seabed, this debate is not relevant for fish species, but can cause problems for lobsters, crabs, etc. (C. Kojima, 'Fisheries, Sedentary' Max Planck Encyclopedia of Public International Law www.mpepil.com §2–3).

[239] UNCLOS (n 226) article 76(8). As of 3 June 2015, the Commission on the Limits of the Continental Shelf (CLCS) had only made recommendations in 22 cases of the 77 submitted to it (Division for Ocean Affairs and the Law of the Sea, 'Submissions, through the Secretary-General of the United Nations, to the Commission on the Limits of the Continental Shelf, pursuant to article 76, paragraph 8, of the United Nations Convention on the Law of the Sea of 10 December 1982' (3 June 2015) http://www.un.org/Depts/los/clcs_new/commission_submissions.htm accessed 6 July 2015), and many states have not even made any submissions yet. Submitting a proposal to the CLCS if their continental shelf continues further than 200 nm would thus be the logical move forward – of course not only in relation to the CITES regime.

b) State of Introduction

The second question to be clarified is which state is the state of introduction and as such responsible for delivering the IFS certificate as well as for establishing the NDF. Determining who is responsible to issue IFS permits is crucial to the efficient implementation of the protection. Indeed, if governmental views do not concur on whether it is the flag state or the port state which is the state of introduction, overlap could happen, with administrative difficulties and lack of mutual recognition possibly causing distrust in permits. On the contrary, if no state takes responsibility for the issuance of permits, a gap is opened in the protection system and a loophole created for illegal trade.

Originally, the question centered on "whether introduction occurs when a fishing vessel takes a specimen of a species of fish, included in Appendices I or II of CITES, on board, or whether that only occurs when the fish is landed in a port of one of the member states."[240] Hence, the CITES Working Group on Introduction from the Sea was trying to determine whether the flag state or the port state was to be considered as the state of introduction, each solution having arguments in its favor and several practical consequences (a). No agreement could be found at CoP15 in 2010, but a new proposal of interpretation was adopted at CoP16, which settled most of the debates (b), although it does not solve all problems and even brings up some difficulties of its own (c).

Hence, it has now been internationally agreed that, when the flag state and the port state are the same state, the provisions related to IFS are applicable, but, when they are different, the flag state is the exporting state and the port state the importing one.

aa) Interpretation

The general method of interpretation is listed in the VCLT. While it is true that interpretation by political organs rarely respects the strict method of the VCLT and usually follows a more general approach[241] the VCLT articles can nevertheless structure the discussion.

i) Ordinary Meaning

First of all, "[a] treaty shall be interpreted in good faith in accordance with the ordinary meaning to be given to the terms of the treaty in their context and in the light of its object and purpose."[242]

According to the ordinary meaning, the concept of 'transportation into a State' conveys a feeling of movement and of national territory. Also, 'which were taken'

[240] Franckx, 'Applications of the term "Introduction from the sea"' (n 232) 12.

[241] Birnie (n 100) 329.

[242] VCLT (n 139) article 31(1); see for example Aust (n 113) 187–188.

is putting a temporal distance between the act of fishing and the IFS.[243] This would rather point to the port state as the state of introduction. Many authors argue in this sense, since "'transportation into a State' requires movement of a CITES-listed specimen from a vessel into a port State."[244]

Indeed, the landing in port represents the transportation movement more accurately than the landing of fish on a deck. Also, whereas 'transportation into a State' probably implies landing on the national territory, a vessel is no extension of a state territory,[245] although that had been argued in the past such as in the *Lotus* case[246] and this 'fictitious assimilation' has been used for certain jurisdictional purposes, in particular to justify the exclusive link between the ship and the flag state.[247] Finally, the need for a specimen to enter into port to be regulated is central, as controls by states usually happen at the border[248] and the concept of IFS has been assimilated to that of import.[249]

However, on the high seas, the flag state's jurisdiction, though maybe not exclusive anymore, is still, at the very least, primary.[250] At the time CITES was drafted, it would have been logical that the flag state was the one state in charge of the specimens caught by its vessels. If the drafters had wanted to move away from the practice of that time, they could have drafted this provision differently.

---

[243] Franckx, 'The Exercise of Jurisdiction over Vessels: Legal Issues Raised by the Relationship Between CITES, FAO and the UN Convention on the Law of the Sea' (n 61) 73.

[244] Wold and Thorson (n 56) 11; see also Favre, *International Trade in Endangered Species* (n 13) 89; Franckx, 'The Relationship between CITES, FAO and Related Agreements: Legal Issues' (n 60) 23; Wijnstekers (n 106) 57.

[245] Calley (n 76) 194; H. Meyers, *The Nationality of Ships: Academisch Proefschrift* (Martinus Nijhoff Publishers 1967) 14.

[246] *Case of the "S.S. Lotus" (France v. Turkey), Judgment,* 7 September 1927, Series A Publications of the Permanent Court of International Justice 4, 25: "All that can be said is that by virtue of the principle of the freedom of the seas, a ship is placed in the same position as national territory [. . .]. It follows that what occurs on board a vessel on the high seas must be regarded as if it occurred on the territory of the State whose flag the ship flies." Also, in their dissenting opinions M. Moore and M. Nyholm made explicit reference to the ship being a (floating) part of the flag state's territory (ibid. 61, 69).

[247] The references to territoriality in the Lotus case were used to determine which jurisdiction applied and how exclusively so. However, this concept of a ship being part of the territory was already criticized at the time of the judgment as a "[f]iction, très généralement admise, mais aussi inutile que malheureuse" (H. Walther, *L'affaire du "Lotus" ou De l'abordage hauturier en droit pénal international* (Les éditions internationales 1928) 157) and was simply used to justify the applicability of the flag state rules on board, so as to avoid lawlessness (ibid. 188).

[248] Franckx, 'The Relationship between CITES, FAO and Related Agreements: Legal Issues' (n 60) 23.

[249] Favre, *International Trade in Endangered Species* (n 13) 89.

[250] FAO, 'Implementation of the International Plan of Action to Prevent, Deter and Eliminate Illegal, Unreported and Unregulated Fishing' FAO Technical Guidelines for Responsible Fisheries 9 (Rome 2002) 21; D. Guilfoyle, *Shipping Interdiction and the Law of the Sea* (Cambridge Studies in International and Comparative Law, Cambridge University Press 2009) 168; Wold and Thorson (n 56) 12.

ii) Object and Purpose

The object and purpose of CITES is to avoid extinction in endangered species, by reducing the harvesting pressure caused by international trade. Hence, the efficiency of either choice in reaching this objective is to be looked into.

The flag state is the entity most likely to show efficiency in issuing IFS certificates. Indeed, it is supposed to control its vessels. As such, it possesses the knowledge and data necessary to determine NDF, particularly if the catches are part of a quota set by an RFMO of which it is a member.[251] Also, for further exports, flag states might be better suited to determine whether the vessel has respected the national laws and/or those of RFMOs to which they are party.[252] The IFS would be taking place upon landing on board, with only one actor involved. As for landing of these catches in a port from another state, it would require an export permit issued from the flag state.[253]

Additionally, if the IFS certificate is to be issued prior to the actual fishing (this aspect has not been settled), the flag state is in a better position to liaise directly with the vessels than the potential future port state would be.[254] Indeed, if the port state is not the same as the flag state, it would have to be informed of the fishing and of the intention of such vessel to come to its port in order to grant an IFS certificate prior to the catch.[255]

As for the port state, "using [. . .it] might help combat IUU fishing, because vessels flying flags of convenience will not be able to rely on lax flag State control to issue IFS certificates".[256] However, one can also point out the existence of a number of ports of convenience.[257] A positive aspect linked to designate the port state as state of introduction is that it avoids the problem caused by fishing entities (mainly Taiwan/Taipei) which are not bound by flag states' obligations.

Regarding the particularly complex question of transshipment (at sea and in port), if, on the one hand, the flag state is the state of introduction, any transshipment to a vessel from another flag state is a type of export, so an export permit would be required.[258] This would signify an administrative burden if the catches are transshipped several times before landing.

---

[251] CITES Secretariat, 'CoP15 Doc. 27 – Interpretation and implementation of the Convention – Trade control and marking – Introduction from the Sea' (n 108) 8–9 annex 3 §22.

[252] E. Franckx, 'Applications of the term "Introduction from the sea"' (n 232) 13; F. Meere and others, 'CITES 'Introduction from the sea' – A practical way forward: Proceedings of a joint workshop' (2008) 3–4.

[253] Meere and others (n 252) 4.

[254] Ibid. 2–4.

[255] Ibid. 3.

[256] Wold and Thorson (n 56) 12.

[257] CITES Secretariat, 'CoP15 Doc. 27 – Interpretation and implementation of the Convention – Trade control and marking – Introduction from the Sea' (n 108) 8 annex 3 §20.

[258] Meere and others (n 252) 4.

On the other hand, if the port state is the state of introduction, transshipment at sea would not influence permits, as it is once in port that the IFS must be issued. For transshipment in port, as the catch does not need to have cleared customs, it would not be considered as introduced from the sea. Such a situation would complicate the NDF if the fish is brought to a distant port, or if the data from the log-books are not transmitted to the port authorities. Indeed, transshipment can take place various times at sea and/or in port, each time driving the fish further away from the waters it was caught in, in which case, "[k]nowing who caught what where, whether a NDF can be made, and whether the product was taken legally [if necessary to determine for export/re-export] will be difficult if not impossible for the port State to determine."[259]

iii) Context

According to the VCLT, in addition to the text, preamble and annexes, one must also take into account any agreement between the parties (all or some of them) related to the conclusion of the treaty or to the treaty in general, or "any subsequent agreement between the parties regarding the interpretation of the treaty or the application of its provisions; [or] any subsequent practice in the application of the treaty which establishes the agreement of the parties regarding its interpretation; or any relevant rules of international law applicable in the relations between the parties."[260]

CITES article XIV (4) makes it clear that when a regime older than CITES exists and affords protection to marine species, a member state of this regime "shall be relieved of the obligations imposed on it under the provisions of [...CITES] with respect to trade in specimens of species included in Appendix II *that are taken by ships registered in that State*." Here, the flag state is mentioned with regard to marine species. The clarity of this paragraph, which is not to be found in the following where reference is made again to the state of introduction, can be interpreted as highlighting a difference between the two types of states. However, the mention of the flag state in paragraph 4 is most likely due to the fact that it would be very unpractical to have an exemption as a result of membership to another institution—which functions on the basis of flags—and then require the port state to declare that catches were taken "in accordance with the provisions of such other treaty, convention or international agreement."[261]

As for other international instruments dealing with fisheries matters and the responsibilities of states, they tend to focus on the flag state (Compliance Agreement, UNFSA), but the port state's role has also increasingly gained in importance, particularly with the adoption of the recent Port State Measures Agreement.[262]

[259] Ibid.

[260] VCLT (n 139) article 31(2)–(3); on this, see for example Aust (n 113) 188–195.

[261] CITES (n 55) article XIV(5).

[262] Franckx, 'The Relationship between CITES, FAO and Related Agreements: Legal Issues' (n 60) 24.

iv) Supplementary Means of Interpretation

Having recourse to the *travaux préparatoires* is envisioned in article 32 of VCLT, inter alia when the means of interpretation provided for in article 31 do not lead to a clarification of the terms.

In the present case, the wording adopted for 'introduction from the sea' came from an IUCN proposal which allowed a compromise in relation to the Convention's coverage of marine species. While the *travaux préparatoires* are not published and hence cannot be examined first hand, Franckx valuable and extensive research on the concept of introduction from the sea uses them. In that respect, it is worth noting that he does not mention any particular discussion on the actual meaning of the terms inserted in article I.[263]

v) Other Authenticated Versions

As referred to in article 33 VCLT, all authenticated versions of a treaty are equally authoritative and are supposed to share the same meaning. If this is still not the case after applying articles 31–32, the interpretation which allows the best reconciliation of meanings, keeping the object and purpose into account, must be preferred.

Here, the French and Spanish versions are also authenticated and read at article I (e): « 'Introduction en provenance de la mer' [signifie] le transport, dans un Etat [. . .]»; "'Introducción procedente del mar' significa el traslado a un Estado [. . .]". Not only do 'le transport dans un Etat' and 'el traslado a un Estado' back up the impression given by 'transportation into' of movement into a state's territory, but the wording 'en provenance de la mer' indicates a physical distance which can hardly be met with IFS happening at the moment of landing on board a vessel.

In conclusion, a clear-cut answer cannot be implied from the text, object and purpose, or other interpretative methods. Moreover, opinions in favor of either the port state or the flag state are quite equally distributed amongst states, academia, and lobbies.[264] However, this is without practical importance as an agreement has been found in 2013.

---

[263] Ibid. 21–22.

[264] For countries' opinions, see CITES, 'CoP15 Com. II Rec. 13 (Rev. 1) – Summary record of the thirteenth session of Committee II' (13–25 March 2010) 1 §27: most of Latin America for example believed that the port state should be the state of introduction, while the European Union was of the opposite opinion.

FAO, 'Report of the Expert Consultation on Legal Issues' (n 4) 3–4 §18; Franckx, 'Applications of the term "Introduction from the sea"' (n 232) 12–14; International Environmental Law Project, 'Understanding Introduction from the Sea' (2010) http://www.lclark.edu/live/files/4356 accessed 6 July 2015. Also, on the practical consequences of either case with regard to permit issuance, see Meere and others (n 252).

bb) Adopted Solution

At the May 2011 meeting of the CITES Standing Committee Working Group on Introduction from the Sea, the Chair and Vice-Chair came up with a practical solution, which was adopted by the Working Group members[265] and then, more importantly, by the states parties at CoP16. Accordingly, the IFS provision only applies if the flag state is also the port state. Otherwise, import/export provisions would need to be used, with the flag state being the exporting state and the port state the importing one.[266]

This solution allows a positive, workable implementation of the CITES documentation system. Not only is this approach practical but it is also a legal interpretation consistent with CITES—a key requirement for the working group.[267] Indeed, the provisions on introduction from the sea never mention the need for two states to be involved.

Also, transshipment at sea can be accommodated by this solution: any state the vessel of which is involved in transshipment just has to ensure that the correct documentation accompanies the fish.[268]

Another fishing practice, chartering, has been addressed in the amendment of the Resolution on IFS adopted at CoP16. Chartering is the situation where a vessel is flagged in one state and chartered by another state: hence, "fishing opportunities allocated to one state [the chartering state] are fished by the vessels of another state",[269] the flag state. When marine species are caught in waters not under the jurisdiction of any state and brought to port in the chartering state, two options are envisioned and can be chosen from by the states in question: the flag state can be the state of export or the chartering state can be the state of introduction. If the marine species are brought back into a third state, the flag state is the export state which can, under the condition that the chartering state has been consulted and gives its agreement, issue the export permit. However, if the chartering state is granted the authorization in the written arrangement between the flag state and the chartering state, it can also be directly considered as the state of export.[270]

---

[265] Committee on Fisheries – Sub-Committee on Fish Trade, 'Update on CITES related activities' presented at the Thirteenth Session of the Sub-Committee on Fish Trade (Hyderabad, 20–24 February 2012) COFI:FT/XIII/2012/8 (2012) 3 §10.

[266] CITES Secretariat (on behalf of the Standing Working Group on Introduction from the Sea), 'SC61 Doc. 32 – Introduction from the sea' (15 August 2011) annex 1 §7, 11.

[267] CITES Secretariat, 'CoP15 Doc. 27 – Interpretation and implementation of the Convention – Trade control and marking – Introduction from the Sea' (n 108) 7 annex 3 §8.

[268] CITES Secretariat 'SC61 Doc. 32 – Introduction from the sea' (n 266) annex 1 §17–19.

[269] E. J. Molenaar, 'Participation, Allocation and Unregulated Fishing: The Practice of Regional Fisheries Management Organisations' (2003) 18 International Journal of Marine and Coastal Law 457, 461.

[270] Resolution Conf. 14.6 (Rev. CoP16) – Introduction from the Sea (n 66).

cc) Remaining Difficulties

Even with the very satisfying solution which was found, a few problems are left. First, when the flag state and the port state are different, the responsibilities' sharing could infer that export can take place without fish clearing customs in the exporting state and/or that the vessel is an extension of the state's territory. Indeed, an export permit is to be issued by the flag state, where, de facto, the fish physically never was, except if the vessel amounts to national territory. The experts' answers to this concern are that the vessel only acts as a transport vessel (i.e. export is *from* the vessel but not *by* the vessel) and that all documentation would be issued by the state's authorities, not by the vessel itself.[271]

Also, this solution has no potential to alleviate the conflict between coastal states and distant water fishing ones. Indeed, as examplified in the debate underlying the sharks listing, an inequality exists between coastal states, such as the EU, being allowed to fish without restriction in their waters, and distant water fleet nations which are required to issue a certificate or permit. This situation was particularly problematic with regard to the EU since it is the principal destination for some products such as shark meat.[272] However, this issue, although linked to the IFS debate, is more due to the focus of CITES on international trade than to a flaw in the IFS system.

Furthermore and mainly, the solution found to address the particularities of chartering is potentially problematic, and this well beyond the CITES framework. While chartering is an increasing practice in fisheries, the obligations of the various states involved still have not been clarified successfully on the international scene. Hence, in current international law, if a vessel is chartered, the allocation of responsibilities remains ambiguous: it is indeed unclear whether the flag state or the state where the operator is based should be in charge and, consequently, responsible for the actions of that vessel.[273]

The solution adopted by the CITES CoP, although it does not intend to apply beyond the Convention,[274] "could set a negative precedent in relation to the exclusive jurisdiction of flag States over vessels fishing on the high seas and could be abused to the benefits of operators involved in illegal, unregulated and unreported fishing."[275] Indeed, the sacrosanct principle of flag state responsibility is potentially at risk. Nevertheless, it seems that the parties were conscious of the particularity of the understanding they reached in Resolution Conf. 14.6 (Rev. 16) and they agreed to provide the Secretariat with information for its monitoring of this practice, as well as to review the chartering provisions at CoP17.[276]

---

[271] CITES Secretariat 'SC61 Doc. 32 – Introduction from the sea' (n 266) annex 1 §13.

[272] China, 'CoP14 Inf. 45 – Shark Issues' (n 3) 3 §1(e).

[273] CITES Secretariat 'SC61 Doc. 32 – Introduction from the sea' (n 266) annex.

[274] Resolution Conf. 14.6 (Rev. CoP16) – Introduction from the Sea (n 66) preamble.

[275] Ireland (on behalf of the European Union), 'CoP16 Inf. 31 – European Union position on Introduction from the Sea Document CoP16 Doc. 32' (3–14 March 2013).

[276] CITES, 'Decisions 16.48–16.51 – Introduction from the sea: chartering' (2013).

Finally, a last aspect that has not been settled by the new proposal is whether the transport from specimens introduced from the sea into another member state is export or re-export.[277] This question has no major practical consequences since the only difference in requirements is that, in the case of re-export, the Scientific Authority is not obliged to undertake an NDF, while it is for export. However, as an NDF is necessary for IFS and will already have been provided by the Scientific Authority of the state of introduction—which is also the (re)exporting state—it makes little difference at the end, whether an export or a re-export permit is to follow IFS.[278]

dd) Conclusions

For commercially-exploited fish species to be listed in Appendix II—where trade can still take place in large quantities—a clear understanding of 'introduction from the sea' and of the parties' corresponding duties regarding marine species caught in areas beyond national jurisdiction was necessary.

This was achieved when an agreement on the matter was found at CoP16. Hence, now, a clear legal framework is set. That might be a sign that states parties want to move forward on the matter and to have the option to afford protection to such species.

3. Non-detriment Finding

In order to deliver an IFS certificate or an export permit, the state of introduction or exporting state must conduct an NDF for both Appendix I and II species[279] and must additionally, in the case of Appendix I species, check that the specimen is not to be used "for primarily commercial purposes".

The Convention does not provide a distinct method to undertake NDF for highly traded species. This might be problematic as the large volume of trade, particularly for species listed in Appendix II, makes a specimen by specimen verification impracticable. In the case of marine species, NDF is further complicated by the fact that such species are moving between states and migrating through waters not under the jurisdiction of any state. Hence, since transboundary, straddling and highly migratory species can be caught in areas under the management of different states or in the high seas, cooperation is necessary for NDF. In order to face these two challenges, methods developed for other species and in other fora can be referred to in the present case.

---

[277] Franckx, 'Applications of the term "Introduction from the sea"' (n 232) 12.

[278] Ibid. 14.

[279] CITES (n 55) respectively articles III(5) and IV(6).

a) Quotas

While not originally planned in the CITES system, quotas progressively became the main standard of NDF.[280] These important management tools serve to make sure that exports of a certain number of specimens every year do not have a detrimental effect on the listed species.[281] Quotas must be set by the Scientific Authority of the exporting state, or on its advice, and can be used instead of determination of non-detriment for each specimen. Hence, they serve to simplify NDF of species traded in high volumes.

Acknowledged as practical since they "may facilitate the issuance of export permits",[282] quotas are also subject to criticisms. Indeed, undesirable effects are possible if decisions are not based on science—let alone with a precautionary principle in mind—and/or if quotas are not well-implemented.[283] It also appears that national consumption is not always taken into account to determine the sustainable levels of specimens entering international trade.[284] Finally, the risk exists that quotas are disrespected, with arguably little consequences.

However, some safeguards are envisioned to attempt to mitigate the problems and risks of abuse. In addition to the requirement to set quotas through an NDF process based on scientific information,[285] guidelines on quotas mention that for specimens caught in the wild, quotas "should be set at a level that takes account of the number or quantity of specimens that are taken from the wild legally or illegally."[286] The need to consider both the volume of illegal and of legal trade

---

[280] Resolution Conf. 14.7 (Rev. CoP15) – Management of nationally established export quotas 2007; CITES, 'CoP11 Inf. 11.3 – Scientific Authorities' Checklist to assist in making Non-detriment Findings for Appendix II Exports' (10–20 April 2000) 13; Resolution Conf. 16.7 – Non-detriment findings (n 168) preamble; L. Little and M. A. Orellana, 'Can CITES Play a Role in Solving the Problem of IUU Fishing?: The Trouble with Patagonian Toothfish' (2005) 16 Colorado Journal of International Environmental Law and Policy 21, 70–71; M. Yeater and J. Vasquez, 'Demystifying the Relationship Between CITES and the WTO' (2001) 10 RECIEL 271, 273–274.

[281] Willock, 'Uncharted Waters/Implementation Issues and Potential Benefits of Listing Toothfish in Appendix II of CITES' (n 231) 22.

[282] Resolution Conf. 14.7 (Rev. CoP15) – Management of nationally established export quotas (n 280) annex §4; E. Thorson, 'IELP White Paper on Export Quotas and Non-Detriment Findings' (2006) 1; Wijnstekers (n 106) 511.

[283] C. Fuchs, 'Convention on International Trade in Endangered Species of Wild Fauna and Flora (CITES) – Conservation Efforts Undermine the Legality Principle' (2008) 9 German Law Journal 1565, 1588; J. B. Murphy, 'Alternative Approaches to the CITES 'Non-detrimental' Finding for Appendix II Species' (2006) 36 Environmental Law 531, 540; C. Raymakers, 'International Trade in Sturgeon and Paddlefish Species – the Effect of CITES listing' (2002) 87 International Review of Hydrobiology 525, 533.

[284] Raymakers, 'International Trade in Sturgeon and Paddlefish Species – the Effect of CITES listing' (n 283) 533.

[285] Resolution Conf. 14.7 (Rev. CoP15) – Management of nationally established export quotas (n 280) annex §6.

[286] Ibid. annex §10.

has also been restated as a guiding principle in the recent Resolution on NDF.[287] To do so, precaution must de facto be used since illegal trade is by nature unreported and the only data available will be constituted of estimates. Hence, the requirements for quota setting are clear and, if they might be disrespected, the same can be said of the conditions for individual specimens' NDF.[288]

Additionally, in some cases, although it is rare and "only for species of special concern",[289] particular quotas can be set directly by the CoP, instead of being determined by parties.[290]

Then, importing parties can and should verify the permits they receive for Appendix II species and have to express their own judgment on NDF for Appendix I species by issuing an import permit. Information is available to all since the state setting a quota is to communicate it to the Secretariat, which posts the relevant values on the CITES website.[291] Furthermore, on each permit issued on the basis of a quota, the quantity already exported compared to the total allowed must appear.[292]

Moreover, any state party which fears that a species is traded at levels that are detrimental to its survival (under a quota management or not) can discuss the matter with the Management Authority of the exporting state, request the help of the Secretariat and decide to apply stricter domestic measures.[293]

Finally, the Animals Committee can exercise some control on the levels of trade for species in Appendix II, through the Significant Trade Review. The idea of this procedure is indeed to verify, for highly traded Appendix II species, that the NDF is made adequately—in other words, that the level of trade is not detrimental to the species.[294] If the Committee is not satisfied, remedial measures can be indicated and, if not implemented, trade suspension can be recommended.[295]

---

[287] Resolution Conf. 16.7 – Non-detriment findings (n 168).

[288] S. Aguilar, 'Regulatory Tools for the Management of Fish and Timber Species through CITES' (2013) 22 RECIEL 281, 283–284.

[289] Fuchs (n 283) 1583.

[290] For the African elephant, African spurred tortoise (in the Appendices); for the leopard (Resolution Conf. 10.14 (Rev. CoP16) – Quotas for leopard hunting trophies and skins for personal use 1997); for the markhor (Resolution Conf. 10.15 (Rev. CoP14) – Establishment of quotas for markhor hunting trophies 1997); for the black rhinoceros (Resolution Conf. 13.5 (Rev. CoP14) – Establishment of export quotas for black rhinoceros hunting trophies 2004). In general on quotas for species in Appendix I, see Resolution Conf. 9.21 (Rev. CoP13) – The interpretation and application of quotas for species included in Appendix I 1994; Yeater and Vasquez (n 280) 274.

[291] Resolution Conf. 14.7 (Rev. CoP15) – Management of nationally established export quotas (n 280) annex §15, 17; Gillespie, *Conservation, biodiversity and international law* (n 101) 87.

[292] Resolution Conf. 12.3 (Rev. CoP16) – Permits and certificates 2002 part VIII Regarding permits and certificates for species subject to quotas.

[293] Resolution Conf. 11.3 (Rev. CoP16) – Compliance and enforcement 2000.

[294] On Significant Trade Review, see supra Chapter 3 C. I. 2. g).

[295] Resolution Conf. 12.8 (Rev. CoP13) – Review of Significant Trade in specimens of Appendix-II species (n 83).

b) Role for RFMOs

Particularly problematic here is that marine species tend to cross borders and that quotas set unilaterally by states are at risk of being inefficient and unsustainable. Indeed, while any national quota will have an effect on the level of stocks in all the range states, unilateral quotas are not usually set with the catches in other waters in mind.

Joint management by all range states is consequently needed if it is to be efficient. For species migrating through the high seas cooperation between all harvesting states is even necessary, a goal which is pursued by RFMOs. It is worth remembering that most high seas areas are under the management of at least one of these bodies, which usually determines TACs and allocates national quotas.[296]

CITES provides, with regard to cooperation in determining NDF, that IFS certificates "may be granted on the advice of a Scientific Authority, in consultation with other national scientific authorities or, when appropriate, international scientific authorities".[297] Hence, reference is made to the possibility to consult with other national authorities or an RFMO, if relevant. In particular, in the recent revision of the Resolution on *Introduction from the sea*, the CoP "recogniz[es] the need for States to consult and cooperate with relevant [...RFMOs] when issuing" the relevant documents.

The quotas set by RFMOs are likely to serve as the benchmark for the determination of NDF, as expertise lies with these organizations. Furthermore, it is their mandate to manage the relevant straddling and highly migratory stocks and to ensure cooperation between range states. Practically speaking, one possibility is for the state of introduction or exporting state to use the data provided by the RFMO: "[r]elying [on its advice...] can minimize the amount of scientific study required by individual parties' Scientific Authorities."[298] It could alternatively nominate the RFMO's scientific body as its Scientific Authority for the purpose of the relevant CITES marine species.

c) Problems in the Absence of Management Regimes

In the absence of a relevant RFMO with numerical CMMs for the species—as in the case of most sharks for example—or in the case of loopholes—such as, for toothfish, in the high seas adjacent to the CCAMLR area—, identifying an adequate

---

[296] In areas of the high seas that are not regulated, it might be more complex at the beginning, with all states involved in the harvesting needing to gather data and agree on TACs – resulting de facto in the actions taken by states within an RFMO.

[297] CITES (n 55) article IV(7).

[298] Murphy, 'Alternative Approaches to the CITES 'Non-detrimental' Finding for Appendix II Species' (n 283) 549.

benchmark against which to undertake the NDF will most likely be difficult.[299] Hence, further practical solutions will be needed, whether through the creation of an RFMO, the extension of the existing ones' jurisdiction or ad hoc cooperation between states involved in the fishery.[300] Indeed, CITES is not a fisheries management body and has neither the resources nor the expertise to be used as such.[301]

Ad hoc cooperation is an option, encouraged for example in the case of sturgeons, fishes which are straddling between waters under the jurisdiction of more than one state. In the CITES regime, export quotas for these species must be adopted collectively by all (or all but one of) the range states of one water basin. Otherwise, international trade will not be allowed.[302] More generally, to avoid the issue of unilateral and incompatible quotas, collaboration between range states is encouraged.[303]

In most cases however, in the absence of RFMOs and management plans, implementation issues can be expected. A risk indeed exists that if a species is listed but no structure for its management is in place, all trade originating from states that cannot undertake the scientific research necessary for an NDF would be considered as prohibited under CITES.[304] In this case, if such international trade represents an important share of a state's (or a community's) resources, it is likely that activities will move into illegality.[305] When illegal trade and illegal fishing become the norm, the potential stigma they carry—which is a driver of compliance[306]—risks being lost; a CITES listing would thus be the cause of a deterioration rather than an improvement of governance. While this problem is not unique to shared species—a state can find it impossible to undertake the necessary research in

---

[299] A. Willock, 'Administrative and Monitoring Implications of listing and down-listing of commercially-exploited aquatic species, including the implications of Annex 4 of Resolution Conf. 9.24' Background paper to CITES Workshop on Introduction from the Sea Issues, 30 November – 2 December 2005, Geneva (Switzerland), http://www.cites.org/eng/news/meetings/ifs-05/IFS05-TRAFFIC-paper.pdf accessed 4 July 2015, 8, 12.

[300] Willock, 'Uncharted Waters/Implementation Issues and Potential Benefits of Listing Toothfish in Appendix II of CITES' (n 231) 9–10, 24.

[301] Little and Orellana (n 280) 77.

[302] Resolution Conf. 12.7 (Rev. CoP16) – Conservation of and trade in sturgeons and paddlefish (n 84); C. Raymakers, 'CITES, the Convention on International Trade in Endangered Species of Wild Fauna and Flora: its role in the conservation of Acipenseriformes' (2006) 22 Journal of Applied Ichthyology 53, 60.

[303] Resolution Conf. 14.7 (Rev. CoP15) – Management of nationally established export quotas (n 280) annex §4; CITES, 'CoP15 Doc. 16.2.2 – Capacity Building – International expert workshop on non-detriment Findings – Report of the Animals and Plants Committees' (13–25 March 2010) 8 §13–14.

[304] It is true that NDF is a sovereign act of the exporting state, but such state should not make an NDF if there is insufficient information (Favre, *International Trade in Endangered Species* (n 13) 63).

[305] Du Plessis (n 183) 23.

[306] U. R. Sumaila, J. Adler and H. Keith, 'Global scope and economics of illegal fishing' (2006) 30 Marine Policy 696, 697–698.

relation to a species only living within its boundaries—it is likely to be more acute for straddling or highly migratory species as coordination in research is furthermore needed.

For sharks for example, "[n]o mechanism for establishing non-detriment findings for these populations is agreed, yet they are required before Parties can issue an Appendix-II export permit or introduction from the sea certificate. An Appendix-II listing may well constitute a '*de facto*' trade ban until the technical aspects of 'non-detriment' have been resolved. It is noted that the Appendix-II listing of great white sharks prompted a total export ban by Australia, even on by-catch specimens, because it was unable to make a non-detriment finding in compliance with CITES."[307]

Nevertheless, a listing can also trigger closer cooperation between range states, as has been the case for many Caribbean states with regard to the queen conch, or, as described above, for the range states of sturgeons.

4. Legality Finding

Before issuing an export permit for a listed species, the Management Authority must be convinced that the specimen has been caught legally, in other words in accordance with its national laws. This is a quite straight-forward requirement when dealing with terrestrial species or species caught in waters under national jurisdiction, because the state sets the standards of what is legal, knows its own laws and exercises—or at least should exercise—actual control over its territory. This legality finding process becomes more complicated for fish under joint management measures or caught on the high seas.

a) Delay in Determination

A fish that is introduced from the sea and later exported—a likely situation as a large proportion of fish enters international trade[308]—does not need a legality check for the former documentation, but will for the latter. As the Convention's rules currently stand, the export permit's issuance is indeed the first stage where a legality check is required.

Problematically, it might not be possible to gather the information necessary for the legality check because of the delay between the harvesting and the permit issuance. Indeed, the later in the process the verification of legality is made, the

[307] China, 'CoP14 Inf. 45 – Shark Issues' (n 3) 2 §1(d); on the export ban, see Willock, 'Administrative and Monitoring Implications of listing' (n 299) 12.

[308] In general terms, between 35 and 40 % of fish and fishery products enter international trade (FAO, 'The State of the World Fisheries and Aquaculture' (Rome 2010) 48; FAO, 'The State of the World Fisheries and Aquaculture' (Rome 2012) 67; FAO, 'The State of the World Fisheries and Aquaculture' (Rome 2014) 47) but can also be a lot higher for species such as Bluefin Tuna.

more difficult it will be to access important data from log-books or information on the respect of quotas.

The state of introduction would indeed "allow [fish] to be landed in its ports, processed, and then [at the time of export] only determine if it had been legally obtained".[309] By that time, the boat that caught the fish will be gone, potentially having taken the information on the location and methods of harvesting with it, data which is crucial for the legality check.

b) Basis to Determine Legality

For fish caught on the high seas but subjected to an import/export certificate, the main issue is related to the source and content of the rules the respect of which must be verified. Indeed, the question of which law is to be checked is an interesting one, especially as many high seas areas are under the management of an RFMO.

The Convention states that the compliance with national laws of the exporting state is to be verified.[310] Indeed, the exporting country's Management Authority must "establish whether a specimen was legally acquired in its country, i.e. not in contravention of its own laws for the protection of fauna and flora."[311] Following the letter of the text, this provision has been applied, so far, as requiring only the national laws of the state to be verified, seemingly forgetting that fish can be caught "in a manner that contravened the conservation and management measures established by a RF[M]O or in a manner that violated international law."[312]

However, it would seem reasonable for international law, even if it is not translated into national law or automatically incorporated therein, to be a source of regulations the respect of which must be verified. CIL, treaties which regulate fishing on the high seas and the relevant CMMs are indeed the logical rules to serve as a point of reference in the legality check. In particular, the obligation to cooperate in the management and conservation of straddling and highly migratory species as well as living resources of the high seas[313] is implemented by RFMOs.

In that respect, although the hard law has not changed, it is worth noting that the rules determined by the relevant RFMO were considered at the third and fourth FAO Expert Advisory Panel meetings, as the basis to determine legality.[314] Furthermore, the CITES website, in the page created to support to implementation of

---

[309] Willock, 'Uncharted Waters/Implementation Issues and Potential Benefits of Listing Toothfish in Appendix II of CITES' (n 231) 24.

[310] Favre, *International Trade in Endangered Species* (n 13) 60.

[311] Wijnstekers (n 106) 122.

[312] Willock, 'Administrative and Monitoring Implications of listing' (n 299) 12.

[313] UNCLOS (n 226) articles 63–64, 118.

[314] FAO, 'Report of the Third FAO Expert Advisory Panel for the Assessment of Proposals to Amend Appendices I and II of CITES Concerning Commercially-exploited Aquatic Species (Rome, 7–12 December 2009)' FAO Fisheries Report 925 (Rome 2010) 126; FAO, 'Report of the Fourth FAO Expert Advisory Panel for the Assessment of Proposals to Amend Appendices I

the regime for shark species, states that "Parties must ensure that specimens in trade have been legally acquired, through monitoring and compliance with applicable laws, *(sub)regional fishery management measures*, etc."[315] as well as put together a list of the regional measures adopted in relation to sharks.[316]

Whether participating in these organizations in order to be allowed to fish is an obligation binding all states still ought to be clarified.[317] Indeed should the UNFSA provisions apply globally, the CITES permitting system would be a tool to curtain IUU fishing and decrease the attractiveness of flags of convenience through trade restrictions applicable in a similar way to all states: if fish had been caught by a vessel disrespecting RFMOs measures or the flag state of which does not cooperate with the relevant RFMO, exports would be illegal under CITES. The fish could consequently only be landed in the flag state and used domestically, as the IFS requirements do not include a legality finding.

c) Improvement Options

aa) Stricter National Requirements

States could modify their national practices towards stricter domestic measures, with or without an international Resolution encouraging them to do so. Indeed, they are allowed to take such measures regarding the conditions to allow trade of species included in the Appendices.[318] It is on the basis of this provision that, following a CoP recommendation, parties are entitled to suspend trade, as a sanction against non-compliant states.[319]

It was suggested that states of introduction could make it the regular procedure to undertake a legality finding at the IFS stage, also with regard to international regulations.[320] That could either be an additional requirement where both NDF and legality verifications have to be undertaken separately, or a process where the latter is included into the former. Indeed, respecting CMMs related to harvesting methods, times of the year open for fishing, minimum fish sizes, etc. are aspects that

---

and II of CITES Concerning Commercially-exploited Aquatic Species (Rome, 3–8 December 2012)' FAO Fisheries and Aquaculture Report R1032 (Rome 2013) 17.

[315] CITES, 'Resources for implementation: Legality' http://cites.org/eng/prog/shark/legality.php accessed 6 July 2015, (italics added).

[316] CITES, 'Regional fisheries management organization measures for shark conservation and management (draft as at March 2014)' (March 2014) http://cites.org/sites/default/files/eng/prog/shark/docs/shark%20RFMO%20measures%20-%20draft%20March%202014.pdf accessed 6 July 2015.

[317] For a discussion on this aspect, see supra Chapter 3 A. II. 3. b).

[318] CITES (n 55) article XIV(1)(a).

[319] Reeve, *Policing International Trade in Endangered Species* (n 8) 91.

[320] FAO, 'Report of the Expert Consultation on Implementation Issues' (n 193) 5 §25.

can count when determining whether trade in a number of specimens is detrimental to the survival of the species.[321]

Furthermore or instead, on the import side, states could put an extra condition to allow trade, requiring that specimens be caught in accordance with all national and international regulations applicable to the flag state. Additional requirements from the importing states are indeed not unheard of. The EU for example, requires an import permit for Appendix II species entering its territory, while the Convention foresees this condition only for Appendix I species.[322]

A last option would be for states to adopt a Lacey Act type of legislation in their domestic system. This federal law of the USA makes it an offense to trade specimens taken in contravention of any national or international laws. Hence, it opens the door to verifying the respect of international regulations under a national law: catching a fish in violation of the RFMO rules and trading it would indeed, in the USA, be a violation of the Lacey Act, hence of the exporting state's laws.

bb) Coordinated by a Resolution?

While states can unilaterally contribute to counterbalancing the system's weaknesses by adopting stricter national measures, there is an added value in coordinating their actions through a Resolution. The adoption of such a document by the CoP appears to be preferred as multilateral measures are promoted. A current line of work at CITES is indeed to get a better overview of the existing stricter measures and assess whether additional multilateral measures should be adopted in order to reduce the need for stricter national measures.[323]

In the present case, the adoption of a Resolution is beneficial as international awareness is raised about the issue at stake, hence making it more likely that

---

[321] Willock, 'Uncharted Waters/Implementation Issues and Potential Benefits of Listing Toothfish in Appendix II of CITES' (n 231) 24; CITES Secretariat, 'Notification to the parties n°2006/023 – Introduction from the sea' (6 April 2006) 5.

[322] 'Council Regulation (EC) No 338/97 of 9 December 1996 on the protection of species of wild fauna and flora by regulating trade therein', *Official Journal of the European Union* (1996) article 4(1)–(2): species listed in Annexes A and B (approximately Appendices I and II) are subject to an import permit (European Community and TRAFFIC Europe, 'Reference Guide – European Community Wildlife Trade Regulations' (2013) http://ec.europa.eu/environment/cites/pdf/2007_referenceguide2_en.pdf accessed 6 July 2015, 21; Bowman, Davies and Redgwell (n 19) 533; Willock, 'Administrative and Monitoring Implications of listing' (n 299) 10). The USA also implements stricter national measures (J. M. Hutton, 'Who Knows Best? Controversy over Unilateral Stricter Domestic Measures' in J. Hutton and B. Dickson (eds), *Endangered Species-Threatened Convention: The Past, Present and Future of CITES* (Earthscan 2000) 63; Yeater and Vasquez (n 280) 273).

[323] Chair of the Standing Committee Working Group on Multilateral Measures, 'CoP16 Doc. 18 (Rev. 1) – Cooperation between parties and promotion of multilateral measures' (3–14 March 2013); CITES, 'Decisions 14.28 (Rev. CoP 15) – 14.29 (Rev. CoP 16) – Cooperation between Parties and promotion of multilateral measures' (2010–2013) and the annex 1 for the Terms of Reference of the requested consultancy.

domestic measures be taken. It also gives such measures legitimacy.[324] Indeed, unilateral measures are at times perceived as questioning the exporting state's abilities to correctly apply CITES rules, if not even threatening its authority.[325] States are also made conscious of the possibility that additional conditions might be required by their trading partners, which is crucial to avoid unnecessary disruptions in trade flows. Finally, having a Resolution is better in relation to WTO law: multilateral trade restrictions are indeed less likely to be successfully challenged in front of the WTO dispute settlement body.[326]

In the revised Resolution 14.6 on IFS, in addition to clarifying the identity of the state of introduction, an operative paragraph recommends that the state of introduction, the state of export and the state of import "take into account whether or not the specimen was or will be acquired and landed: i) in a manner consistent with applicable measures under international law for the conservation and management of living marine resources, including those of any other treaty, convention or agreement with conservation and management measures for the marine species in question; and ii) through any illegal, unreported or unregulated (IUU) fishing activity."

Although this provision recommends states to take into account the aspects above-mentioned, as an additional (non-binding) condition, it does not actually modify the legality findings requirements as these are still based on the national legislation of the exporting state.[327] "Taking into account" means that there is no obligation to forbid IFS if no legality verification is undertaken or to forbid any trade which does not respect international conservation and management measures. A soft wording of this sort is a sign that this provision is not mandatory, in contrast to Resolutions where the CoP 'urges', 'decides', 'agrees' or even, arguably is 'satisfied', the latter being one of the terms proposed at CoP15 for this paragraph.[328] By keeping the language quite general, the opportunity nevertheless arose to refer

---

[324] R. B. Martin, 'When CITES Works and When it Does Not' in J. Hutton and B. Dickson (eds), *Endangered Species-Threatened Convention: The Past, Present and Future of CITES* (Earthscan 2000) 32.

[325] Hutton (n 322) 58–59; R. W. G. Jenkins, 'The Significant Trade Process: Making Appendix II Work' in J. Hutton and B. Dickson (eds), *Endangered Species-Threatened Convention: The Past, Present and Future of CITES* (Earthscan 2000) 21.

[326] Hutton (n 322) 60–62; R. J. McLaughlin, 'Settling Trade-Related Disputes Over the Protection of Marine Living Resources: UNCLOS or the WTO?' (1997–1998) 10 Georgetown International Environmental Law Review 29, 95.

[327] See for example Resolution Conf. 14.6 (Rev. CoP16) – Introduction from the Sea (n 66) annex II.1.1.2 or III.1.1.2.

[328] The revised draft at CoP15 indeed read as: the CoP "recommends that a Management Authority of the State of introduction, prior to issuing a certificate of introduction from the sea, [*be satisfied*] [take into account] not only that the provisions of the Convention are met but also that the specimen was or will be acquired in a manner consistent with applicable measures for the conservation and management of living marine resources, including those of Regional Fisheries Management Organizations or arrangements, and with international law" (CITES Secretariat, 'CoP15 Doc. 27 – Interpretation and implementation of the Convention – Trade control and marking – Introduction from the Sea' (n 108) 4 annex 1 (italics added)).

to *any* relevant RFMO measure—whether the state in question is a party to it or not. Also, a comprehensive application of RFMO measures to all is further reinforced by making references to IUU fishing, since unregulated fishing encompasses fishing in waters managed by an RFMO, without being party to it.

5. Captive-Breeding and Ranching

Captive-breeding is an exception to the CITES permitting system,[329] in which case the Management Authority of the export state simply needs to issue a certificate.[330] The scope of this exception and its practical application are particularly relevant to fisheries, due to the importance of aquaculture in the sector.

Aquaculture is indeed key to the future of fisheries, as it nowadays contributes to 50 % of the fish eaten and is the fastest-growing food producing industry in the world. Technology is also developing rapidly, with some species traditionally not reproducing in captivity now starting to be bred in farms. The term 'aquaculture' encompasses various practices and while some aquaculture operations are based in national waters, some others are in the high seas such as in the Mediterranean—or might even, in the foreseeable future take place in areas beyond 200 nm.[331]

While the practice's ecological advantages can be debated—as, for example, transmission of diseases or escapees have negative impacts—it is also source of interesting legal and practical questions when CITES is involved. Indeed, if a fish species is listed under CITES, fish which can both be wild caught and originate from farms, not only is the normal permitting system relevant for the former group, but two additional types of provisions (captive-breeding and ranching) are possibly applicable to the latter and would facilitate or allow international trade.

Aquaculture practices, although not the panacea, are a way towards more sustainable fisheries and should not be discouraged by a ban on trade. Hence, these provisions are highly relevant to ensuring that the CITES legal regime is adequate for commercially-exploited fish species.

a) Captive-Breeding

aa) Definition

Trade in captive-bred specimens is one of the special cases when, according to the treaty, international trade in listed species is exempted from permits.[332] This

[329] CITES (n 55) article VII.

[330] Ibid. article VII(5); Resolution Conf. 10.16 (Rev) – Specimens of animal species bred in captivity 1997.

[331] J.-F. de Pulvenis Séligny, 'The marine living resources and the evolving law of the sea' (2010) 1 Aegean Review of the Law of the Sea and Maritime Law 61, 88–90.

[332] CITES (n 55) article VII(4)–(5).

exemption was originally created for the very specific case of chinchilla captive-breeding in Argentina but its scope was then widened.[333]

To qualify under CITES as bred-in-captivity, specimens must be born in captivity, from parents who mated in a controlled environment.[334] The criterion related to the place of birth was inserted in the definition in 1979 to make sure that juveniles caught in the wild and reared in captivity would not be exempted from CITES permits.[335]

Additionally, the breeding stock must have been "established in accordance with the provisions of CITES and relevant national laws and in a manner not detrimental to the survival of the species in the wild", must be "maintained without the introduction of specimens from the wild, except" in a few specific cases, and must have produced offspring itself born in captivity or should at least be able to do so.[336] Hence, if the original specimens caught from the wild were imported or introduced from the sea, they must have been so in accordance with the Convention's relevant requirements.

bb) Appendix I Species (for Non-commercial Purposes), Appendices II and III Species

For the specimens fulfilling the requirements outlined above, the Management Authority of the exporting state shall issue a certificate stating that they were bred in captivity, document which shall be accepted as a replacement of the usual permits.[337] This applies clearly to Appendices II and III species (coming from both commercial and non-commercial captive-breeding operations), though the application of the exemption is more complex an issue for Appendix I species, as it is limited to operations with non-commercial purposes.

Specimens of Appendix I species that are bred in captivity are to be considered as specimens of species in Appendix II.[338] According to the current interpretation, this means that specimens of Appendix I species bred in captivity not for commercial purposes can be traded with the relevant certificate and hence that import may be allowed whether the purposes of trade are commercial or not.[339]

---

[333] Favre, *International Trade in Endangered Species* (n 13) 186–187; R. W. G. Jenkins, 'An Overview of the Fundamental Principles of CITES as a Mechanism for Regulating Trade in Listed Species', http://www.cites.org/eng/news/meetings/ifs-05/IFS05-principle.pdf accessed 4 July 2015, 9.

[334] Resolution Conf. 10.16 (Rev) – Specimens of animal species bred in captivity (n 330).

[335] Wijnstekers (n 106) 494.

[336] Resolution Conf. 10.16 (Rev) – Specimens of animal species bred in captivity (n 330); Gillespie, *Conservation, biodiversity and international law* (n 101) 220.

[337] CITES (n 55) article VII(5).

[338] Ibid. article VII(4).

[339] Resolution Conf. 10.16 (Rev) – Specimens of animal species bred in captivity (n 330) fourth preambular; Resolution Conf. 12.10 (Rev. CoP15) – Registration of operations that breed Appendix-I animal species in captivity for commercial purposes (n 13) sixth preambular.

This interpretation is the one decided by the CoP, but raises some issues.[340] It is unclear, at first, how a specimen can be bred in captivity not for commercial purposes and then be traded for commercial purposes. Resolution 5.10 clarifies that point by stating that when surpluses are sold to balance the costs of the captive-breeding program and/or if profits are reinvested in the program, the operation should not be considered as primarily commercial.[341]

The main problem is that CITES article VII (5) states that for "*any* specimen of an animal species [. . .] bred in captivity [. . .] a certificate [. . .] shall be accepted in lieu of any of the permits or certificates required under the provisions of *Article III*, IV or V." The reference to 'any specimen' read together with the mention of article III, which applies to Appendix I species, clearly suggests that all bred-in-captivity Appendix I specimens should be encompassed in this exemption. Nevertheless, the interpretation adopted by the CoP is considered authoritative and is consequently to be applied as it is.

Marking is the only other requirement applicable to bred-in-captivity specimens,[342] even though it is not obvious as to how it could be implemented for fish products.

cc) Appendix I Species (for Commercial Purposes)

The trade in specimens of Appendix I species bred in captivity for commercial purposes is regulated by Appendix II provisions, as developed in article IV, in other words the traditional system of an export permit applies.[343]

Captive-breeding operations of Appendix I species for commercial purposes must be registered by the Secretariat. The decision to apply the article VII (4) exemption is that of the exporting state's Management Authority. However, importing states "shall restrict imports [. . .of such specimens] to those produced by operations included in the Secretariat's Register and shall reject any document granted under Article VII, paragraph 4, if the specimens concerned do not originate from such an operation".[344]

The procedure of registration forces the exporting state to provide information and gives the opportunity to other states to object to such registration.[345] It is a lengthy and complex process, with operations "required to be reviewed by the Parties before being registered with the Secretariat. This cautious approach has been reflected in a series of resolutions, culminating in Resolution Conf. 12.10, that

---

[340] Wijnstekers (n 106) 491–492.

[341] Resolution Conf. 5.10 (Rev. CoP15) – Definition of 'primarily commercial purposes' 1985 annex, e).

[342] Resolution Conf. 10.16 (Rev) – Specimens of animal species bred in captivity (n 330) operative part.

[343] Ibid. third preambular.

[344] Resolution Conf. 12.10 (Rev. CoP15) – Registration of operations that breed Appendix-I animal species in captivity for commercial purposes (n 13).

[345] Ibid. annex 2.

have systematically made registration requirements so onerous that very few captive-breeding operations are registered."[346] Also, an identifying mark must be apposed on the specimen and must be described in the document allowing trade.[347]

For these specimens of Appendix I species bred in captivity for commercial purposes, in addition to the obligation that the breeding stock was "established in accordance with the provisions of CITES and relevant national laws and in a manner not detrimental to the survival of the species in the wild",[348] a proof of legal acquisition of the breeding stock must have been provided in the registration's application.[349] Hence, specimens to start a breeding material must originate from another captive-breeding operation, be ranched, or be wild caught in the same state as the operation takes place. Indeed, "import of wild-caught specimens of Appendix-I species for purposes of establishing a commercial captive-breeding operation is precluded by Article III, paragraph 3(c)".[350]

dd) Adequacy and Relevance for Fish Species

The risks linked to captive-breeding are mainly twofold. First, if a specimen caught in the wild is declared as captive-bred, it will be submitted to lower control requirements. Then, in the case of Appendix I species, the creation of a legal market might simplify smuggling by making enforcement more complex.[351]

In the current state of technology, not many aquaculture farms could hope to see their specimens fit into the definition of captive-breeding.[352] Indeed, for species that could be endangered, such as Bluefin Tuna, fish farmers have not yet been able to hatch specimens on a commercial basis. However, the breakthrough might be at the doorstep.[353]

Also, the complex requirements would make it very expensive to comply with CITES,[354] in particular as aquaculture exists mainly, if not exclusively, for commercial purposes. With regard to aquatic species, there are only very few captive-

---

[346] Jenkins, 'An Overview of the Fundamental Principles of CITES as a Mechanism for Regulating Trade in Listed Species' (n 333) 9.

[347] Resolution Conf. 10.16 (Rev) – Specimens of animal species bred in captivity (n 330); Resolution Conf. 12.10 (Rev. CoP15) – Registration of operations that breed Appendix-I animal species in captivity for commercial purposes (n 13) f).

[348] Resolution Conf. 10.16 (Rev) – Specimens of animal species bred in captivity (n 330).

[349] Resolution Conf. 12.10 (Rev. CoP15) – Registration of operations that breed Appendix-I animal species in captivity for commercial purposes (n 13) annex 1 §5.

[350] Ibid. preamble, as further developed in Resolution Conf. 5.10 (Rev. CoP15) – Definition of 'primarily commercial purposes' (n 341) annex, e).

[351] Favre, *International Trade in Endangered Species* (n 13) 187.

[352] Jenkins, 'An Overview of the Fundamental Principles of CITES as a Mechanism for Regulating Trade in Listed Species' (n 333) 9.

[353] See supra Chapter 2 C. II. 2.

[354] Jenkins, 'An Overview of the Fundamental Principles of CITES as a Mechanism for Regulating Trade in Listed Species' (n 333) 9.

breeding facilities registered by the CITES Secretariat, which is a sign of the restrictiveness of the regime.[355]

As for aquaculture on the high seas, such as in the special area of ecological protection in the Mediterranean Sea, the existing regime of CITES gives no indication as to which state would be responsible to issue the captive-breeding certificate.

b) Ranching

aa) Definition

Following the adoption of the captive-breeding definition, a number of operations, which were based on the raising of wild-caught eggs or juveniles, could not be exempted from a CITES permit anymore. For these operations involving Appendix I species and which were of commercial nature, trade was consequently totally restricted. This is clarified in Resolution Conf. 11.16, with the CoP "noting that the terms of Resolution Conf. 10.16 (Rev.) [. . .] do not allow the entry into trade of specimens of species included in Appendix I that have been taken from the wild and reared in captivity, except in accordance with the provisions of Article III of the Convention".[356]

A solution had to be and was found through the creation of the ranching concept.[357] Ranching is not a terminology which exists in the Convention, but was developed in successive Resolutions, as early as 1981[358]—at the CoP following the adoption of the captive-breeding definition in 1979.[359] Hence, now, under CITES and its Resolution 11.16, ranching is defined as "the rearing in a controlled environment of animals taken as eggs or juveniles from the wild, where they would otherwise have had a very low probability of surviving to adulthood".[360]

The reference to the low probability of surviving to adulthood reflects the fact that certain species have a high number of eggs or juveniles, more likely to die than older specimens,[361] since they are more numerous, competing for limited amounts of food and intrinsically vulnerable to the environment and predators.[362] A number

[355] Willock, 'Administrative and Monitoring Implications of listing' (n 299) 18.

[356] Resolution Conf. 11.16 (Rev. CoP15) – Ranching and trade in ranched specimens of species transferred from Appendix I to Appendix II (n 24) second preambular.

[357] Wijnstekers (n 106) 533–534.

[358] Resolution Conf. 3.15 – Ranching (n 24).

[359] Resolution Conf. 2.12 (Rev) – Specimens Bred in Captivity or Artificially Propagated 1979 (repealed).

[360] Resolution Conf. 11.16 (Rev. CoP15) – Ranching and trade in ranched specimens of species transferred from Appendix I to Appendix II (n 24).

[361] Jenkins, 'An Overview of the Fundamental Principles of CITES as a Mechanism for Regulating Trade in Listed Species' (n 333) 10.

[362] World Ocean Review, *The Future of Fish – The Fisheries of the Future* 2013 (n 184) 97–99.

of these young specimens can be removed without detriment to the wild population,[363] particularly since the important individuals for the species survival are the older reproducing ones.[364] The rationale of ranching is that raising specimens "in a controlled environment [...] maximize[s] the survival rate and rate of growth for the animal."[365]

The particular provisions on ranching are only applicable to Appendix I species; Appendix II species can be ranched as part of standard management programs and international trade in these specimens only needs to fulfill article IV requirements, mainly an NDF.[366] This is rather less demanding than to demonstrate a conservation benefit, as it must be done for ranching.[367]

bb) Procedure and Requirements to Register

The procedure for Appendix I species to be ranched is the following: a population of an Appendix I species is transferred to Appendix II for ranching purposes. The transfer is to be approved by the CoP, on the basis of the conditions developed in Resolution 11.16 and presented here below. The Secretariat must receive the required information at least 330 days before the next CoP and request the necessary scientific expertise to verify the validity of such proposals.

As stated in Resolution 11.16, ranching activities are only possible for "populations of species [...] that occur within the jurisdiction of Parties and are deemed by the Conference of the Parties to be no longer endangered and to benefit by ranching with the intention of trade". This apparently excludes ranching in populations that occur in areas beyond national jurisdictions. Furthermore, there needs to be specific populations of a species which are not any longer considered as endangered—if this was the case for the whole species, then a general down-listing would be more adequate.

To be potentially accepted, ranching activities must bring a benefit to the local population (of animals) and to the wild population in general "through reintroduction or in other ways". The same biological information which is necessary for amending the Appendices is required in the proposal. Also, evidence must be provided "that the taking from the wild will have no significant detrimental impact on wild populations". Adequate identification must be made possible, to avoid confusion with populations which stay in Appendix I. Other monitoring and verification mechanisms are essential such as inventories, harvest-level controls, or

[363] Favre, *International Trade in Endangered Species* (n 13) 206.

[364] R. W. G. Jenkins and others, 'Review of Crocodile Ranching Programs Conducted for CITES by the Crocodile Specialist Group of IUCN/SSC' January – April 2004, AC22 Inf. 2 (CITES 2006) 33.

[365] Favre, *International Trade in Endangered Species* (n 13) 187–188.

[366] Jenkins and others, 'Review of Crocodile Ranching Programs Conducted for CITES by the Crocodile Specialist Group of IUCN/SSC' (n 364) 4, 34.

[367] Ibid. 34.

the compulsory marking system.[368] For programs which include the catches of wild adult specimens, the proposal must be examined more strictly.

If the activities are accepted, then monitoring and controls are quite stringent with annual reports, information to be kept at the disposal of the Secretariat if requested, and the possibility, if both the Standing Committee and the state agree to it, to make a visit to the ranching operation. Finally, if problems arise and it is impossible to solve them, a transfer back to Appendix I may be proposed by the depositary state, at the request of the Standing Committee.[369] In practice, control through reporting and monitoring is not very successful, with, for example in 2004, more than half of the relevant parties failing to report on many required aspects.[370]

cc) Adequacy and Relevance for Fish Species

For fish species listed in Appendix I which are farmed, ranching could be a way to avoid the 'not for primarily commercial purposes' restriction to trade. However, to qualify, the population in question must be within a party's jurisdiction,[371] which rules out high seas farming. Whether this condition also rules out farms receiving juveniles caught on the high seas is unclear, but this would be an additional limitation. Hence, except if an aquaculture operation fulfills the bred-in-captivity requirements, it will be impossible, under the CITES regime, to legally trade the fish farmed on the high seas or maybe even those originally caught as juveniles on the high seas. The adequacy of these restrictions to trade in commercially-exploited fish species would need to be assessed since they have the potential to totally ban international trade for commercial purposes in Appendix I species if no captive-breeding operation can be registered.

Furthermore, even if sourcing juveniles in waters beyond national jurisdiction was allowed, the legal status under CITES of such fish has not been made clear by a Resolution and this scenario is not envisioned directly by the Convention. As mentioned in the Expert Panel report for CoP15, transferring live specimens to a farm "may also be considered international trade, since vessels are not necessarily from the same countries as those in which the farms are located."[372]

Then, it is not evident that juveniles caught for fish farming do benefit from captivity in terms of survival probability. Indeed, they are no larvae anymore, might be beyond the biological bottleneck point and may suffer additional mortality due

---

[368] Gillespie, *Conservation, biodiversity and international law* (n 101) 210–211.

[369] Resolution Conf. 11.16 (Rev. CoP15) – Ranching and trade in ranched specimens of species transferred from Appendix I to Appendix II (n 24).

[370] Jenkins and others, 'Review of Crocodile Ranching Programs Conducted for CITES by the Crocodile Specialist Group of IUCN/SSC' (n 364) 30, 34.

[371] Resolution Conf. 11.16 (Rev. CoP15) – Ranching and trade in ranched specimens of species transferred from Appendix I to Appendix II (n 24): "populations of species included in Appendix I that occur within the jurisdiction of Parties".

[372] FAO, 'Report of the Third FAO Expert Advisory Panel for the Assessment of Proposals' (n 314) 125.

to the transport into cages.[373] If not proven otherwise in the particular case, this specificity would remove farming operations from the ambit of ranching. Also, with the species' survival in mind, and remembering that, in certain regions, a large portion of the fish caught is transferred into farms for fattening purposes,[374] it seems logical that "the potential impacts of the removal of pre-reproductive fish from the wild for mariculture grow-out needs to be assessed in terms of its broader implications before wide-scale promotion or deployment"[375] or before international trade is allowed. Hence, the sustainability of the harvest for the stock and the spawning biomass, as well as the overall benefits of farming for the species must be evaluated.[376]

Fulfilling the requirements for ranching might also be difficult for species straddling between states. Indeed, with transboundary stocks, proving that the taking from the wild is not detrimental requires joint management. This can however be accommodated within CITES, as proven by the special regime developed for the ranching of marine turtles, for which both a national management plan and regional cooperation are needed.[377]

c) Need for Control and Differentiation

The tendency to allow trade under the captive-bred exemption while the specimens do not actually fulfill the requirements exists and is highly problematic. It has been highlighted in Resolution 10.16 where the parties mention their concern that "in spite of the adoption of several Resolutions at various meetings of the Conference of the Parties, much trade in specimens declared as bred in captivity remains contrary to the Convention and to Resolutions of the Conference of the Parties, and may be detrimental to the survival of wild populations of the species concerned".[378] This issue has been mentioned again, more recently, by the EU and the USA in a document suggesting tentative solutions.[379]

---

[373] F. Ottolenghi and others, 'Capture-based aquaculture: The fattening of eels, groupers, tunas and yellowtails' (FAO 2004) 118–121.

[374] Monaco, 'CoP15 Prop. 19 – Proposal to include Atlantic Bluefin Tuna' (n 93) 20 §8.4.

[375] P. J. Mous and others, 'Capture for culture: artificial shelters for grouper collection in SE Asia' (2006) 7 Fish and Fisheries 58, 69.

[376] FAO, 'Aquaculture development – 6. Use of wild fishery resources for capture-based aquaculture' FAO Technical Guidelines for Responsible Fisheries 5 Suppl. 6 (Rome 2011) 3–4, 11–13, 16–22.

[377] Resolution Conf. 9.20 (Rev) – Guidelines for evaluating marine turtle ranching proposals submitted pursuant to Resolution Conf. 11.6 (Rev. CoP15) 1994 annex §1.

[378] Resolution Conf. 10.16 (Rev) – Specimens of animal species bred in captivity (n 330) sixth preambular.

[379] Hungary (on behalf of the European Union) and United States of America, 'SC61 Doc. 27 – Implementation of the Convention relating to captive-bred and ranched specimens' (15 August 2011).

Clearly differentiating between captive-breeding and ranching is also necessary since permits to trade are not based on the same requirements. Problems linked to a lack of differentiation at the national level are well illustrated by the mismanagement of the Arapaima. In the case of this Brazilian fish, sustainability is not enhanced by the application of the CITES regime. This is partly due to the national definition of captive-breeding, which does not specify the need for specimens to be born in captivity. Hence, some Arapaima caught in the wild and raised in captivity are exempted from an NDF—as would usually only be the case for captive-bred specimens.[380] Risks of 'fish laundering' are quite high and wild stocks are possibly endangered by the existence of farms. Indeed, these operations partially or totally rely on specimens caught in their natural habitats, which increases the amount of fish withdrawn from the wild.

Then, an additional differentiation would be probably needed for farmed fish species, between the operations which increase survival rates of juveniles and operations which only increase the specimens' body mass. Indeed, "[t]he concept and definition of 'farmed' specimens varied across the range of taxa. Some aquaculture or mariculture operations are more in the nature of 'enhanced wild' production, as it is understood by botanists."[381] It is indeed necessary to keep the practices' impacts on sustainability in mind. Whereas captive-breeding has the potential to decrease the pressure on wild catches, the positive effects of ranching must be more closely monitored. If the eggs or juveniles caught to be reared in captivity really had little chances of surviving to adulthood, then ranching can be a conservation measure but, otherwise, fattening farms simply increase the demand for wild-caught specimens, some of which might even suffer additional mortality as a result of the transfer into cages.

For species that are highly traded in a captive-bred or ranched form, it could be useful to extend the Significant Trade Review to such specimens. They are indeed currently excluded from it,[382] while it would make sense to independently verify how sustainable major captive-breeding and ranching operations actually are.

d) Overall Sustainability of Aquaculture Practices

Aquaculture has ecological consequences linked to the concentration of fish in a small enclosed area, to the antibiotics given, to the introduction of exotic species in

---

[380] L. Castello and D. J. Stewart, 'Assessing CITES Non-detriment Findings Procedures for Arapaima in Brazil' http://www.conabio.gob.mx/institucion/cooperacion_internacional/TallerNDF/Links-Documentos/WG-CS/WG8-Fishes/WG8-CS1%20Arapaima/WG8-CS1-S.pdf accessed 6 July 2015, 2.

[381] CITES, 'CoP15 Doc. 16.2.2 – Capacity Building – International expert workshop on non-detriment Findings – Report of the Animals and Plants Committees' (n 303) 7 §8.

[382] Hungary and Unites States of America, 'SC61 Doc. 27 – Implementation of the Convention relating to captive-bred and ranched specimens' (n 379) 4.

a region, as well as to the interactions between escapees and specimens living in the wild—usually at the detriment of the latter, but not automatically.

No requirement related to these aspects is explicitly included in the general Resolution regarding captive-breeding. However, the CITES regime for captive-breeding of Appendix I species for commercial purposes mentions some of these factors. It urges parties, "prior to the establishment of captive-breeding operations for exotic species, [to] undertake an assessment of the ecological risks, in order to safeguard against any negative effects on local ecosystems and native species".[383] Also, information about the food production and source is to be provided in the application to register a captive-breeding operation.[384] Finally and most importantly "the Management Authority shall satisfy itself that the captive-breeding operation will make a continuing meaningful contribution according to the conservation needs of the species concerned".[385]

For ranching under CITES, no special requirement is made in relation to more general sustainability questions, but "the programme must be primarily beneficial to the conservation of the local population (i.e., where applicable, contribute to its increase in the wild or promote protection of the species's habitat while maintaining a stable population)".[386] It is recognized that aquaculture practices for fish such as sturgeons may have negative impacts on the wild populations, due to disease transmission or the harvesting of breeding-stocks,[387] but it remains unclear whether such considerations are actually taken into account in the general assessment of the ranching operation.

For all the specimens that are farmed but do not fall under the above categories, a permit including an NDF is needed. Hence, the non-detrimental effect on wild stocks has to be verified not only by assessing the effect of removing specimens from the wild, but arguably also by examining the negative effects of farms on wild stocks—through pollutants, diseases, escapees, etc. It seems however quite illogical that farmed Appendix II specimens should be subject to stronger requirements than Appendix I ranched ones or captive-bred ones.

Overall, there is a need to assess, for each relevant species, the impact of aquaculture on the wild stocks both through negative effects of deterioration of the environment, competition, diseases, introduction of alien species, increased pressure on stocks (both targeted stocks and dependent species used as forage), etc. and through positive effects such as the expected reduced demand for high

---

[383] Resolution Conf. 12.10 (Rev. CoP15) – Registration of operations that breed Appendix-I animal species in captivity for commercial purposes (n 13).

[384] Ibid. annex I §14.

[385] Ibid. j).

[386] Resolution Conf. 11.16 (Rev. CoP15) – Ranching and trade in ranched specimens of species transferred from Appendix I to Appendix II (n 24).

[387] M. Engler and R. Parry-Jones, 'Opportunity or Threat: The Role of the European Union in Global Wildlife Trade' (TRAFFIC Europe 2007) 34.

priced wild-caught specimens.[388] The CITES regime has requirements in relation to most of these factors, although not always in a perfectly coherent manner as conditions differ according to the category (normal protection under Appendices, ranching or captive-breeding) in which the specific specimens fall.

6. Border Control

Border controls must be efficient since it is at that point that national authorities verify the respect of regulations on international trade of certain endangered species. To ensure that, when needed, traded specimens are accompanied by the required export permit, CITES Authorities and border control officials must "monitor the shipments and verify the validity of each CITES permit."[389]

In practice, it is crucial that custom officers are able to single out suspicious shipments and to identify specimens of listed species.[390] If they cannot, the risk is high that "illegally obtained product may be laundered under other species names or products from the listed species may simply enter trade without the necessary documentation having been issued."[391] These deficiencies have been highlighted in practice for example in the case of caviar, which reaches high prices on the market and is hence a flagship species for illegal trade, conducted with the help of fraudulent documents, false declarations, corruption, etc.[392]

In the case at hand, the problems of identification and screening are particularly sensitive since fish usually enter trade in highly processed forms and recognizing a species is consequently deemed difficult. Also, the volume of trade puts an additional strain on the already stretched workload of customs control.

a) Identification

The challenges linked to identification are acute for fish species, as these animals are usually traded in a very processed form, which in some cases makes species

---

[388] On the general interactions and effects, see J. L. Anderson, 'Aquaculture and fisheries: complement or competition' in R. Arthur and J. Nierentz (eds), *Global Trade Conference on Aquaculture*, FAO Fisheries Proceedings 9 (Rome 2007) 231–236; S. M. Fernandez, 'Captive-bred exceptions: an unconventional approach to conservation under the Federal Endangered Species Act' (2003–2004) 15 University of Florida Journal of Law and Public Policy 155, 185–189; Ottolenghi and others (n 373) 211; Wiersema, 'Uncertainty and Markets for Endangered Species under CITES' (n 45) 242–243.

[389] CITES, *CITES trade controls to take effect for mahogany* (11 November 2003) http://www.cites.org/eng/news/pr/2003/031111_mahogany.shtml accessed 6 July 2015.

[390] FAO, 'Report of the Expert Consultation on Implementation Issues' (n 193) 8–9 §46.

[391] Willock, 'Administrative and Monitoring Implications of listing' (n 299) 14.

[392] CITES Secretariat, 'Caviar criminals' *CITES World – Official Newsletter of the Parties* (December 2001) 10.

hardly distinguishable one from the other. A further complicating factor for fish is the existence of both wild-caught and bred-in-captivity specimens, which do not need the same documentation to be allowed into international trade. There is thus a risk to see wild-caught fish smuggled into the market under this treaty exception.[393] However, identification issues are neither something new to CITES nor unique to fish species.[394]

aa) Readily Recognizable Specimen

According to the Convention, a specimen the trade of which can be regulated includes "any readily recognizable part or derivative thereof". The lack of definition of these terms, could have led states to interpret them differently.[395] However, they have been clarified "to include any specimen which appears from an accompanying document, the packaging or a mark or label, or from any other circumstances, to be a part or derivative of an animal or plant of a species included in the Appendices".[396]

Such a very inclusive definition is positive for fish products which are often hard to differentiate or recognize, as labeling is considered sufficient for customs to intervene. However, it is true that this does not guarantee the non-occurrence of mislabeling fraud, which is known to be widespread for fish products in other documentation schemes.[397]

bb) Look-Alike

As envisioned within the Convention framework, one possibility to deal with the difficulty of identifying species is to list their look-alike in Appendix II.[398] However, the application of this provision should be avoided as it would dramatically increase the amount of fish species listed and related paperwork.[399] Hence, states and the CITES community must look into other options to improve identification.

---

[393] As is for example still the case for sturgeons (Engler and Parry-Jones (n 387) 34).

[394] Vincent and others (n 6) 583.

[395] E. M. McOmber, 'Problems in Enforcement of the Convention on International Trade in Endangered Species' (2001–2002) 27 Brooklyn Journal of International Law 673, 692–693.

[396] Resolution Conf. 9.6 (Rev. CoP16) – Trade in readily recognizable parts and derivatives 1994.

[397] J. L. Jacquet and D. Pauly, 'Trade secrets: Renaming and mislabeling of seafood' (2008) 32 Marine Policy 309.

[398] Some species can be listed in order to ensure an efficient control of the protection for an endangered species (CITES (n 55) article II(2)(b)), particularly in the case of species that cannot be distinguished one from the other.

[399] This look-alike provision is probably not implementable for fish species. Indeed, it would potentially mean listing all fish that supply similar looking filets or many sharks because dried fins cannot be easily distinguished. Such a huge listing is not realistic (Willock, 'Administrative and Monitoring Implications of listing' (n 299) 14).

cc) Identification of Different Species

First, identification guides can be very useful to point to the particularities of a species and the details that make one different from another, particularly when dealing with entire fish. Such booklets are for example available for hard coral species or for sturgeons; in that latter case, the guide shows the different types of fishes and the accepted labeling/packaging applicable to them.[400]

Second, a good way to simplify identification of goods entering trade is to use a universally agreed packaging/labeling system. For example, as mentioned, sturgeons can only be traded under a standardized labeling system.[401] Since 2004, CITES parties are indeed to refuse shipments which do not respect the labeling guidelines.[402]

Third, since processing operations usually "bring to the loss of any morphological diagnostic features of the species,"[403] DNA barcoding can be a technical approach to discover species substitutions. It is argued to be a reliable method to improve food traceability, as demonstrated in a study about identification of sharks' meat[404] and in another one about differentiating species of sturgeons.[405] However, in view of the costs involved in DNA tests and the quantity of fish traded, this might stay, for the time being, a method only for exceptional occurrences.[406] It might nevertheless become increasingly mainstream and hence useful.

For example, the implementation of Atlantic Bluefin Tuna listing would have been difficult, because it is not easy to differentiate this species from several other

---

[400] Anonymous, 'A new CITES identification guide for sturgeons and paddlefish' *CITES World – Official Newsletter of the Parties* (December 2001) 15; E. W. T. Cooper and N. Chalifour, 'CITES, Eh?: The Implementation of CITES in Canada since WAPPRIITA' (TRAFFIC North America 2004) 61. These booklets are usually developed by CITES, by states or by partner non-governmental organizations (CITES, 'The CITES Publications' http://www.cites.org/eng/resources/publications.php accessed 6 July 2015).

[401] Originally only for export, these requirements have been extended to domestic trade also (Resolution Conf. 12.7 (Rev. CoP16) – Conservation of and trade in sturgeons and paddlefish (n 84); Raymakers, 'CITES, the Convention on International Trade in Endangered Species of Wild Fauna and Flora: its role in the conservation of Acipenseriformes' (n 502) 62).

[402] Resolution Conf. 12.7 (Rev. CoP16) – Conservation of and trade in sturgeons and paddlefish (n 84): "importing Parties [should] not accept shipments of caviar unless they comply with these provisions [the provisions of the universal labelling system]"; A. Knapp, C. Kitschke and S. von Meibom, 'Proceedings of the International Sturgeon Enforcement Workshop to Combat Illegal Trade in Caviar: Prepared by TRAFFIC Europe for the European Commission' Brussels 2006, SC54 Inf. 6 (CITES 2006) 10.

[403] M. Barbuto and others, 'DNA barcoding reveals fraudulent substitutions in shark seafood products: The Italian case of 'palombo' (Mustelus spp.)' (2010) 43 Food Research International 376, 376.

[404] Ibid. 380.

[405] C. Wolf, P. Hübner and J. Lüthy, 'Differentiation of sturgeon species by PCR-RFLP' (1999) 32 Food Research International 699, 699–700, 704.

[406] Willock, 'Administrative and Monitoring Implications of listing' (n 299) 15.

tunas. But technological developments were considered to tackle this issue,[407] particularly once "genetic testing techniques are easily available, expedient, and cost-effective."[408]

Fourth, training is crucial to improve chances that species are identified. Whether it is on the physical characteristics of the fish, on the required packaging or on the obligatory documentation, expertise exists and should be further shared. CITES already organizes capacity-building seminars. Cooperation with the FAO could be further increased, in order to benefit from the specialized knowledge they have developed in relation to fish species.[409]

dd) Differentiation of Farmed from Wild-Caught Specimens

First, it is crucial that only specimens produced in a manner reflecting CITES understanding of 'bred-in-captivity' are granted a certificate exempting them from permits' requirements by the relevant Management Authority.[410] Otherwise, "there is the risk that wild caught species will be claimed to be captive bred to take advantage of the lesser standards."[411] The definition of aquaculture in national law, at least when referring to trade in endangered species, must indeed reproduce CITES provisions (or can be stricter): only specimens born in a controlled environment, from parents which "were in a controlled environment when development of the offspring began"[412] should benefit from the permit exemption. Otherwise, the implementation of the Convention is doomed to be deficient, opening a door for illegal trade, that no level of custom control can close.

Second, methods to strengthen the chain of custody of products up to their entry into trade are crucial to ensure that all shipments requiring a permit have one.[413] Hence, to help Management Authorities control when they should issue bred-in-captivity certificates, it can be useful to register facilities which produce fish satisfying the relevant requirements. Indeed, such registration increases the possibilities to make verifications and consequently reduces the incentives to mislabel. Currently, only Appendix I captive-breeding facilities and sturgeons producers must be registered,[414] but this practice could be expanded to all fish farms of listed

---

[407] Monaco, 'CoP15 Prop. 19 – Proposal to include Atlantic Bluefin Tuna' (n 93) 21 §9.

[408] Ibid. 3 summary §18.

[409] The FAO already provides training for species such as sharks, queen conch or sea cucumbers (FAO, 'Capacity building information' http://www.fao.org/fishery/topic/16350/en accessed 6 July 2015).

[410] According to the procedure set in CITES (n 55) article VII(5).

[411] Favre, *International Trade in Endangered Species* (n 13) 187.

[412] Resolution Conf. 10.16 (Rev) – Specimens of animal species bred in captivity (n 330).

[413] J. Grogan and P. Barreto, 'Big-Leaf Mahogany on CITES Appendix II: Big Challenge, Big Opportunity' (2005) 19 Conservation Biology 973, 974–975.

[414] For the Appendix I captive-breeding facilities, registration is at the CITES Secretariat (Resolution Conf. 12.10 (Rev. CoP15) – Registration of operations that breed Appendix-I animal species in captivity for commercial purposes (n 13); Jenkins, 'An Overview of the Fundamental Principles

species. Canada for example only implements the captive-bred exemption for registered commercial breeders.[415]

Third, for the cases of alleged inconsistence between products and labels, advancements in DNA-technology could be used to differentiate sturgeons produced in aquaculture facilities from wild-caught specimens. Scientific research is moving forward rapidly with regard to aquaculture, since economic interests encourage the development of a workable and affordable solution. For example, using food additives to distinguish wild from farmed caviar through a DNA analysis has been successfully tested.[416] This technique would "provide a cheap and easy-to-assess tool for labelling caviar from aquaculture, thereby allowing its utilization as an identification system under CITES".[417]

It is nevertheless true that identification methods are only useful if an efficient shipments' screening allows suspicious cargos to be put aside. Indeed, all new techniques "are still reliant on basic visual identification as an initial form of triage. As such, [they] emphasiz[e], rather than eclips[e], the need for effective traditional screening procedures."[418]

b) Screening

Screening wildlife imports/exports in order to identify suspicious cargos is challenging due to the lack of resources compared to the level of trade and because illegal trade in wildlife is a well-organized business.[419] "Often it is very difficult to identify illegal trade, particularly bearing in mind the sheer volume of trade; in the EU alone 'over 7000 seizures involving over 3.5 million CITES-listed specimens' were made in the 2003–4 period."[420] Since fish are traded in large volumes, this problem is likely to be intensified by the listing of such a commercially-exploited species in Appendix II.

---

of CITES as a Mechanism for Regulating Trade in Listed Species' (n 333) 9), whereas registration for the sturgeons is at the national level (Resolution Conf. 12.7 (Rev. CoP16) – Conservation of and trade in sturgeons and paddlefish (n 84)).

415 Cooper and Chalifour (n 400) 30.

416 A. Wuertz and others, 'Identification of caviar from increasing global aquaculture production — Dietary capric acid as a labelling tool for CITES implementation in caviar trade' (2009) 298 Aquaculture Research 51, 51–55.

417 Ibid. 51.

418 S. Clarke, 'Shark Product Trade in Hong Kong and Mainland China and Implementation of the CITES Shark Listings' (TRAFFIC East Asia 2004) 40.

419 U.S. Department of State, 'Against Wildlife Trafficking: Working Together to End the Illegal Trade in Wildlife, Fact Sheet' (November 2007) http://2001-2009.state.gov/r/pa/scp/102487.htm accessed 6 July 2015.

420 Bowman, Davies and Redgwell (n 19) 525 (reference omitted).

aa) Maximization of Existing Resources

It is crucial to have the best and highest possible level of verifications, but there are many competing priorities for limited resources. Hence, ways to maximize the available resources are important, because "[t]here is a definite connection between manpower shortages and the corresponding increase in smuggling activities."[421]

First, designation of ports of entry for wildlife imports concentrates expertise in a few places, is considered helpful in the fight against illegal trade, and is envisioned as an option by CITES.[422] This is the choice made by the USA, which have designated about 20 ports where wildlife trade can enter/leave the country.[423] Any wildlife cargo spotted in another port is illegal. However this also signifies that the expertise on wildlife trade will be concentrated in some ports, leaving the others more prone to be successfully used for smuggling.[424]

Second, for countries such as the USA, where the responsibility to verify wildlife trade lies with professional wildlife inspectors and not with custom officers, there are possibilities, through basic training, to familiarize custom officers with wildlife trade enforcement and thus to increase the available manpower for primary screening, even if the inspection of suspicious shipments is better left to the specialists.[425]

Third, maximum use should be made of the expertise which already exists in certain organizations/states, particularly to assist developing countries. Indeed, when even countries like the USA lack resources and mechanisms to efficiently enforce wildlife trade controls, poorer countries are unlikely to find implementation easy.[426] Cooperation is crucial as "most implementation gaps of environmental regimes are [the result. . .of] institutional and financial constraints".[427]

Training could be organized by the FAO or RFMOs, to use their expertise in fisheries, by Interpol, for information and collaboration on tackling illegal trade, and by partner NGOs—such as TRAFFIC, WWF and the IUCN—which are involved in knowledge dissemination. It is a characteristic feature of CITES to collaborate with other institutions,[428] and states should both take advantage of it and promote it.

---

421 S. Patel, 'The Convention on International Trade in Endangered Species and the Last Unicorn' (1995–1996) 18 Houston Journal of International Law 157, 191 (reference omitted).

422 Bowman, Davies and Redgwell (n 19) 530; CITES (n 55) article VIII(3).

423 S. Cleva, 'Enforcement Starts with Wildlife Inspectors' (2005) 30(2) U.S. Fish & Wildlife Service, Endangered Species Bulletin 18, 19.

424 Wijnstekers (n 106) 483.

425 McOmber (n 395) 698, 701.

426 Fuchs (n 283) 1589; McOmber (n 395) 701.

427 Sand, 'Whither CITES? The Evolution of a Treaty Regime in the Borderland of Trade and Environment' (n 18) 51.

428 Fuchs (n 283) 1573.

bb) Technology

It is clear that not all shipments of wildlife are verified and only few random checks on cargos declared as non-wildlife are undertaken with the objective of finding smuggled species.[429] However, the use of technology could possibly free some resources, in order to give the officers responsible for CITES enforcement at the national level more time to verify cargos, both suspicious ones and randomly.

For example, X-rays are used in mainland China and Hong-Kong to enhance visual controls during the custom controls, in relation to sharks. It allows the determination of the shape of a cargo's content thus improving the chances to find mislabeled/smuggled goods.[430] Related to that, computerized intelligence databases are increasingly used and should be promoted in order to provide specialists and non-specialists with an easy access to the necessary identification information, while visually screening the shipments.[431]

As for e-permitting, this new method might develop "as a way to counterbalance the growing complexity of CITES enforcement and the increasing demand for inspection services."[432] It is currently being developed by CITES and some states have started implementing it.[433] The less paperwork the wildlife officers have to fill in and manually verify, the more time is available to undertake selected and/or random verifications.

cc) Intelligence Gathering and Sharing

Finally, even if more resources are allocated to customs control, if officers benefit from good quality training and if technology improves primary screening efficiency, inspections as such will not detect all smuggled specimens of CITES species.[434] Intelligence gathering between national customs and through international organizations such as Interpol is necessary.[435] Already associated by a Memorandum of Understanding since 1998,[436] CITES and Interpol entered in November 2010 into the International Consortium on Combating Wildlife Crime

[429] Cleva (n 423) 18–19; McOmber (n 395) 698.

[430] Clarke, 'Shark Product Trade in Hong Kong and Mainland China and Implementation of the CITES Shark Listings' (n 418) 39.

[431] Ibid. 43.

[432] Cleva (n 423) 19.

[433] It is meant to assist both the "the handling and processing of CITES applications [and] the issuance of electronic permits" (CITES, 'CITES electronic permitting toolkit' (2010) http://www.cites.org/common/cop/15/doc/E15-30-01T.pdf accessed 6 July 2015, 3, 27).

[434] Willock, 'Administrative and Monitoring Implications of listing' (n 299) 15.

[435] CITES Secretariat, 'Caviar criminals' (n 392) 10.

[436] Memorandum of Understanding between the General Secretariat of ICPO-Interpol and the CITES Secretariat 1998.

with three additional organizations, which agree to "work collaboratively to support national law enforcement agencies."[437]

This type of collaboration has, for example, served to decrease the levels of caviar poaching and smuggling, through issuance, by Interpol and the World Customs Organization, of "intelligence briefings to help target smugglers and illicit dealers."[438] Intelligence gathering and sharing are crucial to improve the processes through which shipments are assessed in order to determine the risk that their content is illegal.[439]

In summary, listing a fish species would have practical consequences on customs' work, by increasing both the workload and complexity of verifications. States and the CITES community could react by adapting their screening and identification methods, as well as the regulations applicable with regard to packaging and registration of captive-bred facilities. Also, maximizing the existing resources and making best use of technology are ways to be pursued—this would not only benefit the efficiency of the CITES regime, but also of any trade documentation scheme. Furthermore, ensuring that national laws reflect the CITES provisions is necessary to avoid, as much as possible, the occurrences of smuggling as well as to be able to sanction them if necessary. Cooperation, here again, is likely to be crucial between states, and with international organizations that have developed useful expertise.

II. Ability of CITES to Address Particularities of a Species Used as Food Supply

Fishery products are very important sources of food as well as livelihood for a large proportion of the world's population. While these characteristics do not as such curtail the use of CITES, it is important to assess whether this regime's provisions and procedures can accommodate them, in particular in relation to the particularities of developing countries.

These countries and communities have the fewest livelihoods alternatives and are those for whom fish contributes most to food security. The fears expressed in various fora that CITES cannot address their specific needs can consequently be a strong argument against using this regime for fish species. Whether or not one considers that livelihood considerations should have a role to play in decisions related to granting protection to particular species, the reality remains that they do;[440] as such they ought to be taken into account.

[437] These are the United Nations Office on Drugs and Crime (UNODC), the World Bank and World Customs Organization (CITES Secretariat, ICPO-Interpol Secretariat, UNODC, World Bank and WCO Secretariat, Letter of Understanding Establishing the International Consortium on Combating Wildlife Crime 2010).

[438] CITES Secretariat, 'Caviar criminals' (n 392) 10.

[439] Cooper and Chalifour (n 400) 49, 52–53.

[440] Cooney and Abensperg-Traun (n 35) 302–307.

1. Specific Relevance of Food Security and Livelihood for Fish Species

Food security is defined by the FAO as the situation "when all people, at all times, have physical, social and economic access to sufficient, safe and nutritious food that meets their dietary needs and food preferences for an active and healthy life."[441] In 2010, undernourishment levels decreased from previous years but still represented an issue for about one billion people.[442]

As mentioned in the Introduction, fish are an important source of food/protein and revenue for a large number of countries, particularly developing ones. They contribute both to direct food security—as well as to the quality of ingested nutrients—and to indirect livelihood as they generate wealth.[443] Being the main source of protein for approximately one billion people, fish products are an important part of a healthy diet worldwide. For low income food deficient countries, they represent 20 % of total animal protein intake.[444] People in developing countries depend on fish for their food intake far more than people in developed countries—not only because of the quantities eaten but also because of the lack of potential ersatz.[445]

Even outside the realm of direct food security, fish products are crucial for a number of communities, particularly in the developing world where they are the most exported and valuable agricultural commodity.[446] The value of international trade in fish products amounted to 78 billion USD in 2005. Also, the estimated number of people directly working in the fishing sector reaches nearly 58 million, nearly all of them in developing countries.[447] As most of the trade in fisheries products goes from developing to developed countries,[448] restrictions to trade would economically affect the former more than the latter.

While, in general, CITES species are almost exclusively native to developing countries,[449] this is not the case for Toothfish or Bluefin Tuna, which are straddling and highly migratory species, difficult to catch and rather high-end market products. Thus their listing would impact many countries but would not directly put food

---

441 FAO, 'The State of Food Insecurity in the World: Addressing food insecurity in protracted crises' (Rome 2010) 8.

442 Ibid.; FAO, 'The State of the World Fisheries and Aquaculture 2014' (n 308) 68.

443 T. Bostock and S. Walmsley, 'Enough to Eat? Fisheries and Food Security' in R. Bourne and M. Collins (eds), *From Hook to Plate: The State of Marine Fisheries: A Commonwealth Perspective* (2009) 109.

444 Ibid. 105.

445 G. Kent, 'Fisheries, food security, and the poor' (1997) 22 Food policy 393, 395–396.

446 Bostock and Walmsley (n 443) 105, 107.

447 FAO, 'The State of the World Fisheries and Aquaculture 2012' (n 308) 10, 41; FAO, 'The State of the World Fisheries and Aquaculture 2014' (n 308) 27, 31–32.

448 F. Asche and F. Khatun, 'Aquaculture: Issues and Opportunities for Sustainable Production and Trade' ICTSD Natural Resources, International Trade and Sustainable Development Series Issue Paper 5 (Geneva 2006) 20; Bostock and Walmsley (n 443) 107–108.

449 Young, 'Contemporary Issues' (n 14) 176.

security in developing countries at risk. Nevertheless, other aquatic species such as the *Strombus* species or the Spiny lobster in the Caribbean are native to and important on an income point of view for developing countries in the region. As for the queen conch, listed in Appendix II, it is considered as the main or an important additional source of revenue for many small-scale fishermen in the region.[450]

The listing of any species is deemed to have economic consequences—and fish are a major commodity representing huge economic assets. The use of CITES here is most likely to have economic impacts through the rise of transaction costs and/or loss of income. It would probably be inacceptable for many countries to restrict the access to species which are crucial food for their population[451] or even to restrict access to trade for species which provide necessary incomes.[452] These considerations do not only matter for Appendix I listing, but also for Appendix II. Indeed, the administrative burden and costs linked to implementing the CITES requirements represent a significant increase in production/trade costs, meaning that some species would not enter into international trade anymore and/or that the costs of fishing would rise.

In conclusion, as the listing of commercially-exploited marine species could be considered a politically inacceptable decision by the developing countries which rely heavily on fisheries, hence triggering a negative vote, it is necessary to assess whether and how CITES takes into account their particularities.

2. A Regime Permitting Sustainable Use

a) Only Targeting International Trade

Regarding direct food security, one must note that CITES only regulates international trade, and in case of marine species, catches beyond waters under national jurisdiction. Hence, coastal fisheries for national consumption would be left untouched by a listing.

---

[450] C. Catarci, 'World Markets and Industry of Selected Commercially Exploited Aquatic Species with an International Conservation Profile' FAO Fisheries Circular 990 (Rome 2004).

[451] Mulliken (n 49) 57–58.

[452] South Africa for example clearly expressed its concern that listing toothfish would set a precedent and that this type of measures towards sustainable development would have negative socio-economic consequences for the country. Such consequences were deemed unacceptable since they would affect a developing country (CCAMLR, 'CoP12 Inf. 20 – Proposal from Australia (CoP12 Prop. 39)' (n 208) 10 §10.42 (South Africa)). Tunisia, the United Arab Emirates and Grenada also argued against the listing of Atlantic Bluefin Tuna in Appendix I, inter alia because it "would affect the livelihoods and food security of Small Island Developing States" (CITES, 'CoP15 Com. I Rec. 8 (Rev. 1) – Summary record of the eighth session of Committee I' (n 95) 2).

However, more than 30 % of the total fish production is internationally traded.[453] Hence, the use of CITES would create additional restrictions and impact national economies. However, this is the case for any species listed by CITES and most were logically an important source of income, since it was partly the pressure exercised by international trade that pushed them towards extinction.

A listing could even improve the local conditions, as it would be more complicated and costly for foreign vessels to come and catch fish in the EEZs of third states. Such vessels would have the choice of either landing the catches in the coastal state's port—which would create employment—or would have to implement import/export permits to land the specimens in other states' ports, making the fishing venture potentially more sustainable and definitely more expensive.

b) No Automatic Ban on Trade

The fear of a preservationist approach taking over in the CITES forum is acute and such apprehension to see a general trade ban implemented is rational, since listed species are difficult to down-/de-list. However, this ban scenario is only a risk in case of listing a species in Appendix I. Under the rules relevant to Appendix II, international trade is possible in accordance with conditions of sustainability and legality, criteria that no one can legitimately criticize. While it might be difficult at first to implement the requirements to allow trade in Appendix II species, the development of management plans and controls is possible, especially with outside support.[454]

Also, some exemptions found in the treaty itself or later developed strengthen the concept of trade under a sustainable use philosophy:[455] captive-breeding, ranching and split-listing of populations allow international trade to continue in species or populations that can put up with it. These practices are well in line with the ability of states to find acceptable compromises relating to sustainable use.[456]

Furthermore, aquaculture is important for food security, as it contributes to food production and creates revenues,[457] and the CITES regime appears to be able to accommodate some of such operations.

---

[453] Bostock and Walmsley (n 443) 105.

[454] See infra Chapter 5 C. II. 3. b) and Chapter 6 D.

[455] See supra Chapter 5 A. I.

[456] McOmber (n 395) 696.

[457] Bostock and Walmsley (n 443) 106; L. Cunningham, 'Assessing the contribution of aquaculture to food security: a survey of methodologies' FAO Fisheries Circular 1010 (Rome 2005) 10–14.

c) Non-charismatic Species

Decisions to amend Appendices are usually not only interest- but also value-based; states vote according to their environmental beliefs as much as to scientific evidence.[458] Some species are emotionally preferred over others, and fish are usually in the latter category.[459] They are indeed not perceived or depicted as charismatic types of animals.

Since the main debate between pro-sustainable use and pro-preservation is mostly centered on the so-called charismatic species,[460] the risk of preservationism is less acute for fish species, in relation to whom the discussions focus on the possibility to implement sustainable trade and the fears about a misuse of a temporary ban to allow species to recover. Hence, a larger margin of maneuver exists to take compromise decisions with regard to non-charismatic species, for which the precautionary principle is less strictly applied.[461] One can also expect quicker reactions to changes in biological statuses, as the criteria are likely to be more closely followed for species which are not listed on value-based arguments.

The concern of fishing states can nevertheless be understood in the view of the evolution of whaling for example: within in a few decades, a traditionally commercial practice[462] has become a taboo and the instruments which were intended to avoid overexploitation were transformed into preservationist treaties.

3. A Regime Taking into Account Particular Needs

The CITES community is increasingly aware of the need to take into account livelihood questions not only while assessing a proposal, but also at the implementation stage.

a) In Assessing a Proposal and Listing

In taking the decision to amend an Appendix, reference is made to the need to take into account socio-economic factors. However, this is only part of the Preamble of Resolution 9.24 and hence does not constitute a formal criterion.

---

[458] Carey (n 8) 1309.

[459] Stoett (n 8) 569.

[460] G. J. W. Webb, 'Are All Species Equal? A Comparative Assessment' in J. Hutton and B. Dickson (eds), *Endangered Species-Threatened Convention: The Past, Present and Future of CITES* (Earthscan 2000) 98–99.

[461] The difference of treatment between crocodiles (many ranching operations have been allowed) and turtles (none has been) can be noticed, even though their relevant characteristics are similar (ibid.).

[462] Whaling used to be a legitimate harvesting activity (on a historical view on the uses of whales and the situation of the so-called fishery, see for example P. C. Jessup, *L'exploitation des richesses de la mer* (Recueil des Cours vol 29, Librairie Hachette 1929) 488–494).

For a listed species, annotations can be used, either as reference or as substantive information to provide clarifications about the scope of the listing of some species.[463] A particular annotation was made regarding the minke whale in order to respect the IWC exemption for aboriginal subsistence whaling. The West Greenland population of the minke whale is indeed in Appendix II instead of Appendix I.[464] This is a particular and unique case under CITES—which does not explicitly state the reason behind this split-listing—but it shows that subsistence issues can be taken into account in the regime, when so desired by the CoP.

Furthermore, the impact of trade restrictions on livelihoods of poorer communities is increasingly recognized as a relevant factor within the CITES community.[465] In order to be able to assess such consequences, Decision 14.3 (now repealed) called for the development of tools to determine the positive and negative impacts of a listing on livelihoods. In practice however, it does not mean that negative impacts are to be avoided in all cases.[466] It also is now explicitly stated that the determination of such impacts is not to serve at the listing stage, but at the implementation one.[467]

It must nevertheless be noted that, as states are the decision-makers and independently weigh the various elements related to a listing, the importance of a species for the livelihoods in a particular region can be presented as an argument against a positive vote to the proposal. That has for example happened in the case of an Indonesian ornamental fish and the proposal did not succeed.[468] Also, at CoP11, when the Devil's claw was proposed for listing, it was strongly opposed on the basis of the threat a listing would put on the livelihood of harvesters, a consideration that is not formally a criterion for/against a listing.

Even though "CITES listing should not be a debate between jobs and species",[469] such considerations can weigh in the balance.[470] In such delicate cases, studies of the relative economic and social importance of a species in a community or state might be useful to clarify the actual impact a listing could have.[471] For some

---

[463] Resolution Conf. 11.21 (Rev. CoP16) – Use of annotations in Appendices I and II 2000; Aguilar (n 288) 282–283.

[464] IWC, 'Aboriginal Subsistence Whaling' http://iwc.int/aboriginal.htm#asw accessed 6 July 2015.

[465] B. Dickson, 'CITES and the livelihoods of the poor' (2008) 42 Oryx 548, 548, 550–551.

[466] Ibid. 551.

[467] Resolution Conf. 16.6 – CITES and livelihoods 2013; on this see infra Chapter 5 C. II. 3 b).

[468] A. A. Vagelli, 'The unfortunate journey of Pterapogon kauderni: A remarkable apogonid endangered by the international ornamental fish trade, and its case in CITES' (2008) 18 SPC Live Reef Fish Information Bulletin 17, 23.

[469] M. B. Sky, 'Getting on the List: Politics and Procedural Maneuvering in CITES Appendix I and II Decisions for Commercially Exploited Marine and Timber Species' (2009–2010) 10 Sustainable Development Law and Policy Review 35, 40.

[470] Cooney and Abensperg-Traun (n 35) 302–307.

[471] Such a study has been undertaken by Catarci (n 450) on sturgeons, Caribbean queen conch, sharks and Patagonian toothfish, although it was not linked to any particular proposal.

species, it is also fundamental to have the local communities as partners rather than opponents of a listing for it to actually improve the conservation status of the species—this might be worth taking into account in the decision to list a species.[472]

b) In Implementing the Regime

Resolution 8.3 on *Recognition of the benefits of trade in wildlife*, as amended at CoP13, "recognizes that implementation of CITES-listing decisions should take into account potential impacts on the livelihoods of the poor."[473] The choice of words such as "should" and "take into account" signifies that the CoP did not want to be bound; states have no firm obligation to take action to deal with such potential impacts.

More recently, the new Resolution 16.6 on *CITES and livelihoods*, adopted at CoP16,[474] includes recognition that the implementation of a CITES listing can potentially have positive effects on livelihoods ("by delivering long-term species conservation and reducing unsustainable and illegal trade"), but that it may also negatively "impact livelihoods of rural communities by restricting access to income, employment and other resources, such as food, materials and medicines". It then recognizes that due to the potential of "short-term negative impacts on rural communities, mitigation strategies should be adopted as appropriate", including through short-term financial support and the promotion of alternative sources of income.[475]

In that same Resolution, parties are invited to make aid more effective through partnerships between the different institutions involved in conservation and development. Furthermore, international financial institutions are encouraged to help in the development of necessary measures to tackle the impacts that CITES listings could have on livelihoods of poorer communities.

It is however explicitly stated that "livelihoods issues and the present Resolution do not pertain to the criteria for the amendment of the Appendices or the requirement to make non-detriment findings".[476] Hence, it is only related to the implementation of the regime, and not to the decision to list or not a species.[477]

Then, two documents have recently been developed: first, the toolkit for the rapid assessment at the national level of the positive and negative impacts of implementing CITES-listing on the livelihoods of the poor presents a range of tools to assess such impacts. Second, voluntary guidelines are provided to the

---

472 Cooney and Abensperg-Traun (n 35) 308.

473 Resolution Conf. 8.3 (Rev. CoP13) – Recognition of the benefits of trade in wildlife (n 26).

474 Resolution Conf. 16.6 – CITES and livelihoods (n 467).

475 Ibid.

476 Ibid.

477 Cooney and Abensperg-Traun (n 35) 304.

parties to address negative impacts.[478] Once fully finalized and if implemented, those documents could be of great use to actually consider and then take care of such problems.

Moreover, capacity building to help developing countries draft and adopt management plans is envisioned in the treaty regime. Indeed, an early Resolution already "call[ed] on all Parties to ensure the inclusion of technical assistance, in matters relating to this Convention, in the bilateral and multilateral programmes of development aid in which they participate; urge[d] Parties to make special funding and qualified staff available, possibly by way of 'associate expert' assignments to the Secretariat and to developing countries, to carry out such technical assistance projects for the benefits of the other Parties; and request[ed] the Secretariat to continue to seek external funding for this purpose, in consultation with the Standing Committee, and to execute the projects so funded."[479] Also, parties are encouraged to offer assistance to the developing countries making such a request in particular with regard to NDF.[480] This is crucial, as implementing a CITES listing has high costs due to the research necessary to undertake NDF, the actual issuance of permits and administrative aspects linked to the reporting obligations, and inspection and enforcement.[481]

For aquaculture, the high requirements that have to be met for an operation to be considered captive-breeding or ranching might prove a difficulty, in particular for developing countries and hence create barriers to trade.[482] Specific financial and technical support will be needed but is not yet envisioned in the regime.

c) In Down-/De-Listing

Finally, an unsuccessful mechanism was developed in relation to the proposal to list Atlantic Bluefin Tuna, in order to mitigate the fears that fish species would be listed without a chance of down-/de-listing. As examined earlier, a Resolution was drafted—but not adopted—in an attempt to provide for easier and more automatic down-/de-listing.[483]

---

[478] Resolution Conf. 16.6 – CITES and livelihoods (n 467); CITES Secretariat (on behalf of the Working Group On CITES and Livelihoods), 'CoP16 Inf. 21 – CITES and livelihoods toolkit and guidelines' (3–14 March 2013).

[479] Resolution Conf. 3.4 – Technical cooperation 1981.

[480] Resolution Conf. 16.7 – Non-detriment findings (n 168).

[481] Willock, 'Administrative and Monitoring Implications of listing' (n 299) 9.

[482] As can already be the case with sanitary and phytosanitary measures as well as technical barriers to trade (Asche and Khatun (n 448) 22–24).

[483] Monaco, 'CoP15 Prop. 19 – Proposal to include Atlantic Bluefin Tuna' (n 93) 3 summary §17; Monaco, 'CoP15 Doc. 52 (Rev. 1) – Species Trade and Conservation, Atlantic Bluefin Tuna, Draft Resolution of the Conference of the Parties' (n 212) operative part; see supra Chapter 5 C. I. 1. c) aa).

Such initiative, even if it had been successful, would not have changed the treaty system, in which the CoP is the sovereign body. The decisions taken in that forum, listing as well as down-/de-listing, reflect the views of various states on the question at stake. A proposal can arguably be directed by such a Resolution, but the parties' votes, i.e. the actual decision-making, cannot. This is the nature of international decisions and is not particular to CITES.

4. Conclusions

The argument that CITES cannot take into account the concerns related to the importance of fish in terms of food and income, as well as those of developing states about food security, or that it is biased against sustainable development appears oversimplified.

In general, it is worth noting that CITES only targets international trade and is no automatic ban on trade, as Appendix II species can be traded as long as an NDF and legality check have been successfully undertaken.

Also, the debate between protectionism and sustainable use has been fierce within CITES for decades, showing the different interests and values of the states parties. Since fish are non-charismatic species, they are less likely to be left on an Appendix while their biological status has recovered. It is would nevertheless be beneficial to address this particular fear, as it is grounded on practical examples of CITES' misuse.

As for specific mechanisms to face the issue of food security or the special needs of developing countries, CITES has adopted at the last CoP a Resolution on livelihoods and some relevant guidelines in relation to the implementation of the regime. The former is quite a milestone as, for the first time, CITES parties acknowledge that "trade regulation has real impacts on people, and that CITES parties have some form of duty (however qualified) to seek to ensure that negative impacts on people are avoided or mitigated."[484]

D. Efficacy of Using CITES

It is quite difficult to empirically assess the effectiveness of CITES. This would require an examination, all other factors remaining equal, of whether the use of this regime helped to reduce the pressure that international trade puts on specific species and hence whether trade restrictions improved their biological status.[485] Such a trade and biology study would be very useful, but is well beyond the scope of this research, and would be complex since CITES measures are usually only one

---

[484] Cooney and Abensperg-Traun (n 35) 304.

[485] Reeve, *Policing International Trade in Endangered Species* (n 8) 7.

amongst others.[486] Also, trade is influenced by many independent mechanisms (such as consumers' preferences or fashion)[487] the importance of which would have to be accounted for. The lack of any existing comprehensive assessment is nevertheless to be noted,[488] but the FAO is currently reviewing the state of commercially-exploited aquatic species listed under CITES,[489] which could, if such exercise is undertaken on a regular basis, serve to consider the impact of protection under this regime.

Hence, that CITES is considered a successful treaty is not backed up by clear-cut facts—and is not uncontroversial.[490] Nevertheless, several practical effects with regard to fisheries can be expected (I). Moreover, this regime functions quite differently than traditional fisheries management and has consequently some potential to positively impact the governance issues identified in Chapter 4 (II).

---

[486] P. H. Sand, 'Commodity or Taboo? International Regulation of Trade in Endangered Species' (1997) Green Globe Yearbook 19, 26.

[487] Mulliken (n 49) viii, 51–52.

[488] R. Reeve, 'The Convention on International Trade in Endangered Species of Wild Fauna and Flora (CITES)' in G. Ulfstein (ed), *Making treaties work: Human Rights, Environment and Arms Control* (Cambridge University Press 2007) 134. In 1996, the states parties commissioned an assessment for 12 species. The conclusion was rather negative: only two species seemed to have improved (Environmental Resources Management, Study on How to Improve the Effectiveness of CITES, 1996 Final Report to the Standing Committee of CITES, Lausanne, as presented in Martin (n 324) 30). As for the study of Sas-Rolfes (n 196), it looks at rhinoceroses, bears, elephants and tigers, species that have little in common with aquatic species.

For example, in relation to sturgeons, the success of CITES' fight against illegal trade is not unanimous: on positive and negative views on how much illegal trade in sturgeons still exists, see CITES Secretariat, 'Caviar criminals' (n 392) 10; Engler and Parry-Jones (n 387) 34. Many species of paddlefish are still imperiled (E. K. Pikitch and others, 'Status, trends and management of sturgeon and paddlefish fisheries' (2005) 6 Fish and Fisheries 233, 251–252).

A study on the effectiveness of CITES for marine species is in preparation by the authors of Vincent and others (n 6), which already provides some comments on the effects of a listing.

[489] Committee on Fisheries – Sub-Committee on Fish Trade, 'Update on CITES related activities' presented at the Fourteenth Session of the Sub-Committee on Fish Trade (Bergen, 24–28 February 2014) COFI:FT/XIV/2014/10 (2014) 5 §17–18.

[490] CITES' effectiveness is indeed put into question by the scope of illegal trade in wildlife – estimated between five and 10 billion USD per year (Fuchs (n 283) 1588) and by the fact that nearly no listed species has recovered so well that it qualified to be delisted (H. Kievit, 'Conservation of the Nile Crocodile: Has CITES Helped or Hindered?' in J. Hutton and B. Dickson (eds), *Endangered Species-Threatened Convention: The Past, Present and Future of CITES* (Earthscan 2000) 88). Also, the effectiveness of CITES is not accepted unanimously: it is considered rather positively for animals such as crocodiles – one of the success stories of the regime, mostly due to the ranching possibilities (Kievit (n 490) 88–89, 91) – and elephants, but less so for rhinoceros or tigers (E. Neumayer, 'WTO Rules and Multilateral Environmental Agreements' in K. P. Gallagher and J. Werksman (eds), *The Earthscan Reader on International Trade and Sustainable Development* (Earthscan 2002) 142, presenting OECD, 'Trade Measures in Multilateral Environmental Agreements: Synthesis Report of Three Case Studies' COM/ENV/TD(98)127/FINAL (Paris 1999); Martin (n 324) 33–34). See also supra Chapter 3 C. I. 3.

I. Practical Effects

1. Against the Main Threats

In the particular case of commercially-exploited fish species, a listing under CITES could have positive effects against unsustainable fishing. Indeed fishing quotas—which are not always set according to scientific advice—are used for NDF and could be reviewed by the Animals Committee during a Significant Trade Review. However, as the main data linked to fisheries comes from RFMOs, one can wonder against which benchmark the review would take place. Indeed, without independent research (i.e. not originating from the RFMOs) it would appear difficult to criticize quotas, except in the case of clearly unsustainable ones. It is worth noting that even the threat of listing has, in certain cases, served its purpose: ICCAT has adopted scientifically-advised quotas for Atlantic Bluefin Tuna ever since the CoP15 proposal to list this species.[491]

With regard to IUU fishing, CITES could help to better control certain aspects of this major issue. Indeed, as a globally applicable TDS, its system of permits would allow improved monitoring on a quasi-universal level.[492] The benefits of a listing would be in terms of reducing demand and consequently catch levels—this would quite probably help rebuilding the stocks. Also, such a trade measure would "provide some additional tools in combating illegal trade and thereby illegal fishing."[493] Illegal fishing itself would not be curtailed, but the accessibility of international markets and hence the expected profits would decrease—except of course if, due to stable demand and poor enforcement at the borders, illegal fishing is then coupled with illegal trade.

The problem of unreported fishing would certainly not be fully solved, since it falls by nature outside of the official framework. However, the probability of determining levels of unreported fishing would increase thanks to the intensification of trade verifications. Finally, regarding unregulated fishing, it is not clear whether CITES could apply RFMOs measures globally, but it seems logical that, everywhere, some regulation would need to be considered applicable in order for the necessary NDF (and arguably also the legality finding) to be undertaken. All states

---

[491] Vincent and others (n 6) 574. However, views differ as to whether ICCAT has adopted scientifically-recommended quotas in 2014 (European Commission – Press release, 'The International Commission for the Conservation of Atlantic Tuna (ICCAT) 2014: Sustainable increase for Bluefin tuna quota supported by science' 17 November 2014 http://europa.eu/rapid/press-release_IP-14-1886_en.htm accessed 6 July 2015; The Pew Charitable Trusts, 'ICCAT Ignores Science and Increases Quota for Atlantic Bluefin Tuna' 20 November 2014 http://www.pewtrusts.org/en/about/news-room/news/2014/11/20/iccat-ignores-science-and-increases-quota-for-atlantic-bluefin-tuna accessed 6 July 2015).

[492] It appears that trade in sturgeon is better monitored and controlled since the relevant species were listed under CITES (Raymakers, 'International Trade in Sturgeon and Paddlefish Species – the Effect of CITES listing' (n 283) 530, 532).

[493] FAO, 'CoP15 Inf. 26 – Statement from the FAO Fisheries and Aquaculture Department on CoP15 Proposal 19 to List the Atlantic Bluefin Tuna in Appendix I' (n 97) annex §9.

parties to CITES would consequently be bound by some CMMs which would have to be determined.

As for by-catch, it seems unlikely that CITES would have much impact on this particular issue. Indeed, most incidental catches are discarded at sea and since "[n]o parts or derivatives are brought into a state's territory [. . .,] CITES does not control this activity."[494] Consequently, for species mostly caught as by-catch and discarded, this regime is of no use, as China mentioned with regard to sawfishes.[495] The only exception would be if the targeted species during the harvesting of which endangered species are incidentally caught were listed, although such use of the listing criteria is probably not to be promoted. Moreover, for these untargeted species that were usually landed, the need for a permit will most probably encourage discard.

It is important to note that CITES' focus on international trade means that it does not regulate domestic trade or protect habitats. However, with the definition of IFS agreed at CoP16, all catches on the high seas are covered by the regime, even if landed in the flag state's ports. Also, external vulnerabilities such as habitat degradation are to be taken into account in the decision-making process. Hence, while this regime does not offer direct habitat protection, it can support the species' recovery by taking into account the increased stress that degradation of the ecosystem represents.[496]

2. Effects of a Ban or Restriction to Trade on Fisheries

A listing in Appendix I is a near total ban for international commercial trade. This would allow the enforcement of clear rules where almost all trade can be identified as illegal. However, it also means the end of available trade data and comes at a high socio-economic cost.

Trade bans have additionally been criticized in particular because they can be counterproductive if trade becomes illegal instead of being brought to a halt. Indeed, they do not necessarily reduce demand and their efficiency will depend on the availability of enforcement mechanisms.[497] "Complete trade bans often merely raise the value of illegal trafficking and render stringent controls more difficult."[498]

Furthermore, it does not make any doubt that the willingness of the range states is crucial to the efficacy of the ban. This depends, inter alia, on whether the decision

[494] Favre, *International Trade in Endangered Species* (n 13) 89.

[495] China, 'CoP14 Inf. 45 – Shark Issues' (n 3) 3 §2.

[496] C. T. Bright, 'The Future of the International Whaling Commission: Can We Save the Whales?' (1992–1993) 5 Georgetown International Environmental Law Review 815, 841.

[497] Carey (n 8) 1307–1308; R. Cooney and P. Jepson, 'The international wild bird trade: what's wrong with blanket bans?' (2006) 40 Oryx 18, 19.

[498] Neumayer (n 490) 142 (reference omitted).

to amend the Appendices is based on clear sustainability arguments, is backed up by the range states (in opposition to being only pushed by rich consumer countries),[499] and is able to react to ameliorations in the species' status (in opposition to keeping the species on the Appendix forever, as is too often the case). Also, without alternative income for the communities who depend on such harvesting[500] and without any further revenue expected, the incentive to protect the habitat of the listed species disappears.

Regarding fish species, some of these pitfalls are important to take into account. The risk of trade continuing and moving underground is quite acute due to the high economic importance of trade in fish products for an important number of communities and countries. Hence, if a listing is perceived as imposed by the consumer countries, the willingness to implement it is likely to be missing. Moreover, the fear of a preservationist approach which would keep a species in Appendix I while it has sufficiently recovered risks putting states on their guard as to using the regime at all.

However, as fish species do not fall under the 'charismatic species' heading, scientific arguments have more chances to serve as a basis for a listing than ideological values do. Also, the argument according to which restrictions to trade, since they reduce the benefits communities could receive from wild species, negatively impact the motivation to protect habitats[501] is of little relevance here. There is indeed less economic incentive to protect or stop protecting a marine habitat in relation to the use which is made of living resources, than would be the case for land-based habitats: there is no competition for space between humans and animals and, usually, no alternative use which can easily be made of the waters.[502] Moreover, habitat loss, which is such a central threat for land-based species,[503] is probably slightly less pronounced for marine species.

A listing in Appendix II causes a restriction to trade but no ban and sustainable trade should be able to continue. However, in countries which do not have the resources to implement the Convention's requirements, this 'nuance' might be lost. Indeed, if no management plan exists or if enforcement is impossible, an NDF cannot take place, and restrictions are thus transformed in a ban.[504] Hence, the risk is considerable that under such circumstances most of the trade becomes illegal. Moreover, in certain countries, an Appendix II listing can represent a real ban if the exporting state does not differentiate Appendix I from II species, such as the

[499] Cooney and Jepson (n 497) 21–22.

[500] Ibid. 20.

[501] J. Hutton and B. Dickson, 'Introduction' in J. Hutton and B. Dickson (eds), *Endangered Species-Threatened Convention: The Past, Present and Future of CITES* (Earthscan 2000) xvi.

[502] M. Bowman, 'Transcending the Fisheries Paradigm: Towards a Rational Approach to Determining the Future of the International Whaling Commission' (2009) 7 New Zealand Yearbook of International Law 85, 119.

[503] Dansky (n 39) 970; Du Plessis (n 183) 16–17.

[504] China, 'CoP14 Inf. 45 – Shark Issues' (n 3) 2 §1(d).

Philippines in the case of sea horses.[505] As there is no interest in CITES causing illegal trade rather than stopping unsustainable trade, these risks must be taken into account in the listing decision and addressed in the implementation, for example by providing support for the development of management plans.

With regard to aquaculture, a listing would complicate matters, make operations more expensive and enormously restrict trade for many farms. Indeed, many aquaculture practices where specimens are born in captivity are still not able to fulfill the criterion of "second-generation offspring" production.[506] However, it would also encourage more sustainable practices, and provide some regulatory framework—or encourage the development thereof—for an activity which is in a phase of rapid but quite unregulated development, at least when moving beyond the domestic level.

II. Added Value Towards Good Governance

1. Legal Clarity, Coherence and Political Unity

The CITES regime has an advantage over the traditional fisheries management regimes in the sense that its regulations are applicable to the high seas, without the need to clarify whether the freedom of fishing still exists or to define the status of the UNFSA. Indeed, nearly all states are parties to CITES and hence bound by the regime. Also, while reservations are possible and a number of the non-parties are also known flags of convenience,[507] the requirement to ask reserving/non-parties for 'comparable documentation' can avoid the creation of too many loopholes.[508]

Moreover, other legal issues, which arose within the CITES regime, such as the questions linked to IFS, chartered vessels, or the standards applicable in a potential legality finding for high seas catches, could be addressed by the CITES CoP. Indeed, CITES provides several fora, such as the Standing Committee and the CoP, where these issues can be discussed and usually multilaterally agreed upon.

Then, the problem of geographical fragmentation and gaps is likely to be improved, since the documentation scheme system under CITES is globally applicable.[509] Closing the geographical loopholes through the use of CITES was already

---

[505] P. Christie, E. G. Oracio and L. Eisma-Osorio, 'Impacts of the CITES listing of seahorses on the status of the species and on human well-being in the Philippines: a case study' FAO Fisheries and Aquaculture Circular 1058 (Rome 2011).

[506] Jenkins, 'An Overview of the Fundamental Principles of CITES as a Mechanism for Regulating Trade in Listed Species' (n 333) 9.

[507] Calley (n 76) 163.

[508] Ibid. 203–204.

[509] M. Lack, 'Catching On?: Trade-related Measures as a Fisheries Management Tool' (TRAFFIC International 2007) 11–12; Willock, 'Uncharted Waters/Implementation Issues and Potential Benefits of Listing Toothfish in Appendix II of CITES' (n 231) 18.

seen by Australia as a major advantage of this instrument when the toothfish proposal was discussed.[510] In the whaling issue, it was also recognized that CITES was useful "to prevent illicit trade by nations the International Whaling Convention could not control."[511]

However, waters under national jurisdiction are left unregulated when the fishing vessel is flagged in the coastal state. Also, several ways to bypass trade restrictions exist in practice, such as reservations and exemptions, which also exacerbate the existing problems regarding permits' verifications at the borders.[512]

Furthermore, the structural fragmentation caused by the number of treaties adopted in relation to fisheries is obviously not solved by the use of an additional instrument. However, cooperation is central in the CITES regime with inter alia shared work programs, attendance to other bodies' meetings and invitations extended to their officers, exchange of information, as well as joint activities.[513] Those collaborative actions were already envisioned in the treaty and have been further developed through Resolutions and Memoranda of Understanding. As such, CITES could serve as a coordination mechanism in the governance of listed species. It has nevertheless been asked whether "[t]he involvement of CITES [...] could raise serious questions about the utility of RFMOs".[514] Such a perceived competition between various institutions does undermine CITES' potential for coordination.

Finally, in terms of philosophical fragmentation, CITES is arguably unlikely to improve the situation. Indeed, a strong lack of political consensus is witnessed at CoPs and the divide between pro-sustainable use and pro-preservationism, which is part of this regime's history, only shows the polarization. However, it must also be noted that the North/South divide which is usually witnessed at CITES could be less pronounced for fisheries than for terrestrial species. Indeed, since these animals straddle and migrate, adopting restrictions is less perceived as a question of developed *versus* developing countries. This and the fact that fish species are not charismatic could actually serve CITES in building up compromises in order to move away from the polarization which undermines it.

---

[510] M. Haward, 'IUU Fishing: Contemporary Practice' in A. G. Oude Elferink and D. R. Rothwell (eds), *Oceans Management in the 21st Century: Institutional Frameworks and Responses* (Martinus Nijhoff Publishers 2004) 102; Australia, 'CoP12 Prop. 12.39 – Consideration of proposals for amendment of Appendices I and II – Patagonian Toothfish' (3–15 November 2002) 2 §6–7.

[511] Bright (n 496) 841 (reference omitted).

[512] Birnie, Boyle and Redgwell (n 39) 689–690.

[513] CITES Secretariat, 'CoP16 Doc. 13 – Cooperation with Organizations and Multilateral Environmental Agreements' (3–14 March 2013); see infra Chapter 6.

[514] Little and Orellana (n 280) 118.

2. Adequacy of Conservation and Management Measures

First, a listing under CITES could improve the scientific data available which is necessary to adopt science-based CMMs. Indeed, trade and permits' information must be reported to the CITES Secretariat on an annual basis. The violation of this requirement can have severe consequences in terms of sanctions. It is nevertheless true, and accepted, that CITES itself does not possess the specialized knowledge to organize fisheries management. Hence, cooperation with the relevant fisheries organizations is necessary and can well be included in the regime.

Second, the CITES regime has a strong precautionary approach, which, in theory, is applicable to all species similarly. However, while it is not automatically triggered and can be influenced by policy arguments, it has unexpected side-effects. Indeed, a consequence of this strong precautionary principle is that it provides many states with a reason to vote in opposition to a listing. The fear that CITES is unable to down-/de-list seems in that respect to be partly justified.[515] In the case at hand, as fish products are valuable commodities a listing which brings about a permanent impossibility to trade those species even if stocks have recovered would be very problematic.

Also, fish is an important source of food having important socio-economic ramifications. Consequently, only endangered species should be targeted by restrictive measures. The look-alike criterion is thus problematic. Moreover, management of species rather than stocks is not traditional in the fisheries sector. The risk to adopt unnecessary limitations to fishing in relation to healthy stocks could be a strong disincentive to use this regime. It does not seem that the need for more flexible 'look-alike' listing and split-listing, which is "critical in increasing the efficacy of CITES as a tool for conservation of commercially-exploited aquatic species",[516] has been dealt with by now.

Third, even if the Fort Lauderdale criteria are supposed to facilitate science-based decisions, the procedure to amend Appendices remains more political than scientific.[517] Many species which should be listed are not or are under-listed—and vice versa. "Decisions on whether to provide protection for commercially exploited species often have more to do with economics than with science, underlining the

[515] FAO, 'Report of the Expert Consultation on Legal Issues' (n 4) 7 §31–33; Jenkins, 'An Overview of the Fundamental Principles of CITES as a Mechanism for Regulating Trade in Listed Species' (n 333) 8.

[516] FAO, 'Report of the Expert Consultation on Legal Issues' (n 4) 9 §43.

[517] However, it is worth noting that the arguments of both preservationists and conservationists refer to the respect or disrespect of science. For example, on the one hand, the states that wanted to allow trade in ivory complained of cultural imperialism of western states using unscientific criteria. On the other hand, "environmentalists argued that the failure to list species such as the Atlantic Bluefin tuna and the Brazilian mahogany resulted from powerful economic interests overruling sound science" (Fuchs (n 283) 1577).

inherent challenge of the Convention: species that are in most need of protection from trade are least likely to get listed because of high levels of demand."[518]

It has been said that the listing in CITES of a commercially-exploited fish species under the unsuccessful management of an RFMO would not take place because the same states act similarly in different structures on the basis of the principle of a state's unity.[519] This is partly a valid opinion but it does not take into account the different scopes of states' membership. Moreover, many RFMOs suffer from IUU fishing which is not only due to their members, who could consequently wish for support to tackle this issue. Finally, different decisions are actually taken in different fora because of the institutional cultures and the ministerial portfolio of the states' representatives.

It is interesting to note that, when the Fort Lauderdale criteria were adopted, a subject of criticism was the preponderant role of science, which risked replacing policy considerations in decision-making.[520] To the contrary, it could now be regarded that less politically influenced decisions and more science-based ones could better fulfill the goals of the regime. Furthermore, political and economic interests have impacts on other processes than amending Appendices: NDF (by the national scientific authorities) and sometimes even Significant Trade Review (by CITES bodies) are indeed also known not to be solely influenced by science.[521] These biases should also be addressed, probably as a matter of even more urgency.

Fourth, as for aspects of procedure, while the Convention provides for a two-third majority vote to amend the Appendices, the rules now state that consensus-based adoption should be preferred. This can complicate amendments in a forum with so many participants, but a back-up exists. Indeed, when adoption or rejection of the proposal is not met by consensus, the CoP moves to voting. A qualified majority can also have side-effects though, in particular as, once listed, species are potentially blocked on an Appendix, until the majority evolves and decides to down-/de-list. Furthermore, the practice of voting through secret ballot represents a threat to transparency This problem will probably not be resolved in the near future in light of the clear rejection suffered by the relevant amendment proposals at CoP16.[522] Finally, CoPs take place every two to three years, which is not a fully adequate decision-making timeframe for fish stocks, whose abundance varies much faster. Nevertheless, at the very least, the postal procedure exists.

---

[518] Sky (n 469) 35.

[519] S. Beslier, 'Pêche et biodiversité marine: complémentarité ou concurrence au sein de l'ordre juridique international? Le cas du thon rouge' (Paris November 2010) Idées pour le débat n°9, 5–6.

[520] Eldridge (n 20) 561–562.

[521] See for example the issuance of export permits for the bottlenose dolphins (E. C. M. Parsons, N. A. Rose and T. M. Telecky, 'The trade in live Indo-Pacific bottlenose dolphins from Solomon Islands: A CITES decision implementation case study' (2010) 34 Marine Policy 384). It has also been remarked that the Significant Trade Review has flaws, such as the lack of uniform standards, poor peer-reviewing, the absence of illegal trade as relevant data or the debatable choice of species to review (Reeve, *Policing International Trade in Endangered Species* (n 8) 181, 184–185).

[522] See supra Chapter 3 C. I. 2. a) and 3.

Fifth, on the positive side, the large range of values and interests represented by states parties to CITES might encourage compromise solutions where conservation is a factor, moving away from the impression that fisheries are managed only by clubs of fishing nations which pursue short-term interests. Also, whereas at RFMOs' meetings the large majority of delegates are from the fishing ministries and industry,[523] the delegates to CITES are likely to come from more varied fields of expertise.

Finally, an important improvement brought by CITES is that any adopted measure is applicable quite globally. Indeed, with 181 parties, and also being indirectly applicable to non-members in their dealings with members, its measures are wide-reaching.[524] This also means that reservations are not a major danger, as long as the party which has made them is trading with parties who have not. Then, since the regime is globally applicable, it has the potential to look into coherent areas beyond politically-defined boundaries. However, only listed species are the focus of this Convention and no ecosystem approach or even considerations are envisioned.

3. Compliance and Enforcement

First, CITES is of major interest inasmuch as it allows a better supervision and monitoring of both states' and RFMOs' decisions and actions. Indeed, rather than being exclusively based on the flag state, its measures are implemented in port. Furthermore, CITES is applicable to fishing on the high seas also when the harvest is landed in the country of the vessel's flag state, which is a major improvement compared to a TDS which only regulates international trade—as understood in a restrictive sense.

Second, with regard to RFMOs, NDF would most probably be based on the quotas that they set. Hence, a review of their decisions by CITES could be possible, either at the NDF stage or during a Significant Trade Review—whether this is actually constructive, since it can be a source of delegitimization for the RFMOs, is still to be determined.[525] In any case, a strong overview system exists: states' actions are indeed coordinated by the Convention Secretariat,[526] the parties have

[523] California Environmental Associates, 'Charting a Course to Sustainable Fisheries' (16 January 2012) http://www.chartingacourse.org/downloads/ accessed 4 July 2015, 57.

[524] CCAMLR for example pointed out that, in 2012, all states which may be involved in the harvest and/or trade of toothfish but are not cooperating with the CDS are parties to CITES (CCAMLR, 'CoP16 Doc. 63 (Rev.1) – Toothfish: Report of CCAMLR' (3–14 March 2013) 2 §7 and 'Comments from the Secretariat' B.).

[525] For a negative view on the question, see Beslier, 'Pêche et biodiversité marine: complémentarité ou concurrence au sein de l'ordre juridique international? Le cas du thon rouge' (n 519) 7.

[526] A. Batchelor, 'The preservation of wildlife habitat in ecosystems: towards a new direction under international law to prevent species' extinction' (1987–1988) 3 Florida International Law Journal 307, 331.

to report annually and NGOs such as TRAFFIC play an important role in monitoring wildlife trade.

Third, while the CITES permits are "an easy and cost-effective system of enforcement"[527] it is crucial not to undermine stricter schemes that already exist. In that regard, it seems that CDS are more efficient than TDS;[528] under TDS the traceability only starts at the point of entrance into trade, national trade is not verified and transshipment is not covered.[529] It must nevertheless be acknowledged that a CITES permit is broader than a simple TDS; as international trade includes IFS, the permitting system actually applies to fish caught on the high seas and landed for domestic purposes in the country of the vessel's flag state. The Convention hence, through this provision, regulates some national trade.[530] The only aspect really left outside of CITES' scope is fishing within one's own territorial sea or EEZ. That weakness is made worse by large domestic markets and by the free internal market of the EU (where no border controls take place within the area),[531] which is the "top global importer by value of many wild animal and plant commodities".[532]

Fourth, it must be noted that implementation and enforcement measures are mainly to be taken by the parties. They include the creation of national authorities to monitor the operations (a Management Authority and one or several Scientific Authority/ies)[533] and the delivery of permits by those authorities. As for enforcement, states must penalize violations of the Convention, seize the specimens and maintain records, as well as are allowed to determine special ports where wildlife shipments can enter.[534] In comparison to other early environmental instruments, CITES imposes clear requirements on states to use permits and to set up two bodies—obligations which are, in general, quite well respected—, provides a

[527] Franckx, 'The Exercise of Jurisdiction over Vessels: Legal Issues Raised by the Relationship Between CITES, FAO and the UN Convention on the Law of the Sea' (n 61) 66.

[528] R. Rayfuse, 'Building sustainable high seas fisheries through certification processes: issues and perspectives' (2009) 35 Océanis 93, 99.

[529] Lack, 'Catching On?' (n 509) 9.

[530] Young, *Trading Fish, Saving Fish* (n 3) 146.

[531] Engler and Parry-Jones (n 387) 12; Sand, 'Whither CITES? The Evolution of a Treaty Regime in the Borderland of Trade and Environment' (n 18) 55–56. It is however important to note that the movement of Annex A species (more or less equivalent to Appendix I species) within the EU is regulated (European Community and TRAFFIC Europe, 'Reference Guide – European Community Wildlife Trade Regulations' (n 322) 96).

[532] Engler and Parry-Jones (n 387) 6. This could be quite critical for sharks for example, as the EU is an important actor in relation to these species and has a large consumption of shark meat and of tunas, partly caught in European waters. Internal market trade (i.e. domestic trade) would only be little influenced by a listing and a small, if any, decrease in demand would hence be witnessed (S. V. Fordham, 'Shark alert: Revealing Europe's impact on shark populations' (August 2006) 5; Wijnstekers (n 106) 731–732).

[533] CITES (n 55) article IX.

[534] Ibid. article VIII(1)(3)(6), as presented in S. Lyster, *International Wildlife Law* (Grotius Publications Limited 1985) 264.

forum to meet regularly and has an overview mechanism in the form of both the Secretariat and the Standing Committee.[535]

However, one must note that compliance with CITES is undermined by lack of resources and limited specialized knowledge, in particular within developing countries. On the ground, a major challenge remains in the actual enforcement at the borders. Additionally, the national fines, required under article VIII, and the level at which the judiciary sets them in particular cases, are often not high enough to serve as a deterrent.[536] Furthermore, however useful the regime requirements are, they are only checked through compulsory annual records, which are unfortunately not well complied with. In that respect, the lack of a separate compliance committee has been considered a weakness[537] although a study commissioned in 2004 found out that the system was not suffering from this situation.[538]

Fifth, a further very useful characteristic of CITES is the existence of sanctions with potentially global applicability. One must nevertheless note that sanctions appear not to always be fairly and impartially decided; that they more easily target developing countries is not only a sign of potential discrimination[539] but also of the lack of financial support given to the states who are struggling with the implementation of the convention to help them make the required changes.[540] Also, the sanctions are not binding[541] and it is not clear how many states actually implement the recommended restrictions to trade.[542] However, the enforcement measures developed to address non-compliant states' behavior or unsustainable significant trade are generally perceived as quite efficient.[543] Indeed, suspension of trade

---

[535] Birnie, Boyle and Redgwell (n 39) 685–687; Bowman, Davies and Redgwell (n 19) 485, 533.

[536] Bowman, Davies and Redgwell (n 19) 527–528, 533; Engler and Parry-Jones (n 387) 14; Reeve, *Policing International Trade in Endangered Species* (n 8) 255–257; R. Torpy, 'If Criminal Offenses Were Added to CITES, Would Nations Be Better Able to Restrict International Trade in Endangered Species and Protect Biodiversity?' (2012) 9 Brazilian Journal of International Law 57, 68.

[537] Reeve, *Policing International Trade in Endangered Species* (n 8) 268–272.

[538] V. Koester, 'Compliance Committees within MEAs and the Desirability and Feasibility of Establishing Special Compliance Bodies under CITES' 20 May 2004, SC54 Inf. 3 (CITES 2006).

[539] Reeve, *Policing International Trade in Endangered Species* (n 8) 324; P. H. Sand, 'Enforcing CITES: The Rise and Fall of Trade Sanctions' (2013) 22 RECIEL 251, 261–262.

[540] Jenkins, 'The Significant Trade Process: Making Appendix II Work' (n 325) 53–54; Neumayer (n 490) 142; P. H. Sand, 'Le rôle des "conférences des parties" aux conventions internationales' (Le droit international face aux enjeux environnementaux, Aix-en-Provence, 4–6 June 2009) 108; CITES, 'Countries currently subject to a recommendation to suspend trade' last update on 19 March 2015 http://www.cites.org/eng/resources/ref/suspend.php accessed 6 July 2015.

[541] Reeve, 'The Convention on International Trade in Endangered Species of Wild Fauna and Flora (CITES)' (n 488) 149; G. Ulfstein, 'Dispute resolution, compliance control and enforcement in international environmental law' in G. Ulfstein (ed), *Making treaties work: Human Rights, Environment and Arms Control* (Cambridge University Press 2007) 130.

[542] Reeve, 'The Convention on International Trade in Endangered Species of Wild Fauna and Flora (CITES)' (n 488) 153.

[543] Birnie, Boyle and Redgwell (n 39) 685.

against a particular state, regarding one or all species, is a strong incentive to change one's behavior and has proven so in several cases.[544]

The sanction system is a clear illustration that, although quite old, the Convention has the ability to evolve, mostly through the adoption of Resolutions, in order to face new challenges, or to better react to old issues. Indeed, while no compliance or sanction mechanism existed in the original treaty regime, one was later developed. Nevertheless, with the difficulty to amend the Convention regime in a binding way, changes are soft law in nature and can thus be perceived by states as suggestions only. Regarding sanctions for example, the Standing Committee recommends their application, but cannot force any state to stop trading with the party found in violation of CITES.

While the risk cannot be excluded that trade sanctions would be considered as conflicting with WTO law, measures adopted within a global regime such as CITES, where consultations with the non-compliant state are formally included in the procedures leading up to sanctions,[545] are not likely to be regarded as hidden discriminatory practices. Nevertheless, the use of such sanctions should ideally be kept as rare as possible; if they were used against too many actors, compliance with the system would be at risk of delegitimization and loopholes would be created. In any case, it is worth noting that a finding of non-compliance by the Standing Committee could have effects beyond the CITES regime,[546] as it could, for example, be the basis to determine state responsibility.

Finally, at the difference of UNCLOS, CITES has no compulsory binding dispute settlement provisions. However, as using CITES puts an emphasis on the environmental side of fisheries (in opposition to the exploitative side), it could encourage the use of ITLOS in relation to Part XII for non-environmentally friendly fishing practices or facilitate the setting of standards defining which behavior respects fisheries regulations under the law of the sea.

III. Conclusions

In conclusion, the expected effects of CITES for a particular species would depend on the importance of each type of threat in its survival. For species mostly victims of by-catch, this regime would not be of much use, but it could offer additional tools to reduce the incentives and/or ability to undertake IUU fishing, especially if most of the catches are intended for international trade. Other threats linked to habitat

---

[544] Reeve, 'The Convention on International Trade in Endangered Species of Wild Fauna and Flora (CITES)' (n 488) 149.

[545] Resolution Conf. 14.3 – CITES compliance procedures 2007 in particular §20, 21, 27.

[546] E. Hey, 'International Institutions' in D. Bodansky, J. Brunnée and E. Hey (eds), *Oxford Handbook of International Environmental Law* (Oxford University Press 2007) 765–766.

destruction and/or other vulnerabilities might be taken into account by the regime, but are not its focus.

Then, the effects to be expected of a trade ban (listing in Appendix I) or of a restriction to trade (listing in Appendix II) in relation to a commercially-exploited fish species should be examined with care and on a case-by-case basis before a listing is adopted. Indeed, the use of Appendix I can only function if enforcement is available in sufficient amount; otherwise the international trade of much demanded species will simply go underground. Furthermore, due to the importance of fish species for some communities or states, the range states must feel some ownership of the decision or else it will not be implemented. As for Appendix II, it allows trade that is sustainable to continue and, as such, does not appear to need much critical scrutiny. However, the efficient implementation of these trade restrictions depends on the existence of a management plan, which some states cannot provide. In those cases, a listing in Appendix II also results in a ban, with the risks outlined above. For species that are raised in fish farms, a listing would importantly increase costs but would also provide incentives for more sustainable practices, and would put in place some regulatory framework.

As for the added value of CITES, it does have characteristics which are needed to improve the governance of fisheries. These are in particular the legal clarity it would bring to the field, its broad membership, ability to evolve, monitoring and enforcement mechanisms, global applicability and the Significant Trade Review procedure. Hence, this regime could generally contribute to the conservation and management of fish stocks, and even serve as a back-up mechanism if the relevant RFMO is facing problems in fulfilling its mandate, applying the precautionary principle, or simply suffers from inadequate geographical boundaries or membership.

However, the fears that the regime is preservationist and that science is not the basis of decisions ought to be addressed. Moreover, CITES can obviously not manage fisheries on its own, since it does not have the necessary scientific expertise or resources. There is also a need to take into account the regional particularities of species/ecosystems as well as to look beyond the protection of species from harvest triggered by international trade towards habitat protection or protection from pollution, areas in which a regional regime might be more efficient. Finally, one cannot deny that a risk of institutional rivalry exists.

As CITES has the potential to complement and support the efforts of RFMOs, it is useful to look at how the cooperation between CITES and other treaty regimes related to fisheries is organized and how it could be improved. Indeed, cooperation could allow the strengths of both CITES and RFMOs to be at play in the governance of fisheries. The examination of existing cooperative partnerships and ways of amelioration is the focus of the next chapter.

Chapter 6
CITES Cooperation with Other Institutions in Relation to Commercially-exploited Fish Species

CITES has a tradition of cooperating with other institutions, be they international organizations, MEAs secretariats or NGOs.[1] Currently, coherency and mutual supportiveness are the explicit focus of one of its Strategic Vision's goal.[2] Cooperation with biodiversity-related conventions has even been the object of a Resolution adopted at CoP16, encouraging parties to continue and strengthen such synergies.[3] In the case of marine species, the legal framework applicable is quite particular (A). Within this framework, collaborative work occurs at several stages: expertise is provided to CITES for specific developments of the regime (B), support is requested and given in relation to listing (C) and some cooperation takes place to support the implementation of the protection under CITES (D).

A. Legal Framework for Cooperation

I. CITES-FAO

1. CITES Convention

According to article XV, the CITES Secretariat has to contact the relevant institutions managing marine species when a proposal concerning them is brought forward. CITES fulfilled this obligation, but did not receive answers from the FAO until 2003. Indeed, "[i]n the past, when CITES ha[s] contacted FAO under Article

[1] For a recent summary, see for example CITES Secretariat, 'CoP16 Doc. 13 – Cooperation with Organizations and Multilateral Environmental Agreements' (3–14 March 2013).

[2] Resolution Conf. 16.3 – CITES Strategic Vision: 2008–2020 2013 goal 3.

[3] Resolution Conf. 16.4 – Cooperation of CITES with other biodiversity-related conventions 2013.

S. Guggisberg, *The Use of CITES for Commercially-exploited Fish Species*, Hamburg Studies on Maritime Affairs 35, DOI 10.1007/978-3-319-23702-2_6

15, FAO has declined to comment because it had not been given a mandate by its members to do so."[4]

In 1998, at the meeting of the Sub-Committee on Fish Trade of COFI, the FAO was requested by a South African proposal to get involved in CITES matters, particularly regarding the potential proposals to amend Appendices and the criteria used to do so.[5] It was only five years later that COFI gave a mandate to the FAO to cooperate with CITES in the assessment of proposals related to marine species and adopted terms of reference thereabout.

Article XV's scope and application will be discussed in detail below since it still provides the framework applicable to the relationship between CITES and RFMOs.[6] For the partnership between CITES and the FAO, the MoU is a more detailed framework which incorporates the obligations included in article XV.

As for article XIV, it is not applicable because the FAO is not an institution which yields actual protection to species. Hence, this provision does not offer a framework for states which are members of the FAO and have also ratified CITES.

2. MoU on CITES-FAO Cooperation

Following years of ad hoc cooperation, a more formal structure was put in place between CITES and the FAO. After long negotiations[7] an MoU was indeed adopted in 2006 by the Sub-Committee on Fish Trade of COFI and signed by

[4] FAO, 'Report of the FAO ad hoc Expert Advisory Panel for the Assessment of Proposals to Amend Appendices I and II of CITES Concerning Commercially-exploited Aquatic Species (Rome 13–16 July 2004)' FAO Fisheries Report 748 (Rome 2004) 13 appendix C 'Welcome speech by Mr. Ichiro Nomura, Assistant Director-General, FIO Fisheries Department'.

[5] Committee on Fisheries – Sub-Committee on Fish Trade, 'Report of the Sixth Session of the Sub-Committee on Fish Trade (Bremen, 3–6 June 1998)' FAO Fisheries Report R589 (Rome 1998) §12. At the following COFI meeting, it was agreed that the FAO's work regarding the CITES criteria should be continued (Committee on Fisheries, 'Report of the Twenty-third Session of the Committee on Fisheries (Rome, 15–19 February 1999)' FAO Fisheries Report 595 (Rome 1999) §78). On a general overview of the history of the early partnership, see Committee on Fisheries – Sub-Committee on Fish Trade, 'Item 5 of the Provisional Agenda – Historical involvement of COFI within the process of amendment to the CITES listing criteria', presented at the Eighth Session of the Sub-Committee on Fish Trade (Bremen, 12–16 February 2002) COFI: FT/VIII/2002/Inf.6 (2002) 6–8 §15-22; E. Franckx, 'The Protection of Biodiversity and Fisheries Management: Issues Raised by the Relationship between CITES and LOSC' in D. Freestone, R. Barnes and D. Ong (eds), *The Law of the Sea – Progress and Prospects* (Oxford University Press 2006) 214–215.

[6] See infra Chapter 6 A. II. 2.

[7] M. A. Young, *Trading Fish, Saving Fish: The Interaction between Regimes and International Law* (Cambridge Studies in International and Comparative Law, Cambridge University Press 2011) 158–169.

both organizations in the same year at the 54th CITES Meeting of the Standing Committee.[8]

According to this document, the FAO and CITES will "communicate and exchange information regularly and bring to each other's attention general information of common interest." They are also to be invited at each other's meetings when topics of relevance are on the agenda.[9] At CITES meetings, such as the CoP, Animals Committee or Standing Committee, a FAO delegation is regularly present and contributes by providing information on activities undertaken in areas of common interest and by taking part in the relevant working groups. As for attendance to the FAO meetings, experts from CITES are invited to join during workshops and at COFI. At the meetings of the Sub-Committee on Fish Trade of COFI, CITES-related items are usually on the agenda. Also, officers from both institutions join workshops and meetings organized by the other institution when relevant.

Then, another common goal is capacity building in developing and transition countries, with regard to issues related to commercially-exploited aquatic species which are listed under CITES.[10]

Last but not least, it is planned for the FAO to provide advice on listing criteria and on proposed amendments of Appendices;[11] these activities, in fact, already took place prior to the signature of the MoU. Noteworthy is that "[i]n order to ensure maximum coordination of conservation measures, the CITES Secretariat *will respect, to the greatest extent possible*, the results of the FAO scientific and technical review of proposals to amend the Appendices, and technical and legal issues of common interest and the responses from all the relevant bodies associated with management of the species in question."[12] While this is no firm obligation since the qualification 'to the greatest extent possible' has been included, this is nevertheless an important undertaking on the CITES Secretariat's side.

This wording 'will respect' was not consensual. On the one hand Japan, Australia and Iceland (on behalf of Europe) made clear that the use of 'respect' was crucial for them (in opposition to the proposals to use 'consider' or 'take into account'), in order to reflect the equality of the two institutions entering into the partnership. On the other hand, Mexico highlighted a discrepancy between the Spanish and the English draft versions, where the former referred to 'take into account' and should be preferred. At the end, the Spanish text was changed to reflect the wording of 'respect',[13] but the terms 'to take due account' of the

[8] WECAFC, 'WECAFC Lesser Antilles Fisheries Committee – An overview of the FAO activities in relation to CITES and commercially-exploited species' WECAFC/XIII/08/10 (21 October 2008) ftp://ftp.fao.org/docrep/fao/meeting/014/aj252e.pdf accessed 5 July 2015, 2 §6.

[9] Memorandum of Understanding between the Food and Agriculture Organization of the United Nations (FAO) and the Secretariat of the Convention on International Trade in Endangered Species (CITES) (CITES-FAO MoU) 2006 §1.

[10] Ibid. §2.

[11] Ibid. §3-4.

[12] Ibid. §6 (italics added).

[13] Standing Committee, 'SC53 Summary Record (Rev. 1) – Summary Record of Fifty-third meeting of the Standing Committee' (2005) 5–6 §10.

Panel's views were still retained in another paragraph, which is not of the clearest effect.[14]

A preamble to the MoU was first envisioned, but since it was difficult to agree on it, it was removed.[15] These paragraphs were an attempt to define respective responsibilities,[16] clearly stating the FAO and RFMOs' primary role in the field of fisheries;[17] this however did not reflect all states' views.[18]

3. Closer Cooperation in Relation to Sharks

The relationship between CITES and the FAO is more intertwined for shark species than for other commercially-exploited aquatic species. Nevertheless, the legal framework for cooperation is the same MoU as for other fish species.

CITES and the FAO have been both very involved in the development of regulations on sharks, as no institution has the exclusive or even primary responsibility in the management of these species. Competences and programs of work are interconnected—and acknowledged as such. On the one hand, in 1994, the CITES parties called for the development of data gathering programs related to sharks and requested the FAO's involvement therefor.[19] On the other hand, in 1999, the FAO adopted the comprehensive IPOA-Sharks and, in Resolution 12.6 on *Conservation and Management of Sharks*, the CITES CoP urged states which have not yet done so to adopt a national plan of action, hence to comply with the IPOA.[20] It also urged COFI to further continue the necessary research for the implementation of the plan of action.[21] This last call, directed at COFI, another international body, is quite

---

[14] E. Franckx and K. van den Bosche, 'The Influence of Environmental Law on the Development of the Law of the Sea: CITES and the International Law of Fisheries' (2011) 54 Japanese Yearbook of International Law 218, 254; M. A. Young, 'Protecting Endangered Marine Species: Collaboration Between the Food and Agriculture Organization and the CITES Regime' (2010) 11 Melbourne Journal of International Law 1, 44–45.

[15] Standing Committee, 'SC53 Summary Record (Rev. 1)' (n 13) 5–6 §10; Committee on Fisheries – Sub-Committee on Fish Trade, 'CITES issues with respect to international fish trade and the CITES/FAO MoU', presented at the Tenth Session of the Sub-Committee on Fish Trade (Santiago de Compostela, 30 May – 2 June 2006) COFI:FT/X/2006/3 (2006) 4 §13-14; Young, *Trading Fish, Saving Fish* (n 7) 168.

[16] Standing Committee, 'SC53 Summary Record (Rev. 1)' (n 13) 6 §10.

[17] See for example the draft version presented in 2004 by Japan: Japan, 'CoP13 Doc. 12.4 – Cooperation with other organizations – Cooperation with the Food and Agriculture Organization of the United Nations' (2–14 October 2004) annex.

[18] Committee on Fisheries – Sub-Committee on Fish Trade, 'CITES issues with respect to international fish trade and the CITES/FAO MoU', presented at the Tenth Session of the Sub-Committee on Fish Trade (n 15) 4 §14.

[19] Resolution Conf. 9.17 – Status of International Trade in Shark Species 1994 (repealed).

[20] Resolution Conf. 12.6 (Rev. CoP16) – Conservation and management of sharks (Class Chondrichthyes) 2002.

[21] Ibid.

noteworthy since it shows how, for sharks matters, CITES feels entitled to urge an independent meeting of states to undertake supplementary research. This is a rare occurrence of the cooperation taking place in this direction, with CITES parties pointing out aspects which need to be developed or improved by the FAO.

In conclusion, the cooperation between the two institutions in relation to sharks is more complex and overlapping than for other fish species, with the CITES CoP even making recommendations to the FAO. This situation is most probably due to the lack of a management institution which specifically targets sharks species. As a result, both the FAO and CITES, trying to fill in the protection gap, interact quite closely.

4. Legal Status of the MoU

It is well accepted that international organizations can enter into MoUs,[22] even though some scope for debate remains on the need for legal personality to do so, on the legal nature of such instruments and on the consequences of them being, or not, binding.[23]

An important aspect to consider in discussing the nature of the MoU between the FAO Fisheries and Aquaculture Department and the CITES Secretariat is the legal status of its participants and the mandate given to them by their founding treaties. Indeed, it is recognized that international organizations can have legal personality and enter into legally-binding obligations through treaties, if provided for by their founding instrument, or if necessary for the fulfilment of their mandates, through the implied powers doctrine developed by the ICJ.[24]

The signatories here are on the one hand a department of an international organization, the FAO, which itself has an international legal personality[25]—but this is not the case for the departments—and, on the other hand, the Secretariat of an

[22] J. Klabbers, *An introduction to international institutional law* (Cambridge University Press 2009) 251, 267. Indeed, depending on the particular rules of the organizations, they can enter into treaties ('Vienna Convention on the Law of Treaties between States and International Organizations or between International Organizations' 1986 A/CONF.129/15 (not yet in force) article 6).

[23] Young, *Trading Fish, Saving Fish* (n 7) 155–156; in favor of non-binding nature are for example R. Wolfrum and N. Matz, *Conflicts in international environmental law* (Beiträge zum ausländischen öffentlichen Recht und Völkerrecht, Springer 2003) 173, while R. R. Churchill and G. Ulfstein leave this to be determined in each particular case ('Autonomous Institutional Arrangements in Multilateral Environmental Agreements: A Little-Noticed Phenomenon in International Law' (2000) 94 American Journal of International Law 623, 651).

[24] A. Aust, *Modern Treaty Law and Practice* (Cambridge University Press 2000) 54; Klabbers, *An introduction to international institutional law* (n 22) 251–252. On the implied powers doctrine, see supra Chapter 3 B. II. 1.

[25] 'Constitution of the Food and Agriculture Organization of the United Nations' 1945 article XVI (1).

MEA in relation to which reference to legal personality has yet to be formally made.[26]

It is generally agreed that institutions created to structure a treaty's implementation are not de jure international organizations and miss some of their characteristics, mainly the international legal personality, which is related to the independence from the states parties to the treaty.[27] However, it has been questioned whether institutions created under CITES might enjoy legal personality on the basis of implied powers, even though it would still not clarify "whether the legal personality resides in the CITES Secretariat, the COP itself or even the international organisation hosting the CITES Secretariat, namely UNEP."[28]

All depends on how the institutions created by MEAs are considered and how legal personality is defined. Schermers and Blokkers acknowledged that "the formal distinction between organizations and organs is far from clear-cut in practice."[29] According to Churchill and Ulfstein, most CoPs have legal personality either by explicit mention of their functions in the treaty or by implied powers stemming from "catch-all phrases relat[ing] to what is necessary for the effective implementation of the MEAs".[30] As seen above, such powers were recognized by the ICJ for international organizations and, because the CoP's structure is very similar to that of an international organization, a parallel reasoning could apply.[31] However, such an approach is debatable since, as representative bodies of the member states, CoPs do not appear to have a will of their own. It is also worth noting that, in the present case, the CITES Convention does not include a provision allowing for an open interpretation of the CoP's competence to enter into binding arrangements with other institutions.

As for the Secretariats, while they work under the mandate given to them by the treaty and the CoP, some of them might benefit from a large textual margin of maneuver.[32] The CITES Secretariat's functions are quite well defined in the

[26] CITES Secretariat, 'SC54 Doc. 8 – Strategic and administrative matters – Legal personality of the Convention and the Secretariat' (2–14 October 2006) 1 §7.

[27] H. G. Schermers and N. Blokker, *International institutional law: Unity within diversity* (5th edn Nijhoff 2011) §38, 44; Wolfrum and Matz, *Conflicts in international environmental law* (n 23) 164; for the characteristics of international organizations, see supra Chapter 3 B. II. 1.

[28] Young, *Trading Fish, Saving Fish* (n 7) 157 (reference omitted).

[29] Schermers and Blokker (n 27) §38.

[30] Churchill and Ulfstein (n 23) 647–649.

[31] P. H. Sand, 'Le rôle des "conférences des parties" aux conventions internationales' (Le droit international face aux enjeux environnementaux, Aix-en-Provence, 4–6 June 2009) 102–103; G. Ulfstein, 'Treaty Bodies' in D. Bodansky, J. Brunnée and E. Hey (eds), *Oxford Handbook of International Environmental Law* (Oxford University Press 2007) 881, 886.

[32] For example 'United Nations Framework Convention on Climate Change' 1992, 1771 *United Nations Treaty Series* 107 article 8(2)(f), as presented in Churchill and Ulfstein (n 23) 649.

treaty,[33] but the Executive Director of the UNEP, who provides the Secretariat, may be, “[t]o the extent and in the manner he considers appropriate, [. . .] assisted by suitable inter-governmental or non-governmental international or national agencies and bodies technically qualified in protection, conservation and management of wild fauna and flora.”[34] Nothing is said as to the nature and form such assistance may take.

Also, the Secretariat is “to perform any other function as may be entrusted to it by the Parties”,[35] a very general statement which could serve as a basis to enter into arrangements if so is the wish of the member states. Moreover, it shall perform the tasks listed in article XV of the Convention, in other words, consult with inter-governmental bodies related to marine species[36]—this could require a formal structure to enable swift consultations. It is worth noting that the Secretariat itself “has always operated under the assumption that is possesses both the legal capacity and international personality (as an autonomous treaty body) needed to perform its functions under the Convention.”[37] However, none of the duties or functions here mentioned strictly requires the ability to enter into binding agreements, as the cooperative activities envisioned are of a technical nature.

The MoU is indeed a technical agreement—its content aims at cooperation on specific points where the regimes’ competences overlap—and, like the majority of such instruments, is hence of a non-binding nature. In particular, the use of ‘the signatories will’ instead of, for example, ‘the parties shall’ is to be noted as a sign of non-binding intent. Other opinions have been voiced, based for example on the existence of a quasi-termination and amendment clause, which could show the binding nature of a document.[38] However, such provision might also have been inserted due to the awareness of the signatories that the partnership could need to evolve and that it would consequently be beneficial to establish procedures for such potential occurrences.

In conclusion, from the nature of its signatories and its wording, the MoU examined here is clearly non-binding. In any case, because of the soft-law wording chosen by the signatories in the actual provisions, the theoretical legal nature of the MoU has no real consequence. Indeed, since the commitments are qualified through the use of ‘as appropriate’ or ‘to the greatest extent possible’, the signatories could hardly be found in non-compliance of this instrument.

[33] ‘Convention on International Trade in Endangered Species of Wild Fauna and Flora’ (CITES) 1973, 993 *United Nations Treaty Series* 243 article XII(2).

[34] Ibid. article XII(1).

[35] Ibid. article XII(2)(i).

[36] Ibid. articles XII(2)(b) and XV(2)(b).

[37] CITES Secretariat, ‘SC54 Doc. 8 – Strategic and administrative matters – Legal personality of the Convention and the Secretariat’ (n 26) 1 §6; R. Caddell, ‘Inter-Treaty Cooperation, Biodiversity Conservation and the Trade in Endangered Species’ (2013) 22 RECIEL 264, 269.

[38] Churchill and Ulfstein (n 23) 651; Young, *Trading Fish, Saving Fish* (n 7) 170–171, 186.

5. Additional General Remarks

Regarding the scope of the cooperation, as already mentioned,[39] the FAO's understanding of 'commercially-exploited aquatic species' for CITES activities includes all marine and large freshwater bodies fish and invertebrate species. Hence it only avoids looking at mammals, birds, reptiles and plants. This mandate is wider than the typical one of the Fisheries and Aquaculture Department of the FAO. Indeed, for CITES activities, ornamental fish are taken into account, while they usually fall outside of the FAO's mandate.

One must finally note that many of the activities undertaken within the 'CITES and commercially-exploited aquatic species' program have been financed by the Japanese Trust Fund project, with a first project until 2010 and a second one, foreseen to run until 2015.[40] Although the Sub-Committee on Fish Trade has recommended that CITES activities should be funded through the Regular Program, little or no core budget is dedicated to this work and only recently have a few countries other than Japan, such as Germany, provided special funding for this project, and the EU for the implementation of the sharks and manta rays listing.[41]

II. CITES-RFMOs

1. Through FAO

As mentioned in Chapter 3, the FAO has no direct role in the management of commercially-exploited aquatic species and only has an indirect overview power over the RFMOs that are entrusted with such a management mandate.

---

[39] See supra Chapter 1.

[40] FAO/Government Cooperative Programme, 'Cites and Commercially-exploited Aquatic Species, Including the Evaluation of Listing Proposals – Terminal Report – Japan – Project Findings and Recommendations' FI:GCP/INT/987/JPN (Report prepared for the Government of Japan by the Food and Agriculture Organization of the United Nations, Rome 2011); FAO/Government Cooperative Programme, 'CITES and Commercially-exploited Aquatic Species, Including the Evaluation of Listing Proposals (Phase 2) – Project Document 2010–2015: GCP/INT/104/JPN'.

[41] Committee on Fisheries – Sub-Committee on Fish Trade, 'Report of the Ninth Session of the Sub-Committee on Fish Trade (Bremen, 10–14 February 2004)' FAO Fisheries Report 736 (Rome 2004) 7 §20; Committee on Fisheries, 'Decisions and Recommendations of the Thirteenth Session of the COFI Sub-Committee on Fish Trade, Hyderabad, Hyderabad, 20–24 February 2012' Thirtieth Session (Rome, Italy, 9–13 July 2012) COFI/2012/4 5 §37; CITES, 'Ongoing projects and activities: EU project' http://cites.org/eng/prog/shark/projects.php accessed 5 July 2015; CITES, 'Strengthening capacity in developing countries for sustainable wildlife management and enhanced implementation of CITES wildlife trade regulations, with particular focus on commercially-exploited aquatic species: Project summary sheet' http://cites.org/sites/default/files/eng/prog/shark/docs/eu_project_summary.pdf accessed 5 July 2015.

While, originally, it was desired that the role of the FAO in liaising with RFMOs be addressed in the MoU,[42] its current version makes no clear mention of the need, for the FAO, to invite RFMOs to the scientific meetings related to the amendments of Appendices or the modification of criteria. Indeed, the only reference to such bodies is in paragraph 6, where the CITES Secretariat is called upon to "respect, to the greatest extent possible, the results of the FAO [. . .] and the responses from all the relevant bodies associated with management of the species in question." Such an enumeration of the sources of information that the CITES Secretariat should respect hints at that it is not the responsibility of the FAO to gather information from the RFMOs, but rather that of the CITES Secretariat. Indeed, the wording does not suggest an inclusion of the relevant regional bodies' recommendations into the FAO report.

However, the terms of reference for the Expert Panel, which were agreed upon at COFI in 2003, clarify that information received from the RFMOs is to be considered. They also envision that it is up to the FAO to send the proposals to the FAO members and to RFMOs as well as to inform them of the deadline to receive comments.[43] This document precedes the adoption of the MoU and is still used for the Panels; hence it seems that the FAO has agreed to assemble the information on listing proposals, including the one coming from other international institutions. As the FAO is to undertake the review of proposals "in a manner it deems appropriate",[44] it can in practice inform RFMOs and consider their comments, if any, on the amendments' proposals. When relevant, representatives from RFMOs are invited to the Expert Panel meetings as observers.

2. Directly

The CITES regime provides a framework for the partnership with RFMOs in the Convention text in relation to IFS, to overlapping competences and to provision of advice at the listing stage, as well as in Resolutions and in guidelines.

a) CITES Convention

aa) For Listing

According to the Convention, in the case of amendments being proposed between two CoPs, in other words when the postal procedure is used, consultation by the

---

[42] United States of America, 'CoP12 Doc. 16.2.2 – Cooperation with other organizations – CITES and FAO: FAO collaboration with CITES through a Memorandum of Understanding' (3–15 November 2002) 'Comments from the Secretariat' B.1.

[43] Terms of Reference, as found for example in FAO, 'Report of the FAO ad hoc Expert Advisory Panel for the Assessment of Proposals' (n 4) appendix D 15–16. When the same information is relevant for all Expert Panels, only the 2004 one will be referred to.

[44] CITES-FAO MoU (n 9) §5.

Secretariat with "inter-governmental bodies having a function in relation to those species" is required.[45] Such bodies include RFMOs, which have conservation and management functions in relation to fish species, but arguably also, more generally, the FAO and Secretariats of MEAs such as the CMS.

It is noteworthy that the wording of article XV differs from that of article XIV, which refers to "any other treaty, convention or international agreement [. . .] under the provisions of which protection is afforded to marine species".[46] Indeed, article XIV mentions a treaty while article XV refers to a body. Moreover, the former provision's focus appears to be restricted to treaties affording protection, while the latter potentially encompasses all institutions with a function in relation to marine species, hence even the ones with an advisory role only.

In particular, the consultation with such institutions, as provided for in article XV (2) (b), shall aim at gathering scientific data and "ensuring co-ordination with any conservation measures enforced by such bodies." Here also, a close look at the choice of words is necessary. Indeed, coordination is needed with regard to conservation measures, which makes one wonder why management measures have not also been mentioned. This omission could suggest that since CITES is a conservation treaty, efforts on the CITES side to efficiently coordinate its trade restrictions with measures taken by other bodies should be limited to those of a conservatory nature—for example when quotas are set with precaution in mind, and only when they are actually enforced. This interpretation can find support in ITLOS which emphasized that "conservation and management cover different aspects".[47]

The procedure to be followed for amendments proposals at the CoP itself is provided for in paragraph 1 of article XV. It indicates the need to consult with 'interested bodies', "in accordance with the provisions of the sub-paragraphs (b) and (c) of paragraph 2", that is, for marine species, the steps presented here above. This cross-reference to a later paragraph and to a little used extraordinary procedure is somewhat confusing.[48] Nevertheless, a reading of the provision makes clear that article XV creates, with regard to proposals related to marine species, an additional obligation for the Secretariat whether a normal or postal procedure is used.[49]

Moreover, it is clear from the Resolution on *Criteria for amendment of Appendices I and II* that CITES parties do not envision a differentiation between the procedures to be followed at the CoP or between CoPs. Their practice in that respect is indeed presented in the Preamble:

---

[45] CITES (n 33) article XV(2)(b).

[46] Ibid. article XIV(4).

[47] *M/V "Virginia G" (Panama/Guinea-Bissau), Judgment,* 14 April 2014, ITLOS case n°19 62 §212.

[48] W. Wijnstekers, *The Evolution of CITES* (9th edn CIC – International Council for Game and Wildlife Conservation 2011) 459.

[49] D. S. Favre, *International Trade in Endangered Species: A Guide to CITES* (Martinus Nijhoff Publishers 1989) 312; A. Gillespie, *Conservation, biodiversity and international law* (Edward Elgar 2011) 96.

> RECOGNIZING that [. . .] all Parties shall be consulted by the Secretariat in accordance with Article XV, paragraph 1 (a), of the Convention;
> RECOGNIZING further that, in accordance with the same Article, the Secretariat shall consult intergovernmental bodies having a function in relation to marine species;[50]

The first Preambular refers to paragraph 1, while the second mentions the actual content of paragraph 2 (b), without hinting to the existence of such a bizarre cross-reference.

bb) For Non-detriment Finding

If a specimen is introduced from the sea, a certificate is to be issued on the basis of an NDF. Since it will be a specimen caught in waters not under the jurisdiction of any state, such finding can require collaboration with other states, or with an international body relevant to this species.

Article IV (7) of the Convention provides in that regard that "[c]ertificates [. . .of IFS] may be granted on the advice of a Scientific Authority, in consultation with other national scientific authorities or, when appropriate, international scientific authorities, in respect of periods not exceeding one year for total numbers of specimens to be introduced in such periods." It refers to internationally agreed annual quotas, on the basis of which the NDF could be made. The international scientific authority of reference could be a specialized body such as ICES, the scientific committee of an RFMO or even a FAO Panel—however this has not been envisioned in the partnership.

It is interesting to note that no such reference to an 'international scientific authority' is found in the article applicable to trade in Appendix I species. Taking into account that trade in Appendix I species will be a rare occurrence (as opposed to a regular one for Appendix II species), this omission appears reasonable. It also seems to be voluntary, as the same wording is used in an operative part of the Resolution on *Introduction from the Sea*, with explicit reference to Appendix II species only.[51]

Similarly, the Convention does not mention international scientific authorities in relation to specimens not introduced from the sea, but submitted to the normal import/export permits. This silence is probably due to the long-lasting—and only recently revoked—assumption that all specimens caught in the high seas would, at some point, be introduced from the sea. To ensure a coherent and efficient regime, the Resolution on *Introduction from the Sea* recommends that "the Scientific Authority of the State of export, in making its non-detriment finding, consult with other national scientific authorities or, when appropriate, international scientific authorities".

[50] Resolution Conf. 9.24 (Rev. CoP16) – Criteria for amendment of Appendices I and II (Resolution Conf. 9.24 (Rev. CoP16) – Lauderdale criteria) 1994 preamble.

[51] Resolution Conf. 14.6 (Rev. CoP16) – Introduction from the Sea 2007 operative part.

cc) To Avoid Implementation Overlap

CITES envisions practical cooperation with regimes already existing at the time it entered into force, in order to avoid overlapping implementation measures. As discussed in Chapter 5, according to article XIV (4)–(5) which applies to marine species, states which were parties on 1 July 1975 to another treaty protecting a species listed in Appendix II, such as one founding an RFMO, are relieved of applying the CITES permitting system for specimens of the relevant species.[52]

In that case the state of introduction will only have to issue a certificate assuring that the "specimen was taken in accordance with the provisions of the other treaty, convention or international agreement in question."[53] This provision creates a simplified mechanism for the implementation of the protection system for Appendix II species already protected by another regime. It hence provides for automatic overlap avoidance in specific situations.

b) Existing Frameworks

aa) CCAMLR

For the 2002 CITES CoP, Australia, a member of CCAMLR, sent to the CITES Secretariat a proposal to list Patagonian Toothfish in Appendix II. This proposal caused many debates as to the respective roles of CITES and RFMOs, the nature of their relationship and the suitability of CITES for marine species. A Resolution on the *Cooperation between CITES and the CCAMLR regarding trade in toothfish* was adopted following the withdrawal of the Australian listing proposal. It envisions exchange of information and recommends a broadening of the CCAMLR CDS to all CITES parties.

The Resolution indeed 'recommends' parties to adopt the CDS, 'invites' states and the FAO to provide relevant information to the CCAMLR Secretariat and 'encourages' CCAMLR to share information of common interest. It also 'urges' parties to further study trade in toothfish and cooperate with the Secretariat in the gathering of such information as well as 'requests' the Secretariat to transmit information on illegal trade to the CCAMLR Secretariat.

Except for these last two demands, one of which is directed at its own Secretariat and hence binding,[54] and the other one which is vague, the ordinary meanings of the other calls for action hint at a non-binding intent—particularly when the CoP directs these calls to relationships involving third parties such as between the

[52] See supra Chapter 5 B. III. 1.

[53] CITES (n 33) article XIV(5).

[54] Indeed, under the Convention, the Secretariat shall "perform any other function as may be entrusted to it by the Parties" (CITES (n 33) article XII(2)(i)).

FAO and the CCAMLR Secretariat.[55] Also, later practice showed that neither the exchange of information from the parties nor the broadening of the CDS mechanism to all CITES parties was successfully implemented.[56]

Consequently, the CITES parties are under no binding obligation regarding toothfish, and only the Secretariat has to transmit information to the CCAMLR Secretariat.

bb) ICCAT

While the adoption of a Resolution on CITES and ICCAT was proposed at CoP15, it was rejected together with the listing of Atlantic Bluefin Tuna. Since part of the draft Resolution was linked to a special down-listing procedure—only applicable if the species was listed—this rejection was to be expected.[57] This document also intended to direct "the Secretariat to enter into negotiations with ICCAT in order to develop a Memorandum of Understanding between both conventions".[58] The Secretariat commented on this draft Resolution by highlighting that it would only have value if the Bluefin Tuna was listed and that the MoU should go beyond mere exchange of information.[59]

Regarding this call for further cooperation between the institutions, no Resolution or Decision has been adopted to provide a framework to the development of an MoU. However, *Guidelines for Cooperation between ICCAT and the CoP of CITES* have been adopted, respectively in November 2011 and in July 2012.[60] They have been endorsed by the contracting parties of ICCAT and by the parties of CITES.

---

[55] To recommend is defined as to "advise as a course of action"; to invite is defined as to "request (something) formally or politely from someone"; to encourage is defined as to "give support [. . .] to", "help or stimulate the development of" (C. Soanes and A. Stevenson (eds), *Concise Oxford English Dictionary* (11th edn Oxford University Press 2008)). Note however that D. S. Calley considers that Resolution 12.4 actually "oblig[es] CITES parties to apply the CDS" (*Market Denial and International Fisheries Regulation: The Targeted and Effective Use of Trade Measures Against the Flag of Convenience Fishing Industry* (Martinus Nijhoff Publishers 2012) 180).

[56] CITES Secretariat, 'CoP13 Doc. 36 (Rev. 1) – Conservation of and trade in Dissostichus species' (2–14 October 2004) 1 §3; CCAMLR, 'CoP14 Doc. 61 – Toothfish: Report of CCAMLR' (3–15 June 2007) 2 §8-9; CCAMLR, 'CoP16 Doc. 63 (Rev.1) – Toothfish: Report of CCAMLR' (3–14 March 2013) 1 §4.

[57] Monaco, 'CoP15 Prop. 19 – Proposal to include Atlantic Bluefin Tuna (*Thunnus thynnus* (Linnaeus, 1758)) on Appendix I of CITES in accordance with Article II 1 of the Convention' (13–25 March 2010) 3 summary §17; Monaco, 'CoP15 Doc. 52 (Rev. 1) – Species Trade and Conservation, Atlantic Bluefin Tuna, Draft Resolution of the Conference of the Parties' (13–25 March 2010) operative part.

[58] Monaco, 'CoP15 Doc. 52 (Rev. 1) – Species Trade and Conservation, Atlantic Bluefin Tuna, Draft Resolution of the Conference of the Parties' (n 57) operative part.

[59] Ibid. 'Comments from the Secretariat' A.

[60] CITES Secretariat, 'CoP16 Doc. 13 – Cooperation with Organizations and Multilateral Environmental Agreements' (n 1) 6 §35; ICCAT – CITES CoP, 'SC62 Doc. 14.6 – Annex Guidelines for cooperation between the International Commission for the Conservation of Atlantic Tunas (ICCAT) and the Conference of the Parties to the Convention on International Trade in

The contemplated cooperation includes invitation at meetings of common interest as well as communication of relevant information on species and issues of common interest.[61] However, exchange of information is limited in scope, as, unless the Secretaries decide otherwise in consultation with the parties, only information published by the said institutions is to be shared.[62]

Another area of cooperation is the exchange of information to increase the mutual understanding about each other's activities and goals,[63] which is undoubtedly meaningful after the tense relations surrounding the proposals to list Atlantic Bluefin Tuna. The Secretariats are also "encouraged to facilitate communication, collaboration, and information exchange among national representatives to ICCAT and national CITES Authorities as feasible and appropriate."[64] Such promotion of cooperation at the national level can be most helpful to increase coherency between institutions and consequently mutual supportiveness.

The cooperation to evaluate listing proposals, to address technical and legal issues linked to such listing and to implement the protection, which was included in an earlier draft version,[65] has been removed. Similarly, the 2011 draft recognized the shared objective of CITES and ICCAT regarding the conservation of tunas and tuna-like species,[66] but this statement did not make it to the adopted version. It is nevertheless worth mentioning that the CITES Secretariat mentioned, together with the information on the endorsement of the Guidelines, that "[a]s provided in Article XV of the Convention, ICCAT and other inter-governmental bodies having a function in relation to marine species will be consulted by the Secretariat on any amendment proposals involving such species".[67]

As for the legal nature of the Guidelines, at least one of two bodies who endorsed them could agree to binding commitments under international law: ICCAT is indeed an international organization. This is less clear for the CITES CoP, which, as examined above, might dispose of a legal personality on the basis of implied powers but probably does not. Moreover, a quasi-termination and amendment

---

Endangered Species of Wild Fauna and Flora (CITES)' http://www.cites.org/common/com/sc/62/E62-14-06-A.pdf accessed 5 July 2015; CITES Secretariat, 'SC62 Sum. 2 (Rev. 1) – 62nd meeting of the Standing Committee – Executive Summary 23.07.2012 Afternoon' (23 July 2012) 2 §14.6.; CITES Secretariat, 'SC62 Doc. 14.6 – Cooperation with other organizations – International Commission for the Conservation of Atlantic Tuna' (23 July 2012) recommendation §8.

[61] ICCAT – CITES CoP, 'SC62 Doc. 14.6 – Annex Guidelines for cooperation between the ICCAT and the CoP to CITES' (n 60) §1-2.

[62] Ibid. §2.

[63] Ibid. §3-4.

[64] Ibid. §5.

[65] CITES Secretariat, 'SC61 Doc. 15.7 – Cooperation with other organizations – International Commission for the Conservation of Atlantic Tunas: Proposed guidelines for cooperation between the ICCAT and CITES' (15 August 2011) annex.

[66] Ibid.

[67] CITES Secretariat, 'CoP16 Doc. 13 – Cooperation with Organizations and Multilateral Environmental Agreements' (n 1) 6 §35.

clause has been included. However, the CITES CoP and ICCAT parties did not actually sign this instrument (it was endorsed only), they agreed to highly qualified undertakings ('is/are encouraged to'; 'as necessary and appropriate', 'may include'), and the final paragraph mentions that the 'Guidelines will become operative' as well as 'unless discontinued'; all those are strong indicators of an intention not to formally be bound. These guidelines are consequently to be treated as a document promoting technical cooperation and mutual understanding, but not as creating hard obligations on the CITES CoP or ICCAT.

cc) IWC

The CITES-IWC relations are determined by the Convention and by Resolution 11.4 on the *Conservation of cetaceans, trade in cetacean specimens and the relationship with the International Whaling Commission.*

The provisions of the Convention relevant to marine species were drafted with the particular relationship between CITES and the IWC in mind. Hence, it is clear that articles XIV and XV, the content and nature of which have been discussed here above,[68] are applicable to and binding on the partnership between CITES and the IWC.

As mentioned in Chapter 5, the history of CITES and the IWC has been intertwined for a number of decades.[69] At CoP4, CITES harmonized Appendices with the IWC by listing in Appendix I all species protected by the IWC. Following the 1986 moratorium, the IWC requested the listing under CITES Appendices of all cetaceans species under its management. CITES parties agreed to this move. Since then, it has kept the status quo in terms of decisions.

The IWC's central position regarding whales has been recognized early on by the CITES community.[70] This role takes several forms, ranging from provision of advice for the amendment of Appendices to broadening the scope of application of IWC decisions to all CITES members by tightly coordinating measures.[71] Resolution 11.4 makes soft recommendations to CITES parties to join the IWC, to refrain from issuing permits, to cooperate in preventing and monitoring illegal trade in whales, as well as encourages the IWC to provide relevant information to CITES parties. It also uses stronger wording ('urges', 'directs') to require its Secretariat to share data with the IWC, its parties 'to explore the issue of illegal trade in whale meat' and 'every country concerned' to submit information in relation to the

---

[68] See supra Chapter 6 A. II. 2. a).

[69] See supra Chapter 5 B. II. 1. a).

[70] Resolution Conf. 2.7 (Rev) – Relationship with the International Whaling Commission 1979 (repealed); Resolution Conf. 2.8 – Introduction from the Sea 1979 (repealed); Resolution Conf. 11.4 (Rev. CoP12) – Conservation of cetaceans, trade in cetacean specimens and the relationship with the International Whaling Commission 2000.

[71] Resolution Conf. 11.4 (Rev. CoP12) – Conservation of cetaceans, trade in cetacean specimens and the relationship with the International Whaling Commission (n 70) preamble and operative part.

inventory of whales parts and derivatives. While the effect of some parts of this Resolution will be discussed below,[72] it is, at the exception of the order made to 'every country concerned', which appears to extend beyond CITES parties, and arguably of the order made to its parties on potential research (although exploring an issue is by far not an obligation of result), an instrument worded in soft and overall non-binding terms.

c) Within the UNFSA

It might be worth mentioning that states which are parties to the UNFSA and wish to propose a listing under CITES have first to consult with the relevant RFMO if any exists. Indeed, article 8 (6) of the UNFSA calls for consultation:[73] "[a]ny State intending to propose that action be taken by an intergovernmental organization having competence with respect to living resources should, where such action would have a significant effect on conservation and management measures already established by a competent sub-regional or regional fisheries management organization or arrangement, consult through that organization or arrangement with its members or participants. To the extent practicable, such consultation should take place prior to the submission of the proposal to the intergovernmental organization."[74]

However, it is not clear whether this consultation has to be formal or if a notification with a call for comment could suffice. Also, the wording 'to the extent practicable' strongly qualifies this requirement.

III. CITES-CMS

The CITES and the CMS have a specific MoU which, in quite general and qualified terms, requires institutional cooperation, exchange of information and mentions the possibility to "decide on joint activities".[75] This might become of interest to

[72] See infra Chapter 6 C. IV. 1. and D. IV. 1.

[73] FAO, 'Report of the Expert Consultation on Legal Issues Associated to CITES and Commercially-exploited Aquatic Species (22–25 June 2004)' FAO Fisheries and Aquaculture Report 746 (Rome 2004) 6 §27.

[74] 'Agreement for the Implementation of the Provisions of the United Nations Convention on the Law of the Sea of 10 December 1982 Relating to the Conservation and Management of Straddling Fish Stocks and Highly Migratory Fish Stocks' (UNFSA) 1995, 2167 *United Nations Treaty Series* 3 article 8(6).

[75] Memorandum of Understanding between the Secretariat of the Convention on International Trade in Endangered Species of Wild Fauna and Flora (CITES Secretariat) and the Secretariat of the Convention on the Conservation of Migratory Species of Wild Animals (CMS Secretariat) (CITES-CMS MoU) 2002 articles 2–4.

develop cooperation with regard to the management of migratory sharks, four of which (whale shark, great white shark, basking shark, and porbeagle) are listed under CITES and are also the object of a CMS MoU.

Sharks are very present in the activities planned in concert by the two conventions regimes (the relevant work program is still in a draft version but was endorsed by the CITES Standing Committee in 2012[76]), as the Secretariats agree to "develop joint programmes for the conservation and/or sustainable use of these species. Each Secretariat tak[es] the lead where appropriate with CITES focusing on sustainable use and CMS on restoration of species."[77] If implemented, this would be a very practical and potentially fruitful partnership.[78]

IV. Comparison: CITES and Forestry Matters

Moving the discussion beyond marine species is useful for comparison purposes. Some parallels can be drawn with the strong cooperation between CITES and the FAO in relation to timber species and some lessons learnt from it. As the FAO's mandate includes forestry matters while CITES is also applicable to flora species, it is indeed interesting to look at the relationship between these institutions and regimes, as well as to examine the role of regional organizations for listed tree species.

Three commercially-exploited tree species have been listed on CITES Appendix II: the afrormosia in 1992, the bigleaf mahogany in 2002 and the ramin in 2004.[79] Originally, such proposals were highly controversial. Concerns of the industry and range states were expressed, the scientific basis for listing was questioned, CITES was argued not to be the best forum to deal with forestry matters (as opposed to intergovernmental organizations) and the possibility of using CITES for sustainable use (rather than as a trade ban) was disputed.[80]

However, with time, negotiations, and the deconstruction of myths, the listing of timber species has become accepted, is now considered one of the regime's very active fields of work and CITES plays an important role in the regulation of the

---

[76] CITES Secretariat, 'SC62 Sum. 2 (Rev. 1) – 62nd meeting of the Standing Committee – Executive Summary 23.07.2012 Afternoon' (n 60) 2 §14.3.

[77] CITES Secretariat and CMS Secretariat, 'SC61 Doc. 15.4 (Rev. 1) Cooperation with other organizations – Convention on the conservation of migratory species of wild animals' (15 August 2011) 12 annex 1; CITES Secretariat, 'SC62 Doc. 14.3 – Cooperation with other organizations – Convention on the Conservation of Migratory Species of Wild Animals' (23 July 2012).

[78] Caddell, 'Inter-Treaty Cooperation, Biodiversity Conservation and the Trade in Endangered Species' (n 37) 275.

[79] For a full list of tree species' proposals, see T. Mulliken, 'The Role of CITES in Controlling the International Trade in Forest Products: Implications for Sustainable Forest Management' Non-Wood Forest Products Working Documents 7 (Rome 2009) 8.

[80] Ibid. vi.

international trade in some timber species, in collaboration with partners whose support could all but be taken for granted a couple of decades ago.[81] A sign of the positive attitude towards CITES is that, at CoP16, all proposals of timber species were adopted by consensus and that range states were encouraging such listings.[82]

1. FAO

a) In the Convention

In contrast to the case of marine species there is no mention in the Convention itself of a need for partnerships or for requesting information from institutions dealing with timber species.

b) Memorandum of Cooperation: Under Development

CITES and the FAO have not, yet, agreed on an MoU of any sort regarding forestry. However, they were for a time in the process of doing so, which would have constituted a further step into the global FAO-CITES cooperation. Indeed, CITES Decision 15.18 read as: "The Standing Committee shall analyze the current Memorandum of Understanding (MoU) between CITES and FAO, determine the cooperative scheme between CITES and FAO on forestry issues, and ensure that future cooperation between FAO and CITES is carried out under the framework of the MoU." It is worth noting that this Decision has been deleted at CoP16 and not replaced by another one related to the cooperation with the FAO—the status and future prospects of the draft examined here below are thus unclear.

The development of a formal partnership, already considered by the CITES Secretariat in 2002,[83] was later called upon by the CITES parties, in the form of an MoU. However, further discussions between the two institutions rather led to the drafting of a global Memorandum of Cooperation (MoC), to avoid that each FAO department signs its own MoU with CITES. The MoC draft is a framework instrument the format of which is based on the MoC between the FAO and the Convention on Biological Diversity—with the difference that the model version contains a Preamble.[84]

[81] S. F. Oldfield, 'The Evolving Role of CITES in Regulating the International Timber Trade' (2013) 22 RECIEL 291, 291, 293, 297–298, 300.

[82] J. E. Scanlon, 'CITES at Its Best: CoP16 as a 'Watershed Moment' for the World's Wildlife' (2013) 22 RECIEL 222, 223.

[83] United States of America, 'CoP12 Doc. 16.2.2 – Cooperation with other organizations – CITES and FAO' (n 42) 'Comments from the Secretariat' C-D, already suggesting a global instrument not only focused on fisheries or timber.

[84] Memorandum of Cooperation between the FAO and CBD Secretariat 2005.

Its draft articles are not overly surprising, with provisions on representation at each other's meetings, exchange of information, joint activities and specific cooperation through the development of MoUs to be added as annexes. The MoU on fisheries would constitute Annex I.

A number of CITES parties were not satisfied with the MoC solution which was developed by the international organizations. A few states expressed a strong opposition to a general agreement with the FAO, favoring ad hoc agreements with the relevant departments. Also, paragraphs (g)-(h) of draft article IV were perceived as problematic, with some countries expressing their concerns at having the FAO interfering with their decisions taken within CITES.[85] These paragraphs read as:

> Cooperation under this Memorandum of Cooperation shall include: [. . .]
> g. Involvement by FAO in revisions of the CITES listing criteria for species of interest to both Parties and provision of advice by FAO to CITES regarding the CITES listing criteria.
> h. Cooperation between the Parties to ensure adequate consultations in the scientific and technical evaluation of proposals for including, transferring or deleting species of interest to both organizations in the CITES Appendices based on the criteria agreed by the Parties to CITES, and [in the efforts of] both Parties [to] address scientific, technical, socio-economic and legal issues relating to the listing [of species] and implementation of such listings.[86]

Hence, while such provisions do not transform the FAO into a decision-maker at CITES, they seem to acknowledge a role for the FAO which goes beyond that of advice provider: 'involvement' for revision of the listing criteria and 'cooperation' for the amendment of Appendices. It is interesting to note that, regarding the criteria revision, a similar formulation had already been agreed upon in the 2006 MoU: "FAO will continue to provide advice to CITES on, and be involved in any future revision of, the CITES listing criteria."[87] It seems that these problematic paragraphs would in fact not give any additional rights to the FAO.

In any case, a new draft presented at the Standing Committee in 2012 fully removed paragraph (g), replacing it by a capacity building provision and the chapeau changed from 'shall' to 'should'.[88]

. As for the legal nature of the MoC, if signed, it could have, at one stage of the drafting been subject to discussion. Indeed, the FAO would be one of the parties

[85] Committee on Fisheries – Sub-Committee on Fish Trade, 'Update on CITES related activities' presented at the Thirteenth Session of the Sub-Committee on Fish Trade (Hyderabad, 20–24 February 2012) COFI:FT/XIII/2012/8 (2012) 4 §13-14.

[86] CITES Secretariat, 'SC61 Doc. 15.5 – Cooperation with other organizations – Food and Agriculture Organization of the United Nations' (15 August 2011) annex 4 (the [. . .] in paragraph h. are replacing the underlined text still under negotiation).

[87] CITES-FAO MoU (n 9) §3.

[88] FAO and CITES Secretariat, 'SC62 Doc. 14.4 Annex 1 – Draft Memorandum for Cooperation between the Food and Agriculture Organization of the United Nations (FAO) and the Secretariat of the Convention on International Trade in Endangered Species of Wild Fauna and Flora (clean version)' (23 July 2012) item IV.

and, as an international organization with legal personality, it could enter into a binding agreement with the potential consequences it involves in case of non-compliance. Also, some terms in this particular early draft MoC with the FAO were suggesting a binding intent: 'shall', 'the Parties', 'entry into force' were used instead of 'will', 'the signatories' and 'will take effect' in the MoU.[89]

However, major changes took place in the wording between 2011 and 2012: the 'parties' have become the 'participants', 'articles' are renamed 'items', 'entered into' has become 'initiated' and past references to 'shall' now read 'should'.[90] Furthermore, the other party, the CITES Secretariat, according to the view presented here above,[91] can probably not enter into a binding agreement.

A draft MoU between the Forestry Department of the FAO and the CITES Secretariat regarding forestry has been presented at the Standing Committee in 2012.[92] In generally termed provisions and quite similar to the guidelines for cooperation with ICCAT, the participants are encouraged to exchange information and may develop joint activities.[93] No mention of the cooperation during the listing stage is made. Also, it is interesting that an explicit statement clarified the non-binding nature of the instrument.[94] The potential relationship of this instrument with the MoC is left unaddressed.

2. International Tropical Timber Organization (ITTO) and Regional Organizations

Two Resolutions provide CITES and institutions active in the management of timber species with a framework for their collaboration: one is on the implementation of the convention regarding such species and the other focuses on the cooperation between CITES and the International Tropical Timber Organization (ITTO) regarding trade in tropical timber.

ITTO is an intergovernmental organization, with a membership representing 80 % of the world's tropical forests and 90 % of the global trade in such species. Its goal is to support "the conservation and sustainable management, use and trade of

[89] Aust (n 24) 27; P. Gautier, 'Non-Binding Agreements' Max Planck Encyclopedia of Public International Law www.mpepil.com §12.

[90] FAO and CITES Secretariat, 'SC62 Doc. 14.4 Annex 1 – Draft Memorandum for Cooperation between the Food and Agriculture Organization of the United Nations (FAO) and the Secretariat of the Convention on International Trade in Endangered Species of Wild Fauna and Flora (clean version)' (n 88) chapeau, item IV.

[91] See supra Chapter 6 A. I. 4.

[92] Forestry Department of the FAO and CITES Secretariat, 'SC62 Doc.14.4 Annex 3 – Draft Memorandum of Understanding' (2012) http://www.cites.org/common/com/sc/62/E62-14-04-A3.pdf accessed 5 July 2015.

[93] On this, see infra Chapter 6 D. IV. 2.

[94] Forestry Department of the FAO and CITES Secretariat, 'SC62 Doc.14.4 Annex 3 – Draft Memorandum of Understanding' (n 92) 3.

tropical forest resources."[95] The other institutions of interest here, the regional timber organizations, have mandates ranging from general competence, to focus on scientific research or trade aspects.

In the Preamble of Resolution 10.13 on the *Implementation of the Convention for timber species*, the CoP recognizes the importance of biological and trade information in a listing proposal and admits that such information is often in the hands of other international organizations. It also recognizes both the beneficial and negative aspects that international trade can have on wildlife. It finally mentions the unfavorable impact that misconceptions about the effects of a listing—i.e. that it would automatically result in a ban on trade—can have on the industry.[96] This Resolution also establishes a procedure of consultation with relevant global and regional organizations before a listing proposal can be submitted to CITES.[97]

With ITTO, CITES adopted a Resolution on cooperation, against the advice of the Secretariat, which would have preferred an MoU.[98] This Resolution highlights in its Preamble the overlapping goals and interests of CITES and ITTO. It also welcomes the multi-level cooperation between the two institutions and the support provided by ITTO in various occasions towards the efficient implementation of CITES listings. Then, in the operative part, the CoP urges and encourages its parties to further cooperate with ITTO.[99]

V. Remarks and Conclusions

The legal bases for cooperation between CITES and other institutions involved with marine species are varied, with provisions in the Convention, two MoUs (one with the FAO and one with the CMS), two Resolutions (one with the IWC and one with CCAMLR) and guidelines for cooperation with ICCAT.

The existence of such frameworks for cooperation is positive as their adoption has triggered discussions, debates and clarification of responsibilities. For example the IWC-CITES Resolution made explicitly clear that the procedures to be followed to ask the IWC for advice are the same whether the proposal is made at a CoP or between CoPs. However, these instruments also entail some ambiguous provisions and reflect compromise solutions on tense subjects.

---

[95] ITTO, 'About ITTO' http://www.itto.int/about_itto/ accessed 5 July 2015.

[96] Resolution Conf. 10.13 (Rev. CoP15) – Implementation of the Convention for timber species 1997 preamble.

[97] Ibid. 'Regarding international organizations', a)-b); see infra Chapter 6 C. IV. 2.

[98] United States of America, 'CoP14 Doc. 18.2 – Cooperation between CITES and ITTO Regarding Trade in Tropical Timber' (3–15 June 2007) 'Comments from the Secretariat' B-C.

[99] Resolution Conf. 14.4 – Cooperation between CITES and ITTO regarding trade in tropical timber 2007 preamble and operative part.

The legal nature of the CITES-FAO MoU can be discussed, but, in practice, has only a very limited impact, if any. Indeed, commitments are phrased in a general manner and reflect the cooperation which was already taking place before 2006. Also, it is worth highlighting that it is the CITES Secretariat which undertook some commitments, not the CITES CoP. Hence the sovereign decision-making powers are left untouched and the fears which have been expressed with regard to the MoC are, as such, unfounded.

As can be illustrated by the partnership of CITES with organizations related to timber species, the formality of the framework is irrelevant: these relationships are not mentioned in the Convention and only take the form of Resolutions (instead of an MoU). They are thus, structurally speaking, looser than the ones for marine species, but also seem to function better. The aspects which prove crucial are the willingness to cooperate and the acknowledgement of mutual interest; these characteristics are nowadays very present with regard to timber[100] and arguably less so for marine species.

Indeed, in the case of timber, the potential of CITES to strengthen the regulatory framework for sustainable use seems now to be recognized—although evidently not by all ITTO members.[101] This is the result of a process, which started with critical questions of adequacy being asked, until mentalities progressively evolved; the main issue was that range states perceived the use of CITES as a ban on trade and this problem was addressed through discussions between national Management Authorities, the public, the industry and the relevant governmental and non-governmental organizations.[102] A similar development might be unfolding for fish species, although, for the time being, arguments are still heard on CITES' irrelevance and inadequacy.

In particular, the fears of a ban as well as both the positive and negative impacts of trade on the status of timber species are actually mentioned in the Preamble of the Resolution on the *Implementation of the Convention for timber species*. Reaching an agreement on such aspects and explicitly acknowledging the existing worries could help to depolarize the discussion related to fisheries. Recognizing the benefits of trade in marine products could equally be beneficial.[103]

Finally, an important feature that CITES has not yet developed sufficiently with organizations involved with marine species is the inclusive and formal involvement of RFMOs, in contrast to what has been reached with the forestry management institutions. Already in 2002, FAO members pointed to "[t]he need for the existing process within CITES for scientific evaluation of proposals for listing, transfer and

---

[100] ITTO, 'Making the mahogany trade work: Report of the workshop on capacity-building for the implementation of the CITES Appendix-II listing of mahogany' ITTO Technical Series 22 (2004) 14–16.

[101] Mulliken (n 79) ix, 13–15.

[102] M. Engler and R. Parry-Jones, 'Opportunity or Threat: The Role of the European Union in Global Wildlife Trade' (TRAFFIC Europe 2007) 24; Mulliken (n 79) 13.

[103] FAO, 'Report of the Expert Consultation on Legal Issues' (n 73) 6 §28.

de-listing to be strengthened through more explicit involvement of fisheries agencies and Regional Fishery Management Organizations (RFMOs) [... as well as indicated t]hat regional fisheries management organizations should be more involved in scientific evaluation of proposals concerning species falling within their mandates."[104]

B. External Provision of Expertise

Since the FAO has expertise in relation to some precise aspects of the CITES regime which were under consideration for revision, the UN specialized agency provided relevant information on topics such as the adequacy of CITES general listing criteria for aquatic species, particular listing situations, and the introduction from the sea questions.

This type of cooperation, originally of an ad hoc nature, already took place before the adoption of the MoU. Since 2006, provision of general expertise (to be distinguished from scientific expertise for specific listing proposals)[105] is governed by the MoU, but has occurred more rarely, as many issues have already been addressed.

I. Listing Criteria

1. Development in Relation to Commercially-exploited Aquatic Species

As examined above, a first set of criteria, the Bern criteria were adopted in 1976 and replaced in 1994 by the Fort Lauderdale criteria.[106] Later on, the FAO made recommendations to CITES regarding the adequacy of the existing listing criteria in relation to commercially-exploited aquatic species. Two technical consultations took place,[107] in order, amongst other goals, to "[r]efine the criteria and guidelines,

[104] Committee on Fisheries – Sub-Committee on Fish Trade, 'An analysis of the CITES listing criteria applied to commercially exploited aquatic species', presented at the Eighth Session of the Sub-Committee on Fish Trade (Bremen, 12–16 February 2002) COFI:FT/VIII/2002/3 (2002) §5, making reference to the FAO, 'Second technical consultation on the suitability of the CITES criteria for listing commercially-exploited aquatic species – A background analysis and framework for evaluating the status of commercially-exploited aquatic species in a CITES context' FI:SLC2/2001/2 (22 October 2001) http://www.fao.org/docrep/MEETING/003/Y1455E.HTM accessed 5 July 2015 §77, 80.

[105] For the latter, see infra Chapter 6 C.

[106] Resolution Conf. 9.24 (Rev. CoP16) – Lauderdale criteria (n 50); see supra Chapter 5 C. I.

[107] FAO, 'Technical consultation on the suitability of the CITES criteria for listing commercially-exploited aquatic species – The key points from an appraisal of the suitability of the CITES criteria for listing commercially-exploited aquatic species' FI:SLC/2000/2 (28 June 2000) http://www.fao.

as appropriate, considering the specific characteristics of commercially-exploited aquatic resources, and remove any possible ambiguities in the application of such criteria and guidelines which might lead to false alarms and misses" and to "[r]econcile language regarding species, populations, and population-units used respectively by CITES and the community of fisheries science and management, with corresponding terms clearly identified."[108] This led to a report endorsed in 2002 by the Sub-Committee on Fish Trade, on behalf of COFI.[109]

The specific modifications of criteria for commercially-exploited aquatic species were adopted at CoP13, in 2004. The main advice from the FAO technical consultations was followed.[110] Of particular importance was the inclusion of different benchmarks to define decline for commercially-exploited aquatic species—the so-called Annex 5 footnote—which take into account such species' particular characteristics and their various life histories.[111]

2. Recent Debate About Interpretation of the Appendix II Listing Criteria

The debate regarding the interpretation of Annex 2a (listing in Appendix II) has brought the FAO and CITES together again to discuss the criteria.[112] The contro-

org/docrep/meeting/x4894e.htm accessed 5 July 2015; FAO, 'Report of the Technical Consultation on the Suitability of the CITES Criteria for Listing Commercially-exploited Aquatic Species (Rome, 28–30 June 2000)' FAO Fisheries Report 629 (Rome 2000); FAO, 'Second technical consultation on the suitability of the CITES criteria for listing commercially-exploited aquatic species – A background analysis and framework for evaluating the status of commercially-exploited aquatic species in a CITES context' (n 104); FAO, 'Report of the Second Technical Consultation on the Suitability of the CITES Criteria for Listing Commercially-exploited Aquatic Species (Windhoek, Namibia 22 October 2001)' FAO Fisheries Report 667 (Rome 2002).

[108] Committee on Fisheries, 'Review of the CITES listing criteria for species exploited by fisheries in marine and large freshwater bodies, at the Twenty-fourth Session of the Committee on Fisheries (Rome, 26 February – 2 March 2001)' COFI/2001/6 (Rome 2001) §11(a), (c).

[109] Committee on Fisheries – Sub-Committee on Fish Trade, 'Report of the Eighth Session of the Sub-Committee on Fish Trade (Bremen, 12–16 February 2002)' FAO Fisheries Report 673 (Rome 2002) 7 §16.

[110] WECAFC, 'WECAFC Lesser Antilles Fisheries Committee – An overview of the FAO activities in relation to CITES and commercially-exploited species' (n 8) 2 §4-5.

[111] As highlighted as important in FAO, 'Report of the Second Meeting of FAO and non-FAO regional fishery bodies or arrangements (Rome, 20–21 February 2001)' FAO Fisheries Report 645 (Rome 2001) 8 §44. For suggested changes, see FAO, 'Second technical consultation on the suitability of the CITES criteria for listing commercially-exploited aquatic species – A background analysis and framework for evaluating the status of commercially-exploited aquatic species in a CITES context' (n 104); FAO, 'Report of the Second Technical Consultation on the Suitability of the CITES Criteria for Listing Commercially-exploited Aquatic Species' (n 107).

[112] See supra Chapter 5 C. I. 1. c) bb).

versial aspects have been examined at a 2011 workshop[113] and have been discussed at the CITES Animals Committee meeting in March 2012. The text of Resolution 9.24 could be modified in the future, in order to clarify the criteria applicable and to ensure coherency between the two institutions' interpretations. However, while a common understanding would be ideal such an agreement seems unlikely, in the view of the current positions.

II. Specific Listing Situations

1. Split-Listing

Even though split-listing would be quite positive with regard to commercially-exploited fish species as it would avoid unnecessary restrictions being applied to healthy stocks, it is not encouraged under CITES and total split-listing is not possible.[114]

This situation is not ideal and, already in 2004, the FAO Expert Consultation "suggested that it would be useful to review the text covering split-listing to determine if there is adequate flexibility for commercially-exploited aquatic species."[115] However, such a review is still to take place.

2. Look-Alike

Listing species under the look-alike criterion can be extremely costly and ineffective for fish species, which, in many cases, look similar and/or are traded in highly processed forms.[116]

Parts of this issue were addressed successfully at CoP13 when the advice of the FAO Expert Consultation on Legal Issues was followed and Resolution Conf. 9.24 (then Rev.CoP12) was modified. Hence, Annex 2b was changed, from "species should be included" to "species may be included".

Using more split-listing and listing less look-alike species does in turn increase enforcement issues. Hence, already in 2004, it was recognized by the FAO that new or improved implementation mechanisms would be needed.[117] However, the workshop to address enforcement and identification issues, thus facilitating the use of split-listing and decreasing the use of look-alike listing, which was planned for

---

[113] FAO, 'Report of the FAO Workshop to Review the Application of CITES Criterion Annex 2 a B to Commercially-exploited Aquatic Species (Rome, 19–21 April 2011)' FAO Fisheries and Aquaculture Report 976 (Rome 2011).

[114] See supra Chapter 5 C. I. 1. b) bb) and c) aa).

[115] FAO, 'Report of the Expert Consultation on Legal Issues' (n 73) 9 §44.

[116] See supra Chapter 5 C. I. 1. b) cc).

[117] FAO, 'Report of the Expert Consultation on Legal Issues' (n 73) 9–10 §46.

2009, had to be cancelled due to budgetary restrictions. The 2010–2015 CITES activities program of the FAO has this item on the agenda again; it intends to give advice on the application of these criteria to aquatic species, to further produce identification guides, as well as to develop tools for easier differentiation between species (genetic tools, labeling schemes, etc.).

3. By-Catch

As examined, the wording of Annex 2b could suggest that, should fishing a targeted species cause the by-catch of endangered species, these target species could possibly be listed. This would make the application of CITES extremely difficult and should be avoided.[118]

The FAO advised that "the practice of listing [. . .] should not be extended to non-threatened target species whose exploitation endangers other species."[119] However, the practical rewording "to control effective trade", which would have clarified paragraph B was not adopted by the CoP.[120]

III. Introduction from the Sea

The concept of introduction from the sea, crucial for aquatic species, has been clarified recently both with regard to 'from the sea' and to the identification of the state of introduction.[121]

The FAO provided an analysis of the issues at stake in the 2004 legal consultation. It advised that, for the geographical component of the issue, an interpretation which would place the limits of waters under national jurisdiction where they were in 1973 would contravene UNCLOS and the law of fisheries developed thereupon.[122] In 2005, a CITES Workshop on introduction from the sea suggested a draft Resolution. Its adoption at the following CoP clarified the "marine environment not under the jurisdiction of any State" definition, which is to be understood "as reflected in the United Nations Convention on the Law of the Sea."[123]

As for the question of which state qualifies as state of introduction, it remained unresolved until March 2013 and caused much discussion. No clear advice had been

[118] See supra Chapter 5 C. I. 1. b) dd) and c) bb).

[119] FAO, 'An Appraisal of the Suitability of the CITES Criteria for Listing Commercially-exploited Aquatic Species' FAO Fisheries Circular 954 (Rome 2000) 33–35 §4.2.

[120] FAO, 'Report of the Expert Consultation on Legal Issues' (n 73) 9 §45.

[121] See supra Chapter 5 C. I. 2.

[122] FAO, 'Report of the Expert Consultation on Legal Issues' (n 73) 4 §24.

[123] Resolution Conf. 14.6 (Rev. CoP16) – Introduction from the Sea (n 51).

given by the FAO, although research was undertaken on this question.[124] At CoP16, a solution was adopted distinguishing between the situation where the flag and port state is the same state and the one where two states are involved.

When not organizing consultations itself, the FAO Secretariat has been invited throughout the process and at meetings, as an observer. It was tasked with providing technical information as requested in the CITES Working Group on Introduction from the Sea.[125]

IV. Conclusions

The FAO, acting as a consultant for CITES, provided expert information on legal questions, such as whether the criteria for listing were adapted for fish species and on the introduction from the sea questions. Through this mechanism, expert advice is externalized to the institution most able to develop it. Such guidance was overall well-followed since the Fort Lauderdale criteria were modified, the listing of look-alike species is now possible but not automatic and the IFS provision is clarified. It is also noteworthy that, since the FAO put an important emphasis on avoiding criteria which encourage false alarms, crucial socio-economic interests might be better protected.[126] However, the recommendations about split-listing and about the wording of Annex 2b which could affect by-catch have not been followed. Furthermore, interpretation problems related to Annex 2a remain, with important practical consequences.[127]

With regard to the FAO's advice for the criteria, one could have hoped that, by going to the specifics and looking at when a species should scientifically be afforded protection, the two polarized viewpoints on the use of CITES for marine species could be brought together. This worked for a time, but stalled at CoP15 because Atlantic Bluefin Tuna was too sensitive a proposal for science to be automatically followed. This problem is similar to the one suffered by the criteria in general: while it was hoped that the Fort Lauderdale criteria would make the listing process a science-based and neutral one, listing remained a political

[124] FAO, 'Report of the Expert Consultation on Legal Issues' (n 73) 3–4 §18-19; E. Franckx, 'The Relationship between CITES, FAO and Related Agreements: Legal Issues' FAO Fisheries and Aquaculture Circular 1062 (Rome 2011) 23–26.

[125] CITES Secretariat, 'CoP14 Doc. 18.1 – Cooperation with other organizations – Cooperation with the Food and Agriculture Organization of the United Nations' (3–15 June 2007) 4 §25; Committee on Fisheries – Sub-Committee on Fish Trade, 'Report of the Twelfth Session of the Sub-Committee on Fish Trade (Buenos Aires 26–30 April 2010)' FAO Fisheries and Aquaculture Report R939 (Rome 2010) 10 §45.

[126] FAO, 'Technical consultation on the suitability of the CITES criteria for listing commercially-exploited aquatic species – The key points from an appraisal of the suitability of the CITES criteria for listing commercially-exploited aquatic species' (n 107) §23.

[127] See infra Chapter 6 C. I. 2.

decision.[128] States are indeed the ones not only to choose the criteria, but also to vote, and they do so defending their interests.

As for the aspects of the criteria that have been identified as requiring further research—mainly the questions of look-alike and split-listing—they hopefully will be addressed in the near future. Whether RFMOs should be involved in such provision of expertise does not seem to be much a topic of discussion. Indeed, while they have the practical expertise 'on the ground' the FAO is the institution with the strongest ability to have an overview of the issues and was requested to "play a facilitating role in improving dialogue and communications among member States, RFMOs and CITES."[129] Also, it is worth noting that, at some meetings where RFMOs gather, which take place at the FAO, CITES matters were discussed.[130] Hence, the FAO has the ability to remain informed of RFMOs' views on CITES-related questions and to take them into account if needed.

C. Cooperation in Listing

I. FAO

1. Cooperation

a) Terms of Reference

For CITES CoP13, 14, 15 and 16 a FAO Expert Advisory Panel provided advice on whether the species proposed to be added to the Appendices fulfilled the CITES criteria.[131] To do so, the FAO gathered a group of experts, informed its member states who could send information and requested input from relevant regional organizations.[132]

---

[128] S. Young, 'Contemporary Issues of the Convention on International Trade in Endangered Species of Wild Fauna and Flora (CITES) and the Debate Over Sustainable Use' (2003) 14 Colorado Journal of International Environmental Law and Policy 167, 188 (reference omitted).

[129] FAO, 'Report of the Second Meeting of FAO and non-FAO regional fishery bodies or arrangements' (n 111) 8 §44.

[130] Ibid. 8–10 §43-50.

[131] FAO, 'Report of the FAO ad hoc Expert Advisory Panel for the Assessment of Proposals' (n 4); FAO, 'Report of the Second FAO ad hoc Expert Panel for the Assessment of Proposals to Amend Appendices I and II of CITES Concerning Commercially-exploited Aquatic Species (Rome, 26–30 March 2007)' FAO Fisheries Report 833 (Rome 2007); FAO, 'Report of the Third FAO Expert Advisory Panel for the Assessment of Proposals to Amend Appendices I and II of CITES Concerning Commercially-exploited Aquatic Species (Rome, 7–12 December 2009)' FAO Fisheries Report 925 (Rome 2010).

[132] FAO, 'Report of the FAO ad hoc Expert Advisory Panel for the Assessment of Proposals' (n 4) 2 §8 and 15 appendix D 'Terms of Reference' §6.

The terms of references for the Expert Panels were adopted at the 25th session of COFI. The Expert Panels are organized by the FAO but are constituted of experts working in their personal capacity, chosen on the basis of equitable geographical representation[133] and supported by a few FAO Secretariat staff as well as a CITES observer.

The mandate of such panels is to "assess each proposal from a scientific perspective in accordance with the CITES biological listing criteria, taking account of the recommendations on the criteria made to CITES by FAO [and] comment, as appropriate, on technical aspects of the proposal in relation to biology, ecology, trade and management measures, as well as, to the extent possible, the likely effectiveness for conservation."[134]

b) Content of the Advice

The advice of the Panel is based on the information included in the proposal and any available scientific data of relevance, which is then judged on its reliability. Indeed, research papers are noted from zero to five according to a scale agreed upon by the Expert Panel (zero as no value for flawed data to five for "[s]tatistically designed, fishery-independent survey of abundance")[135]. Such a qualitative scientific exercise is highly commendable in a field where data's reliability and independence are regularly questioned.

The reports incorporated several important sections allowing to better take into account the particularities of commercially-exploited aquatic species, whether on a biological, implementation or trade-related level. Since the second report indeed, a section on whether fluctuations in abundance could be due to natural fluctuations—to which fish species are known to be exposed—is included.[136] Also, particular vulnerabilities and mitigating factors are examined.[137]

Then, in accordance with the terms of reference, already the first report incorporated a section on a listing's likely effectiveness for conservation; this section includes remarks on issues of implementation such as look-alike, on the special

[133] Ibid. 15 appendix D 'Terms of Reference' §2-3.

[134] Ibid. 15 appendix D 'Terms of Reference' §5.

[135] Ibid. 3–4 §16.

[136] FAO, 'Report of the Second FAO ad hoc Expert Panel for the Assessment of Proposals' (n 131) 24, 43, 73; FAO, 'Report of the Third FAO Expert Advisory Panel for the Assessment of Proposals' (n 131) 22, 46, 64, 91, 124; FAO, 'Report of the Fourth FAO Expert Advisory Panel for the Assessment of Proposals to Amend Appendices I and II of CITES Concerning Commercially-exploited Aquatic Species (Rome, 3–8 December 2012)' FAO Fisheries and Aquaculture Report R1032 (Rome 2013) 14, 38, 70, 111.

[137] FAO, 'Report of the Second FAO ad hoc Expert Panel for the Assessment of Proposals' (n 131) 24, 43, 73; FAO, 'Report of the Third FAO Expert Advisory Panel for the Assessment of Proposals' (n 131) 22–23, 46–47, 64–65, 91–92, 124–125; FAO, 'Report of the Fourth FAO Expert Advisory Panel for the Assessment of Proposals' (n 136) 7–8, 19, 42, 61–62, 76, 107.

situation of small-island states and on the identification of threats.[138] Since the second report, additionally, the applicability of the IFS provision is analyzed, and the basis for NDF and for legality finding, the possibility of split-listing as well as issues of identification in trade are addressed.[139] For the NDF, interesting remarks were for example for porbeagle, that "advice is available on which NDF could be based although a closer alignment between management measures and scientific advice would be required",[140] and for the spiny dogfish, that the issues of straddling stocks and discards should be taken into account.[141] Such comments hint that a listing under CITES could question the adequacy of the existing measures and suggest improvements. The lack of adequate information and management to determine NDF was also highlighted when relevant.[142] All in all, the likely effectiveness of a listing is discussed,[143] taking into account the level of legal trade, the rate of incidence of by-catch and the level of domestic trade.[144] Hence, even though the criteria used by the CoP have been criticized for not analyzing the likely effectiveness of the protection, the Expert Panel does examine this aspect.[145]

The importance of trade as a threat is also taken into account, as required by the criteria. However, due to the lack of specialized knowledge of the experts on trade matters, they have to rely on the proponents' information, which has not always been deemed sufficient.[146] The Expert Panel requested that states provide more precise and unbiased information in their proposals and "quantitative information on the impact of international trade on species status".[147] In the second report, trade

[138] FAO, 'Report of the FAO ad hoc Expert Advisory Panel for the Assessment of Proposals' (n 4) 23, 37–38.

[139] FAO, 'Report of the Second FAO ad hoc Expert Panel for the Assessment of Proposals' (n 131) 25–26, 45–47, 74–75; FAO, 'Report of the Third FAO Expert Advisory Panel for the Assessment of Proposals' (n 131) 24–25, 47–49, 66–67, 93–95, 125–127; FAO, 'Report of the Fourth FAO Expert Advisory Panel for the Assessment of Proposals' (n 136) 16–18, 41–42, 72–74, 113–114.

[140] FAO, 'Report of the Second FAO ad hoc Expert Panel for the Assessment of Proposals' (n 131) 26.

[141] Ibid. 46.

[142] FAO, 'Report of the Third FAO Expert Advisory Panel for the Assessment of Proposals' (n 131) 48; FAO, 'Report of the Fourth FAO Expert Advisory Panel for the Assessment of Proposals' (n 136) 17, 41, 113.

[143] Which was later on expressly requested by COFI Members at Committee on Fisheries – Sub-Committee on Fish Trade, 'Report of the Twelfth Session of the Sub-Committee on Fish Trade' (n 125) 9 §43.

[144] FAO, 'Report of the Second FAO ad hoc Expert Panel for the Assessment of Proposals' (n 131) 27, 47–48, 76; FAO, 'Report of the Third FAO Expert Advisory Panel for the Assessment of Proposals' (n 131) 25, 49, 67, 94, 126–127; FAO, 'Report of the Fourth FAO Expert Advisory Panel for the Assessment of Proposals' (n 136) 18, 42, 62, 74–76, 114–115.

[145] B. Dickson, 'International Conservation Treaties, Poverty and Development: the Case of CITES' (2002) 74 Natural Resource perspectives http://www.odi.org.uk/publications/2069-international-conservation-treaties-poverty-development-case-cites accessed 5 July 2015, 3.

[146] FAO, 'Report of the FAO ad hoc Expert Advisory Panel for the Assessment of Proposals' (n 4) 25.

[147] FAO, 'Report of the Second FAO ad hoc Expert Panel for the Assessment of Proposals' (n 131) 5–6 §15, §18-19.

considerations were included with, for example, important mentions of how internal trade within the EU would not be affected.[148]

Furthermore, the potential socio-economic impacts of a listing have been addressed since 2007, in terms of additional costs, restrictions to trade or bans in the absence of a management plan for a listed species, and effects on vulnerable local communities active in the fishing or farming of the relevant species.[149]

As for aquaculture-related matters, a few remarks were made, mostly regarding the effects of aquaculture on demand, but also, more recently, on the impacts of a listing on aquaculture. Indeed, in its report for CoP14, the Expert Panel mentioned that with regard to eels, the demand from Asian aquaculture for elvers or live young eels had increased the prices on European markets and could cause smuggling of young specimens.[150] Also, in its report for CoP15, the Expert Panel mentioned that sourcing Atlantic Bluefin Tuna juveniles for farms in the Mediterranean was an aggravating factor to the already high level of fishing pressure.[151] The implications of a listing on aquaculture have been briefly mentioned: under an Appendix I listing, "[n]o commercial harvesting of Atlantic bluefin on the high seas, either for direct use or for supplying tuna farms, would be allowed."[152] Finally, remarks on the potential conservation benefits stemming from aquaculture were included in the advice for the humphead wrasse.[153]

c) CITES Secretariat's Recommendations

The CITES Secretariat also gives parties advice on the amendments' proposals. While Wijnstekers considers that this was originally not envisioned for the normal procedure (as opposed to the postal procedure),[154] the reference in article XV (1) to the procedures of article XV (2) (b) rather implies the opposite conclusion.[155] In practice, states have anyway endorsed the role of the CITES Secretariat in providing its own views. Indeed, while Resolution Conf. 5.20 could have clarified limitations in the Secretariat's role, it did not, and the Secretariat makes recommendations for all proposals.[156] It is responsible to present both its views and those of the FAO to the parties.

---

[148] Ibid. 24–25, 45, 73–74.

[149] Ibid. 26, 47, 75, 86, 111–112.

[150] Ibid. 84.

[151] FAO, 'Report of the Third FAO Expert Advisory Panel for the Assessment of Proposals' (n 131) 124. A similar comment was made in relation to the humphead wrasse (FAO, 'Report of the FAO ad hoc Expert Advisory Panel for the Assessment of Proposals' (n 4) 38).

[152] FAO, 'Report of the Third FAO Expert Advisory Panel for the Assessment of Proposals' (n 131) 125.

[153] FAO, 'Report of the FAO ad hoc Expert Advisory Panel for the Assessment of Proposals' (n 4) 38.

[154] CITES (n 33) article XV(1)(2); Wijnstekers (n 48) 460.

[155] For the presentation of this cross-reference, see supra Chapter 6 A. II. 2. a) aa).

[156] Wijnstekers (n 48) 461.

Overall, the Secretariat followed the Expert Panel's advice in its recommendations to the Parties, at the major exception of CoP14. Indeed, then, out of seven proposals on which the Expert Panel had to provide expertise, the CITES Secretariat recommendations dissented on four occasions.[157] It is worth keeping in mind, that, according to the MoU, the Secretariat must "respect, to the greatest extent possible" the FAO's conclusions and the responses from the relevant RFMOs but that this remains a qualified requirement.

Out of the 21 pieces of advice from the FAO Expert Panel, 13 were clearly followed by the CoPs (six recommending a listing and seven stating that the evidence did not support the listing); three led to listings while the Expert Panel admitted that it was unable to evaluate the effectiveness of a listing and/or evidence could bring different types of conclusions; and five pieces of advice were clearly rejected (leading to four CoP decisions not to list and one to add a species against the expert advice). In two cases, the Expert Panel recommended a listing but expressed a lack of consensus, respectively stating that "on balance, the available evidence supports the proposal"[158] and that "[a] majority of the FAO Expert Advisory Panel considered that the available evidence supported the proposal"[159]—in both cases, the species was not listed. In average, recommendations were followed in 62 % of the cases and disregarded in 24 %—the remaining percent representing cases where recommendations were lacking or not based on a consensus.[160]

2. Problems

As mentioned, not only does the CoP sometimes take political decisions against the Expert Panel's advice, but the CITES Secretariat also does not always follow the FAO in its recommendations. A main explanation to this situation is the problem in the interpretation of the criteria, particularly those in Annex 2aB.[161] This issue leads to an incompatibility not only of language and results but also of methodology. Indeed, the benchmarks which serve to advise on a listing are not perfectly similar, since the FAO Expert Panel applies the decline definition and the CITES Secretariat does not use a set scientific basis to assess 'reducing'. This interpretative inconsistency underlies the differences between a maybe slightly preservationist CITES Secretariat and a pro-sustainable use FAO.

[157] See table in Annex 3 for an overview of the FAO Expert Panels' pieces of advice, the CITES Secretariat recommendations and the CoP decisions.

[158] FAO, 'Report of the Third FAO Expert Advisory Panel for the Assessment of Proposals' (n 131) 41.

[159] Ibid. 117.

[160] See Annex 3.

[161] CITES Secretariat, 'CoP14 Inf. 26 – Correspondence between the FAO and the CITES Secretariat concerning proposals to amend the appendices' (3–15 June 2007); see supra Chapter 6 B. I. 2.

While both Secretariats reaffirmed their intent to collaborate, such opposing advice certainly does not strengthen the credibility of the Expert Panel in the eyes of the CITES parties. The FAO addressed its concerns in that regard to the CITES Secretariat which apparently shared this perception.[162] Also, recurrent diverging recommendations could bring up doubts, at the FAO, on the usefulness of providing scientific advice if it seems not to be followed by the partner institution. Overall, if the experience of CoP14 was to occur again, a problem of mistrust could develop, on the one hand from the CITES parties towards the scientific accuracy of the advice provided by both the FAO and the Secretariat and, on the other hand from the FAO members who might see the work undertaken as not sufficiently valued or profitable.

Such a situation has been criticized in clear terms in the case of the porbeagle shark and spiny dogfish: "the Secretariat's final recommendations largely ignored these difficulties [i.e. technical and implementation ones, recognized by the CITES Animals Committee and included in the provisional Secretariat recommendations], contradicted advice by FAO, and appears to be based on the unspecified belief that listing would be beneficial to conservation of the two species. The apparent reversal in the Secretariat's position has the potential to seriously undermine the cooperative spirit that should underpin the MOU with FAO."[163] One should note that this opinion stems from China, a country generally opposed to the use of CITES for commercially-exploited marine species.

However, the possibility to express a range of views is a sign of transparency, independence, as well as trust,[164] and, hence hopefully of good scientific practices. Nevertheless, the problem here is not related to a disagreement on data but to the definition of 'reduction', which influences the final conclusions in a problematic manner. One must finally note that states' representatives are no biology specialists and rely on expert advice which should be clear if it is to have a real impact on the decision-making processes. One FAO member state for example commented on the need to have clear recommendations from the Expert Panel.[165] The FAO members requested a review of how the Panel's advice was followed and why,[166] but this has not yet been undertaken.

---

[162] CITES Secretariat, 'CoP14 Inf. 26 – Correspondence between the FAO and the CITES Secretariat concerning proposals to amend the appendices' (n 161). This concerns remains to this date (Committee on Fisheries – Sub-Committee on Fish Trade, 'Update on CITES related activities' presented at the Fourteenth Session of the Sub-Committee on Fish Trade (Bergen, 24–28 February 2014) COFI:FT/XIV/2014/10 (2014) 3 §7).

[163] China, 'CoP14 Inf. 45 – Shark Issues' (3–15 June 2007) 2 §1(c).

[164] FAO, 'Report of the Second FAO ad hoc Expert Panel for the Assessment of Proposals' (n 131) 2 §6.

[165] Ibid. 3 §12.

[166] Ibid. 16 appendix C 'Welcome speech by Ichiro Nomura, Assistant Director-General, FAO Fisheries and Aquaculture Department'; Committee on Fisheries – Sub-Committee on Fish Trade, 'Report of the Tenth Session of the Sub-Committee on Fish Trade (Santiago de Compostela, 30 May–2 June 2006)' FAO Fisheries Report 807 (Rome 2006) 8 §26.

II. RFMOs

1. Cooperation in Practice

In practice, since the Expert Panels are organized by the FAO, the cooperation with RFMOs mainly takes place through this UN specialized agency. For the First Panel, the RFMOs were notified but no comment was received;[167] the FAO had to recontact some of the regional bodies to discuss white sharks' management. For the Second Panel, relevant RFMOs were also notified and three organizations answered.[168] For the Third Panel, RFMOs were notified and comments were received from ICCAT, GFCM and the Federation of European Aquaculture Producers Tuna Commission.[169] The ICCAT's report on the biological status of the porbeagle and Atlantic Bluefin Tuna was over 400 pages long[170] and the SCRS report was used to determine decline.[171] For the Fourth Panel, information was received from many more RFMOs and bodies (IATTC; NEAFC, WCPFC; ICES, Southeast Asian Fisheries Development Center Policy and from the SRFC) and an ICCAT staff member attended.[172] The information provided was related to available data and the biological status of the species.[173]

In accordance with article XV, the CITES Secretariat also requests comments from intergovernmental bodies having a function in relation to marine species.[174] However, answers were only rarely received from RFMOs—the exception being from ICCAT for CoP15—until CoP16 when several regional fishery bodies sent return letters. Most of these letters were simply acknowledging the request for

---

[167] FAO, 'Report of the FAO ad hoc Expert Advisory Panel for the Assessment of Proposals' (n 4) 2 §8.

[168] SEAFDEC, ASSOCORAL and GFCM (FAO, 'Report of the Second FAO ad hoc Expert Panel for the Assessment of Proposals' (n 131) 3 §12).

[169] FAO, 'Report of the Third FAO Expert Advisory Panel for the Assessment of Proposals' (n 131) 3 §13.

[170] ICCAT, 'CoP15 Doc. 68 Annex 4 – Submission of documents by ICCAT to the FAO for the Panel' (11 January 2010).

[171] FAO, 'Report of the Third FAO Expert Advisory Panel for the Assessment of Proposals' (n 131) 120–122.

[172] FAO, 'Report of the Fourth FAO Expert Advisory Panel for the Assessment of Proposals' (n 136) 3.

[173] FAO, 'Report of the FAO ad hoc Expert Advisory Panel for the Assessment of Proposals' (n 4); FAO, 'Report of the Second FAO ad hoc Expert Panel for the Assessment of Proposals' (n 131); FAO, 'Report of the Third FAO Expert Advisory Panel for the Assessment of Proposals' (n 131); FAO, 'Report of the Fourth FAO Expert Advisory Panel for the Assessment of Proposals' (n 136).

[174] CITES, 'CoP13 Doc. 60 – Amendments of the Appendices – Proposals to Amend Appendices I and II' (2–14 October 2004) 1–2 §7; CITES, 'CoP14 Doc. 68 – Amendments of the Appendices – Proposals to Amend Appendices I and II' (3–15 June 2007) 1–2 §6; CITES, 'CoP15 Doc. 68 – Amendments of the Appendices – Proposals to Amend Appendices I and II' (13–25 March 2010) 1 §6; CITES, 'CoP16 Doc. 77 – Amendments of the Appendices – Proposals to Amend Appendices I and II' (3–14 March 2013) 1–2 §6.

information, stating that no scientific data was available, presenting the relevant measures applicable in their areas of competence or pointing out mistakes in the listing proposals.[175]

It is also worth mentioning that, following the proposals to list Patagonian Toothfish and Atlantic Bluefin Tuna, both CCAMLR and ICCAT keep on providing information to the CITES CoP on the measures they have taken in the inter-sessional period. CCAMLR is in a formal relationship with CITES (although it does not oblige its Secretariat to anything) and the ICCAT Secretariat, under the framework of the guidelines, is encouraged to provide information on matters of common interest, which it did by sending, for CoP16, a letter summarizing the progress made in the Eastern Atlantic.[176]

2. Problems

a) Perceived or Real Competition

The proposals to list Patagonian Toothfish and Atlantic Bluefin Tuna caused tensions with members of respectively CCAMLR and ICCAT due to a perceived competition between CITES and the RFMOs. Indeed, in 2002, all CCAMLR members, except Australia and New Zealand, were strongly opposed to a listing,[177] arguing that it would undermine CCAMLR's efficiency by decreasing its competence,[178] weakening its reputation[179] and hence putting its credibility and authority at risk.[180]

Similarly, in the run up to the vote for Atlantic Bluefin Tuna, the tension between CITES and ICCAT was high and arguments put into question the competence of these institutions in a mutually exclusive way.[181] In particular, the naming and shaming of ICCAT by certain states and NGOs during the debates at CoP15—using words such as "last chance for Atlantic Bluefin Tuna"; "gross mismanagement";[182]

[175] CITES, 'CoP16 Doc. 77 – Amendments of the Appendices – Proposals to Amend Appendices I and II' (n 174) annexes.

[176] Respectively CCAMLR, 'CoP16 Doc. 63 (Rev.1) – Toothfish: Report of CCAMLR' (n 56) for toothfish; and ICCAT, 'CoP16 Inf. 42 – Measures taken by the ICCAT on Eastern Bluefin Tuna' (3–14 March 2013) for Bluefin Tuna.

[177] CCAMLR, 'CoP12 Inf. 20 – Proposal from Australia (CoP12 Prop. 39)' (3–15 November 2002) 7–8 §10.26-10.28 (only Australia, New-Zealand, and to some extent Sweden agreed on the interest of a listing); on this general opposition, see Calley (n 55) 175–181.

[178] CCAMLR, 'CoP12 Inf. 20 – Proposal from Australia (CoP12 Prop. 39)' (n 177) 4 §10.4 (Japan).

[179] Ibid. 4 §10.6 (Japan), 15–16 §10.68 (UK).

[180] Ibid. 5 §10.13 (Norway), 14 §10.58 (Italy).

[181] FAO, 'CoP15 Inf. 26 – Statement from the FAO Fisheries and Aquaculture Department on CoP15 Proposal 19 to List the Atlantic Bluefin Tuna in Appendix I' (13–25 March 2010) annex §2–3.

[182] Greenpeace, 'CITES – Last Chance for Atlantic Bluefin Tuna' (2010) http://www.greenpeace.org/international/en/publications/reports/cites-last-chance-for-bluefin/ accessed 5 July 2015, 1.

"abject failure";[183] or "a dismal failure"[184]—is of questionable value and risks feeding competition as well as a sense of dispossession rather than improving accountability. While neither ICCAT nor CITES directly attacked each other's potential role, the FAO stated that deciding to list or not Atlantic Bluefin Tuna on CITES Appendices "requires, in part, a choice as to whether the relevant regional fisheries management organization, ICCAT, should retain its sole mandate to do so or whether CITES is also required."[185] It further mentioned that moving away from RFMOs' primary management functions was possible, particularly if they do not meet the legal obligations incumbent upon them, such as the adoption of sustainable management plans.[186] This remark can be understood as suggesting that listing a fish species under CITES would be a punishment for mismanagement of resources by an RFMO and kept as a threat equivalent to shaming the management regimes for inadequate work. Any decision regarding a listing would then be perceived as a vote of confidence for/against the relevant RFMO.

However, opinions were not unanimous in assuming that the interrelationship was a competitive one, with Sweden agreeing that, should CCAMLR measures not be sufficient, using CITES was an option to consider.[187] Other countries also stated that a listing would not put into question CCAMLR's primary competence to decide catch levels, adopt CMMs on fishing methods and implement the CDS. Moreover, the above mentioned FAO's comment made during the discussions related to Atlantic Bluefin Tuna asked "whether CITES is *also* required", not whether CITES should be used instead of ICCAT. More recently, following the listing of shark species at CoP16, the FAO and CITES jointly issued a brochure stating that "CITES can *complement* fishery management to encourage fishery stakeholders—managers and fishermen alike—to implement and follow sustainable fishing practices."[188]

---

[183] Ibid. 2.

[184] The Pew Environment Group, 'CITES Proposal 19 – Atlantic Bluefin Tuna' (2010) http://www.pewtrusts.org/~/media/legacy/uploadedfiles/peg/publications/fact_sheet/CITES20201020Fact20SheetsAllEnglishpdf.pdf accessed 5 July 2015, 1.

[185] FAO, 'CoP15 Inf. 26 – Statement from the FAO Fisheries and Aquaculture Department on CoP15 Proposal 19 to List the Atlantic Bluefin Tuna in Appendix I' (n 181) annex §1.

[186] The FAO, in CoP15 Inf. 26 (ibid. annex §2 (reference omitted)), stated that "[i]n accordance with the UN Law of the Sea and the UN Fish Stocks Agreement, FAO recognises and respects the mandate of regional fisheries management organisations (RFMOs) for conservation and management of living resources in the areas of the high seas. Fundamentally therefore, FAO is supportive of these bodies having the primary management function for straddling fish stocks and highly migratory fish stocks. At the same time, the rights granted to these bodies by international law brings with it legal obligations including, in particular, the need to 'adopt measures to ensure the long-term sustainability of straddling fish stocks and highly migratory fish stocks and promote the objective of their optimum utilization'."

[187] CCAMLR, 'CoP12 Inf. 20 – Proposal from Australia (CoP12 Prop. 39)' (n 177) 8 §10.28 (Sweden).

[188] CITES and FAO, 'A brief introduction to CITES listing of shark species' http://cites.org/sites/default/files/eng/prog/shark/docs/sharks_brochure_v8.pdf accessed 5 July 2015, 2 (italics added).

One can wonder whether formal consultations with the RFMO prior to proposing a listing could have avoided this perceived competition. In the case of Patagonian Toothfish, the proposal mentioned that range states had been approached for consultation, including CCAMLR members but that only few responded. Following this consultation, Australia included a note saying that "[t]he concerns expressed by those parties that do not support a nomination at this time were largely due to a perception that a CITES listing would undermine the role of CCAMLR. We have structured the nomination to ensure that the strength of CCAMLR management is ensured."[189] Whether the proposal actually succeeded in reflecting this intent is debatable, but, in any case, CCAMLR member states were of the opinion that their RFMO's role was being disregarded. A formal discussion within CCAMLR or between the proponent and the institution might have been positive. In that regard, Norway highlighted CCAMLR's article XXIII on the Commission's cooperation with other institutions. It inferred from this provision, rightly or not, that, since CCAMLR's individual members cannot act on behalf of the Commission, they have to refrain from any independent contact with institutions whose areas of competence overlap with that of CCAMLR.[190]

b) Lack of Clarity in the Processes

The procedures used to coordinate information gathering and advice from the relevant RFMOs in relation to amendments' proposals lack clarity. CITES explicitly provides for the need to consult with the relevant international organizations in the postal and normal procedures and such direct consultation with RFMOs does take place. However, it runs in parallel to the FAO's effort to contact RFMOs in order to gather information. It is unclear whether those two procedures have different goals as they are undertaken at the same stage of the listing process, that is after the proposal has been submitted.

As for the FAO, it is noteworthy that the MoU does not provide it with a clear mandate to coordinate information gathering from the relevant RFMOs or to inform them about the proposals. The only reference to RFMOs in the MoU specifies a responsibility of the CITES Secretariat at the end part of the process, i.e. once recommendations are made. The pertinence of mentioning third parties and their interactions with the CITES Secretariat in the MoU between CITES and the FAO is questionable. Moreover, the lack of consistency between the explicit mandate and the reality of FAO's coordinating role is problematic, particularly since the FAO Expert Panel is criticized whatever conclusions it reaches, whether by pro-conservation or by pro-fishing states. Its role should be clarified and agreed

---

[189] Australia, 'CoP12 Prop. 12.39 – Consideration of proposals for amendment of Appendices I and II – Patagonian Toothfish' (3–15 November 2002) 22 attachment A.

[190] CCAMLR, 'CoP12 Inf. 20 – Proposal from Australia (CoP12 Prop. 39)' (n 177) 5 §10.9, 15 §10.66 (Norway).

upon by CITES and the relevant RFMOs in a manner that would entitle the Expert Panel to cooperate in an all-inclusive scientific review. Alternatively, the most advisable option might be for the FAO to keep clear from the somehow tense relationships between CITES and RFMOs and to let direct partnerships follow their course.

This situation is probably not the cause of the perceived competition but does not help to decrease the tension. It also tends to put the FAO in an in-between position which is potentially harmful to its advice-providing role.

III. CMS

The CMS has its own scientific body, the mandate of which includes the provision of scientific advice to the Secretariat and the CoP, research on the status of migratory species and review of the Appendices.[191]

The CMS, as an intergovernmental body having a function in relation to marine species, is requested to provide information and did so for CoP16. It briefly presented the protection it affords to the species included in the relevant proposals, commented on their biological status and need for additional protection.[192]

The CMS could further serve as a partner institution to CITES with regard to migratory sharks, in particular if the CMS gathers data under the ambit of its MoU.[193] Indeed, the conservation plan adopted under this recent instrument includes as one of its goals the promotion of "practical and enforceable conservation recommendations based on the best available science within relevant [...] biodiversity-related MEAs".[194] This is however not yet implemented.

IV. Comparison

1. IWC

The delivery of advice and information from another institution has a long history in CITES in relation to cetaceans. It was already envisioned at CoP2 with a

---

[191] 'Convention on the Conservation of Migratory Species of Wild Animals' (CMS) 1979, 1651 *United Nations Treaty Series* 333 article VIII(5).

[192] CITES, 'CoP16 Doc. 77 – Amendments of the Appendices – Proposals to Amend Appendices I and II' (n 174) annex 4 'CMS comments on CITES listing proposals' (10 January 2013).

[193] Caddell, 'Inter-Treaty Cooperation, Biodiversity Conservation and the Trade in Endangered Species' (n 37) 279.

[194] Annex 3 to the MoU on the Conservation of Migratory Sharks: Conservation Plan 27 September 2012 http://cms.eaudeweb.ro/fr/document/conservation-plan accessed 5 July 2015, 1 §1, 5 §6.3, 8 §15.1.

Resolution providing for consultation with the IWC regarding listing and de-listing. The IWC was accepted as more able, scientifically, to gather data on the relevant species' status.[195] This reflects the requirement in article XV (2) (b) that the Secretariat "consult[s] with inter-governmental bodies having a function in relation to those species especially with a view to obtaining scientific data these bodies may be able to provide." A Resolution on the *Conservation of cetaceans, trade in cetacean specimens and the relationship with the International Whaling Commission* was also adopted in 2000, recognizing "that the IWC is the major source of information on whale stocks around the world".[196]

It has been widely debated whether CITES is actually independent in its choice of measures towards whales or whether it is obliged to adapt them to the IWC's decisions. The hierarchical preeminence of the IWC could be argued for,[197] in particular since CITES has, so far, always supported the other institution's decisions. However, legally speaking, it is under no obligation to always follow the IWC's advice as to amending the Appendices.[198] Indeed, the Convention only requires the Secretariat to consult the IWC to gather information and to coordinate conservation measures. This obligation lies with the Secretariat, not with the CoP. Moreover, it is an obligation to consult, not to comply. Finally, Resolution 11.4, in its operative part, says nothing about considering the IWC's decisions as binding, or as the unique source of science.

While it is true that the CITES CoP listed whale species which were in fact not fulfilling the listing criteria,[199] this does not necessarily signify, as Arbour and Lavallée see it, that states parties definitely handed over their decision-making power to the IWC.[200] Rather, the majority of the parties took a politically motivated decision in order, at a given moment, to coordinate CITES measures with the IWC. The CoP has taken political decisions regularly, also in matters in which no other

---

[195] Resolution Conf. 11.4 (Rev. CoP12) – Conservation of cetaceans, trade in cetacean specimens and the relationship with the International Whaling Commission (n 70) preamble; C. P. Carlarne, 'Saving the Whales in the New Millennium: International Institutions, Recent Developments and the Future of International Whaling Policies' (2005–2006) 24 Virginia Environmental Law Journal 1, 24–25; Favre, *International Trade in Endangered Species* (n 49) 91.

[196] Resolution Conf. 11.4 (Rev. CoP12) – Conservation of cetaceans, trade in cetacean specimens and the relationship with the International Whaling Commission (n 70) preamble.

[197] M. Arbour and S. Lavallée, *Droit international de l'environnement* (Bruylant 2006) 411.

[198] A. Proelß, 'Marine Mammals' Max Planck Encyclopedia of Public International Law www.mpepil.com §18.

[199] Such as the minke whale (A. Gillespie, *Whaling Diplomacy: Defining Issues in International Environmental Law* (New Horizons in Environmental Law, Edward Elgar 2005) 338; or in general Arbour and Lavallée (n 197) 412).

[200] Arbour and Lavallée (n 197) 411, as recognized in Resolution Conf. 11.4 (Rev. CoP12) – Conservation of cetaceans, trade in cetacean specimens and the relationship with the International Whaling Commission (n 70) and, earlier, in Resolution Conf. 9.12 – Illegal Trade in Whale Meat 1994 (repealed), where it was said that the IWC is the main source of information regarding whales.

institution is involved. The majority can decide to modify these decisions, sovereignly and whenever they wish.

For example, some CITES parties and even the Secretariat have 'rebelled' at CoP10 in Harare against the obedience CITES had so far shown towards the IWC's decisions. Norway's proposal to down-list the minke whale[201] has been recommended by the Secretariat[202] and a majority of 57 to 51 agreed to it.[203] Although this proposal did not reach the required two-third majority, it clearly showed that the criticized lack of scientific basis for some measures as well as the debated hardline protectionism defended at the IWC—and sometimes at CITES—were considered valid arguments by several countries.

Also, at CoP12, Mexico unsuccessfully suggested the adoption of a new Resolution in which the CoP would "[decide] to retain whale species in the Appendices in which they are currently listed, considering the results of the 54th Annual Meeting of the IWC".[204] The Secretariat expressly commented that this proposal goes against the amendment procedure to be found in article XV and negates the existence of the Fort Lauderdale criteria.[205] That this proposal was not adopted shows a willingness to formally retain, and maybe practically regain, decision-making power for species under the CITES regime, even if they are managed by another institution.

At CoP13, it was Japan's turn to propose down-listing of the minke whale and to come up with a new draft Resolution criticizing the non-adoption of the revised management scheme and the negative effects this had on CITES;[206] this document was partly supported by the Secretariat after some changes, but both proposals failed. According to Japan as well as to Norway, the failure of the IWC to adopt the revised management whaling regime promised for so long showed the regime's dysfunctionality[207] and consequently the need for the protection under CITES to free itself from its influence.

In conclusion, so far, the CITES CoP has followed the IWC's advice, as a matter of principle, not of science. Some states try to challenge the status quo in both institutions by requesting the end of the IWC moratorium and trying to down-list some species. Legally speaking, a review of the listing would be possible under

---

[201] Norway, 'CoP10 Prop. 10.22 – Proposal to transfer the Northeast Atlantic and North Atlantic Central stocks of minke whale (*Balaenoptera acutorostrata*) from Appendix I to Appendix II' (1997).

[202] CITES Secretariat, 'CoP10 Doc. 10.89 (Rev) Annex 1 – Comments from the Parties and Comments and Recommendations from the Secretariat' (9–20 June 1997) 911–912.

[203] CITES, 'CoP10 Com.I 10.9 – Summary Report of the Committee I Meeting' (9–20 June 1997) 194.

[204] Mexico, 'CoP12 Doc. 16.4 – Cooperation between CITES and the International Whaling Commission' (3–15 November 2002) 3 draft Resolution's operative part.

[205] Ibid. 1 'Comments from the Secretariat' B.

[206] Japan, 'CoP13 Doc. 12.2 – CITES listing of whale stocks and the International Whaling Commission' (2–14 October 2004) 2 draft Resolution.

[207] Gillespie, *Whaling Diplomacy* (n 199) 340–341.

CITES as nothing in the Convention or in a Resolution makes the CITES CoP dependent on another institution for listing decisions.

Should a down-listing be accepted by the required majority of CITES parties, the whale species in Appendix II would arguably, for states parties to the IWC, fall under the simplified procedure provided for in article XIV (4)–(5): the flag state would just have to issue a certificate stating that the specimen has been caught in accordance with the provisions of the other treaty, here the ICRW.[208] This would be a typical situation where the parties to CITES while formally allowed to trade specimens of a species, would not be allowed to catch them under the IWC and hence would have to refrain from any harvesting in order to fulfil article XIV (5)'s criterion. As examined earlier, although far from ideal, this situation would not result in a *stricto sensu* conflict of rules as the CITES provisions are permissive of trade, not making it an obligation.[209]

The tension witnessed in the whaling question is not so much between the organizations but rather between the states forming them and trying to instrumentalize each meeting to express their opposing views which are based on philosophical differences. Neither institution has a mandatory dispute settlement mechanism;[210] it has been said that "as with the IWC, differing perspectives concerning the treatment of whales can cause frequent ripples in CITES' institutional stability."[211] Without going that far it cannot be denied that the tension linked to the use of CITES for commercially-exploited aquatic species is influenced or maybe even stems from the philosophical conflict which underlies the whaling debate.

2. Forestry

When requested, the FAO advises the CITES Secretariat about listing timber species, but there is no established bilateral framework, such as the MoU, yet, for its role.

According to the Resolution on the *Implementation of the Convention for Timber Species*, before submitting a proposal for a listing, the future proponent states have to consult with at least four organizations specialized in timber issues (two on trade matters and two on biological questions), from a list provided in the Resolution. This list includes the FAO, regional timber organizations, research institutions and NGOs such as TRAFFIC or the WWF. The information so gathered has then to be

[208] See supra Chapter 5 B. III. 1. a).

[209] See supra Chapter 5 B. III. 2. c) aa).

[210] The case brought by Australia against Japan in front of the ICJ was so on the basis of article 36(2) of the 'Annex to the United Nations Charter: Statute of the International Court of Justice' (Statute of the ICJ) 1945, 1 *United Nations Treaty Series* xvi, both countries having made a declaration of acceptance of the Court's jurisdiction.

[211] Carlarne (n 195) 24.

incorporated in the proposal.[212] States that are parties to the 1994 International Tropical Timber Agreement (ITTA) or its successor treaty have to consult with ITTO if they intend to submit a proposal.[213] For proposals to amend Appendices in relation to timber species, to species of which all specimens in trade are artificially propagated, the Secretariat has to seek the views of ITTO, FAO and IUCN and share them at the CoP.[214]

Hence roles are well divided: states willing to submit a proposal are the ones who have to contact and consult relevant organizations, including some regional ones. As for the CITES Secretariat it sends its request for opinions on certain species to some key institutions, one of which being the FAO. This two-steps cooperation with international organizations might be perceived as an overlap, but it might also help reduce possible tensions as well as separate the political from the scientific advice. Indeed, as the state has to be in direct contact with the relevant organizations some room and time is set for discussion at this early stage. Then, while the state might disregard the opinion of the specialized institutions in its proposal—a political move—the CITES Secretariat is able, when needed, to gather neutral advice to present to the CoP. Finally, keeping the FAO as an information provider instead of requesting it to coordinate the information can increase its legitimacy as a neutral adviser.

V. Conclusions

1. Division of Roles and Independence of Science

The current cooperation with regard to listing commercially-exploited fish species is quite efficient. Indeed, requesting scientific information from the specialized institutions is positive and necessary due to the complexity of the data to analyze and the lack of CITES' specialized knowledge in fisheries. Also, more generally, "[n]o one [party] could possibly have the scientific expertise to be able to independently evaluate all the plant and animal species that might be proposed. In fact, many Parties do not even have reasonably available to them access to individuals capable of doing scientific evaluations for all species. [...] The opinions of the

[212] Resolution Conf. 10.13 (Rev. CoP15) – Implementation of the Convention for timber species (n 96) 'Regarding international organizations', a).

[213] Resolution Conf. 14.4 – Cooperation between CITES and ITTO regarding trade in tropical timber (n 99) operative part.

[214] Resolution Conf. 10.13 (Rev. CoP15) – Implementation of the Convention for timber species (n 96) 'Regarding international organizations', b), with reference to Resolution Conf. 9.24 (Rev. CoP16) – Lauderdale criteria (n 50) first RESOLVES, h). The Secretariat did so, for example, for CoP16 (CITES, 'CoP16 Doc. 77 – Amendments of the Appendices – Proposals to Amend Appendices I and II' (n 174) 2 §7).

Secretariat is [sic] one very important source of information."[215] However, even the Secretariat must refer to other bodies for specialized knowledge.

The fact that CITES externalizes the provision of advice is mostly constructive. First of all, it is necessary to cooperate with the specialized institutions due to resources' limitations. Nevertheless, the FAO Expert Panel might be speaking another institutional language and has goals that are different from the aims of the CITES Secretariat. This is made clear by the lack of similar interpretation of Annex 2aB.

However, it does not seem that the disagreements with the FAO's recommendations are due to the problem of having an outside institution providing advice—at least not beyond the issue of interpretation mentioned. In fact, as will be highlighted below, the Expert Panel, with its focus on sustainable use rather than preservation, seems to share the philosophy of the CoP more than the Secretariat does.

Two other positive aspects of the current situation are externalization of advice and the fact that the scientists invited to the Expert Panel participate in their personal capacity and not as the representatives of their state or institution. Indeed, provision of scientific advice in-house can create problems of its own, in particular if scientists come from the countries involved in the decision-making process and do not act in their personal capacity, if somehow funding is dependent on the conclusions reached, if policy makers can review the advice or if they are even involved in giving scientific advice.[216] "In some cases, such externalisation of research collaboration can help insulate it from sensitive political issues which otherwise might obstruct problem solving."[217]

Furthermore, the FAO Expert Panel brings quality science-based advice to the discussion and, in most cases, the arguments contradicting the findings of the Panels are also developed in scientific terms.[218] This situation provides for a better chance of finding a compromise than when arguments are purely political or ideological.

Moreover, that the Expert Panel's recommendations are free from political interests is quite accepted. For example, Japan, which is financing the CITES-activities of the FAO and which is the ultimate supporter of the FAO, was disillusioned by the Expert Panel when they advised to list the Atlantic Bluefin Tuna, but did not cancel its funding for the project. To be able to displease the main funder is always a good sign. One must nevertheless note that financial resources are limited, and CITES is not a top priority item on the FAO's agenda. Even though funding the CITES related activities through extra-budgetary resources originating

[215] Favre, *International Trade in Endangered Species* (n 49) 311.

[216] T. Polacheck, 'Politics and independent scientific advice in RFMO processes: A case study of crossing boundaries' (2012) 36 Marine Policy 132, 139.

[217] O. S. Stokke, 'Governance of High Seas Fisheries: The Role of Regime Linkages' in D. Vidas and W. Ostreng (eds), *Order for the oceans at the turn of the century* (Kluwer Law International 1999) 161.

[218] Germany (on behalf of the European Union), 'CoP14 Inf. 48 – Comments on the FAO Assessment of the CITES Amendment – Proposals to list the porbeagle (lamna nasus) and the spiny dogfish (squalus acanthias) on Appendix II of CITES' (3–15 June 2007).

from a single donor does not directly affect the independence of the research, it is not an ideal situation. Indeed, one way or another, "[f]unding will always be a limiting factor and a potential mechanism for control or explicit or implicit influence on science process."[219]

As for the negative aspects of the partnership, opinions differ as to whether overlap and duplication with regard to scientific and analytical activities are desirable or not.[220] While it might provide an opportunity to reach a consensual scientific basis for decisions and understanding, it also entails risks of delegitimization of scientific advice if institutions or people cannot agree. The opposing conclusions reached by the Expert Panel and the Secretariat, although positive scientifically speaking, allow a *prima facie* scientific argument to be made on both opinions and could be contributing to a stalemate.

Then, to have the assessment both of whether a species meets the CITES criteria and of the potential effectiveness of a listing discussed in the same meeting is not necessarily the most adequate format. Indeed, while the former aspect is clearly based on science, the second includes approximations and relies on assumptions; it could thus be beneficial to separate these issues in the report. Furthermore, the qualifications of the Expert Panel members might not be optimal to discuss questions of international trade. Hence, it would be recommendable to organize a second panel of experts in wildlife trade matters. In any case, it is positive that studies, independent of the CITES-related program, are undertaken by the FAO to determine the socio-economic importance of trade in a species for a community or a state,[221] in that such research's existence simplifies the examination of a listing's potential effectiveness. The FAO is in any case aware that the provision of advice on the technical aspects (in opposition to the biological ones) of a proposal could be improved. It envisions, funding permitting, to organize a workshop in order to identify issues, in fisheries management and international trade, which are of importance in relation to a listing in CITES, as well as the factors having an influence on how successful a listing can be.[222]

Moreover, early inclusiveness is an important feature to avoid competition and misunderstandings. In that regard the current situation is not fully satisfying as the role given to RFMOs is an *ex post facto* one. This can only fuel the perceived competition. While it might not be desirable to require the support of the relevant RFMO in order to be able to submit a proposal, consultation at an early stage, such as is envisioned for timber species, would be an improvement.

---

219 Polacheck, 'Politics and independent scientific advice in RFMO processes: A case study of crossing boundaries' (n 216) 140.

220 M. Ivanova and J. Roy, 'The Architecture of Global Environmental Governance: Pros and Cons of Multiplicity' http://www.centerforunreform.org/node/251 accessed 5 July 2015, 2–3, 11.

221 Such as that by C. Catarci, 'World Markets and Industry of Selected Commercially Exploited Aquatic Species with an International Conservation Profile' FAO Fisheries Circular 990 (Rome 2004).

222 Committee on Fisheries – Sub-Committee on Fish Trade, 'Update on CITES related activities' presented at the Fourteenth Session (n 162) 8 §35-36.

Finally, in addition to being involved too late in the listing procedure, RFMOs are contacted through two channels at same time; the logic behind such parallel processes remains unclear. Also, to have the FAO coordinate the request of information from RFMOs entails a risk for the FAO's legitimacy both towards RFMOs which could feel relegated to a secondary role, and towards CITES which could think that the transmission of the information gathered from the RFMOs reflects the FAO's views. This issue was already highlighted at the second technical consultation on the criteria.[223] Most representatives of RFMOs also "expressed the view that CITES must take direct responsibility for liaising with the appropriate intergovernmental body and that this responsibility should not be delegated to FAO."[224] This recommendation appears to be followed in practice, although only in the sense that RFMOs are also contacted by the CITES Secretariat; this does not resolve the uncomfortable middle-man position of the FAO.

2. Listing Remains a Political Decision

It should not be forgotten that decision-making under CITES often remains of a political nature. One (quite harmless) example is that of listing the date mussel listing at CoP13. The species is not threatened because of international trade, but because harvesting destroys its habitat.[225] Its listing was a feel-good decision at the end of a CoP, although it does not fulfil the CITES criteria.

The decisions to list commercially-exploited aquatic species decreased from CoP13 to CoP15—none was taken in 2009—but experienced a regain in 2013. One must note the suspicion related to such proposals; Japan commented on all proposals and is known to have been wary of Western states trying to get as many species of sharks on the list as possible.

When the Expert Panel's advice was not backing up a listing, the CITES Secretariat dissented, most probably due to the different interpretation of Annex 2aB. However, in such cases the CoP reached the same decision as was advised by the Expert Panel. In contrast, when the Expert Panel's conclusions supported the listing of species, its advice was widely followed by the CITES Secretariat but not by the majority of the CoP. The way these recommendations are taken into account gives a clear picture of the different values underlying the CITES Secretariat's and the Expert Panel's methodology as well as the goals pursued by the CoP. A majority of states is indeed happy to follow science when it advises not to list a species and to put it aside when it does. Science appears not to be the decisive factor in decisions

[223] FAO, 'Report of the Second Technical Consultation on the Suitability of the CITES Criteria for Listing Commercially-exploited Aquatic Species' (n 107) §28.

[224] FAO, 'Report of the Second Meeting of FAO and non-FAO regional fishery bodies or arrangements' (n 111) 9 §47.

[225] FAO, 'Report of the FAO ad hoc Expert Advisory Panel for the Assessment of Proposals' (n 4) 47–49.

and, currently, a majority of states may not only feel unconvinced by the science, but also uncomfortable with the listing of further commercially-exploited fish species.

Lack of consensus between the CITES Secretariat and the Expert Panel does not help but appears not to be the determining factor in the success or failure of a proposal. Even a high level of consensus does not ensure that decision-makers will follow science. For example, at CoP15, not only did the Expert Panel and the CITES Secretariat agree on the need to list the spiny dogfish, but studies from the IUCN and TRAFFIC also reached the same conclusions.[226] However, this shark species was not listed.

As a matter of comparison however, it must be noted that, for the bigleaf mahogany (also a controversial but successful listing), it was widely accepted that the criteria were met, that trade is a major (if not the only) driver of exploitation, and that illegal logging is very important.[227] That this proposal was finally adopted shows that, in certain cases, the better the scientific data and the larger the consensus, the more likely a listing will take place.[228]

In conclusion, the provision of advice from a third party, ideally an independent and specialized institution, is desirable as well as useful. The earlier the relevant RFMOs are contacted, the higher is the likelihood that they feel involved with the proposal. In any case, clarifying the roles of the relevant actors and rationale of various procedures is important, even though it does not have the potential to resolve the underlying political issues at stake.

D. Cooperation in Implementing Protection

I. FAO

1. Permit Issuance and Control

In order to deliver a permit, an NDF is necessary, which in turn requires the exporting state or state of introduction to have a management system in place. In that regard, the CITES-FAO MoU provides that "[t]he signatories will cooperate as appropriate to facilitate capacity building in developing countries and countries with economies in transition on issues relating to commercially-exploited aquatic species listed on the CITES Appendices [and that. . .] both signatories will address technical and legal issues relating to the listing and implementation of such listings."[229]

---

[226] Spain (on behalf of the European Union), 'CoP15 Inf. 36 – Frequently asked questions related to the listing proposals of porbeagle and spiny dogfish' (13–25 March 2010) 5 §16.

[227] A. Blundell, 'A Review of the CITES Listing of Big-leaf Mahogany' (2004) 38 Oryx 84, 84–86.

[228] Also noteworthy is the example of the humphead wrasse as opposed to the white great shark (T. Gehring and E. Ruffing, 'When Arguments Prevail Over Power: The CITES Procedure for the Listing of Endangered Species' (2008) 8 Global Environmental Politics 123, 139–141, 142–144).

[229] CITES-FAO MoU (n 9) §2, 4.

For some of the commercially-exploited aquatic species listed in CITES Appendices, the FAO has, even before the MoU, undertaken capacity-building activities, mostly workshops and case studies, to help member states comply fully with their obligations.[230] Workshops and technical guidelines for the sustainable management of sea cucumbers were organized and developed, a multi-dimensional case-study for the situation in the Eritrean Red Sea coast implemented and a catalogue for the identification of this species produced.[231]

With regard to the humphead wrasse, the FAO and IUCN developed a stock assessment model to be used by trading countries in their NDF.[232] The FAO also more recently worked on monitoring and management of this species and on reef habitat evaluation.[233]

The FAO, with the help of local and regional partners in the Caribbean, developed a 'Manual for the monitoring and management of queen conch'.[234] Then as a consequence of the CITES Significant Trade Review in that species, it provided advice to the countries identified as facing difficulties on ways to better manage their fisheries of queen conch.[235] It also contributed to and organized capacity-building meetings in 2005 and 2006.[236]

As for sharks, since no RFMO has a direct management mandate over them—although many have adopted CMMs in the recent years—the data available through the FAO might be necessary. In any case, its expertise is crucial. The FAO can indeed be of particular use, by helping states which have not done so, to develop a national plan of action for sharks. In 2005, an FAO Expert Consultation was organized to review the implementation of the IPOA-Sharks.[237] More recently, a

---

[230] For a review of such early activities, see CITES Secretariat, 'CoP14 Doc. 18.1 – Cooperation with other organizations – Cooperation with the Food and Agriculture Organization of the United Nations' (n 125) 2–3 §12-22.

[231] FAO, 'Capacity Building – Sea Cucumbers' http://www.fao.org/fishery/topic/16360/en accessed 5 July 2015.

[232] FAO, 'Stock Assessment Approach for the Napoleon Fish, Cheilinus Undulatus in Indonesia, A tool for quota-setting for data-poor fisheries under CITES Appendix II Non-Detriment Finding requirements' FAO Fisheries Circular 1023 (Rome 2007).

[233] R. Gillett, 'Monitoring and management of the humphead wrasse, Cheilinus undulatus' FAO Fisheries and Aquaculture Circular 1048 (Rome 2010); A. Oddone and others, 'Estimating reef habitat coverage suitable for the humphead wrasse, Cheilinus undulatus, using remote sensing' FAO Fisheries and Aquaculture Circular 1057 (Rome 2010).

[234] P. Medley, 'Manual for the monitoring and management of queen conch' FAO Fisheries Circular 1012 (Rome 2005).

[235] FAO, 'Capacity Building – Caribbean Queen conch' http://www.fao.org/fishery/topic/16390/en accessed 5 July 2015.

[236] Ibid. The 2006 workshop resulted in a report: FAO, 'Regional Workshop on the Monitoring and Management of Queen Conch, Strombus gigas (Kingston, 1–5 May 2006)' FAO Fisheries Report 832 (Rome 2007).

[237] FAO, 'Report of the FAO Expert Consultation on the Implementation of the FAO International Plan of Action for the Conservation and Management of Sharks (Rome, 6–8 December 2005)' FAO Fisheries Report 795 (Rome 2006).

technical workshop examined the ways to improve the monitoring of shark fisheries and trade.[238] Also, a FAO/CITES workshop took place to scrutinize the application and effectiveness of international regulatory measures for the conservation and sustainable use of Elasmobranchs.[239]

Following the CoP16 decision to list additional shark species, the FAO and CITES have, under the auspices of an EU-funded project,[240] intensified their cooperation in order to ensure the efficient implementation of such listings after the 18 months delay period. In particular, regional consultative meetings have been organized to support states in the implementation of CITES requirements.[241]

For sturgeons, the FAO and CITES have also organized technical workshops on stock assessment and TACs methodologies as well as on ways to combat illegal fishing of sturgeons.[242] The former meeting, held in 2008, responded to a request of the CITES Animals Committee, and targeted the methodology used by the five Caspian countries in their management of sturgeons. The latter, held in 2009, was also convened by both institutions to "identify the various steps towards the development of a regional strategy and action plan to fight crime in the sturgeon fisheries sector".[243]

Continued support in the implementation of CITES regulations is on the program of the FAO activities. This is needed, since already now, with the limited range of commercially-exploited aquatic species listed, "[a] number of countries, especially developing countries and countries in transition, are experiencing

[238] FAO, 'Report of the Technical Workshop on the Status, Limitations and Opportunities for Improving the Monitoring of Shark Fisheries and Trade (Rome, 3–6 November 2008)' FAO Fisheries and Aquaculture Report 897 (Rome 2009).

[239] FAO, 'Report of the FAO/CITES workshop took place to review the application and effectiveness of international regulatory measures for the conservation and sustainable use of Elasmobranchs (Genazzano, 19–23 July 2010)' FAO Fisheries and Aquaculture Report R984 (Rome 2012).

[240] CITES, *CITES getting ready for sharks and rays* (2013) http://www.cites.org/eng/news/pr/2013/20130914_shark_ray.php accessed 5 July 2015; Scanlon, *FAO-CITES related activities on commercially exploited aquatic species*, CITES Secretary General's intervention presented at the Fourteenth Session of the Sub-Committee on Fish Trade (Bergen, 24–28 February 2014) http://cites.org/eng/news/sg/2014/20140226_cofi-ft.php accessed 5 July 2015; CITES, 'Strengthening capacity in developing countries for sustainable wildlife management and enhanced implementation of CITES wildlife trade regulations, with particular focus on commercially-exploited aquatic species: Project summary sheet' (n 41).

[241] Scanlon, *FAO-CITES related activities on commercially exploited aquatic species* (n 240); Committee on Fisheries – Sub-Committee on Fish Trade, 'Update on CITES related activities' presented at the Fourteenth Session (n 162) 8 §34.

[242] FAO and CITES, 'Report of the FAO and CITES Technical Workshop – Stock Assessment and TAC Methodologies (Rome, 11–13 November 2008)' TCP/INT/3101 (Rome 2008); FAO and CITES, 'Report of the FAO and CITES Technical Workshop on Combating Illegal Sturgeon Fishing and Trade (Antalya, Turkey, 28–30 September 2009)' TCP/INT/3101 (Rome 2010).

[243] FAO and CITES, 'Report of the FAO and CITES Technical Workshop on Combating Illegal Sturgeon Fishing and Trade' (n 242) 1.

substantial difficulties in meeting CITES requirements, for example in the case of queen conch, humphead wrasse and the Caspian Sea sturgeons."[244]

In addition to the above mentioned activities linked to NDF, such as the development of a management plan, or linked to controls, such as identification booklets, the FAO, by its privileged relation with RFMOs, could serve as a centralized source of information. Indeed, scientific data is not only needed to take decisions on the amendment of the Appendices, but also at the stage of permit issuance—RFMOs have evidently a role to play in both phases.

With regard to legality findings, no specific activity has been undertaken. As long as states only have to check that catches are made in compliance with their national laws, they are in the best position to make this verification. FAO cannot help beyond supporting the development of efficient national frameworks for the management of fisheries. However, now that the CoP recommends that the legality of fishing should also be taken into account at the IFS stage and on the basis of international regulations, the FAO could have a role in helping to clarify the relevant rules and whether they are respected.

As for implementation at the borders, in addition to the above-mentioned identification guides, a shark fin software and guide have respectively been developed by the FAO[245] and by Pew.[246] However, the workshop planned in order to address enforcement and identification issues and consequently to avoid unnecessary listing of look-alike species while also facilitating the implementation of split-listing was cancelled due to budget reduction and has not been rescheduled.

2. Aquaculture

The potential impacts of a CITES listing on aquaculture could be addressed, at least partly, by captive-breeding and ranching. However, these particular frameworks within the regime have, so far, only been acknowledged briefly and incidentally in the CITES-related activities of the FAO. Also, while the Sub-Committee on Fish Trade of COFI drew attention, in 2012, to the need to include information on the role of aquaculture for the species examined, the discussion remained open as to

[244] FAO/Government Cooperative Programme, 'CITES and Commercially-exploited Aquatic Species, Including the Evaluation of Listing Proposals (Phase 2) – Project Document 2010-2015' (n 40) 4 (on file with the author).

[245] FAO, 'iSharkFin' http://www.fao.org/fishery/ipoa-sharks/iSharkFin/en accessed 5 July 2015; Committee on Fisheries – Sub-Committee on Fish Trade, 'Update on CITES related activities' presented at the Fourteenth Session (n 162) 4 §12.

[246] Honduras, 'CoP16 Inf. 40 – Shark fin identification guide in English, Spanish and French: Identifying Shark Fins: Oceanic Whitetip, Porbeagle and Hammerheads by the School of Marine and Atmospheric Sciences and The Pew Environmental Group' (3–14 March 2013).

whether this review should take place together with the existing Expert Panel procedure or whether it should be separated.[247]

Of the aquatic species listed or once proposed to be listed under CITES, some are reared in aquaculture programs (eel), some are fattened in farms (Bluefin Tuna), and some are bred in captivity (sturgeons). While both technical workshops on sturgeons, here above described, have mentioned the role and existence of aquaculture, no further detail was given. This is surprising as it was recognized that "[s]ince aquaculture facilities are suspected of playing a major role in laundering illegal sturgeon products, the actual output of the aquaculture facilities should therefore be assessed, compared to the representations being made about the production from the facilities, and monitored on a regular basis."[248]

Questions of definition will need to be answered,[249] ways to determine sustainability developed, traceability improved and identification guides put together. The FAO could also help build capacity with regard to sustainable management of aquaculture, in order for specimens originating from farms to qualify as CITES bred-in-captivity or ranched ones. This would avoid unnecessary restrictions to international trade in fish products, an aspect very important to FAO members.

II. RFMOs

1. Non-detriment Finding and Legality Finding

If a species managed by an RFMO is listed on a CITES Appendix, cooperation with the RFMO would be necessary in many regards and under different forms. A first situation arises if the RFMO is older than CITES and the species is listed in Appendix II: in that case, the RFMO parties would not have to apply the CITES permit system. Article XIV (4)–(5) is then indeed the framework for cooperation in implementing protection and it avoids doubling of work. As mentioned earlier several aspects—in particular the dating and meaning of protection—would need to be clarified for this provision to be easily applicable.[250]

[247] Committee on Fisheries – Sub-Committee on Fish Trade, 'Report of the Thirteenth Session of the Sub-Committee on Fish Trade (Hyderabad, India, 20–24 February 2012)' FAO Fisheries and Aquaculture Report 996 (Rome 2012) 12 §53.

[248] FAO and CITES, 'Report of the FAO and CITES Technical Workshop on Combating Illegal Sturgeon Fishing and Trade' (n 242) 13.

[249] In particular since a wider range of practices are included under the term 'aquaculture' at the FAO than would be falling under the CITES category of captive-breeding (A. Willock, 'Administrative and Monitoring Implications of listing and down-listing of commercially-exploited aquatic species, including the implications of Annex 4 of Resolution Conf. 9.24' Background paper to CITES Workshop on Introduction from the Sea Issues (30 November – 2 December 2005, Geneva (Switzerland)) http://www.cites.org/eng/news/meetings/ifs-05/IFS05-TRAFFIC-paper.pdf accessed 4 July 2015, 17–18).

[250] See supra Chapter 5 B. III. 1. a).

In all other situations the relevant RFMO would play a crucial role in terms of NDF and legality finding. For example, the Second Expert Panel indicated that for the spiny dogfish, which is likely to be caught in waters not under national jurisdiction, GFCM "could facilitate the development of agreed criteria for making non-detriment findings."[251] Also, as mentioned by the Third Expert Panel with reference to Atlantic Bluefin Tuna, the SCRS stock assessment could be used for NDF, and compliance with ICCAT and GFCM rules for legality findings.[252] Following the listing of several sharks at CoP16, a compendium of regional measures applicable to such species was put together and is available online, in relation to the legality finding requirement.[253] With regard to aquaculture, cooperation between the relevant RFMO and CITES parties would also be useful, if not even necessary in order to determine legality and the non-detrimental character of trade.

Since the RFMOs are the primary management bodies for fisheries and since they are the ones most likely to have information on the sustainability of catches and on the measures applicable to a specific fishery, CITES parties will not be able to determine non-detriment[254] or legality finding without their collaboration. That states consult and cooperate with RFMOs in issuing the relevant permit or certificate is explicitly recognized as necessary by the CoP.[255] Some types of custom-made partnerships would probably have to be developed to fulfill the specific requirements of each particular case. In a more general manner, it is acknowledged within the CITES-activities program of the FAO that "the extent and form of cooperation with regional fishery management organizations [. . .] need to be addressed."[256]

Whether the FAO is to serve as a clearing house mechanism or not (and whether it can) is a political decision that might have to be taken in time and place. Also, before any RFMO can be given the powers to determine NDF, the situation of

---

[251] FAO, 'Report of the Second FAO ad hoc Expert Panel for the Assessment of Proposals' (n 131) 45.

[252] FAO, 'Report of the Third FAO Expert Advisory Panel for the Assessment of Proposals' (n 131) 126.

[253] CITES, 'Resources for implementation: Legality' http://cites.org/eng/prog/shark/legality.php accessed 5 July 2015; CITES, 'Regional fisheries management organization measures for shark conservation and management (draft as at March 2014)' (March 2014) http://cites.org/sites/default/files/eng/prog/shark/docs/shark%20RFMO%20measures%20-%20draft%20March%202014.pdf accessed 5 July 2015.

[254] L. Little and M. A. Orellana, 'Can CITES Play a Role in Solving the Problem of IUU Fishing?: The Trouble with Patagonian Toothfish' (2005) 16 Colorado Journal of International Environmental Law and Policy 21, 78.

[255] Resolution Conf. 14.6 (Rev. CoP16) – Introduction from the Sea (n 51) preamble.

[256] FAO/Government Cooperative Programme, 'CITES and Commercially-exploited Aquatic Species, Including the Evaluation of Listing Proposals (Phase 2) – Project Document 2010-2015' (n 40) 4.

non-members to the RFMO and the applicability of the RFMO's finding to them would have to be clarified.[257]

2. Catch-Documentation Schemes

When a CDS or a TDS is in place it could be used instead of CITES permits and hence be applicable to all CITES parties, avoiding costly duplication of work. This was envisioned in the case of the Patagonian Toothfish proposal,[258] in parallel to giving CCAMLR the responsibilities usually entrusted to the Management and/or Scientific Authorities.[259] It seems that most of the information needed for a CITES permit is available or could easily be inserted in the existing CDS, which is rather more than less comprehensive.[260]

In the current situation already, where toothfish is not listed, a tentative to broaden the CDS to all CITES parties can be observed: the CITES CoP indeed recommends to its parties, whether member of CCAMLR or not, to adopt the CDS developed under CCAMLR for toothfish.[261]

This proposed widening of the CDS to all CITES parties has not been a success. At CoP13, the meeting which followed the adoption of this Resolution, the CITES Secretariat concluded that no state that was not already implementing the CDS started doing so as a consequence of Resolution Conf. 12.4.[262] The Secretariat commented that "the implementation of Resolution Conf. 12.4 is seriously hampered by the fact that the species concerned are not included in CITES Appendix II."[263] The same sentiment has probably underlined the negotiations at CoP16: CCAMLR requested the CITES parties to consider an amendment to Resolution Conf. 12.4 to insert a further call for CITES parties to cooperate with CCAMLR through the CITES Secretariat, but no modification of the Resolution was adopted.[264]

---

[257] E. Franckx, 'The Exercise of Jurisdiction over Vessels: Legal Issues Raised by the Relationship Between CITES, FAO and the UN Convention on the Law of the Sea' in E. Franckx and P. Gautier (eds), *The Exercise of Jurisdiction over Vessels: New Developments in the Fields of Pollution, Fisheries, Crimes at Sea and Trafficking of Weapons of Mass Destruction* (Bruylant 2010) 78.

[258] A. Willock, 'Uncharted Waters/Implementation Issues and Potential Benefits of Listing Toothfish in Appendix II of CITES' (TRAFFIC 2002) 27–31.

[259] International Environmental Law Project, 'Understanding Introduction from the Sea' (2010) http://www.lclark.edu/live/files/4356 accessed 6 July 2015, 14, 27–28.

[260] Willock, 'Uncharted Waters/Implementation Issues and Potential Benefits of Listing Toothfish in Appendix II of CITES' (n 258) 31.

[261] Resolution Conf. 12.4 – Cooperation between CITES and the Commission for the Conservation of Antarctic Marine Living Resources regarding trade in toothfish 2002.

[262] CCAMLR, 'CoP14 Doc. 61 – Toothfish: Report of CCAMLR' (n 56) 2 §8-9; CCAMLR, 'CoP16 Doc. 63 (Rev.1) – Toothfish: Report of CCAMLR' (n 56) 1 §4.

[263] CCAMLR, 'CoP14 Doc. 61 – Toothfish: Report of CCAMLR' (n 56) 5 'Comments from the Secretariat' B.

[264] CCAMLR, 'CoP16 Doc. 63 (Rev.1) – Toothfish: Report of CCAMLR' (n 56).

The situation would have been different had toothfish been listed. In that case, following a few adaptations, the CDS could have been used as a CITES permit, either between the CCAMLR parties only, or for all CITES parties.[265] The proposal to list toothfish explicitly stated that CITES would complement CCAMLR efforts:[266] it included the idea of further using the CDS between parties to CCAMLR and cooperating parties as an equivalent to a permit under CITES. As the CDS would not automatically have been applicable as a permit ersatz for all CITES parties, if such a wide application had been the preferred solution, a Resolution would have been required to modify the normal regime. In such case, a more binding wording than the one found in Resolution 12.4 would be required, at the risk, otherwise, to see some states applying the CDS and others the CITES permits.

Moreover, it is not clear whether toothfish caught beyond the CCAMLR Convention area could have been covered by an extended CDS[267] or whether only CITES permits would have been applicable to these areas.[268] Hence, the geographical scope of application of a CDS would have had to be defined. Also, additional quotas would have been needed for areas beyond the scope of application of the RFMO, in order to issue the CDS or the CITES permit. The identification of the forum where such quotas would have been adopted would have required a clarification.

Finally, it would have been necessary to ensure that the strength of the CDS was not weakened by its implementation through CITES—a fear behind much of the resistance within CCAMLR.

III. CMS

In the case of sharks, the CMS could become the specialized institution, providing the necessary data for NDF and a central point of reference for legality findings. Joint programs can also enhance the management abilities of states and hence the capacity to implement CITES.[269]

---

[265] Willock, 'Uncharted Waters/Implementation Issues and Potential Benefits of Listing Toothfish in Appendix II of CITES' (n 258) 29–31.

[266] Australia, 'CoP12 Prop. 12.39 – Consideration of proposals for amendment of Appendices I and II – Patagonian Toothfish' (n 189) 2 §7, 17 §6.2.

[267] Little and Orellana are of the opinion that the CDS could not be applicable for fish caught beyond the CCAMLR area, because these fisheries are not covered by the Convention ((n 254) 102–103).

[268] Australia, 'CoP12 Prop. 12.39 – Consideration of proposals for amendment of Appendices I and II – Patagonian Toothfish' (n 189) 1.

[269] CITES Secretariat, 'SC62 Doc. 14.3 – Cooperation with other organizations – Convention on the Conservation of Migratory Species of Wild Animals' (n 77) annex 4.

IV. Comparison

1. The IWC's Use of CITES as an Enforcement Mechanism

In parallel to the debate about CITES' alleged deference to the IWC in listing decisions, one can discuss whether CITES is obliged to coordinate its measures with the IWC's decisions for permit issuance. This would transform the regime into a mere (although extremely useful) enforcement mechanism.[270]

It is indeed questioned whether CITES can implement its own normal regime. According to the Convention, it is the Secretariat which is required to "consult [. . .] with a view [. . .] to ensuring co-ordination with any conservation measures enforced by such bodies";[271] this wording implies an obligation of means, not necessarily of results. As such, if measures cannot be coordinated, as long as this was tried in good faith, the obligation is still fulfilled. As for article XIV (4)–(5) it only applies to Appendix II species, while nearly all whales are in Appendix I.

Then, the relevant Resolution "recommends that the Parties agree not to issue any [. . . permit . . .] for any specimen of a species or stock protected from commercial whaling by the International Convention for the Regulation of Whaling".[272] Hence CITES parties are requested not to internationally trade products from whales protected by the IWC, whether they are parties to that regime or not.

The wording 'recommend' is not as such binding on parties. Nevertheless, as this provision is indeed implemented by parties, it does remove the decision-making about permit issuance away from states, puts it into the IWC's hands and modifies the CITES regime; "the Parties have allowed the judgment of the I.W.C. to replace that of the individual Scientific Authorities who would have to make a non-detrimental finding".[273] Indeed, while within CITES some exemptions are envisioned, the Resolution acts as a total ban. It also goes further than article XIV (4)–(5), which is only applicable to Appendix II species, while the Resolution makes no distinction.[274]

In conclusion, one can say that CITES parties, by their willingness to coordinate the amending decisions with the IWC and their voluntary agreement to modify the regime, "ha[ve] gradually evolved into an external enforcement mechanism for IWC policies."[275] More than cooperation, it is a hierarchical relation. While, in the short term, it avoids inconsistencies between the implementation of the two

---

[270] A. D'Amato and S. K. Chopra, 'Whales: Their Emerging Right to Life' (1991) 85 American Journal of International Law 21, 47.

[271] CITES (n 33) article XV(2)(b).

[272] Resolution Conf. 11.4 (Rev. CoP12) – Conservation of cetaceans, trade in cetacean specimens and the relationship with the International Whaling Commission (n 70) operative part.

[273] Favre, *International Trade in Endangered Species* (n 49) 92.

[274] Gillespie, *Whaling Diplomacy* (n 199) 337; Favre, *International Trade in Endangered Species* (n 49) 92.

[275] Carlarne (n 195) 24.

regimes, this also comes at a cost, for example in terms of decreased legitimacy for CITES. That regime, by handing over the decision-making powers to another institution, has indeed also relinquished control over the measures adopted and their compliance with its own mandate.

Nevertheless, legally speaking, CITES parties are not under an obligation to automatically enforce the IWC decisions, since the particular Resolution is not formulated in binding terms and the general provisions of the treaty do not require an actual result.[276]

2. Forestry

The ITTO-CITES partnership, even before the adoption of Resolution 14.4, was a fruitful one, with ITTO technically and financially supporting reunions of the CITES Bigleaf Mahogany Working Group, providing funds for a capacity-building workshop on the implementation of Bigleaf Mahogany listing and organizing a meeting on the effective implementation of the regime following the inclusion of ramin in Appendix II.[277]

It was recognized that, for the mahogany, the implementation of the listing requirements was going to be a challenge, because controlling harvest and trade is so difficult.[278] Hence, a delay of 12 months between the listing and the implementation was agreed upon, in order to resolve some of the issues. Consequently, an implementation strategy was developed in that period of time and a workshop on capacity building was held under the sponsorship of ITTO shortly after the listing decision.[279] This workshop focused on supporting exporting states in developing methods for the NDF of mahogany.[280]

In general, ITTO projects, in collaboration with CITES and funded by some member countries, have served to facilitate the successful implementation of CITES listings. Indeed, on the basis of a multi-donor project for afrormosia, bigleaf mahogany and ramin, the range states' capacities are being strengthened, in particular through the support in gathering relevant data, developing management plans

[276] P. H. Sand, 'Japan's 'Research Whaling' in the Antarctic Southern Ocean and the North Pacific Ocean in the Face of the Endangered Species Convention (CITES)' (2008) 17 RECIEL 56, 61. This is however not the conclusion reached by Gillespie who considers, in a rather vague interpretation of article XV, that the adoption of "a listing which [is] effectively different from the IWC recommendations" would fail the coordination required by the Convention and hence be an *ultra vires* action (Gillespie, *Whaling Diplomacy* (n 199) 344).

[277] United States of America, 'CoP14 Doc. 18.2 – Cooperation between CITES and ITTO Regarding Trade in Tropical Timber' (n 98) 1 §3-4; ITTO, 'Making the mahogany trade work' (n 100) 14–16.

[278] Blundell (n 227) 88.

[279] J. Grogan and P. Barreto, 'Big-Leaf Mahogany on CITES Appendix II: Big Challenge, Big Opportunity' (2005) 19 Conservation Biology 973, 973.

[280] ITTO, 'Making the mahogany trade work' (n 100) 3, 16–17.

in order to be able to undertake NDF, drafting national legislation, improving enforcement mechanisms, and increasing trust in the range states.[281]

V. Conclusions

In conclusion, the existing cooperation related to the implementation of protection is less intensive than the other forms of cooperation. This is most probably due to the fact that, currently, only few commercially-exploited aquatic species are listed on CITES Appendices.

However, if the trend set by the listing at CoP16 of several shark species is anything to judge from, cooperation to help states undertaking NDF and determining legality is on the rise in relation to commercially-exploited fish species. Such a tendency was also witnessed in the case of timber. Indeed, a major increase in the number of activities took place following the listing of relevant species. Hence, in a way, such listings already proved beneficial as they forced the development of management plans.

The FAO's activities meant to develop national management plans support the effective implementation of the CITES protection system, since they allow states to control the sustainability of trade. As for the cooperation with RFMOs, it will need to be organized, in order to establish the most efficient mechanisms possible for NDF and legality finding, as well as to clarify the possibility of using the existing CDS/TDS in place of the CITES permits. It is worth noting that the status of RFMOs as 'key project partners' has been recognized in relation to the on-going project related to supporting the implementation of the sharks and manta rays' listing.[282]

The partnership between the IWC and CITES is an example of utmost collaboration where the CITES parties apply the IWC decisions. Since using an existing CDS/TDS would have some parallels with that situation, it would be useful, if such decision needs to be made, to keep in mind the advantages but also the problems linked to handing over power to another institution. Indeed, on the one hand it

---

[281] CITES Secretariat, 'PC17 Doc. 16.2 – Timber Issues – Progress Report on the Joint ITTO-CITES Timber Project: First Regional Meeting' (15 April 2008) 1 §3 and annex; Phase I of the ITTO-CITES program (2006–2011) is considered to have been a success and Phase II is under development (CITES Secretariat, 'CoP16 Doc. 13 – Cooperation with Organizations and Multilateral Environmental Agreements' (n 1) 3–5 §16-25); ITTO, 'ITTO-CITES Program for Implementing CITES Listings of Tropical Timber Species' http://www.itto.int/cites_programme/ accessed 5 July 2015; CITES, 'ITTO-CITES programme on tree species' http://www.cites.org/eng/prog/itto.php accessed 5 July 2015; Engler and Parry-Jones (n 102) 26.

[282] CITES, 'Strengthening capacity in developing countries for sustainable wildlife management and enhanced implementation of CITES wildlife trade regulations, with particular focus on commercially-exploited aquatic species: Project summary sheet' (n 41) 2.

avoids inconsistencies, but on the other hand it submits one regime to the vagaries of the other, putting the legitimacy of the former at risk.

E. Conclusions

I. Need for Cooperation

In the case of the listing of commercially-exploited fish species in CITES Appendices, cooperation with other institutions is necessary for several reasons. First, CITES lacks expertise—and resources—with regard to marine species. The Secretariat thus does not necessarily have sufficient knowledge to present all the arguments in favor or against a structural change in the regime or to single-handedly advise parties on a listing proposal.

Also, the CITES regime is not a management regime, but a trade regulation one. Hence, in order to efficiently apply its provisions, states must have sufficient scientific information on which to base decisions. Furthermore, a cooperative management framework is needed for shared, straddling or highly migratory species, in which measures can be decided collectively. As many management regimes exist in the field of fisheries, cooperation to avoid a doubling of work is desirable.

Finally, the implementation of the Convention is a costly business and some requirements are difficult to comply with, in particular for developing countries. Indeed, management plans must exist in order to determine NDF—otherwise trade restrictions might become trade bans—and controls at the borders can be challenging with respect to marine species. Any partnership which can support countries in implementing CITES is evidently beneficial.

II. Appraisal of the Partnerships

1. To Provide Expertise

The cooperation linked to the external provision of expertise only takes place between the FAO and the CITES Secretariat and is rather a success. Many pieces of advice were followed even though some problems remain. In particular the different interpretations of the criteria sometimes cause divergent recommendations regarding the listing proposals.

It is worth noting that such collaboration predates the development of the MoU and that the existing political will to cooperate seems not to have been triggered or reinforced by the adoption of this instrument.

2. To Provide Science-Based Advice

The second type of cooperation, linked to the actual listing of species, has a slightly more nuanced track record. The FAO organizes Expert Panels to assess whether species proposed to be listed meet the CITES criteria and to discuss the potential effects of a listing. The CITES Secretariat also provides the parties with advice, which, in two of the last CoPs, did not concur perfectly or even to a large extent with the Expert Panel's conclusions. This might have had an effect on the effectiveness of this partnership but the CoP has not followed consistently one or the other advice provider and not even the scientific advice when both recommendations coincided.

Receiving scientific advice prepared by an outside institution can be an issue in certain cases. Indeed, the institutional mentality and language might differ and create mistrust. However, in the present situation, it appears that while the Expert Panel has a slightly stricter understanding of the criteria (and hence supports less proposals) than the CITES Secretariat, this dissonance has no automatic consequence on the decisions of the CoP. Whether the existence of opposed opinions is positive—because it reflects different views with transparency—or rather a source of confusion for the parties is unclear.

One must note that it is not the provision of science by an external institution or its quality which is the real issue at stake within CITES, but the actual role of science in the decision-making processes. It appears that, for many species, states follow their values and interests rather than the scientific advice. Hence, the scientific advice is not necessarily respected and its origin (Secretariat recommendations or FAO Expert Panel) does not impact this. It has also even been suggested that the absence of an in-house scientific committee in CITES might be positive, as it avoids "internal conflicts along the science/politics divide".[283]

The problem of how science and politics interact is nevertheless not resolved. While similar in many fields, it might be particularly acute here due to the imperfection of data, the very different agendas of institutions and states, and the potential socio-economic impacts of any type of decision.

The cooperation between the FAO and the CITES Secretariat is nevertheless a positive exercise, with independent and quality information provided by an (or even by the) expert institution. At the very least, it can provide a concrete basis for the debate and somehow force the CITES parties who would be willing to disregard the advice to express their reasons for doing so.

That the Secretariat shall "respect, to the greatest extent possible" the FAO conclusions as well as the responses from the relevant RFMOs does not create a strong obligation and does not, in any case, bind the CoP. Consequently, the legal nature of the MoU has no influence with regard to the effectiveness of the partnership.

---

[283] P. H. Sand, 'Whither CITES? The Evolution of a Treaty Regime in the Borderland of Trade and Environment' (1997) 8 European Journal of International Law 29, 37.

The role of RFMOs in the process of giving advice is neither fully streamlined nor is beyond improvement. As it stands, both the FAO Expert Panel and the CITES Secretariat take the responsibility to gather information from the relevant management institutions. While this inclusiveness shall be commended for existing, running two similar procedures in parallel is not necessarily an ideal situation in terms of clarity. It also leaves the FAO in a complex middle-man position. Ideally, instead of being the recipients of two *ex post facto* requests for comments on a proposal, RFMOs could be directly and formally involved before a proposal is submitted; this would have the potential to decrease the feeling of being put aside or subjected to a vote of confidence. The example of the process developed for listing timber should be followed.

3. To Help Implementing the Protection Under CITES

The last type of cooperation which includes ways of supporting the implementation of CITES requirements consists mainly in capacity-building. The help provided to develop management plans in order to fulfill the requirements imposed by the CITES regime and the publication of identification guides can be very helpful, in particular for developing countries which would otherwise be unable to trade in Appendix II species and/or to enforce the restrictions. As CITES is a relatively poor MEA and hence has problems organizing enough technical assistance,[284] cooperation with the FAO can only be positive and beneficial.

This third aspect of the partnership is both the least controversial and the least developed, due not only to funding limitations but also to the novelty of listing commercially-exploited fish species. It is unfortunate since 'on the ground' joint actions could actually strengthen the cooperation, allow the development of management plans and even demystify the effects of a listing on trade opportunities. Financial constraints could partly be remedied once a species is listed. One can indeed observe, in the forestry sector, that incentives to fund capacity-building and strategic plans meetings closely followed the listing of mahogany. The current development of programs on sharks seems to concur with this precedent.[285]

The type of partnership which would have to be developed with an RFMO, if a species under its management was listed in CITES, would probably include some support for NDF and legality finding. In practice, it would depend, on a case-by-case basis, on the measures already existing (for example a CDS/TDS) and on the willingness of the CITES parties to defer some powers to another institution.

[284] R. Reeve, *Policing International Trade in Endangered Species* (Earthscan 2002) 238.

[285] CITES, 'Ongoing projects and activities: EU project' (n 41); CITES, 'Strengthening capacity in developing countries for sustainable wildlife management and enhanced implementation of CITES wildlife trade regulations, with particular focus on commercially-exploited aquatic species: Project summary sheet' (n 41).

In conclusion, the partnerships of CITES with other institutions concerned with marine species are necessary, positive in making use of the existing and specialized bodies and quite successful, but they nevertheless suffer from some flaws. They attained more than the—already useful—traditional exchange of information; for expertise, listing advice or implementation support, partnerships have indeed been developed and could be further strengthened. The synergies are in most cases constructive but some feeling of competition remains, for example with RFMOs. While this could be decreased by formally involving these organizations earlier on, such rivalry arguably rather reflects the RFMOs member states' unwillingness to have a larger community of states involved in the (mis)management of the resources under their mandates than a real issue of the RFMO as such. As the FAO declared: "we regret the situation in which the effectiveness of RFMOs to achieve their goals is having to be questioned by the international community."[286] Yet, if unwise decisions keep on being taken in regional foras, the international community might want to step in to ensure the protection of the global commons.

[286] FAO, 'CoP15 Inf. 26 – Statement from the FAO Fisheries and Aquaculture Department on CoP15 Proposal 19 to List the Atlantic Bluefin Tuna in Appendix I' (n 181) annex §3.

Part III
General Conclusions and Recommendations

Chapter 7
Desirability of Using CITES

A. Legal and Legitimate Use of CITES

I. Within the CITES Mandate

While some states have expressed that CITES is not the right forum for fisheries and hence tried to exclude commercially-exploited fish species from its mandate, or at least to reduce the options to use this regime to the cases where a management regime is absent, nothing in the Convention can legally support such proposals.

Commercially-exploited fish species, whether under the protection of another regime or not, can indeed fall under the ambit of CITES, if they meet the biological and trade-related requirements. This is illustrated by several provisions of the Convention, which refer to marine species, by Resolutions and by the listing of various fish species such as sharks or sturgeons.

II. No Automatic Threat to RFMOs' Reputation or Efficiency

The threat to the relevant RFMO's reputation or efficiency, to which member states to CCAMLR made reference in 2002, is not an automatic feature of the listing in a CITES Appendix of a species under the management of another regime. Indeed, it depends on how the member states of the relevant RFMO approach the involvement of CITES: it can be perceived as a menace and competition, or as a chance to benefit from a globally applicable enforcement mechanism. The evolution of ITTO members' opinions towards a CITES listing for example illustrates the usefulness of discussing the common goals and cooperation opportunities.

Furthermore, such a listing does not automatically represent a doubling of work. Indeed, CITES already envisions overlap avoidance for Appendix II species protected by a treaty pre-dating 1975. As for species in Appendix I and species

S. Guggisberg, *The Use of CITES for Commercially-exploited Fish Species*,
Hamburg Studies on Maritime Affairs 35, DOI 10.1007/978-3-319-23702-2_7

managed by regimes set up after 1975, coordination mechanisms could be put in place. In order to avoid the existence of two documentation schemes one on top of the other, the TDS/CDS of the RFMO could for example be adapted, as was suggested for toothfish.

B. Adequate and/or Modifiable Framework

I. Adequate Regime

That the CITES regime is not perfectly adapted for marine species is rather a consequence of its limited use therefor than of a structural flaw. The Convention text is termed quite generally and many definitions and implementation aspects were later developed through Resolutions in order to respond to practical needs—these are only now arising in relation to fisheries.

The progressive adoption of quotas as the main NDF and of sanction mechanisms are illustrations of this regime's flexibility. Also more recently, the experience with the mahogany shows that on-the-spot solutions can be developed and problems addressed when they actually appear, through cooperation or the work of a specialized working group.[1]

With regard to commercially-exploited aquatic species, the listing criteria have been modified, the IFS provision has been clarified and the legality-finding requirement stream-lined in order to enable the regime's efficient implementation. For the NDF of highly traded species, quotas will be necessary and are already in place for a large number of listed species. As for transboundary, straddling or migratory species, the cooperation of all states involved is possible through the existing RFMOs. Moreover, partnerships can be further promoted and developed. Issues related to the definition of captive-breeding and ranching will need to be addressed in order to accommodate the specificities of aquaculture.

II. Cooperation for Specialized Expertise and Management

Another concern raised is directed at CITES' lack of particular expertise in relation to marine species. Whereas this is a correct assessment, nothing prevents that regime from developing specialized knowledge or externalizing some scientific aspects of its work to partner institutions.

[1] ITTO, 'Making the mahogany trade work: Report of the workshop on capacity-building for the implementation of the CITES Appendix-II listing of mahogany' ITTO Technical Series 22 (2004) 11.

It is worth keeping in mind that, with regard to marine species, cooperation is envisioned in the Convention and has already started to take place. For example, the Secretariat entered into a partnership with the FAO for the provision of specialized information on listing criteria and partly relies on the Expert Panel for the assessment of amendments' proposals.

CITES has also been criticized as inadequate because it is not a management regime. It should be noted that it has never assumed such a role and it is not within its mandate. Such a situation is only an issue if no management regime exists or if the use of CITES and of an RFMO are perceived as mutually exclusive. While the former scenario is problematic and ought to be addressed by establishing RFMOs where a geographical loophole remains, the latter does not need to occur. A partnership with the relevant RFMO is indeed not only possible but even advisable in order to assess the species status, to determine NDF and to examine the legality of catches.

III. Enforcement Issues

It is true that several identification and enforcement issues are likely to arise with regard to commercially-exploited fish species listed under CITES. The high numbers in which the specimens are traded, the processed form in which they enter markets, the inadequacy of look-alike listing, the probable temptation to use split-listing and the existence of both wild-caught and farmed specimens are indeed complicating factors to the implementation of efficient controls.

However, for fisheries, some traceability schemes, identification tools and training already exist. Moreover, it must be noted that identification and enforcement issues at the border are the same for many other species under CITES, such as timber,[2] and that this has not impaired their listing. In the case of forestry, these issues were identified before the listing but only addressed in the 12-month run-up to the implementation of protection, through workshops and capacity-building.

IV. Bias Towards Preservationism

Several states expressed their apprehension about using CITES due to the alleged bias of this regime towards preservationism. It is true that down-/de-listing seems difficult under CITES and that legitimate concerns of stakeholders need to be

[2] ITTO, 'Making the mahogany trade work' (n 1) 8; T. Mulliken, 'The Role of CITES in Controlling the International Trade in Forest Products: Implications for Sustainable Forest Management' Non-Wood Forest Products Working Documents 7 (Rome 2009) viii.

addressed. As abundance varies faster in fish stocks than it does in many other species and as fish plays an important role in food security as well as in livelihoods, the prospect of a precautionary non-delisting is certainly a serious disincentive to use the protection regime.

Nevertheless, CITES is not an automatic ban on trade and sustainability is at the basis of this regime: under Appendix II, the main requirement is indeed that trade should not be detrimental to the species. Also, only international trade is targeted, leaving artisanal fisheries undertaken close to the coasts and for domestic consumption unaffected. Moreover, solutions to questions related to livelihoods are currently developed within CITES, showing an increased awareness to poorer countries' interests. In the case of timber, an argument against the listing was that CITES imposes restrictions—usually on developing states—but does not provide any support to help these states meet their obligations; this concern was acted upon[3] and the same appears to be happening for sharks. It has also been pointed out that, while CITES listings might have short-term negative impacts on rural communities, they can actually contribute, in the longer term, to sustainable livelihoods, as recognized in Resolution 16.6 on *CITES and livelihoods*. Finally, as fish species are not particularly charismatic, decisions related to amending Appendices, while remaining of an inherently political nature, might be more science-based than in the case of megafaunic mammals.

C. Benefits in Terms of Governance

I. Governance Benefits for Fisheries

CITES can have a beneficial role towards the protection of commercially-exploited fish species in several regards. First, it applies to all international trade involving at least one of its 181 member states. Hence, as a quasi-universal enforcement mechanism it can support the usually regional management measures by making them globally applicable. Thanks to the IFS provision fishing operations in the high seas and landing in the flag state's port can also be controlled, hence putting a sustainability restriction to the freedom of the high seas. Geographical fragmentation and loopholes are consequently more likely to be closed than under a regional framework.

Second, while it is no stranger to politically motivated decision, the CITES regime has the potential to improve the adequacy of measures applied to fish stocks. While it itself lacks the necessary scientific data and has to rely on other institutions' resources, in particular on the FAO Expert Panel, this externalization has positive sides. This independent panel delivers nonpartisan analyses on the amend-

[3] Mulliken (n 2) 13–14.

ment proposals, is transparent in the sense that differences of opinions are expressed in the reports and assesses the quality of the data available. As a group made of specialists, it has proved able to provide advice on most aspects highlighted as relevant or problematic by stakeholders in the fisheries field, such as natural fluctuations, particular vulnerabilities and mitigation factors, or the impact of a listing on livelihoods.

Third, compliance with and enforcement of CMMs might be improved through the use of CITES measures. Indeed, the regime is based on verifications at the customs; the strengthened role of the port state decreases incentives to use flags of convenience. Also, the Significant Trade Review permits an assessment of the sustainability of quotas decided by the Scientific Authorities: TACs set at too high a level might come under scrutiny, putting necessary pressure on some states or groups of states, sometimes acting through an RFMO. Finally, the sanction mechanism developed within CITES strongly encourages compliance with the regime in order to avoid trade suspensions.

II. Example of Good Practices

While some authors consider that the use of both CITES and a regional organization is the sign of a regime's weakness,[4] it is also possible to approach this multiplicity of fora as a strength. Indeed, while RFMOs manage fisheries, they cannot regulate trade as effectively as a quasi-global institution. Hence, in an imperfect international structure, where laws and institutions have been developed in a piecemeal approach, cooperation and complementarity between institutions focused on different aspects of a same issue are a possible solution.[5]

CITES has a long history and practice of entering into partnerships with related institutions but the information exchange with the FAO has taken this feature to a new level: the specialist body indeed puts together the Expert Panel which acts as an independent consultant. Also, the practical workshops organized do not only support the implementation of CITES but also more generally are beneficial to the management of the relevant natural resources.

Furthermore, CITES is a good example of positive developments in inter-treaties cooperation and of cross-fertilization, mostly in the form of systemic interpretation and rules of reference. Indeed, the CoP interpreted 'waters not under the jurisdiction of any State' in a manner which simplifies the relations between this treaty regime, adopted pre-UNCLOS, and the law of the sea under UNCLOS. By deciding that

---

[4] D. R. Rothwell, 'Building on the Strengths and Addressing the Challenges: The Role of Law of the Sea Institutions' (2004) 35 Ocean Development and International Law 131, 147.

[5] P. Birnie, A. Boyle and C. Redgwell, *International Law and the Environment* (3rd edn Oxford University Press 2009) 697.

these terms were to be understood as in the 1982 Convention, it indeed avoided incoherence between the conservation treaty and the international law of fisheries. Then, the CITES regime requires a verification of catches' legality. Hence, since other regimes such as the law of the sea and the law of fisheries provide for the rules applicable in the relevant maritime zones, this creates an indirect rule of reference. Such mutual supportiveness is very positive.

Chapter 8
Recommendations

A. Conditions Under Which the Use of CITES Can Be Recommended

In general, a proposal only has chances to succeed if enough political will is present. Even if the required qualified majority is reached, if the fishing states and main importing states are not on board, the regime will suffer, at worst from reservations and at best from unwilling and hence imperfect implementation. Also relevant is the condition that international trade is a major factor in the species depletion; otherwise indeed, the regime will be of little use and such feel-good decisions can be dangerous for its credibility.

I. In the Presence of an Institution Responsible for Management and Conservation

If an RFMO exists, a CITES listing can be recommended when the species in question is endangered, or at risk of being so, and when international trade is an important driving factor.

If the unsustainable practices are driven by free riders to the RFMO or by loopholes of some sorts, it seems (even though the toothfish proposal illustrates the contrary) that CITES would be a very valued and likely to be used tool. Indeed, the issues faced by CCAMLR, mainly IUU fishing in the high seas adjoining the RFMO's mandate area, ports of convenience undermining the CDS and vessels flying a flag of convenience involved in fishing within the RFMO area, are important problems which cannot be directly solved by the RFMO. In such cases, CITES could, thanks to its global applicability, serve as an enforcement and enlargement mechanism.

S. Guggisberg, *The Use of CITES for Commercially-exploited Fish Species*, Hamburg Studies on Maritime Affairs 35, DOI 10.1007/978-3-319-23702-2_8

Ideally, the RFMO members should, if not lead, at least support such proposal. Indeed, antagonized RFMO members are likely to make the listing unsuccessful or the implementation of the protection useless: CITES will most certainly be ineffective if range states feel obliged by others to use CITES permits. Hence, the need to work with the range states (here both geographical range states and relevant flag states), not against them, cannot be highlighted enough.[1]

It is unclear whether one permitting system would overtake the other or whether they would work in parallel at great bureaucratic cost. In the former situation, it would be crucial not to undermine the strongest measure—usually the CDS—but also not to put CITES' legitimacy at risk by handing over power to another institution.

If the unsustainable practices which cause the deterioration of the species' status are decided, carried out or accepted by the members of the relevant RFMO (TACs set well above the MSY and/or catches in violation of the quotas), the situation is more complex. In that case, listing under CITES is, first of all, quite unlikely to happen. Indeed, the members of the RFMO will certainly oppose it and strongly lobby in that direction. However, as CITES benefits from a far broader membership than any RFMO, a qualified majority could still be reached without the support of these states.

Reservations exist and it is true that RFMO members who are totally opposed to regulations will probably enter into them. Nevertheless, when trading with a CITES member, similar documentation is required from the reserving state/non-party. Hence, CITES is only inapplicable in the relationship between non-members and/or members who made reservations. In that respect, high seas fisheries for domestic consumption could also fall under a simplified loophole, as the IFS involves only one state and hence a single reservation would suffice.

Some problems of implementation would moreover appear, as the data on which to determine NDF would be in the hands of a non-partner institution. Moreover, the TACs, being unsustainable, would not be of much use for CITES and it is not clear on which basis the NDF would be undertaken. Also, the risk that fish enters the markets illegally would be quite high if demand is not somehow reduced or strong enforcement implemented.

Nevertheless, such a listing should potentially be considered if a group of fishing states, acting through an RFMO, are failing the international community by mismanaging the stocks under their responsibility. Indeed, while coordinated action between institutions is definitely to be preferred, the 'failure' of an institution should not cause all the others to wait and see. CITES has its own mandate to fulfil and should not refrain from acting in accordance with it, simply because another institution exists but has evolved or gone astray. One could wonder in that regard

---

[1] R. B. Martin, 'When CITES Works and When it Does Not' in J. Hutton and B. Dickson (eds), *Endangered Species-Threatened Convention: The Past, Present and Future of CITES* (Earthscan 2000) 31–32.

whether it is really beneficial that CITES accepts to be subjected to IWC decisions, whether or not they conform to the CITES mandate.

In all cases, the threat of a CITES listing—as long as it is taken seriously—provides public scrutiny and can be a leverage tool to force the RFMOs to move towards better management. However, to deliver its full potential, a threat must be credible and not be a one-off, without follow-up. Indeed, history shows that engagements are not always held: a proposal to list Bluefin Tuna was already made in 1992 by Sweden before being withdrawn against a pledge by ICCAT to better manage stocks. The arguments against the listing were similar to those put forward in 2010, in other words that ICCAT was the appropriate forum to manage those stocks, even if it had so far failed.[2] Promises were made towards sustainability, and indeed, improvements in the quotas voted could be noted. Nevertheless, "[b]y 1994, the scientific quota for western bluefin was reset to pre-1991 levels",[3] which were not based on scientific advice.

In a fashion similar to the development of monitoring and enforcement mechanisms in the 1990s, ICCAT reacted to the announced decision to propose a listing in 2009 by creating a penalty mechanism for members who disrespect their quotas. It also began to think about reducing fishing capacity, reorganized its compliance committee and "started a compliance task force to improve monitoring and enforcement of all regulations related to bluefin tuna."[4] So far, it seems that the promises of improvement are longer-lived than in 1994, but, should they not remain so, one can hope that it will not take another 17 years before the international community reacts again.

II. In the Absence of an Institution Responsible for Management and Conservation

The absence of an institution responsible for managing and conserving species which might be protected under CITES is problematic. Indeed, this regime can be an adequate complementary mechanism, but has not been designed to manage fisheries alone. It is indeed no replacement for effective measures 'on the ground'.[5] Hence it would be difficult—and this is the case for sharks for example—to make the most of this trade regulation regime when a management regime is missing.

---

[2] D. G. Webster, 'The irony and the exclusivity of Atlantic bluefin tuna management' (2011) 35 Marine Policy 249, 251.

[3] Ibid.

[4] Ibid.

[5] M. 't Sas-Rolfes, 'Assessing CITES: Four Case Studies' in J. Hutton and B. Dickson (eds), *Endangered Species-Threatened Convention: The Past, Present and Future of CITES* (Earthscan 2000) 69.

It can nevertheless be used if a species is endangered, or risks to be, and if international trade is an important driving factor. A major requirement is the availability of some data, in order for CITES or partner institutions such as the FAO or the CMS to be able to assess threats and the level of trade that could be sustainable. In any case, a listing proposal and/or an actual listing provide publicity and can be a powerful trigger towards the development of much needed management plans for largely unregulated species.

III. Recommended Changes and Additions

Some modifications within CITES would greatly improve its adequacy for commercially-exploited fish species and hence the likelihood of this regime being used therefor.

First, as fish stocks have other life cycles than terrestrial species and as their harvest is globally important in trade as well as locally crucial for food and livelihoods, the precautionary principle as it exists under CITES should be reconsidered. Indeed, down-/de-listing must be possible when the species has recovered and should maybe be rendered more flexible. Also, the decision-making process should be adapted to coincide with the faster cycle used in fisheries management.

Then, while the IFS provision might need to be further reviewed in relation to chartering, it is otherwise now fully applicable. As for the legality finding requirement, it might be positive to ensure in more binding terms that it is undertaken as early in the process as possible, in other words already at the IFS stage, as well as that it includes relevant rules of international law. In that regard, a more general clarification of the legal framework applicable to high seas fisheries, in particular the actual content of the obligation to cooperate in the conservation of living resources, will be necessary. Indeed, whether any state fishing on the high seas has to respect the measures set by the relevant RFMO or only needs to cooperate to some extent will ultimately determine the benchmark for the legality finding.

Moreover, it might be useful to look at the feasibility and sustainability of adapting the captive-breeding criteria to the specificities of fish species, in order to both better frame this activity and avoid unnecessary restrictions to trade. Indeed, considering the important consequences that the listing in CITES may have on aquaculture if a species can be both caught in the wild and potentially farmed, it is necessary to address the technical, legal and practical impacts of such a trade restriction. On the one hand, questions of efficient labeling and clear distinction between hatchery and wild-caught juvenile practices need to be solved so as to avoid the creation of a loophole calling for corruption and 'fish-laundering' practices. On the other hand, captive-breeding practices, although not the panacea, are nevertheless a way towards sustainable fisheries and should not be discouraged by a blanket ban on trade.

Furthermore, some implementation and enforcement challenges such as identification issues still need to be tackled for the regime to be efficient and flexible enough, for example by permitting measures to be taken at the stocks' rather than the species' level. In that regard, as the necessary compliance requires investment that can be beyond the reach of developing countries, mitigation measures and implementation support are needed respectively to counteract short-term negative impacts and to avoid closing the markets to communities which considerably rely on fish trade. Sharing costs or at least supporting poorer countries in technical aspects of implementation is recommended—as it would also be for RFMO-based trade measures.[6]

Finally, it would be extremely beneficial to undertake a thorough assessment of CITES' effects in order to be able to project, beyond mere hypotheses, the added value of a listing under different circumstances. This would require resources but could serve the regime in the future by tying the arguments for or against a listing on actual past experiences.

B. Changes Needed in the Partnerships

I. Cooperation with FAO

The partnership between the FAO and CITES has been, so far, a very fruitful one, as has been expressed by the following words: "[i]t may not yet be a perfect union, as divergences in opinion have arisen, but the two organizations have achieved a great deal under their cooperative framework and are fully committed to its effective implementation."[7]

The fact that the FAO is an external advisor, in opposition to an in-house one, does not seem to be of considerable importance in terms of the trust given by states parties to the recommendations. This might actually even be a positive feature, as the FAO's advice, although not always followed, is perceived as independent.

1. Structure of the Partnership

Regarding the structure of the partnership, the development of the MoU has allowed discussions about the different roles of the organizations and a clarification of the various positions. The non-binding nature of this instrument is of little

[6] C. A. Roheim and J. G. Sutinen, 'Trade and Market-Related Instruments to Promote Sustainable Fishing Practices' International Trade and Sustainable Development Series Issue Paper 3 (Geneva 2006) 37.

[7] CITES Secretariat (in consultation with the FAO), 'CoP15 Doc. 10.2 – Cooperation with other organizations – Cooperation with the Food and Agriculture Organization of the United Nations' (13–25 March 2010) 1 §3.

significance, as it provides a framework for technical cooperation between institutions and does not create obligations for any decision-maker.

Ideally, an important aspect which should be addressed is the disagreement about criteria which results in different recommendations by the CITES Secretariat and the FAO Expert Panel. In such cases, the lack of clarity for states is not to be underestimated and it is conceivable that the FAO could question the practical value of its recommendations, if they are regularly not acted upon by the CITES Secretariat. In that regard, it might be useful for both partners to keep trying to find a compromise solution for the understanding of Annex 2aB. Moreover, the FAO should assess—as requested by its member states—when and why recommendations are, or not, followed by both the CITES Secretariat and the CoP. As to the CITES community, it should maybe give more vocal recognition to the important work provided by the FAO, acting as a free-of-charge consultant.

Finally, as emphasized by the Sub-Committee on Fish Trade of COFI, the CITES activities of the FAO should be financed through the Regular Program Funds. If this does not happen, FAO members other than Japan will be required to contribute substantial amounts. Indeed, budgetary issues have far from only theoretical consequences, as a workshop planned for 2009 on enforcement and identification issues was cancelled due to a reduction in the allowance.

2. Content of Reports

a) Two Reports

It might be beneficial to have two Expert Panels and consequently two reports, one on whether the species meets the biological criteria and the other one on trade-related aspects and the expected effects of a listing.

This would allow the first section not to become as politicized as the second always is. Also, the specialists are not the same and it is unreasonable to ask fish biology experts to make precise trade projections. Finally, this would offer the possibility to further develop the parts related to food security and impacts on livelihoods.

b) Development of Aspects on Aquaculture

The FAO seems to be the adequate institution to address the questions linked to CITES and aquaculture. Broadening the CITES-activities program of the FAO to aquaculture questions would certainly fall under the mandate of the CITES-FAO partnership. Mention of these issues has already been made in the past, such as in the 2004 Consultation on Implementation Issues.

First, several aspects related to the application of captive-breeding or ranching operations' exemptions to commercially-exploited aquatic species remain either unclear or would potentially benefit from an expert analysis. It would in particular be useful to discuss whether the relevant CITES provisions are adequate in that regard or whether they should be ameliorated.

Such questions include, for example, the fact that, for many aquaculture practices where specimens are born in captivity, the requirement of ‘second-generation offspring’ production is still unattainable. It would be helpful to look at the feasibility and sustainability of adapting the captive-breeding criteria to the specificity of fish species.

Regarding ranching, the reference to the population having to be under the jurisdiction of a state seemingly rules out both farms in the high seas and farms receiving high seas caught juveniles. The adequacy of these restrictions for commercially-exploited fish species would need to be assessed since it has the potential to totally ban international trade for commercial purposes in Appendix I species if no captive-breeding operation can be registered and if the species is not found in the wild in the waters of the state wanting to set up a farm.

Even if transfer of fish from the high seas to a farm was allowed under CITES, the legal status of such operation should be clarified as this scenario is not yet envisaged by the regime. As it stands now, this could be considered as international trade, with the permitting obligations this qualification triggers.

Second, the aspects related to aquaculture which were mentioned in the Expert Panel reports have been very general so far and mostly scrutinizing the impacts of such activities on the status of species. However, the possible implications of a listing on aquaculture operations—total ban or exemption through the captive-breeding exception or, more likely, a ranching operation—have not been analyzed in much detail, if at all. Whether examining such consequences should be done during the existing process of the Expert Panel which so far focuses on science, or through a parallel one is unclear. In any case, it is necessary to look in detail at the possible positive and negative impacts of a listing on these activities.

Third, capacity-building could improve the chances that farmed fish enters the markets as captive-bred or ranched, hence with lower restrictions to trade. This technical help to make operations more sustainable could be supported by the FAO.

A particularly important difficulty related to the existence of the bred-in-captivity exemption and of ranching is linked to the risk of ‘laundering’ wild-caught fish using one of these special cases, when otherwise the fish specimens would not be allowed to enter international trade at all. As examined, there are, for example, strong suspicions that this happens for sturgeons. Since the problem of differentiating wild-caught, captive-bred and farmed fish one from another is very similar to that of differentiating look-alike species, solutions to it should be searched at the CITES enforcement and identification consultation. Additionally, questions of practicability in the required marking of captive-bred and ranched specimens would need to be looked at.

II. With RFMOs

If CITES is to come into play for commercially-exploited fish species under the management of other institutions, RFMOs' roles need to be clarified in advance and most probably increased. The existence of an MoU or guidelines providing an agreed framework of cooperation would indeed be beneficial as it would give RFMOs the position of explicit partners—beyond the role of advice provider accorded under article XV—and could hence decrease, or even hopefully avoid, any perceived competition if a fish species is proposed to be listed.

In any case, RFMOs should be involved earlier in the listing process and the procedure should be redesigned. The current situation where RFMOs are formally contacted only after the proposal has been submitted is not the most adequate. The procedure used for timber, where a certain number of institutions have to be consulted by the proponent before it can send its amendment's proposal seems a possible and advisable solution to apply to fisheries. Moreover, that both the CITES Secretariat and the FAO contact RFMOs at approximately the same time and for the same information is confusing and without any clear added value. This also puts the FAO in an intermediary role without having officially been requested to do so by CITES. A streamlining of procedures can be recommended so as to avoid doubling of work and unclarified functions.

Regarding the implementation phase, RFMOs have a crucial role to play in the procedures towards the NDF and the legality finding processes. Also, the enlargement of a regional documentation through CITES could be developed on an ad hoc basis, but could also be subject to prior consultation. The CMS should also be a cooperating partner in relation to shark species which fall under both regimes.

While handing decision-making powers over (de jure or de facto) to another regime dramatically increases coherence, such coordination comes at a price. As illustrated by the IWC-CITES relation indeed, no trade permit under CITES is issued for species under the protection of the IWC, hence avoiding discrepancies. However, this situation entails several risks, in particular for the legitimacy of CITES; the IWC has evolved towards preservationism, and decisions are not always based on science. Hence, CITES, as a de facto mechanism of the IWC, has adopted these measures and is consequently not fulfilling its own mandate anymore. From the experience they had with whales, Japan and other states have become extremely wary of listing a commercially-exploited fish species under CITES—due to the perceived risk of never being able to de-list it again. Polarization can create tensions and a paralysis of the regimes, and it is not a lie to say that the IWC is at a stalemate and is dysfunctional.[8]

[8] M. Fitzmaurice, 'Divided we stand: the legal issues concerning the international whaling commission' in H. P. Hestermeyer and others (eds), *Coexistence, Cooperation and Solidarity: Liber Amicorum Rüdiger Wolfrum* (Martinus Nijhoff Publishers 2011) 851.

In conclusion, while close cooperation is needed, one cannot change another institution, nor control its evolution. Hence, safeguards must be implemented in order to avoid loss of legitimacy and/or failure towards one's own purpose.

III. Depolarization

Opinions are currently quite polarized towards the CITES regime because states have different perceptions of its role. In very simplified terms, the pro-CITES countries are afraid that most of the other institutions and even scientific advice are biased towards unsustainable use and that species might disappear before action is taken.[9] As for the anti-CITES countries, they fear that any listing would create a precedent[10] and that CITES' goal is preservationism rather than sustainable use. In general, the latter states consider that CITES should not be used for fisheries, in particular when a management regime exists.

While the situation of fisheries is different from that of whales—the IWC has evolved into a preservationist group whereas the management of many commercially-exploited fish species is not even sustainable—the polarization of debates at the IWC has certainly spilled over to CITES[11] first in relation to the whales and now to many other aquatic species. In particular, the parallel between the ever-lasting moratorium and the quasi-impossibility to down-/de- list species under CITES shows that some fears of the pro-sustainable use group are founded and that some decisions are based on ideology rather than science.[12]

While states' views cannot be forced in any direction, the evolution of opinions witnessed for timber could be of importance and act as a model for fish species. Indeed, only a decade ago, a commentator mentioned that "[i]f CITES were to take on the regulation of the trade in endangered commercial timber and fish species it would bring about a significant change in its character, in the direction of greater acceptance of the concept of sustainable resource management. But it is not clear that the Parties to CITES are ready for this."[13] Undoubtedly, parties changed their minds for timber and might currently be doing so for fish species.

---

[9] D. S. Favre, 'Debate within the CITES Community: What Direction for the Future?' (1993) 33 Natural Resources Journal 875, 915.

[10] China, 'CoP12 Inf. 30 – "Sharks" and COP12 – A Case for Caution' (3–15 November 2002) 2–3 §5.

[11] A. Gillespie, *Whaling Diplomacy: Defining Issues in International Environmental Law* (New Horizons in Environmental Law, Edward Elgar 2005) 340–341.

[12] E. J. Molenaar, 'Marine Mammals: The Role of Ethics and Ecosystem Considerations' (2003) 6 Journal of International Wildlife Law and Policy 31, 41.

[13] B. Dickson, 'International Conservation Treaties, Poverty and Development: the Case of CITES' (2002) 74 Natural Resource perspectives http://www.odi.org.uk/publications/2069-international-conservation-treaties-poverty-development-case-cites accessed 4 July 2015, 2 Box 1.

A first and much needed transformation is to put more focus on sustainable use.[14] Indeed, this is a compromise solution between various interests. While CITES and RFMOs traditionally have different practices, one being at times more oriented towards conservation/preservation and the other more towards exploitation, both aims can be accommodated through the compromise solution that is sustainable use.

Furthermore, promoting sustainability instead of pure preservation allows some margin of negotiation as to where sustainability is set, while preservation, being absolute, tends to polarize views for or against it. Sustainable use also makes sense from an environmental and economic point of view (and consequently from a social one in the mid-/long-term), provides some economic incentive for good management, and gives an important decision-making role to science.

More practically, it entails the recognition, by the CITES community, that both trade and restrictions thereto can be beneficial to conservation and that the use of CITES does not result in an automatic prohibition of trade. In the case of timber, the evolution of states' opinions, which culminated with their agreement to a listing, is fairly recent. The Resolution on the *Implementation of the Convention for timber species*, after stating that trade can be beneficial to the conservation of species and ecosystems when carried out at levels that are not detrimental to the survival of the species in question, rightly and explicitly points out that a listing in Appendix II or III is often wrongly perceived as a trade ban, and that such misconception has negative consequences.

Similarly, positive discussions could take place and be encouraged for fish species. Moreover, a listing proposal for a commercially-exploited fish species would be a chance to address broader issues typical of the CITES regime, such as decisions taken in total disregard of the listing criteria or problems of down-/de-listing. Indeed, since fish are non-charismatic species, the debate could be less heated than, for example, for whales or elephants.

A second modification which is required to depolarize the situation is an earlier and clearer consultation procedure with the relevant RFMOs. This could indeed avoid or at least decrease confrontation and has the potential to address feelings of competition as well as to promote mutual interests.

In the case of timber, the ITTO-CITES relation appears to be non-confrontational, with the international organization and the biodiversity treaty's CoP knowing their respective roles and strengths but not feeling endangered by the other's involvement on overlapping matters. While the Secretariat originally had suggested a clarification of their roles stating "the primary role of ITTO and regional forestry management organizations in forestry management and the role of CITES in regulating international trade"[15] this formulation was not

[14] C. Huxley, 'CITES: The Vision' in J. Hutton and B. Dickson (eds), *Endangered Species-Threatened Convention: The Past, Present and Future of CITES* (Earthscan 2000) 11.

[15] United States of America, 'CoP14 Doc. 18.2 – Cooperation between CITES and ITTO Regarding Trade in Tropical Timber' (3–15 June 2007) 'Comments from the Secretariat' B.

retained. Instead, the Resolutions adopted are nuanced, mentioning the positive role that CITES can have, through trade restrictions, in the conservation of timber species while also noting the ITTO's present and future role regarding international trade in tropical timber species.

Third, the effectiveness of CITES needs to be actually assessed or projected. Indeed, in order to move away from the politically motivated polarization and towards an honest weighing of the potential value added in using CITES, one needs more than hypotheses. CITES is only a tool towards conservation of endangered species;[16] its implementation has evidently a cost and it should consequently solely be encouraged when and if it brings—or can be expected to bring—positive results. Generally, this result-focused approach needs to be acknowledged in order to avoid that the fear of overprotective measures continues to provide a practical excuse not to undertake any action towards long-term sustainability of resources.

[16] Huxley (n 14) 3–4.

Annexes

S. Guggisberg, *The Use of CITES for Commercially-exploited Fish Species*, Hamburg Studies on Maritime Affairs 35, DOI 10.1007/978-3-319-23702-2

Annex 1a—General Presentation of RFMOs: Non-tuna RFMOs

	CCAMLR	GFCM	NAFO	NEAFC	SEAFO	SIOFA	SPRFMO
Full name	Commission for the Conservation of Antarctic Marine Living Resources	General Fisheries Commission for the Mediterranean	Northwest Atlantic Fisheries Organization	North East Atlantic Fisheries Commission	South East Atlantic Fisheries Organisation	South Indian Ocean Fisheries Agreement	South Pacific Regional Fisheries Management Organisation
Founding treaty	CCAMLR Convention entered into force on 07/04/1982, as part of the Antarctic Treaty System	Agreement for the Establishment of the GFCM, adopted under article XIV FAO Constitution, approved by FAO Conference in 1949//entered into force 20/02/1952//amended 1963, 1976, 1997	Convention on Future Multilateral Cooperation in the Northwest Atlantic Fisheries, signed 24/10/1978// entered into force 01/01/1979. Amendment to the Convention on Future Multilateral Cooperation in the Northwest Atlantic Fisheries adopted 28/09/2007//not yet in force	Convention on Future Multilateral Cooperation in Northeast Atlantic Fisheries signed 18/11/1980//entered into force 17/03/1982. Amendments to the 1982 Convention adopted in 2004 and 2006// contracting parties have agreed to use the "new" Convention on a provisional basis pending ratification	Convention on the Conservation and Management of Fishery Resources in the South-East Atlantic Ocean signed 20/04/2001//entered into force 13/04/2003	Southern Indian Ocean Fisheries Agreement, adopted 07/07/2006//entered into force 21/06/2012	Convention on the Conservation and Management of High Seas Fishery Resources in the South Pacific Ocean, adopted 14/11/2009//entered into force 24/08/2012
Objectives	Find balance between "conservation" and	Promote the development, conservation,	Contribute through consultation and cooperation to the	Ensure the long-term conservation and optima	Ensure long-term conservation and	Ensure the long-term conservation and	Through the application of the

	"rational use" of marine living resources to ensure that stocks are harvested sustainably, existing ecological relationships between harvested, dependent and related species are maintained and depleted populations are restored to levels at which their biological productivity is greatest.	rational management and best utilization of living marine resources, as well as the sustainable development of aquaculture in the Mediterranean, Black Sea and connecting waters.	optimum utilization, rational management and conservation of the fishery resources of the Convention Area amended Convention adopted in 2007, still to be ratified, the objective had been revised as follows: "... ensure the long term conservation and sustainable use of the fishery resources in the Convention Area and, in so doing, to safeguard the marine ecosystems in which these resources are found".	utilization of the fishery resources in the Convention Area, providing sustainable economic, environmental and social benefits.	sustainable use of fishery resources in the management area, except highly migratory species.	sustainable use of fishery resources other than tuna in areas that fall outside national jurisdictions.	precautionary approach and an ecosystem approach to fisheries management, to ensure the long-term conservation and sustainable use of fishery resources and, in so doing, to safeguard the marine ecosystems in which these resources occur.
Species targeted	Antarctic marine living resources comprise the populations of fin fish, molluscs, crustaceans and all other species of	All living resources in waters under its mandate	Most fishery resources of the Northwest Atlantic except salmon, tunas/marlins, mammals as these are already under the responsibility	All fish, molluscs, crustaceans and including sedentary species within its area of competence, excluding, in so	Species in the management area, including sedentary/discrete and straddling species such as alfonsino,	Species in the management area, except tunas	All fish within the Convention area, including molluscs and crustaceans but excluding sedentary species, highly

(continued)

	CCAMLR	GFCM	NAFO	NEAFC	SEAFO	SIOFA	SPRFMO
	living organisms, including birds, found south of the Antarctic Convergence, but excluding whales and seals		of other intergovernmental regional fisheries management bodies. It also does not apply to sedentary species such as many shellfish over which coastal States exercise sovereign rights	far as they are dealt with by other international agreements, highly migratory species listed in Annex I of the United Nations Convention on the Law of the Sea of 10 December 1982, and anadromous stocks	orange roughy, oreo dories, armourhead, sharks, deepwater hake and red crab		migratory species, anadromous and catadronomous species, marine mammals, reptiles and se birds
Area of applicability	Southern Ocean – high seas and waters under national jurisdiction	Mediterranean Sea and Black Sea – high seas and waters under national jurisdiction	Northwest Atlantic Ocean – high seas and waters under national jurisdiction, but so-called regulatory area only in waters beyond 200 nm	Northeast Atlantic Ocean – high seas and waters under national jurisdiction, but so-called regulatory area only in waters beyond 200 nm	Southeast Atlantic Ocean – high seas only	South Indian Ocean – high seas only	South Pacific Ocean – high seas only
Examples of measures	Measures on fish stocks, by-catch reduction, gear; closed areas/seasons; IUU vessels list and lists of authorised vessels; port	Measures on fish stocks, by-catch reduction, gear, fishing capacity, aquaculture; compliance; IUU vessels list;	TACs; by-catch reduction; measures on gear; effort allocation schemes; vessel registers; IUU vessels list; closed areas/seasons;	TACS; closed areas/seasons; VMS; IUU vessels list; port state measures	TACs; by-catch reduction; measures on level of fishing effort, gear; lists of authorised vessels and of authorised ports;	n/a	TACs; measures on gear, minimum sizes; by-catch reduction; port state measures; VMS; inspection and

	inspections; VMS; regulation of transshipment	closed areas/ seasons	VMS; observer programme; port inspection programme; closures		IUU vessel list; MCS		observer programme; list of authorised vessels; IUU vessels list
TDS/CDS	CDS	–	–	–	–	–	–
Measures against non-parties	Scheme to promote compliance by non contracting parties vessels	–	Scheme to promote compliance by non contracting parties	Scheme to promote compliance by non contracting parties	Contact with non-parties; IUU list coupled with port state measures	Yes, measures consistent with international law	Yes, individual or collective measures consistent with international law
Reference to Precautionary Approach	Yes	Yes	Made use of, but not in the Convention	Yes	Yes	Yes	Yes
Reference to Ecosystem Approach	Yes	Yes	No	Yes	Yes	Yes	Yes
CMMs adoption procedure	Consensus for substantive matters	(Qualified) majority vote	(Qualified) majority vote	(Qualified) majority vote	Consensus	Consensus for substantive matters	Consensus
Opt-out possibility	Yes	Yes	Yes	Yes	Yes but restricted	No	Yes but restricted
Possibility for non-members to join	n/a	2/3 majority of existing members	Unqualified openness	3/4 majority of existing members	Unqualified openness	Unqualified openness	Unqualified openness
Explicit legal personality	Yes	–	Yes	Yes	Yes	–	Yes
Dispute settlement provision	Yes	Yes	No	Yes	Yes	Yes	Yes

Annex 1b—General Presentation of RFMOs: Tuna RFMOs

	CCSBT	IATTC	ICCAT	IOTC	WCPFC
Full name	Commission for the Conservation of Southern Bluefin Tuna	Inter-American Tropical Tuna Commission	International Commission for the Conservation of Atlantic Tunas	Indian Ocean Tuna Commission	Western and Central Pacific Fisheries Commission
Founding treaty	Convention for the Conservation of Southern Bluefin Tuna, signed 10/05/1993//entered into force 20/05/1994	Convention for the Establishment of an IATTC, signed 31/05/1949//entered into force 03/03/1950. Amended by Convention for the strengthening of IATTC established the the 1949 Convention, opened for signature in 2003// entered into force 27 August 2010	International Convention for the Conservation of Atlantic Tunas, signed 14/05/1966//entered into force 21/03/1969// amended 1984, 1992	Agreement for the Establishment of the IOTC, adopted under article XIV FAO Constitution, approved by FAO Council 11/1993// entered into force 27/03/1996	Convention for the Conservation and Management of Highly Migratory Fish Stocks in the Western and Central Pacific Ocean, signed 04/09/2000//entered into force 19/06/2004
Objectives	Ensure through appropriate management conservation and optimum utilization of the global Southern Bluefin Tuna fishery	Ensure long-term conservation and sustainable use of fish stocks under the management of the Convention, in accordance with rules of international law	Ensure the conservation of tunas and tuna-like species in the Atlantic ocean and adjacent seas; prevent bycatch (sharks)	Promote cooperation in order to ensure, through appropriate management, the conservation and optimum utilization of tunas and tuna-like species and encourage sustainable development of these fisheries	Ensure long-term conservation and sustainable use of highly migratory fish stocks under its mandate, in accordance with UNCLOS and UNFSA

Species targeted	Southern Bluefin Tuna	Tunas and tuna-like species; other species taken in the tuna fishery	Tunas and tuna-like species	Tunas and tuna-like species	All stocks of highly migratory species in the area (except sauries). In particular, this means skipjack tuna, yellowfin tuny, albacore tuna, bigeye tuna
Area of applicability	Whole range of Southern Bluefin Tuna – high seas and waters under national jurisdiction	Eastern Pacific Ocean – high seas and waters under national jurisdiction	Atlantic Ocean and adjacent seas – high seas and waters under national jurisdiction	Indian Ocean and adjacent seas, north of the Antarctic Convergence – high seas and waters under national jurisdiction	Western and Central Pacific Ocean – high seas and waters under national jurisdiction
Examples of measures	TACs; by-catch mitigation; lists of vessels; VMS; monitoring of transhipment at sea; IUU vessels list	Measures on fish stocks, on fishing capacity, on gear; by-catch reduction; closed areas/seasons; observers; IUU vessels list	TACs; by-catch reduction; measures on fishing capacity; closed areas/seasons; records of vessels, farming facilities, traps, ports; IUU vessels list; regional observer programme; joint schemes of international inspection	Measures on fish stocks; by-catch reduction; control of fishing capacity; list of authorised vessels; IUU vessels list; VMS; control and inspection scheme; regional observer scheme; inspection in port programme	Measures on fish stocks, on gear, on fishing capacity; by-catch reduction; VMS; regulation of transshipment; record fo fishing vessel; IUU vessels list
TDS/CDS	CDS	TDS	Bluefin Tuna CDS	TDS	–
Measures against non-parties	Contact with non parties fishing in the area and possible trade-restrictive measures	–	–	Scheme to promote compliance by non contracting parties	Contact with non-parties

(continued)

	CCSBT	IATTC	ICCAT	IOTC	WCPFC
Reference to Precautionary Approach	Some reference to it, but not in a binding text	Yes	No	Yes	Yes
Reference to Ecosystem Approach	No	Yes	No	No	Yes
CMMs adoption procedure	Unanimity	Consensus	(Qualified) majority vote	(Qualified) majority vote	Consensus
Opt-out possibility	No	No	Yes	Yes	Yes
Possibility for non-members to join	Unqualified openness	Consensus of existing members	Unqualified openness	Unqualified openness (for FAO members)	Invitation decided by consensus of existing members
Explicit legal personality	Yes	Yes	–	–	Yes
Dispute settlement provision	Yes	Yes	No	Yes	Yes

Annex 2a—Measures Related to Sharks Taken by RFMOs: Non-tuna RFMOs

	CCAMLR	GFCM	NAFO	NEAFC	SEAFO
Requirement of a NPOA	No, but reference to it				No, just recalling its existence
Prohibition of a direct fishery	Fishing of sharks in the Convention area for other than scientific research purposes is forbidden (could be resumed on the advice of the Scientific Committee).	Vessels should not undertake direct fishery for thresher shark.		16 species of deep sea sharks; porbeagle until 2014; spiny dogfish; basking shark until 2014.	Deep-water shark directed fisheries in the SEAFO Convention Area until additional information becomes available to identify sustainable harvesting levels
Obligation to use in full		x			x
Prohibited to discard before the first point of offloading (except head, guts, skin)		x	x		x
Prohibited to have shark fins totalling more than 5% of the weight of sharks onboard		x	x		x
Ban on finning and transhipment and landing of shark fins			x		

(continued)

	CCAMLR	GFCM	NAFO	NEAFC	SEAFO
Prohibited to retain on board, transship or land any fins harvested in contravention of the RFMO regulations		x	x		x
Prohibition to retain onboard, transship, land, store, sell, or offer for sale any part or whole carcass of a species		Bigeye thresher sharks, hammerhead sharks.			
Research to make fishing gear more selective	Fishing of sharks in the Convention area for other than scientific research purposes forbidden	x	x		x
Research to identify nursery areas		x	x		x
Release	Sharks (taken accidentally as by-catch).	Juveniles (whenever possible); Bigeye thresher sharks and hammerhead sharks (except if caught by a developing country for the latter).	Sharks (in case of by-catch; not used for subsistence).	Sharks (any incidental catches shall be promptly released; to the extent possible).	Sharks (in fisheries not directed at sharks; release of sharks, especially juveniles; to the extent possible; not used for food or subsistence).
Report		Fishing vessels' logbooks to include information on daily number of sharks caught; report on releases/discards (hammerhead); States to provide data for Atlantic shortfin mako.	Catches.	Data to be submitted to International Council for the Exploration of the Sea for further evaluation of stocks.	Catches (annually).

Relevant document(s)	Conservation measure 32-18 – Conservation of Sharks	Recommendation 2005/3 – ICCAT Recommendation [04-10] Concerning the Conservation of Sharks Caught in Association with Fisheries Managed by ICCAT(E); Recommendation 2006/8 (B) – ICCAT Recommendation [Rec. 05-05] to Amend Recommendation [04-10] Concerning the Conservation of Sharks Caught in Association with Fisheries Managed by ICCAT; Recommendation 34/2010/4 (C) – ICCAT Recommendation [09-07] on the Conservation of Thresher Sharks Caught in Association with Fisheries in the ICCAT Convention Area; Recommendation 35/2011/7 (B) ICCAT recommendation [10-06] on Atlantic Shortfin Mako sharks	Conservation and Enforcement Measures NAFO/FC Doc. 12/1, article 12 Conservation and Management of Sharks	Recommendation 7: 2012 for Conservation and Management Measures for Deep Sea Sharks in the NEAFC Regulatory Area for 2012; Recommendation 6: 2012 for Conservation and Management Measures for Porbeagle (Lamna Nasus) in the NEAFC Regulatory Area from 2012 to 2014; Recommendation 5: 2012 for Conservation and Management Measures for Spurdog (Squalus Acanthias) in the NEAFC Regulatory Area from 2012 to 2014; Recommendation 4: 2012 for Conservation and Management Measures for Basking Shark (Cetorhinus Maximus) in the NEAFC Regulatory Area from 2012 to 2014	Conservation Measure 04/06 on the Conservation of Sharks Caught in Association with Fisheries Managed by SEAFO; Recommendation 1/2008 Banning of Deep-Water Shark Catches.

(continued)

	CCAMLR	GFCM	NAFO	NEAFC	SEAFO
		caught in association with fisheries managed by ICCAT; Recommendation 35/2011/7 (C) – ICCAT Recommendation [10-08] on Hammerhead Sharks (family Sphyrnidae) Caught in Association with Fisheries Managed by ICCAT; Recommendation 35/2011/1 Concerning the Establishment of a GFCM Logbook, Amending Recommendation GFCM/34/2010/1, Annex 1			
Measures applicable to	Not specified	Contracting parties and cooperating non-contracting parties	Contracting parties	Contracting parties. Encouragement to take similar measures for waters under national jurisdiction.	Contracting parties

Annex 2b—Measures Related to Sharks Taken by RFMOs: Tuna RFMOs

	CCSBT	IATTC	ICCAT	IOTC	WCPFC
Requirement of a NPOA	No, but strong encouragement to comply with IPOA-Sharks	x	x		Implementation of the IPOA; NPOAs have to include measures to minimize waste and discards and encourage the live release of incidental catches.
Prohibition of a direct fishery	No binding measure developed by CCSBT with regards to sharks, but strong encouragement to comply with measures adopted by IOTC, WCPFC and ICCAT.		Vessels should not undertake direct fishery for thresher shark; States that have not reported their data for Atlantic shortfin mako sharks shall be prohibited from retaining this species, from 2013.		
Obligation to use in full			x	x	x
Prohibited to discard before the first point of offloading (except head, guts, skin)		x	x	x	x
Prohibited to have shark fins totalling more than 5% of the weight of sharks onboard		x	x	x	x

(continued)

	CCSBT	IATTC	ICCAT	IOTC	WCPFC
Ban on finning and transhipment and landing of shark fins					
Prohibited to retain on board, tranship or land any fins harvested in contravention of the RFMO regulations		x	x	x	x
Prohibition to retain onboard, transship, land, store, sell, or offer for sale any part or whole carcass of a species		Oceanic whitetip sharks.	Bigeye thresher sharks, oceanic whitetip sharks, hammerhead sharks.	Family Alopiidae.	
Research to make fishing gear more selective	No binding measure developed by CCSBT with regards to sharks, but strong encouragement to comply with measures adopted by IOTC, WCPFC and ICCAT.	x	x		
Research to identify nursery areas		x	x	x	
Release		Sharks not used for subsistence (in particular juveniles, as much as possible); oceanic whitetip sharks (to the extent possible).	Juveniles (whenever possible); Bigeye thresher sharks and hammerhead sharks (except if caught by a developing country for the latter).	Juveniles (incidentally-caught ones, whenever possible); females (incidentally-caught ones, in particular pregnant ones, whenever possible); Family Alopiidae (to the extent practicable specimens).	Sharks (in fisheries non directed at sharks; for sharks not used for food or other purposes).

Report		Implementation of resolution; report on discards/releases (oceanic whitetip sharks).	Catches (annually); report on releases/discards (oceanic whitetip and hammerhead sharks); States to provide data for Atlantic shortfin mako sharks.	Catches (annually)	Implementation of IPOA-Shark, including the results of their assessment and the status of the NPOA; data on catches, fishing effort and key shark species (annually).
Relevant document(s)	Recommendation to Mitigate the Impact on Ecologically Related Species of Fishing for Southern Bluefin Tuna	Resolution C-05-03 Resolution on the Conservation of Sharks Caught in Association with Fisheries in the Eastern Pacific Ocean; Resolution C-11-10 Resolution on the Conservation of Oceanic Whitetip Sharks Caught in Association with Fisheries in the Antigua Convention Area	Resolution [03-10] on the Shark Fishery; Recommendation [04-10] Concerning the Conservation of Sharks Caught in Association with Fisheries Managed by ICCAT; Recommendation [05-05] to Amend ICCAT Recommendation [04-10]; ICCAT Supplemental Recommendation [07-06] Concerning Sharks or ICCAT Resolution [08-08] on Porbeagle Shark; Recommendation [09-07] on the Conservation of Thresher Shark Caught in Association with Fisheries Managed in	Resolution 05/05 Concerning the Conservation of Sharks Caught in Association with Fisheries Managed by IOTC; Resolution 10/12 On the Conservation of Thresher Sharks (Family Alopiidae) Caught in Association with Fisheries in the IOTC Area of Competence	Conservation and Management Measures 2010-07 for Sharks

(continued)

	CCSBT	IATTC	ICCAT	IOTC	WCPFC
			the ICCAT Convention Area; Recommendation [10-06] on Atlantic Shortfin Mako Sharks Caught in Association with ICCAT Fisheries; Recommendation [10-07] on the Conservation of Oceanic Whitetip Shark Caught in Association with Fisheries Managed by ICCAT; Recommendation [10-08] on Hammerhead Sharks Caught in Association with Fisheries Managed by ICCAT		
Measures applicable to		Members and cooperating non-members	Contracting parties and cooperating non-contracting parties, entities or fishing entities	Contracting parties and cooperating non-contracting parties	Members, Cooperating non-Members, and participating Territories. Measures apply to sharks caught in association with fisheries managed under the WCPF Convention, and to sharks listed in Annex 1 of the 1982 Convention.

Annex 3—Comparison of FAO Advice/CITES Secretariat Recommendation/CoP Decision

CITES CoP13	Species	Proposal number and Appendix	Reason for proposal	Recommendation of FAO	Recommendation of the CITES Secretariat	Status now
2004				FAO Fisheries Report No. 748	CoP 13 Doc. 60	
1	Carcharodon carcharias (white shark)	Prop n 32 – Ap II	The proposal states that white shark meets the criteria in Resolution Conf. 9.24 (Rev. CoP 12) criteria A and B i) and ii) of Annex 2a (AC19 Doc. 9) because of the significant and ongoing population declines reported in literature and unpublished data.	The available evidence could support a range of hypotheses, and that it was not possible to confirm or exclude the possibility that the species as a whole meets the criteria for listing in Appendix II.	Secretariat recommends that this proposal be adopted if it is amended to remove the annotation indicating a zero annual export quota.	Listed in Ap II
2	Cheilinus undulatus (humphead wrasse)	Prop n 33 – Ap II	Inclusion of Cheilinus undulatus in Appendix II in accordance with Article II, paragraph 2(a) of the Convention and Resolution Conf. 9.24 (Rev. CoP 12), Annex 2a, paragraph B.	Humphead wrasse meets the Annex 2a criterion B and possibly also criterion A for inclusion in CITES Appendix II. Regulation of trade as a result of CITES listing could make a significant contribution to the conservation of this species.	Secretariat recommends that this proposal be adopted.	Listed in Ap II

(continued)

CITES CoP13	Species	Proposal number and Appendix	Reason for proposal	Recommendation of FAO	Recommendation of the CITES Secretariat	Status now
2004				FAO Fisheries Report No. 748	CoP 13 Doc. 60	
3	Lithophaga lithophaga (Mediterranean date mussel)	Prop n 35 – Ap II	Inclusion in Appendix II, in accordance with Article II, paragraph 2(a).	No present risk of extinction. While there are destructive practices though, illegal trade will not be tackled by a listing under CITES. Also, an Appendix II listing would be less strict than existing measures.	Secretariat recommends that this proposal be rejected.	Listed in Ap II
4	Helioporidae spp., Tubiporidae spp., Scleractinia spp., Milleporidae spp., and Stylasteridae spp. (corals)	Prop n 36 – amendment of annotation		Unable to evaluate the effectiveness of the proposed annotation.	Secretariat recommends that this proposal be adopted.	Annotation included

CITES CoP14	Species	Proposal number and Appendix	Reason for proposal	Recommendation of FAO	Recommendation of the CITES Secretariat	Status now
2007				FAO Fisheries Report No. 833	CoP14 Doc. 68	
5	Lamna lasus (porbeagle)	Prop n 15 – Ap II	Regulation of trade is necessary to avoid the species becoming eligible for inclusion in Ap I in the near future.	The available evidence does not support the proposal.	Secretariat recommends that this proposal be adopted.	Not listed until CoP16
6	Squalus acanthias (spiny dogfish)	Prop n 16 – Ap II	Regulation of trade is necessary to avoid that the species be eligible for listing in Ap I in the near future.	The available evidence does not support the proposal.	Secretariat recommends that this proposal be adopted.	Not listed
7	All species of the family Pristidae (sawfishes)	Prop n 17 – Ap I	All known species in the family Pristidae have severely declined. A listing preventing trade would benefit the wild populations of this low-productivity species.	The available evidence supports the proposal.	Secretariat recommends that this proposal be adopted.	All listed in Ap I except Pristis microdon in Ap II
8	Anguilla anguilla (European eel)	Prop n 18 – Ap II	Long and steady decline and the stock is outside safe biological limits.	The available evidence supports the proposal.	Secretariat recommends that this proposal be adopted.	Listed in Ap II
9	Pterapogon kaudemi (Banggai cardinal fish)	Prop n 19 – Ap II	Regulation of trade required to ensure that the harvest of specimens from the wild is not reducing the wild population to a level where survival might be threatened by continued harvesting or other influences.	The Banggai cardinalfish does not meet the biological criteria for Appendix II listing.	Secretariat recommends that this proposal be adopted.	Withdrawn by proponent; not listed

(continued)

CITES CoP14	Species	Proposal number and Appendix	Reason for proposal	Recommendation of FAO	Recommendation of the CITES Secretariat	Status now
2007				FAO Fisheries Report No. 833	CoP14 Doc. 68	
10	Species of Panulirus argus and laevicauda of the Brazilian lobster population	Prop n 20 – Ap II		The available evidence does not support the proposal.	Secretariat recommends that this proposal be rejected.	Withdrawn by proponent; not listed
11	All species in the genus Corallium (red and pink corals)	Prop n 21 – Ap II	Regulation of trade required to ensure that the harvest of specimens from the wild is not reducing the wild population to a level where survival might be threatened by continued harvesting or other influences.	The available evidence does not support the proposal to include all species.	Secretariat recommends that this proposal be adopted.	Not listed; some in Ap III (China)

CITES CoP15	Species	Proposal number and Appendix	Reason for proposal	Recommendation of FAO	Recommendation of the CITES Secretariat	Status now
2010				FAO Fisheries and Aquaculture Report No. 925	CoP15 Doc. 68 Annex 2	
12	Sphyrna lewini (scalloped hammer-head); Shpyrna mokarran (great hammerhead); Sphyrna zygaena (smooth hammer-head); Carcharhinus plumbeus (sandbar shark); Carcharhinus obscurus (dusky shark)	Prop n 15 – Ap II	Sphyrna Lewini is over-exploited for its fins and also is victim of bycatch. According to projection, it will be eligible for Ap I listing within a 10-year period if not regulated. Others are look-alike.	The available evidence supports the proposal to include scalloped hammerhead, as well as great hammerhead and smooth hammerhead because of look-alike. But not sandbar and dusky sharks.	Secretariat recommends that this proposal be adopted for S. lewini, S. mokarran and S. zygaena but rejected for C. plumbeus and C. obscurus.	Not listed but three species listed in CoP16
13	Carcharhinus longimanus (oceanic whitetip shark)	Prop n 16 – Ap II	Carcharhinus longimanus is over-exploited for its fins and also is victim of bycatch. Regulation of trade in the species is necessary to avoid it becoming eligible for inclusion in Appendix I in the near future.	On balance, the available evidence supports the proposal.	Secretariat recommends that this proposal be adopted.	Not listed until CoP16
14	Lamna lasus (porbeagle)	Prop n 17 – Ap II	Some populations of lamna nasus have experienced marked decline. Others can be projected to decrease unless international trade is regulated. Finally look-alike problems could arise. Listing necessary to avoid the species becoming eligible for inclusion in Appendix I in the near future.	The available evidence supports the proposal.	Secretariat recommends that this proposal be adopted.	Not listed until CoP16

(continued)

CITES CoP15	Species	Proposal number and Appendix	Reason for proposal	Recommendation of FAO	Recommendation of the CITES Secretariat	Status now
2010				FAO Fisheries and Aquaculture Report No. 925	CoP15 Doc. 68 Annex 2	
15	Squalus acanthias (spiny dogfish)	Prop n 18 – Ap II	Some stocks have experienced marked decline. Others' survival might be at risk if continued harvesting or other influences. Look-alike issues could arise.	The available evidence does not support the proposal.	Secretariat recommends that this proposal be adopted.	Not listed
16	Thunnus thynnus (Atlantic bluefin tuna)	Prop n 19 – Ap I	Genetically effective population size of subpopulations in the Mediterranean (400-700 individuals) is close to or below the minimum threshold related to the maintenance of genetic diversity and evolutionary potential in the long term; decline criteria for low productivity species met and the species also has added vulnerabilities such as its strong aggregating behaviour.	A majority of the Panel considered that the available evidence supported the proposal; but agreed with ICCAT SCRS that the decline criteria for Ap I was not met. Consensus for Ap II listing.	Secretariat recommends that this proposal be adopted.	Not listed
17	All species in the family Coralliidae (red and pink corals)	Prop n 21 – Ap II	Intensively harvested; vulnerability to over-exploitation.	The available evidence does not support the proposal.	Secretariat recommends that this proposal be adopted.	Not listed; some under Ap III (China)

CITES CoP16	Species	Proposal number and Appendix	Reason for proposal	Recommendation of FAO	Recommendation of the CITES Secretariat	Status now
2013				**FAO Fisheries and Aquaculture Report No. 1032**	**CoP16 Doc.77 Annex 1**	
18	Carcharhinus longimanus (oceanic whitetip shark)	Prop n 42 – Ap II	The greatest threats to this low-productivity species worldwide are harvest for the international fin trade and bycatch, which have led to significant declines. Based upon rates of exploitation, this species may become threatened with extinction unless international trade regulation provides an incentive to introduce or improve monitoring and management measures to provide a basis for non-detriment and legal acquisition findings.	As at CoP15, on the basis of available evidence and on the basis of new data, the oceanic whitetip shark meets the biological criteria for Appendix II listing.	Secretariat recommends that this proposal be adopted.	Listed in Ap II
19	Sphyrna lewini (scalloped hammerhead); Sphyrna mokarran (great hammerhead); and Sphyrna zygaena (smooth hammerhead)	Prop n 43 – Ap II	The greatest threats to Sphyrna lewini, a low-productivity species, worldwide are due to the international fin trade and bycatch, which have caused historic declines of at least 15-20% from the baseline for long-term time series in multiple ocean basins. Sphyrna mokarran and Sphyrna zygaena as look-alike.	The Panel concluded that based on the available evidence Sphyrna lewini meets the biological criteria for listing on CITES Appendix II. The other two proposed species, S. mokarran and S. zygaena fulfil the criteria for inclusion under CITES Appendix II stipulated in Article II, paragraph 2b (“look-alike clause”).	Secretariat recommends that this proposal be adopted.	Three listed in Ap II

(continued)

CITES CoP16	Species	Proposal number and Appendix	Reason for proposal	Recommendation of FAO	Recommendation of the CITES Secretariat	Status now
2013				FAO Fisheries and Aquaculture Report No. 1032	CoP16 Doc.77 Annex 1	
20	Lamna lasus (porbeagle)	Prop n 44 – Ap II	North and Southwest Atlantic and Mediterranean stocks of Lamna nasus qualify for listing Annex 2a A because their marked decline in population size meets CITES' guidelines. It also qualifies under Annex 2a B, as, due to shifting of effort from Northeast to Northwest Atlantic stocks, and a rapid recent rate of decline in catch per unit effort data for some southern stocks, it can be projected that southern hemisphere populations have experienced or are likely to undergo similar declines, triggering consideration for Appendix 1.	The majority of Panel members considered that the species as a whole meets the decline criteria for Appendix II.	Secretariat recommends that this proposal be adopted.	Listed in Ap II
21	Genus Manta	Prop n 46 – Ap II	All Manta species qualify for inclusion in Appendix II under Annex 2a A, meeting CITES' definition of decline for low productivity aquatic species. Significant rates of decline are due to increasing fishing pressure driven by international trade. Manta spp. also qualify for inclusion in Appendix II under criterion B, because of their small and highly fragmented populations, extremely low productivity, and known aggregating behaviour, that cause them to be highly vulnerable to exploitation.	The Panel, due to paucity of information, was unable to identify reliable information to assess against the decline criteria or the small population criteria.	Secretariat recommends that this proposal be adopted.	Genus listed in Ap II

About the International Max Planck Research School for Maritime Affairs at the University of Hamburg

The International Max Planck Research School for Maritime Affairs at the University of Hamburg was established by the Max Planck Society for the Advancement of Science, in co-operation with the Max Planck Institute for Foreign Private Law and Private International Law (Hamburg), the Max Planck Institute for Comparative Foreign Public Law and International Law (Heidelberg), the Max Planck Institute for Meteorology (Hamburg) and the University of Hamburg. The School's research is focused on the legal, economic, and geophysical aspects of the use, protection, and organization of the oceans. Its researchers work in the fields of law, economics, and natural sciences. The School provides extensive research capacities as well as its own teaching curriculum. Currently, the School has 22 Directors who determine the general work of the School, act as supervisors for dissertations, elect applicants for the School's PhD-grants, and are the editors of this book series:

Prof. Dr. Dr. h.c. mult. Jürgen Basedow is Director of the Max Planck Institute for Foreign Private Law and Private International Law; *President and Professor Monika Breuch-Moritz* is the President of the German Federal Maritime and Hydrographic Agency; *Prof. Dr. Dr. h.c. Peter Ehlers* is the Director ret. of the German Federal Maritime and Hydrographic Agency; *Prof. Dr. Dr. h.c. Hartmut Graßl* is Director emeritus of the Max Planck Institute for Meteorology; *Dr. Tatiana Ilyina* is the Leader of the Research Group "Ocean Biogeochemistry" at the Max Planck Institute for Meteorology in Hamburg; *Prof. Dr. Florian Jeßberger* is Head of the International and Comparative Criminal Law Division at the University of Hamburg; *Prof. Dr. Lars Kaleschke* is Junior Professor at the Institute of Oceanography of the University of Hamburg; *Prof. Dr. Hans-Joachim Koch* is Director emeritus of the Seminar of Environmental Law at the University of Hamburg; *Prof. Dr. Robert Koch* is Director of the Institute of Insurance Law at the University of Hamburg; *Prof. Dr. Doris König* is the President of the Bucerius Law School; *Prof. Dr. Rainer Lagoni* is Director emeritus of the Institute of Maritime Law and the Law of the Sea at the University of Hamburg; *Prof. Dr. Gerhard Lammel* is Senior Scientist and Lecturer at the Max Planck Institute

S. Guggisberg, *The Use of CITES for Commercially-exploited Fish Species*, Hamburg Studies on Maritime Affairs 35, DOI 10.1007/978-3-319-23702-2

for Chemistry, Mainz; *Prof. Dr. Ulrich Magnus* is Managing Director of the Seminar of Foreign Law and Private International Law at the University of Hamburg; *Prof. Dr. Peter Mankowski* is Director of the Seminar of Foreign and Private International Law at the University of Hamburg; *Prof. Stefan Oeter* is Managing Director of the Institute for International Affairs at the University of Hamburg; *Prof. Dr. Marian Paschke* is Managing Director of the Institute of Maritime Law and the Law of the Sea at the University of Hamburg; *PD Dr. Thomas Pohlmann* is Senior Scientist at the Centre for Marine and Climate Research and Member of the Institute of Oceanography at the University of Hamburg; *Dr. Uwe A. Schneider* is Assistant Professor at the Research Unit Sustainability and Global Change of the University of Hamburg; *Prof. Dr. Detlef Stammer* is Professor in Physical Oceanography and Remote Sensing at the Institute of Oceanography of the University of Hamburg; *Prof. Dr. Jürgen Sündermann* is Director emeritus of the Centre for Marine and Climate Research at the University of Hamburg; *Prof. Dr. Rüdiger Wolfrum* is Director emeritus at the Max Planck Institute for Comparative Foreign Public Law and International Law and a judge at the International Tribunal for the Law of the Sea; *Prof. Dr. Wilfried Zahel* is Professor emeritus at the Centre for Marine and Climate Research of the University of Hamburg.

At present, *Prof. Dr. Dr. h.c. Jürgen Basedow* and *Prof. Dr. Ulrich Magnus* serve as speakers of the Research School.

Bibliography

References

J. Adler and others, 'Forage fish: from ecosystems to markets' (2008) 33 Annual Review of Environmental Resources 153.

D. J. Agnew, 'The illegal and unregulated fishery for toothfish in the Southern Ocean, and the CCAMLR catch documentation scheme' (2000) 24 Marine Policy 361.

D. J. Agnew and others, 'Estimating the Worldwide Extent of Illegal Fishing' (2009) http://www.plosone.org/article/info:doi/10.1371/journal.pone.0004570 accessed 4 July 2015.

S. Aguilar, 'Regulatory Tools for the Management of Fish and Timber Species through CITES' (2013) 22 RECIEL 281.

M. Ahmed, 'Market Access and Trade Liberalisation in Fisheries' ICTSD Natural Resources, International Trade and Sustainable Development Series Issue Paper 4 (Geneva 2006).

M. Akehurst, 'The Hierarchy of Sources in International Law' (1974–1975) 47 British Yearbook of International Law 273.

C. F. Amerasinghe, *Principles of the Institutional Law of International Organization* (Cambridge Studies in International and Comparative Law, 2nd edn Cambridge University Press 2005).

R. P. Anand, 'Changing Concepts of Freedom of the Seas: A Historical Perspective' in J. M. van Dyke, D. Zaelke and G. Hewison (eds), *Freedom for the Seas in the 21st Century: Ocean Governance and Environmental Harmony* (Island Press 1993).

C. N. K. Anderson and others, 'Why fishing magnifies fluctuations in fish abundance' (2008) 452 Nature 835.

D. Anderson, 'Port States and Environmental Protection' in A. Boyle and D. Freestone (eds), *International Law and Sustainable Development – Past Achievements and Future Challenges* (Oxford University Press 1999).

D. H. Anderson, 'The Straddling Stocks Agreement of 1995: An Initial Assessment' (1996) 45 International and Comparative Law Quarterly 463.

J. L. Anderson, 'Aquaculture and fisheries: complement or competition' in R. Arthur and J. Nierentz (eds), *Global Trade Conference on Aquaculture*, FAO Fisheries Proceedings 9 (Rome 2007).

—— 'Sustainable Aquaculture: What Does It Mean and How Do We Get There?' in P. Leung, C.-S. Lee and P. J. O'Bryen (eds), *Species and System Selection for Sustainable Aquaculture* (Blackwell Publishing 2007).

G. Andreone and G. Cataldi, 'Regards sur les évolutions du droit de la mer en Méditerranée' (2010) 66 Annuaire français de droit international 1.

S. Guggisberg, *The Use of CITES for Commercially-exploited Fish Species*, Hamburg Studies on Maritime Affairs 35, DOI 10.1007/978-3-319-23702-2

S. Andresen, 'The International Whaling Regime: Order at the Turn of the Century' in D. Vidas and W. Ostreng (eds), *Order for the oceans at the turn of the century* (Kluwer Law International 1999).

Anonymous, 'A new CITES identification guide for sturgeons and paddlefish' *CITES World – Official Newsletter of the Parties* (December 2001).

—— 'Document numbering' *CITES World – Official Newsletter of the Parties* (July 2004).

—— 'The CITES Fort Lauderdale Criteria: The Uses and Limits of Science in International Conservation Decisionmaking' (2001) 114 Harvard Law Review 1769.

D. K. Anton, 'Dispute Concerning Japan's JARPA II Program of "Scientific Whaling" (Australia v. Japan)' (2010) http://www.asil.org/insights/volume/14/issue/20/dispute-concerning-japan%E2%80%99s-jarpa-ii-program-%E2%80%9Cscientific-whaling%E2%80%9D accessed 4 July 2015.

—— "Treaty congestion' in contemporary international environmental law' in S. Alam and others (eds), *Routledge Handbook of International Environmental Law* (Routledge 2013).

M. Arbour and S. Lavallée, *Droit international de l'environnement* (Bruylant 2006).

M. Ariji, Y. Senda and A. Takahara, 'Achievements and Problems of Aquaculture in Japan' in P. Leung, C.-S. Lee and P. J. O'Bryen (eds), *Species and System Selection for Sustainable Aquaculture* (Blackwell Publishing 2007).

F. Asche and F. Khatun, 'Aquaculture: Issues and Opportunities for Sustainable Production and Trade' ICTSD Natural Resources, International Trade and Sustainable Development Series Issue Paper 5 (Geneva 2006).

S. Ásmundsson, 'Whaling' in M. H. Nordquist and others (eds), *Law, Science & Ocean Management* (Martinus Nijhoff Publishers 2007).

A. Aust, *Modern Treaty Law and Practice* (Cambridge University Press 2000).

K. Baakman, *Testing Times: The Effectiveness of Five International Biodiversity-Related Conventions* (Wolf Legal Publishers 2011).

D. A. Balton, 'The Compliance Agreement' in E. Hey (ed), *Developments in International Fisheries Law* (Kluwer Law International 1999).

D. A. Balton and H. R. Koehler, 'Reviewing the United Nations Fish Stocks Treaty' (2006–2007) 7 Sustainable Development Law and Policy Review 5.

M. Barbuto and others, 'DNA barcoding reveals fraudulent substitutions in shark seafood products: The Italian case of 'palombo' (Mustelus spp.)' (2010) 43 Food Research International 376.

L. D. Barchue, 'Making a case for the Voluntary IMO Member State Audit Scheme: Paper delivered at a seminar on 'Auditing Flag States: New Directions for Smaller States' in Malmö, World Maritime University' (2005) http://www.imo.org/OurWork/Safety/Implementation/Documents/Voluntary.pdf accessed 6 July 2015.

R. Barnes, 'The Convention on the Law of the Sea: An Effective Framework for Domestic Fisheries Conservation?' in D. Freestone, R. Barnes and D. Ong (eds), *The Law of the Sea – Progress and Prospects* (Oxford University Press 2006).

R. Barston, 'The Law of the Sea and Regional Fisheries Organisations' (1999) 14 International Journal of Marine and Coastal Law 333.

A. Batchelor, 'The preservation of wildlife habitat in ecosystems: towards a new direction under international law to prevent species' extinction' (1987–1988) 3 Florida International Law Journal 307.

J. K. Baum and others, 'Collapse and Conservation of Shark Populations in the Northwest Atlantic' (2003) 299 Science 389.

A. J. Bederman, 'CCAMLR in crisis: a case study of marine management in the southern oceans' in H. N. Scheiber (ed), *Law of the Sea: The Common Heritage and Emerging Challenges* (Martinus Nijhoff Publishers 2000).

J. Beer-Gabel and V. Lestang, *Les Commissions de Pêche et leur Droit: La Conservation et la Gestion des Ressources Marines Vivantes* (Bruylant 2003).

N. Bernasconi-Osterwalder and others, *Environment and Trade: A Guide to WTO Jurisprudence* (Earthscan 2006).

S. Beslier, 'The exercise of jurisdiction over vessels: new developments in the sector of fisheries' in E. Franckx and P. Gautier (eds), *The Exercise of Jurisdiction over Vessels: New Developments in the Fields of Pollution, Fisheries, Crimes at Sea and Trafficking of Weapons of Mass Destruction* (Bruylant 2010).

—— 'Pêche et biodiversité marine: complémentarité ou concurrence au sein de l'ordre juridique international? Le cas du thon rouge' (Paris November 2010) Idées pour le débat n°9.

M. Bettati, 'Création et personnalité juridique des organisations internationales' in R.-J. Dupuy (ed), *Manuel sur les organisations internationales – A Handbook on International Organizations* (2nd edn Martinus Nijhoff Publishers 1998).

U. Beyerlin and J. Grote Stoutenburg, 'Environment, International Protection' Max Planck Encyclopedia of Public International Law www.mpepil.com.

U. Beyerlin and T. Marauhn, *International Environmental Law* (Hart Publishing; Verlag CH Beck 2011).

D. Bialek, 'Sink or Swim: Measures Under International Law for the Conservation of the Patagonian Toothfish in the Southern Ocean' (2003) 34 Ocean Development and International Law 105.

F. Biermann, 'Reforming Global Environmental Governance: The Case for a United Nations Environment Organisation (UNEO)' (2011) http://www.stakeholderforum.org/fileadmin/files/WEO%20Biermann%20FINAL.pdf accessed 6 July 2015.

—— 'The Emerging Debate on the Need for a World Environmental Organization' (2001) 1 Global Environmental Politics 45.

E. A. Bilsky, 'Conserving Marine Wildlife Through World Trade Law' (2008–2009) 30 Michigan Journal of International Law 599.

P. W. Birnie, 'Are Twentieth-Century Marine Conservation Conventions Adaptable to Twenty-First Century Goals and Principles?: Part I' (1997) 12 International Journal of Marine and Coastal Law 305.

P. Birnie, A. Boyle and C. Redgwell, *International Law and the Environment* (3rd edn Oxford University Press 2009).

T. Bjorndal, 'Overview, roles, and performance of the North East Atlantic fisheries commission (NEAFC)' (2009) 33 Marine Policy 685.

N. Blokker, 'International Organizations and Their Members: International Organizations Belong to All Members and to None' (2004) 1 International Organizations Law Review 139.

A. Blundell, 'A Review of the CITES Listing of Big-leaf Mahogany' (2004) 38 Oryx 84.

L. Boisson de Chazournes, 'Precaution in International Law: Reflection on its Composite Nature' in T. M. Ndiaye and R. Wolfrum (eds), *Law of the sea, environmental law and settlement of disputes: Liber amicorum Judge Thomas A. Mensah* (Nijhoff 2007).

L. Boisson de Chazournes and M. M. Mbengue, 'A 'footnote as a principle' Mutual supportiveness and its relevance in an era of fragmentation' in H. P. Hestermeyer and others (eds), *Coexistence, Cooperation and Solidarity: Liber Amicorum Rüdiger Wolfrum* (Martinus Nijhoff Publishers 2011).

C. J. Borgen, 'Resolving Treaty Conflicts' (2005) 37 George Washington International Law Review 573.

T. Bostock and S. Walmsley, 'Enough to Eat? Fisheries and Food Security' in R. Bourne and M. Collins (eds), *From Hook to Plate: The State of Marine Fisheries: A Commonwealth Perspective* (2009).

M. Bothe, 'Compliance' Max Planck Encyclopedia of Public International Law www.mpepil.com.

S. Boutillon, 'The Precautionary Principle: Development of an International Standard' (2001–2002) 23 Michigan Journal of International Law 429.

M. Bowman, 'Transcending the Fisheries Paradigm: Towards a Rational Approach to Determining the Future of the International Whaling Commission' (2009) 7 New Zealand Yearbook of International Law 85.

—— 'A Tale of Two CITES: Divergent Perspectives upon the Effectiveness of the Wildlife Trade Convention' (2013) 22 RECIEL 228.

M. Bowman, P. Davies and C. Redgwell, *Lyster's International Wildlife Law* (2nd edn Cambridge University Press 2010).

A. Boyle, 'Environmental Dispute Settlement' Max Planck Encyclopedia of Public International Law www.mpepil.com.

—— 'Dispute Settlement and the Law of the Sea Convention: Problems of Fragmentation and Jurisdiction' (1997) 46 International and Comparative Law Quarterly 37.

—— 'The Southern Bluefin Tuna Arbitration' (2001) 50 International and Comparative Law Quarterly 447.

A. Boyle and C. Chinkin, *The Making of International Law* (Oxford University Press 2007).

D. Brack, 'Environmental Treaties and Trade: Multilateral Environmental Agreements and the Multilateral Trading System' in G. P. Sampson and W. B. Chambers (eds), *Trade, Environment and the Millennium* (United Nations University Press 2002).

J. Braig, 'Whaling' Max Planck Encyclopedia of Public International Law www.mpepil.com.

R. Bratspies, 'Finessing King Neptune: Fisheries Management and the Limits of International Law' (2001) 25 Harvard Environmental Law Review 213.

C. T. Bright, 'The Future of the International Whaling Commission: Can We Save the Whales?' (1992–1993) 5 Georgetown International Environmental Law Review 815.

I. Brownlie, 'State responsibility: the problem of delegation' in K. Ginther and K. Zemanek (eds), *Völkerrecht zwischen normativem Anspruch und politischer Realität: Festschrift für Karl Zemanek zum 65. Geburtstag* (Duncker & Humblot 1994).

—— *International law at the fiftieth anniversary of the United Nations: general course on public international law* (Recueil des Cours vol 255, Martinus Nijhoff Publishers 1995).

—— 'The Responsibility of States for the Acts of International Organizations' in M. Ragazzi (ed), *International Responsibility Today: Essays in Memory of Oscar Schachter* (Brill 2005).

J. Brunnée, 'International Law and Collective Concerns: Reflections on the Responsibility to Protect' in T. M. Ndiaye and R. Wolfrum (eds), *Law of the sea, environmental law and settlement of disputes: Liber amicorum Judge Thomas A. Mensah* (Nijhoff 2007).

—— 'COPing with Consent: Law-making Under Multilateral Environmental Agreements' (2002) 15 Leiden Journal of International Law 1.

W. T. Burke, 'Unregulated High Seas Fishing and Ocean Governance' in J. M. van Dyke, D. Zaelke and G. Hewison (eds), *Freedom for the Seas in the 21st Century: Ocean Governance and Environmental Harmony* (Island Press 1993).

—— *The New International Law of Fisheries* (Clarendon Press 1994).

—— 'Compatibility and Precaution in the 1995 Straddling Stock Agreement' in H. N. Scheiber (ed), *Law of the Sea: The Common Heritage and Emerging Challenges* (Martinus Nijhoff Publishers 2000).

W. C. Burns, 'The International Whaling Commission and the Future of Cetaceans: Problems and Prospects' (1997) 8 Colorado Journal of International Environmental Law and Policy 8.

D. Butterworth, 'Science and Fisheries Management Entering the New Millennium' in M. H. Nordquist and G. Moore (eds), *Current Fisheries Issues and the Food and Agriculture Organization of the United Nations* (Kluwer Law International 2000).

D. S. Butterworth, 'Possible interpretation problems for the current CITES listing criteria in the context of marine fish species under commercial harvest' (2000) 42 Population Ecology 29.

R. Caddell, 'International Law and the Protection of Migratory Wildlife: An Appraisal of Twenty-Five Years of the Bonn Convention' (2005) 16 Colorado Journal of International Environmental Law and Policy 113.

—— 'Inter-Treaty Cooperation, Biodiversity Conservation and the Trade in Endangered Species' (2013) 22 RECIEL 264.

California Environmental Associates, 'Charting a Course to Sustainable Fisheries' (16 January 2012) http://www.chartingacourse.org/downloads/ accessed 5 June 2014.

D. S. Calley, *Market Denial and International Fisheries Regulation: The Targeted and Effective Use of Trade Measures Against the Flag of Convenience Fishing Industry* (Martinus Nijhoff Publishers 2012).

L. K. Camenzuli, 'The Development of International Law and the Multilateral Environmental Agreements' Conference of the Parties and Its Validity' (2007) http://cmsdata.iucn.org/downloads/cel10_camenzuli.pdf accessed 4 July 2015.
J. Cameron and J. Abouchar, 'The Precautionary Principle: A Fundamental Principle of Law and Policy for the Protection of the Global Environment' (1991) 14 Boston College International & Comparative Law Review 1.
H. Caminos, 'The International Tribunal for the Law of the Sea and its Alleged Threat to the Unity of International Law' in R. Casado Raigón and G. Cataldi (eds), *L'évolution et l'état actuel du droit international de la mer: Mélanges de droit de la mer offerts à Daniel Vignes* (Bruylant 2009).
E. Cannizzaro and P. Palchetti, 'Ultra vires acts of international organizations' in J. Klabbers and A. Wallendahl (eds), *Research Handbook on the Law of International Organizations* (Edward Elgar 2011).
J. E. Carey, 'Improving the Efficacy of CITES by Providing the Proper Incentives to Protect Endangered Species' (1999) 77 Washington University Law Quarterly 1291.
C. P. Carlarne, 'Saving the Whales in the New Millennium: International Institutions, Recent Developments and the Future of International Whaling Policies' (2005–2006) 24 Virginia Environmental Law Journal 1.
D. D. Caron, 'The International Whaling Commission and the North Atlantic Marine Mammal Commission: The Institutional Risks of Coercion in Consensual Structures' (1995) 89 The American Journal of International Law 154.
R. T. Carson and others, 'Fisheries Management Under Cyclical Population Dynamics' (2009) 42 Environmental and Resource Economics 379.
C. F. de Casadevante Romani, 'Objective Regime' Max Planck Encyclopedia of Public International Law www.mpepil.com.
L. Castello and D. J. Stewart, 'Assessing CITES Non-detriment Findings Procedures for Arapaima in Brazil' http://www.conabio.gob.mx/institucion/cooperacion_internacional/TallerNDF/Links-Documentos/WG-CS/WG8-Fishes/WG8-CS1%20Arapaima/WG8-CS1-S.pdf accessed 6 July 2015.
C. Catarci, 'World Markets and Industry of Selected Commercially Exploited Aquatic Species with an International Conservation Profile' FAO Fisheries Circular 990 (Rome 2004).
R. D. Cavanagh and C. Gibson, 'Overview of the Conservation Status of Cartilaginous Fishes (Chondrichthyans) in the Mediterranean Sea' (IUCN 2007).
M. Ceo and others, 'Performance Reviews by Regional Fishery Bodies: Introduction, summaries, synthesis and best practices, Volume I: CCAMLR, CCSBT, ICCAT, IOTC, NAFO, NASCO, NEAFC' FAO Fisheries and Aquaculture Circular 1072 (Rome 2012).
J. I. Charney, *Is International Law Threatened by Multiple International Tribunals?* (Recueil des Cours vol 271, Martinus Nijhoff Publishers 1998).
S. Charnovitz, 'A World Environment Organization' (2002) 27 Columbia Journal of Environmental Law 323.
L. A. Chaves, 'Illegal, Unreported and Unregulated Fishing: WTO-consistent Trade related Measures to Address IUU Fishing' AUS:IUU/2000/16 (Background paper for the Expert Consultation on Illegal, Unreported and Unregulated Fishing Organized by the Government of Australia in Cooperation with FAO, 2000) http://www.fao.org/docrep/005/Y3274E/y3274e0i.htm#bm18 accessed 6 July 2015.
C. Chen, *Fisheries Subsidies under International Law* (Springer 2010).
W. W. L. Cheung and others, 'Intrinsic vulnerability in the global fish catch' (2007) 333 Marine Ecology Progress Series 1.
C. Chevalier, 'Governance in the Mediterranean Sea – Legal Regime and Prospectives", IUCN Centre for Mediterranean Cooperation' http://cmsdata.iucn.org/downloads/legalspects_en_1.pdf accessed 6 July 2015.
V. Christensen and others, 'Hundred-year decline of North Atlantic predatory fishes' (2003) 4 Fish and Fisheries 1.

D. R. Christie, 'The Conservation and Management of Stocks Located Solely within the Exclusive Economic Zone' in E. Hey (ed), *Developments in International Fisheries Law* (Kluwer Law International 1999).

P. Christie, E. G. Oracio and L. Eisma-Osorio, 'Impacts of the CITES listing of seahorses on the status of the species and on human well-being in the Philippines: a case study' FAO Fisheries and Aquaculture Circular 1058 (Rome 2011).

R. R. Churchill, 'Legal uncertainties in international high seas fisheries management' (1998) 37 Fisheries Research 225.

—— 'The Management of Shared Fish Stocks: the Neglected 'other' Paragraph of Article 63 of the UN Convention on the Law of the Sea' in A. Strati, M. Gavouneli and N. Skourtos (eds), *Unresolved Issues and New Challenges to the Law of the Sea: Time Before and Time After* (Martinus Nijhoff Publishers 2006).

—— 'The Jurisprudence of the International Tribunal for the Law of the Sea Relating to Fisheries: Is There Much in the Net?' (2007) 22 International Journal of Marine and Coastal Law 383.

R. R. Churchill and A. V. Lowe, *The law of the sea* (3rd edn Juris Publishing – Manchester University Press 1999).

R. R. Churchill and G. Ulfstein, 'Autonomous Institutional Arrangements in Multilateral Environmental Agreements: A Little-Noticed Phenomenon in International Law' (2000) 94 American Journal of International Law 623.

CITES Secretariat, 'A brief history of sturgeons & CITES' *CITES World – Official Newsletter of the Parties* (December 2001).

—— 'Caviar criminals' *CITES World – Official Newsletter of the Parties* (December 2001).

P. J. Clapham and others, 'The whaling issue: Conservation, confusion, and casuistry' (2007) 31 Marine Policy 314.

S. Clarke, 'Shark Product Trade in Hong Kong and Mainland China and Implementation of the CITES Shark Listings' (TRAFFIC East Asia 2004).

—— 'Best Practice Study of Fish Catch Documentation Schemes' (23 August 2010) http://www.m2cms.com.au/uploaded/5/Final%20CDS%20Report%20-August%2023.pdf accessed 4 July 2015.

S. C. Clarke and others, 'Global estimates of shark catches using trade records from commercial markets' (2006) 9 Ecology Letters 1115.

S. C. Clarke and others, 'Identification of Shark Species Composition and Proportion in the Hong Kong Shark Fin Market Based on Molecular Genetics and Trade Records' (2006) 20 Conservation Biology 201.

M. Clark, 'Are deepwater fisheries sustainable? The example of orange roughy (*Hoplostethus atlanticus*) in New Zealand' (2001) 51 Fisheries Research 123.

S. Cleva, 'Enforcement Starts with Wildlife Inspectors' (2005) 30(2) U.S. Fish & Wildlife Service, Endangered Species Bulletin 18.

C. Clover, *The end of the line* (Ebury Press 2005).

K. L. Cochrane, 'Complexity in fisheries and limitations in the increasing complexity of fisheries management' (1999) 56 ICES Journal of Marine Science 917.

—— 'Reconciling sustainability, economic efficiency and equity in fisheries: the one that got away?' (2000) 1 Fish and Fisheries 3.

—— 'A fishery manager's guidebook. Management measures and their application' FAO Fisheries Technical Paper 424 (Rome 2002).

K. L. Cochrane and others, 'Management procedures in a fishery based on highly variable stocks and with conflicting objectives: experiences in the South African pelagic fishery' (1998) 8 Reviews of Fish Biology and Fisheries 177.

Commission on Global Governance, *Our global neighbourhood: the report of the Commission on Global Governance* (Oxford University Press 1995).

B. J. Condon, *Environmental Sovereignty and the WTO: Trade Sanctions and International Law* (Transnational Publishers 2006).

R. Cooney and M. Abensperg-Traun, 'Raising Local Community Voices: CITES, Livelihoods and Sustainable Use' (2013) 22 RECIEL 301.

R. Cooney and P. Jepson, 'The international wild bird trade: what's wrong with blanket bans?' (2006) 40 Oryx 18.

E. W. T. Cooper and N. Chalifour, 'CITES, Eh?: The Implementation of CITES in Canada since WAPPRIITA' (TRAFFIC North America 2004).

J. Copeland Nagle, 'Why Chinese Wildlife Disappears as CITES Spreads?' (1996–1997) 9 Georgetown International Environmental Law Review 435.

M. Cordonier Segger, 'Sustainable development in international law' in D. Armstrong (ed), *Routledge Handbook of International Law* (Routledge 2009).

H. Corell, 'Oversight of the Implementation of the Global Ocean Regime: The Role of the United Nations' in D. Vidas and W. Ostreng (eds), *Order for the oceans at the turn of the century* (Kluwer Law International 1999).

T. Cottier and J. P. Müller, 'Estoppel' Max Planck Encyclopedia of Public International Law www.mpepil.com.

E. Couzens, 'CITES at Forty: Never Too Late to Make Lifestyle Changes' (2013) 22 RECIEL 311.

J. R. Crawford, 'State Responsibility' Max Planck Encyclopedia of Public International Law www.mpepil.com.

—— 'The General Assembly, the International Court and self-determination' in V. Lowe and M. Fitzmaurice (eds), *Fifty years of the International Court of Justice: Essays in honour of Sir Robert Jennings* (Cambridge University Press 1996).

—— *Brownlie's Principles of Public International Law* (8th edn Oxford University Press 2012).

—— 'The International Court of Justice and the Law of State Responsibility' in C. J. Tams and J. Sloan (eds), *The Development of International Law by the International Court of Justice* (Oxford University Press 2013).

J. P. Croxall and S. Nicol, 'Management of Southern Ocean fisheries: global forces and future sustainability' (2004) 16 Antarctic Science 569.

S. Cullis-Suzuki and D. Pauly, 'Failing the high seas: A global evaluation of regional fisheries management organizations' (2010) 34 Marine Policy 1036.

L. Cunningham, 'Assessing the contribution of aquaculture to food security: a survey of methodologies' FAO Fisheries Circular 1010 (Rome 2005).

A. D'Amato and S. K Chopra, 'Whales: Their Emerging Right to Life' (1991) 85 American Journal of International Law 21.

D. J. Dankel, D. W. Skagen and Ø. Ulltang, 'Fisheries management in practice: review of 13 commercially important fish stocks' (2008) 18 Reviews of Fish Biology and Fisheries 201.

S. Dansky, 'The CITES "Objective" Listing Criteria: Are They Objective Enough to Protect the African Elephant?' (1999) 73 Tulane Law Review 961.

J. d'Aspremont, 'Abuse of the Legal Personality of International Organizations and the Responsibility of Member States' (2007) 4 International Organizations Law Review 91.

B. Davis and B. Worm, 'The International Plan of Action for Sharks: How does national implementation measure up?' (2013) 38 Marine Policy 312.

J. Delbrück, '"Laws in the Public Interest" – Some Observations on the Foundations and Identification of erga omnes Norms in International Law' in V. Götz, P. Selmer and R. Wolfrum (eds), *Liber amicorum, Günther Jaenicke – Zum 85. Geburtstag* (Springer 1998).

E. DeLone, 'Improving the Management of the Atlantic Tuna: The Duty to Strengthen the ICCAT in Light of the 1995 Straddling Stocks Agreement' (1998) 6 New York University Environmental Law Journal 656.

B. Dickson, 'Precaution at the Heart of CITES?' in J. Hutton and B. Dickson (eds), *Endangered Species-Threatened Convention: The Past, Present and Future of CITES* (Earthscan 2000).

—— 'CITES and the livelihoods of the poor' (2008) 42 Oryx 548.

——'International Conservation Treaties, Poverty and Development: the Case of CITES' (2002) 74 Natural Resource perspectives, 3 http://www.odi.org.uk/publications/2069-international-conservation-treaties-poverty-development-case-cites accessed 4 July 2015.

C. Dominicé, 'A la recherche des droits erga omnes', *Droit du pouvoir, pouvoir du droit: Mélanges offerts à Jean Salmon* (Bruylant 2007).

E. Druel, R. Billé and J. Rochette, ‘Getting to yes? Discussions towards an Implementing Agreement to UNCLOS on biodiversity in ABNJ’ (2013) 10 Biodiversity http://www.iddri.org/Publications/Collections/Syntheses/PB1013_ABNJ%20IA_druel%20bille%20rochette.pdf accessed 6 July 2015.

M. A. Du Plessis, ‘CITES and the Causes of Extinction’ in J. Hutton and B. Dickson (eds), *Endangered Species-Threatened Convention: The Past, Present and Future of CITES* (Earthscan 2000).

N. K. Dulvy and others, ‘Methods of assessing extinction risk in marine fishes’ (2004) 4 Fish and Fisheries 255.

N. K. Dulvy and others, ‘You can swim but you can’t hide: the global status and conservation of oceanic pelagic sharks and rays’ (2008) Aquatic Conservation: Marine and Freshwater Ecosystems, http://www.iotc.org/documents/you-can-swim-you-cant-hide-global-status-and-conservation-oceanic-pelagic-sharks-and-rays accessed 4 July 2015, 459.

N. K. Dulvy, Y. Sadovy and J. D. Reynolds, ‘Extinction vulnerability in marine populations’ (2003) 4 Fish and Fisheries 25.

P. Dupuy, ‘Le principe de précaution et le droit international de la mer’ in Mélanges offerts à Laurent Lucchini et Jean-Pierre Quéneudec (ed), *La mer et son droit* (Pedone 2003).

—— ‘Evolutionary Interpretation of Treaties: Between Memory and Prophecy’ in E. Cannizzaro (ed), *The Law of Treaties Beyond the Vienna Convention* (Oxford University Press 2011).

R. D. Eckert, *The Enclosure of Ocean Resources: Economics and the Law of the Sea* (Hoover Institution Press 1979).

W. Edeson, ‘The International Plan of Action on Illegal, Unreported and Unregulated Fishing: The Legal Context of a Non-Legally Binding Instrument’ (2001) 16 International Journal of Marine and Coastal Law 603.

W. R. Edeson, ‘Current Legal Developments – Food and Agriculture Organization of the UN: The Code of Conduct for Responsible Fisheries: An Introduction’ (1996) 11 International Journal of Marine and Coastal Law 233.

H. Edwards, ‘When predators become prey: the need for international shark conservation’ (2006–2007) 12 Ocean and Coastal Law Journal 305.

K. Eldridge, ‘Whale For Sale?: New Developments in the Convention on International Trade in Endangered Species of Wild Fauna and Flora’ (1994–1995) 24 Georgia Journal of International and Comparative Law 549.

J. Ellis, ‘The Straddling Stocks Agreement and the Precautionary Principle as Interpretive Device and Rule of Law’ (2001) 32 Ocean Development and International Law 289.

E. R. Englebrecht, ‘Can Aquaculture continue to circumvent the regulatory net of the Magnuson-Stevens Fishery Conservation and Management Act?’ (2002) 51 Emory Law Journal 1187.

M. Engler and R. Parry-Jones, ‘Opportunity or Threat: The Role of the European Union in Global Wildlife Trade’ (TRAFFIC Europe 2007).

Environmental Justice Foundation, ‘Lowering the Flag – Ending the use of Flags of Convenience by Pirate Fishing Vessels’ (2009) http://ejfoundation.org/sites/default/files/public/Lowering%20the%20flag.pdf accessed 6 July 2015.

—— ‘Transhipment at sea: The need for a Ban in West Africa’ (2013) http://ejfoundation.org/sites/default/files/public/ejf_transhipments_at_sea_web_0.pdf accessed 5 July 2015.

C. Epstein, ‘The Making of Global Environmental Norms – Endangered Species Protection’ (2006) 6 Global Environmental Politics 32.

European Community and TRAFFIC Europe, ‘Reference Guide – European Community Wildlife Trade Regulations’ (2013) http://ec.europa.eu/environment/cites/pdf/2007_referenceguide2_en.pdf accessed 6 July 2015.

A. Fabra and V. Gascón, ‘The Convention on the Conservation of Antarctic Marine Living Resources (CCAMLR) and the Ecosystem Approach’ (2008) 23 International Journal of Marine and Coastal Law 567.

D. S. Favre, *International Trade in Endangered Species: A Guide to CITES* (Martinus Nijhoff Publishers 1989).

—— 'Debate within the CITES Community: What Direction for the Future?' (1993) 33 Natural Resources Journal 875.

S. M. Fernandez, 'Captive-bred exceptions: an unconventional approach to conservation under the Federal Endangered Species Act' (2003–2004) 15 University of Florida Journal of Law and Public Policy 155.

F. Ferretti and others, 'Loss of Large Predatory Sharks from the Mediterranean Sea' (2008) 22 Conservation Biology 952.

J. Firestone and J. Lilley, 'Aboriginal Subsistence Whaling and the Right to Practice and Revitalize Cultural Traditions and Customs' (2005) 8 Journal of International Wildlife Law & Policy 177.

J. Fischer and others, 'Review of the Implementation of the International Plan of Action for the Conservation and Management of Sharks' FAO Fisheries and Aquaculture Circular 1076 (Rome 2012).

M. Fitzmaurice, 'Environmental protection and the International Court of Justice' in V. Lowe and M. Fitzmaurice (eds), *Fifty years of the International Court of Justice: Essays in honour of Sir Robert Jennings* (Cambridge University Press 1996).

—— *International Protection of the Environment* (Recueil des Cours vol 293, Kluwer Law International 2002).

—— 'Third Parties and the Law of Treaties' (2002) 6 Max Planck Yearbook of United Nations Law 37.

—— *Contemporary issues in international environmental law* (Edward Elgar 2009).

—— 'Divided we stand: the legal issues concerning the international whaling commission' in H. P. Hestermeyer and others (eds), *Coexistence, Cooperation and Solidarity: Liber Amicorum Rüdiger Wolfrum* (Martinus Nijhoff Publishers 2011).

—— 'The International Court of Justice and International Environmental Law' in C. J. Tams and J. Sloan (eds), *The Development of International Law by the International Court of Justice* (Oxford University Press 2013).

—— 'Treaties' Max Planck Encyclopedia of Public International Law www.mpepil.com.

S. Flothmann and others, 'Closing Loopholes: Getting Illegal Fishing Under Control' (2010) 328 Science 1235.

S. V. Fordham, 'Shark alert: Revealing Europe's impact on shark populations' (Shark Alliance August 2006).

S. Fordham and C. Dolan, 'A Case Study in International Shark Conservation: The Convention on International Trade in Endangered Species and the Spiny Dogfish' (2004) 34 Golden Gate University Law Review 531.

H. Fox, 'Jurisdiction and immunities' in V. Lowe and M. Fitzmaurice (eds), *Fifty years of the International Court of Justice: Essays in honour of Sir Robert Jennings* (Cambridge University Press 1996).

R. I. C. C. Francis and R. Shotton, "Risk' in fisheries management: a review' (1997) 54 Canadian Journal of Fisheries and Aquaculture Science 1699.

E. Franckx, 'Pacta Tertiis and the Agreement for the Implementation of the Straddling and Highly Migratory Fish Stocks Provisions of the United Nations Convention on the Law of the Sea' (2000) 8 Tulane Journal of International and Comparative Law 49.

—— 'Fisheries Enforcement: Related legal and institutional issues: national, subregional or regional perspectives' FAO Legislative Study 71 (Rome 2001).

—— (ed) *Vessel-source Pollution and Coastal State Jurisdiction: The Work of the ILA Committee on Coastal State Jurisdiction Relating to Marine Pollution (1991–2000)* (Kluwer Law International 2001).

—— 'The Protection of Biodiversity and Fisheries Management: Issues Raised by the Relationship between CITES and LOSC' in D. Freestone, R. Barnes and D. Ong (eds), *The Law of the Sea – Progress and Prospects* (Oxford University Press 2006).

—— 'The Exercise of Jurisdiction over Vessels: Legal Issues Raised by the Relationship Between CITES, FAO and the UN Convention on the Law of the Sea' in E. Franckx and P. Gautier (eds), *The Exercise of Jurisdiction over Vessels: New Developments in the Fields of Pollution, Fisheries, Crimes at Sea and Trafficking of Weapons of Mass Destruction* (Bruylant 2010).

—— 'The Relationship between CITES, FAO and Related Agreements: Legal Issues' FAO Fisheries and Aquaculture Circular 1062 (Rome 2011).

—— 'Applications of the term 'Introduction from the sea" (31 May 2014) http://www.cites.org/eng/news/meetings/ifs-05/term_IFS.pdf accessed 5 July 2015.

E. Franckx and K. van den Bosche, 'The Influence of Environmental Law on the Development of the Law of the Sea: CITES and the International Law of Fisheries' (2011) 54 Japanese Yearbook of International Law 218.

S. Freeland and J. Drysdale, 'Co-Operation or Chaos? – Article 65 of United Nations Convention on the Law of the Sea and the Future of the International Whaling Commission' (2005) 2 Macquarie Journal of International and Comparative Environmental Law 1.

D. Freestone, 'Implementing Precaution Cautiously: the Precautionary Approach in the Straddling and Highly Migratory Fish Stocks Agreement' in E. Hey (ed), *Developments in International Fisheries Law* (Kluwer Law International 1999).

—— 'International Fisheries Law since Rio: the continued Rise of the precautionary Principle' in A. Boyle and D. Freestone (eds), *International Law and Sustainable Development – Past Achievements and Future Challenges* (Oxford University Press 1999).

—— 'Protecting Our Oceans: New Challenges, New Solutions An Overview of This Issue' (2006–2007) 7 Sustainable Development Law and Policy Review 2.

—— 'Principles Applicable to Modern Oceans Governance' (2008) 23 International Journal of Marine and Coastal Law 385.

—— 'Problems of High Seas Governance' in D. Vidas and P. J. Schei (eds), *The World Ocean in Globalisation: Climate Change, Sustainable Fisheries, Biodiversity, Shipping, Regional Issues* (Martinus Nijhoff Publishers 2011).

C. Frid, 'The role of marine science in participatory fisheries governance' in T. S. Gray (ed), *Participation in Fisheries Governance* (Springer 2005).

C. Frid, O. Paramor and C. Scott, 'Ecosystem-based fisheries management: progress in the NE Atlantic' (2005) 29 Marine Policy 461.

J. Friedrich, 'Legal Challenges of Nonbinding Instruments: The Case of the FAO Code of Conduct for Responsible Fisheries' (2008) 9 German Law Journal 1539.

C. Fuchs, 'Convention on International Trade in Endangered Species of Wild Fauna and Flora (CITES) – Conservation Efforts Undermine the Legality Principle' (2008) 9 German Law Journal 1565.

The Future We Want – endorsed in UNGA Resolution A/RES/66/288 27 July 2012 (Document adopted at Rio+20).

G. Gaja, 'Obligations and Rights Erga Omnes in International Law: Preparatory Work' (2005) 71-I Annuaire de l'Institut de Droit International 117.

S. M. Garcia, 'The Precautionary Principle: its Implications in Capture Fisheries Management' (1994) 22 Ocean and Coastal Management 99.

—— 'The Ecosystem Approach to Fisheries: on the Way to Implementation' in M. H Nordquist and others (eds), *Law, Science & Ocean Management* (Martinus Nijhoff Publishers 2007).

S. M. Garcia and K. L. Cochrane, 'Ecosystem approach to fisheries: a review of implementation guidelines' (2005) 62 ICES Journal of Marine Science 311.

S. M. Garcia and C. Newton, 'Current situation, trends, and prospects in world capture fisheries' in E. K. Pikitch, D. D. Huppert and M. P. Sissenwine (eds), *Global Trends: Fisheries Management: Proceedings from the symposium held in Seattle, Washington, USA, June 14–16, 1994* (American Fisheries Society 1997).

J. L. Garrison, 'The Convention on International Trade in Endangered Species of Wild Fauna and Flora (CITES) and the Debate Over Sustainable Use' (1994) 12 Pace Environmental Law Review 301.

P. Gautier, 'Comments on flag state's responsibility, enforcement measures against IUU fishing activities, and the settlement of disputes' in E. Franckx and P. Gautier (eds), *The Exercise of Jurisdiction over Vessels: New Developments in the Fields of Pollution, Fisheries, Crimes at Sea and Trafficking of Weapons of Mass Destruction* (Bruylant 2010).

—— 'Non-Binding Agreements' Max Planck Encyclopedia of Public International Law www.mpepil.com.

T. Gazzini, 'Personality of international organizations' in J. Klabbers and A. Wallendahl (eds), *Research Handbook on the Law of International Organizations* (Edward Elgar 2011).

M. W. Gehring, 'WTO law and sustainable development' in D. Armstrong (ed), *Routledge Handbook of International Law* (Routledge 2009).

T. Gehring and E. Ruffing, 'When Arguments Prevail Over Power: The CITES Procedure for the Listing of Endangered Species' (2008) 8 Global Environmental Politics 123.

A. Gillespie, 'Forum Shopping in International Environmental Law: The IWC, CITES, and the Management of Cetaceans' (2002) 33 Ocean Development and International Law 17.

—— 'The Search for a New Compliance Mechanism Within the International Whaling Commission' (2003) 34 Ocean Development and International Law 349.

—— *Whaling Diplomacy: Defining Issues in International Environmental Law* (New Horizons in Environmental Law, Edward Elgar 2005).

—— 'The Precautionary Principle in the Twenty-First Century: A Case Study of Noise Pollution in the Ocean' (2007) 22 International Journal of Marine and Coastal Law 61.

—— *Conservation, biodiversity and international law* (Edward Elgar 2011).

R. Gillett, 'Monitoring and management of the humphead wrasse, Cheilinus undulatus' FAO Fisheries and Aquaculture Circular 1048 (Rome 2010).

M. Giordano, 'The Internationalization of Wildlife and Efforts Towards its Management: A Conceptual Framework and the Historic Record' (2001–2002) 14 Georgetown International Environmental Law Review 607.

K. M. Gjerde, 'High-seas Fisheries Management under the Convention on the Law of the Sea' in D. Freestone, R. Barnes and D. Ong (eds), *The Law of the Sea – Progress and Prospects* (Oxford University Press 2006).

—— 'High Seas Fisheries Governance: Prospects and Challenges in the 21st Century' in D. Vidas and P. J. Schei (eds), *The World Ocean in Globalisation: Climate Change, Sustainable Fisheries, Biodiversity, Shipping, Regional Issues* (Martinus Nijhoff Publishers 2011).

M. J. Glennon, 'Has International Law Failed the Elephant?' (1990) 84 The American Journal of International Law 1.

F. González-Laxe, 'The precautionary principle in fisheries management' (2005) 29 Marine Policy 495.

M. Gorina-Ysern, 'World Ocean Public Trust: High Seas Fisheries After Grotius Towards a New Ocean Ethos?' (2004) 34 Golden Gate University Law Review 645.

J. Grogan and P. Barreto, 'Big-Leaf Mahogany on CITES Appendix II: Big Challenge, Big Opportunity' (2005) 19 Conservation Biology 973.

A. Gros, *La Convention sur la pêche et la conservation des ressources biologiques de la haute mer* (Recueil des Cours vol 97, A. W. Sijthoff 1960).

H. Grotius, *The Freedom of the Sea or the Right Which Belongs to the Dutch to Take Part in the East Indies Trade: Brown Scott, J. (ed.) and Van Daman Magoffin, R. (translation)* (Elsevier/Oxford University Press 1609/1916).

D. Guilfoyle, *Shipping Interdiction and the Law of the Sea* (Cambridge Studies in International and Comparative Law, Cambridge University Press 2009).

P. M. Haas, 'Introduction: epistemic communities and international policy coordination' (1992) 46 International Organization 1.

—— 'When Does Power Listen to Truth? A Constructivist Approach to the Policy Process' (2004) 11 Journal of European Public Policy 569.

G. Hafner, 'Pros and Cons Ensuing from Fragmentation of International Law' (2003–2004) 25 Michigan Journal of International Law 849.

—— 'The Division of the Commons? The Myth of the Commons: Divide or Perish' in H. P. Hestermeyer and others (eds), *Law of the sea in dialogue* (Beiträge zum ausländischen öffentlichen Recht und Völkerrecht vol 221, Springer 2011).

J. Hampton and others, 'Decline of Pacific tuna populations exaggerated?' (2005) 434 Nature.

R. Hannesson, 'Aquaculture and Fisheries' (2003) 27 Marine Policy 169.

G. Hardin, 'The Tragedy of the Commons' (1968) 3859 Science 1243.

D. Harland, *Killing Game: International Law and the African Elephant* (Praeger 1994).

J. Harrison, *Making the Law of the Sea: A Study in the Development of International Law* (Cambridge University Press 2011).

S. Hart, 'Elements of a Possible Implementation Agreement to UNCLOS for the Conservation and Sustainable Use of Marine Biodiversity in Areas beyond National Jurisdiction' Marine Series 4 (Gland 2008) https://portals.iucn.org/library/efiles/edocs/EPLP-MS-4.pdf accessed 6 July 2015.

M. Haward, 'IUU Fishing: Contemporary Practice' in A. G. Oude Elferink and D. R. Rothwell (eds), *Oceans Management in the 21st Century: Institutional Frameworks and Responses* (Martinus Nijhoff Publishers 2004).

M. Hayashi, 'The 1995 Agreement on the Conservation and Management of Straddling and Highly Migratory Fish Stocks: Significance for the Law of the Sea Convention' (1995) 29 Ocean and Coastal Management 51

—— 'Enforcement by Non-Flag States on the High Seas Under the 1995 Agreement on Straddling and Highly Migratory Fish Stocks' (1996–1997) 9 Georgetown International Environmental Law Review 1.

—— 'Regional Fisheries Management Organisations and Non-Members' in T. M. Ndiaye and R. Wolfrum (eds), *Law of the sea, environmental law and settlement of disputes: Liber amicorum Judge Thomas A. Mensah* (Nijhoff 2007).

T. Henriksen, 'Revisiting the Freedom of Fishing and Legal Obligations on States Not Party to Regional Fisheries Management Obligations' (2009) 40 Ocean Development and International Law 80.

T. Henriksen, G. B. Hønneland and A. K. Sydnes, *Law and Politics in Ocean Governance: The UN Fish Stocks Agreement and Regional Fisheries Management Regimes* (Publications on Ocean Development, Martinus Nijhoff Publishers 2006).

R. Herr, 'The International Regulation of Patagonian Toothfish: CCAMLR and High Seas Fisheries Management' in O. S. Stokke (ed), *Governing High Seas Fisheries: The Interplay of Global and Regional Regimes* (Oxford University Press 2001).

E. Hey, 'International Institutions' in D. Bodansky, J. Brunnée and E. Hey (eds), *Oxford Handbook of International Environmental Law* (Oxford University Press 2007).

M. J. Hickey, 'Acceptance of sustainable use within the CITES community' (1998–1999) 23 Vermont Law Review 861.

B. L. Hicks, 'Treaty Congestion in International Environmental Law: the Need for Greater International Coordination' (1998–1999) 32 University of Richmond law review 1643.

R. Higgins, 'The Legal Consequences for Member States of the Non-Fulfilment by International Obligations of their Obligations towards Third Parties' (1995) 66-I Annuaire de l'Institut de Droit International 251.

—— 'Respecting Sovereign States and Running a Tight Courtroom' (2001) 50 International and Comparative Law Quarterly 121.

R. Hilborn and others, 'State of World Fisheries' (2003) 28 Annual Review of Environment and Resources 359.

R. Hilborn, 'Faith-based Fisheries' (2006) 31 Fisheries Research 554.

—— 'Defining success in fisheries and conflicts in objectives' (2007) 31 Marine Policy 153.

—— 'Reinterpreting the State of Fisheries and their Management' (2007) 10 Ecosystems 1362.

K. Hirata, 'Why Japan Supports Whaling' (2005) 8 Journal of International Wildlife Law & Policy 129.

A. H. Hoel, 'Political uncertainty in international fisheries management' (1998) 37 Fisheries Research 239.

G. Hubold, 'Fishery and Sustainability' in P. Ehlers, E. Mann-Borgese and R. Wolfrum (eds), *Marine Issues: From a Scientific, Political and Legal Perspective* (Kluwer Law International 2002).

T. Huntington and M. R. Hasan, 'Fish as feed inputs for aquaculture – practices, sustainability and implications: a global synthesis' in M. R. Hasan and M. Halwart (eds), *Fish as feed inputs for aquaculture: Practices, sustainability and implications,* FAO Fisheries and Aquaculture Technical Paper 518 (Rome 2009).

J. A. Hutchings and J. D. Reynolds, 'Marine fish population collapses: consequences for recovery and extinction risk' (2004) 54 BioScience 297.

J. A. Hutchings, C. Walters and R. L. Haedrich, 'Is scientific enquiry incompatible with government information control?' (1997) 54 Canadian Journal of Fisheries and Aquatic Sciences 1198.

J. Hutton and B. Dickson, 'Introduction' in J. Hutton and B. Dickson (eds), *Endangered Species-Threatened Convention: The Past, Present and Future of CITES* (Earthscan 2000).

J. M. Hutton, 'Who Knows Best? Controversy over Unilateral Stricter Domestic Measures' in J. Hutton and B. Dickson (eds), *Endangered Species-Threatened Convention: The Past, Present and Future of CITES* (Earthscan 2000).

C. Huxley, 'CITES: The Vision' in J. Hutton and B. Dickson (eds), *Endangered Species-Threatened Convention: The Past, Present and Future of CITES* (Earthscan 2000).

ICTSD, 'Fishing, International trade and Sustainable Development: Policy Discussion Paper' Natural Resources, International Trade and Sustainable Development Studies (2006) http://www.ictsd.org/themes/environment/research/fisheries-international-trade-and-sustainable-development accessed 4 July 2015.

J. B. Ingerowski, *Exclusive Fishery Zones (EFZ): Legal Concept & State Practice* (MV Wissenschaft 2009).

International Environmental Law Project, 'CITES and Marine Species: An Analysis of the Draft Resolutions on Sharks, Sea Turtles, Toothfish and Whales' (2002) http://www.lclark.edu/live/files/154 accessed 6 July 2015.

—— 'Understanding Introduction from the Sea' (2010) http://www.lclark.edu/live/files/4356 accessed 6 July 2015.

IUCN (World Conservation Union), UNEP (United Nations Environment Programme) and WWF (World Wide Fund for Nature), *Caring for the earth: A strategy for sustainable living* (1991).

M. Ivanova and J. Roy, 'The Architecture of Global Environmental Governance: Pros and Cons of Multiplicity' http://www.centerforunreform.org/node/251 accessed 5 July 2015.

A. Jackson, 'The Convention on the Conservation and Management of Fishery Resources in the South East Atlantic Ocean, 2001: an introduction' (2002) 17 International Journal of Marine and Coastal Law 33.

J. B. C. Jackson and others, 'Historical overfishing and the recent collapse of coastal ecosystems' (2001) 293 Science 629.

J. Jackson and J. Jacquet, 'The shifting baselines syndrome: perception, deception, and the future of our oceans' in V. Christensen and J. MacLean (eds), *Ecosystem Approaches to Fisheries: A Global Perspective* (Cambridge University Press 2011).

J. Jacquet, 'Beyond food: fish in the twenty-first century' in V. Christensen and J. MacLean (eds), *Ecosystem Approaches to Fisheries: A Global Perspective* (Cambridge University Press 2011).

J. L. Jacquet and D. Pauly, 'Trade secrets: Renaming and mislabeling of seafood' (2008) 32 Marine Policy 309.

L. Jenkins, 'Trade Sanctions: An Effective Enforcement Tool' (1993) 2 RECIEL 362.

R. W. G. Jenkins, 'The Significant Trade Process: Making Appendix II Work' in J. Hutton and B. Dickson (eds), Endangered Species-Threatened Convention: The Past, Present and Future of CITES (Earthscan 2000).

—— 'An Overview of the Fundamental Principles of CITES as a Mechanism for Regulating Trade in Listed Species' http://www.cites.org/eng/news/meetings/ifs-05/IFS05-principle.pdf accessed 4 July 2015.

R. W. G. Jenkins and others, 'Review of Crocodile Ranching Programs Conducted for CITES by the Crocodile Specialist Group of IUCN/SSC' January – April 2004, AC22 Inf. 2 (CITES 2006).

C. W. Jenks, 'Conflict of Law-Making Treaties' (1953) 30 British Yearbook of International Law 401.

C. Jessup, *L'exploitation des richesses de la mer* (Recueil des Cours vol 29, Librairie Hachette 1929).

L. Juda, 'The 1995 United Nations agreement on straddling fish stocks and highly migratory fish stocks: A critique' (1997) 28 Ocean Development and International Law 147.

L. Juda and R. H. Burroughs, 'The prospects for comprehensive ocean management' (1990) 14 Marine Policy 23.

M. P. Karns and K. A. Mingst, *International Organizations: The Politics and Processes of Global Governance* (Lynne Rienner Publishers 2004).

S. B. Kaye, *International Fisheries Management* (Kluwer Law International 2001).

R. Kearney and others, 'Questionable interpretation of the Precautionary Principle in Australia's implementation of 'no-take' marine protected areas' (2012) 36 Marine Policy 592.

S. J. Keller, 'Is the International Ban on the Importation of Ivory Saving the African Elephant?' (1992) 3 Colorado Journal of International Environmental Law and Policy 381.

G. Kent, 'Fisheries, food security, and the poor' (1997) 22 Food policy 393.

H. Kievit, 'Conservation of the Nile Crocodile: Has CITES Helped or Hindered?' in J. Hutton and B. Dickson (eds), *Endangered Species-Threatened Convention: The Past, Present and Future of CITES* (Earthscan 2000).

E. A. Kirk, 'Maritime Zones and the Ecosystem Approach: A Mismatch?' (1999) 8 RECIEL 67.

A. Kiss and D. Shelton, *Guide to International Environmental Law* (Martinus Nijhoff Publishers 2007).

J. Klabbers, *An introduction to international institutional law* (Cambridge University Press 2009).

—— 'Beyond the Vienna Convention: Conflicting Treaty Provisions' in E. Cannizzaro (ed), *The Law of Treaties Beyond the Vienna Convention* (Oxford University Press 2011).

M. Klinkhardt, 'The blue revolution – feed alternatives for aquaculture' in R. Arthur and J. Nierentz (eds), *Global Trade Conference on Aquaculture,* FAO Fisheries Proceedings 9 (Rome 2007).

M. Klinkhardt and B. Myrseth, 'New aquaculture candidates' in R. Arthur and J. Nierentz (eds), *Global Trade Conference on Aquaculture,* FAO Fisheries Proceedings 9 (Rome 2007).

A. Knapp, C. Kitschke and S. von Meibom, 'Proceedings of the International Sturgeon Enforcement Workshop to Combat Illegal Trade in Caviar: Prepared by TRAFFIC Europe for the European Commission' Brussels 2006, SC54 Inf. 6 (CITES 2006).

K.-H. Kock, 'Understanding CCAMLR's Approach to Management' (2000) www.ccamlr.org/en/system/files/am-all.pdf accessed 4 July 2015.

V. Koester, 'Compliance Committees within MEAs and the Desirability and Feasibility of Establishing Special Compliance Bodies under CITES' 20 May 2004, SC54 Inf. 3 (CITES 2006).

C. Kojima, 'Fisheries, Sedentary' Max Planck Encyclopedia of Public International Law www.mpepil.com.

C. Kojima and V. S. Vereshchetin, 'Implementation Agreements' Max Planck Encyclopedia of Public International Law www.mpepil.com.

D. König, 'Flags of Convenience' Max Planck Encyclopedia of Public International Law www.mpepil.com.

—— 'Flag of Ships' Max Planck Encyclopedia of Public International Law www.mpepil.com.

—— 'Marine Environment, International Protection' Max Planck Encyclopedia of Public International Law www.mpepil.com.

—— 'Tacit Consent/Opting Out Procedures' Max Planck Encyclopedia of Public International Law www.mpepil.com.

E. M. Kornicker Uhlmann, 'State Community Interests, Jus Cogens and Protection of the Global Environment: Developing Criteria for Peremptory Norms' (1998) 11 Georgetown International Environmental Law Review 101.

C. L. Krieps, 'Sustainable use of endangered species under CITES: is it a sustainable alternatives' (1996) 17 University of Pennsylvania Journal of International Law 461.

J. C. Kunich, 'Losing Nemo: The Mass Extinction Now Threatening the World's Ocean Hotspots' (2005) 30 Columbia Journal of Environmental Law 1.

B. Kwiatkowska, 'The Southern Bluefin Tuna (New Zealand v Japan; Australia v Japan) Cases' (2000) 15 International Journal of Marine and Coastal Law 1.

L. de La Fayette, 'Access to Ports in International Law' (1996) 11 International Journal of Marine and Coastal Law 1.

—— 'The Role of the UN in International Oceans Governance' in D. Freestone, R. Barnes and D. Ong (eds), *The Law of the Sea – Progress and Prospects* (Oxford University Press 2006).

M. Lack, 'Catching On?: Trade-related Measures as a Fisheries Management Tool' (TRAFFIC International 2007).

—— 'Continuing CCAMLR's Fight Against IUU Fishing for Toothfish' (WWF Australia and TRAFFIC International 2008).

M. Lack and G. Sant, 'The Future of Sharks: A Review of Action and Inaction' (TRAFFIC January 2011).

M. Lack, K. Short and A. Willock, 'Managing risk and uncertainty in deep-sea fisheries: lessons from Orange Roughy' (2003) http://www.mar-eco.no/learning-zone/__data/page/89/WWF-orange_roughy.pdf accessed 4 July 2015.

R. Lagoni, 'Ports' Max Planck Encyclopedia of Public International Law www.mpepil.com.

P. A. Larkin, 'Concepts and issues in marine ecosystem management' (1996) 6 Reviews of Fish Biology and Fisheries 139.

B. Le Gallic and A. Cox, 'An economic analysis of illegal, unreported and unregulated (IUU) fishing: Key drivers and possible solutions' (2006) 30 Marine Policy 689.

M. Lemke, *Erfüllungsdefizite des Flaggenstaats* (Nomos 2011).

S: Lieberman, 'Science versus Politics: Tales from CITES' (Elisabeth Mann Borgese Lecture, 2010) http://internationaloceaninstitute.dal.ca/EMBlecturetext_2010.pdf accessed 5 July 2015.

A. Lindroos, 'Addressing Norm Conflicts in a Fragmented Legal System: The Doctrine of Lex Specialis' (2005) 74 Nordic Journal of International Law 27.

L. Little and M. A. Orellana, 'Can CITES Play a Role in Solving the Problem of IUU Fishing?: The Trouble with Patagonian Toothfish' (2005) 16 Colorado Journal of International Environmental Law and Policy 21.

Y. Liu, J. O. Olaussen and A. Skonhoft, 'Wild and farmed salmon in Norway – A review' (2011) 35 Marine Policy 413.

T. Lobach, 'Port State control of foreign fishing vessels' FAO Fisheries Circular 987 (Rome 2003).

M. W. Lodge and others, 'Recommended Best Practices for Regional Fisheries Management Organizations: Report of an independent panel to develop a model for improved governance by Regional Fisheries Management Organizations' (Chatham House 2007).

A. Longhurst, *Mismanagement of Marine Fisheries* (Cambridge University Press 2010).

E. Louka, *International Environmental Law: Fairness, Effectiveness, and World Order* (Cambridge University Press 2006).

L. Lucchini, 'La loi canadienne du 12 mai 1994: la logique extrême de la théorie du droit préférentiel de l'Etat côtier en haute mer au titre des stocks chevauchants' (1994) 40 Annuaire français de droit international 864.

G. Lugten, 'The role of international fishery organizations and other bodies in the conservation and management of living aquatic resources' FAO Fisheries and Aquaculture Circular 1054 (Rome 2010).

S. Lyster, *International Wildlife Law* (Grotius Publications Limited 1985).

J. R. Mack, 'International Fisheries Management: How the U.N. Conference on Straddling and Highly Migratory Fish Stocks Changes the Law of Fishing on the High Seas' (1995–1996) 26 California Western International Law Journal 313.

A. Mahiou, *Le droit international ou la dialectique de la rigueur et de la flexibilité: Cours général de droit international public* (Recueil des Cours vol 337, Martinus Nijhoff Publishers 2008).

—— 'De quelques incertitudes institutionnelles et normatives en matière d'environnement' (Le droit international face aux enjeux environnementaux, Aix-en-Provence, 4–6 June 2009).

S. Marr, *The Precautionary Principle in the Law of the Sea: Modern Decision Making in International Law* (Martinus Nijhoff Publishers 2003).

G. S. Martin and J. W. Brennan, 'Enforcing the International Convention for the Regulation of Whaling: The Pelly and Packwood-Magnuson Amendments' (1989) 17 Denver Journal of International Law and Policy 293.

R. B. Martin, 'When CITES Works and When it Does Not' in J. Hutton and B. Dickson (eds), *Endangered Species-Threatened Convention: The Past, Present and Future of CITES* (Earthscan 2000).

E. Masood, 'Scientific Caution blunts efforts to conserve fish stocks' (1996) 379 Nature 481.

N. Matz-Lück, 'Treaties, Conflicts between' Max Planck Encyclopedia of Public International Law www.mpepil.com.

N. Matz-Lück and R. Wolfrum, 'Article XX lit. g GATT' in R. Wolfrum, P.-T. Stoll and A. Seibert-Fohr (eds), *WTO – Technical Barriers and SPS Measures* (Martinus Nijhoff Publishers 2007).

N. Matz, 'The Interaction between the Convention on Biological Diversity and the UN Convention on the Law of the Sea' in P. Ehlers, E. Mann-Borgese and R. Wolfrum (eds), *Marine Issues: From a Scientific, Political and Legal Perspective* (Kluwer Law International 2002).

—— *Wege zur Koordinierung völkerrechtlicher Verträge: Völkervertragsrechtliche und institutionelle Ansätze* (Springer 2005).

—— 'Chaos or Coherence? – Implementing and Enforcing the Conservation of Migratory Species through Various Legal Instruments' (2005) 65 Zeitschrift für ausländisches öffentliches Recht und Völkerrecht 197.

M. L. McConnell, 'Observations on Compliance and Enforcement and Regional Fisheries Institutions: Overcoming the Limitations of the Law of the Sea' in D. A. Russell and D. Vanderzwaag (eds), *Recasting Transboundary Fisheries Management Arrangements in Light of Sustainability Principles: Canadian and International Perspectives* (Legal Aspects of Sustainable Development vol 8, Martinus Nijhoff Publishers 2010).

T. L. McDorman, 'Iceland, Whaling and the U.S. Pelly Amendment: The International Trade Law Context' (1997) 66 Nordic Journal of International Law 453.

—— 'Fisheries Conservation and Management and International Trade Law' in E. Hey (ed), *Developments in International Fisheries Law* (Kluwer Law International 1999).

—— 'Implementing Existing Tools: Turning Words Into Actions – Decision-Making Processes of Regional Fisheries Management Organisations (RFMOs)' (2005) 20 International Journal of Marine and Coastal Law 423.

C. McLachlan, 'The Principle of Systemic Integration and Article 31(3)(C) of the Vienna Convention' (2005) 54 International and Comparative Law Quarterly 279.

R. J. McLaughlin, 'Settling Trade-Related Disputes Over the Protection of Marine Living Resources: UNCLOS or the WTO?' (1997–1998) 10 Georgetown International Environmental Law Review 29.

E. M. McOmber, 'Problems in Enforcement of the Convention on International Trade in Endangered Species' (2001–2002) 27 Brooklyn Journal of International Law 673.

K. Mechlem, 'Food and Agriculture Organization of the United Nations (FAO)' Max Planck Encyclopedia of Public International Law www.mpepil.com.

P. Medley, 'Manual for the monitoring and management of queen conch' FAO Fisheries Circular 1012 (Rome 2005).

F. Meere and others, 'CITES 'Introduction from the sea' – A practical way forward: Proceedings of a joint workshop' (2008).

E. Meltzer, 'Global Overview of Straddling and Highly Migratory Fish Stocks: Maps and Charts Detailing RFMO Coverage and Implementation' (2005) 20 International Journal of Marine and Coastal Law 571.

T. A. Mensah, 'Marine Pollution from Ships, Prevention of and Responses to' Max Planck Encyclopedia of Public International Law www.mpepil.com.

J. G. Merrills, *International Dispute Settlement* (5th edn Cambridge University Press 2011).

H. Meyers, *The Nationality of Ships: Academisch Proefschrift* (Martinus Nijhoff Publishers 1967).

E. J. Molenaar, 'Port State Jurisdiction' Max Planck Encyclopedia of Public International Law www.mpepil.com.

—— 'Marine Mammals: The Role of Ethics and Ecosystem Considerations' (2003) 6 Journal of International Wildlife Law and Policy 31.

—— 'Participation, Allocation and Unregulated Fishing: The Practice of Regional Fisheries Management Organisations' (2003) 18 International Journal of Marine and Coastal Law 457.

—— 'Managing Biodiversity in Areas Beyond National Jurisdiction' (2007) 22 International Journal of Marine and Coastal Law 89.

—— 'Port State Jurisdiction: Towards Mandatory and Comprehensive Use' in D. Freestone, R. Barnes and D. Ong (eds), *The Law of the Sea – Progress and Prospects* (Oxford University Press 2006).

—— 'Port State Jurisdiction: Towards Comprehensive, Mandatory and Global Coverage' (2007) 38 Ocean Development and International Law 225.

—— 'Non-Participation in the Fish Stocks Agreement: Status and Reasons' (2011) 26 International Journal of Marine and Coastal Law 195.

J. Montgomery and L. Carter, 'Marine Science in the Past 25 Years: Main Findings and Trends' in D. Vidas and P. J. Schei (eds), *The World Ocean in Globalisation: Climate Change, Sustainable Fisheries, Biodiversity, Shipping, Regional Issues* (Martinus Nijhoff Publishers 2011).

G. Moore, 'The FAO Compliance Agreement' in M. H. Nordquist and G. Moore (eds), *Current Fisheries Issues and the Food and Agriculture Organization of the United Nations* (Kluwer Law International 2000).

T. Morato and others, 'Fishing down the deep' (2006) 7 Fish and Fisheries 24.

D. L. Morgan, 'Implications of the Proliferation of International Legal Fora: The Example of the Southern Bluefin Tuna Cases' (2002) 43 Harvard International Law Journal 541.

J. Morishita, 'Multiple analysis of the whaling issue: Understanding the dispute by a matrix' (2006) 30 Marine Policy 802.

—— 'What is the ecosystem approach for fisheries management?' (2008) 32 Marine Policy 19.

P. J. Mous and others, 'Capture for culture: artificial shelters for grouper collection in SE Asia' (2006) 7 Fish and Fisheries 58.

T. Mulliken, 'The Role of CITES in Controlling the International Trade in Forest Products: Implications for Sustainable Forest Management' Non-Wood Forest Products Working Documents 7 (Rome 2009).

C. Mullon, P. Fréon and P. Curry 'The dynamics of collapse in world fisheries' (2006) 6 Fish and Fisheries 111.

S. A. Murawski, 'Ten myths concerning ecosystem approaches to marine resource management' (2007) 31 Marine Policy 681.

J. B. Murphy, 'Alternative Approaches to the CITES 'Non-detrimental' Finding for Appendix II Species' (2006) 36 Environmental Law 531.

J. K. Murphy, 'FAO's general fisheries commission for the Mediterranean: a new role, a new role model' (2007) 12 Drake Journal of Agricultural Law 391.

J. B. Mus, 'Conflicts Between Treaties in International Law' (1998) 45 Netherlands International Law Review 208.

R. Myers and B. Worm, 'Rapid worldwide depletion of predatory fish communities' (2003) 423 Nature 280.

A. Najam, 'The Case Against a New International Environmental Organization' (2003) 9 Global Governance 367.

K. Nakatani, 'Responsibility of Member States towards Third Parties for an Internationally Wrongful Act of the Organization' in M. Ragazzi (ed), *Responsibility of international organizations: Essays in Memory of Sir Ian Brownlie* (Martinus Nijhoff Publishers 2013).

S. N. Nandan, S. Rosenne and N. R. Grandy, *The United Nations Convention on the Law of the Sea: A Commentary: Volume II* (Martinus Nijhoff Publishers 1993).

R. L. Naylor and others, 'Effect of aquaculture on world fish supplies' (2000) 405 Nature 1021.

R. L. Naylor and others, 'Feeding aquaculture in an era of finite resources' (2009) 106 Proceedings of the National Academy of Science 15103.

E. Neumayer, 'WTO Rules and Multilateral Environmental Agreements' in K. P. Gallagher and J. Werksman (eds), *The Earthscan Reader on International Trade and Sustainable Development* (Earthscan 2002).

P. A. Nickler, 'A tragedy of the commons in coastal fisheries: contending prescriptions for conservation, and the case of the Atlantic Bluefin Tuna' (1998–1999) 26 Environmental Affairs Law Review 549.

M. H. Nordquist and others (eds), *United Nations Convention on the Law of the Sea 1982: A Commentary: Volume IV* (Martinus Nijhoff Publishers 1991).

A. Oddone and others, 'Estimating reef habitat coverage suitable for the humphead wrasse, Cheilinus undulatus, using remote sensing' FAO Fisheries and Aquaculture Circular 1057 (Rome 2010).

OECD, 'Trade Measures in Multilateral Environmental Agreements: Synthesis Report of Three Case Studies' COM/ENV/TD(98)127/FINAL (Paris 1999).

—— 'Globalisation in Fisheries and Aquaculture: Opportunities and Challenges' (Paris 2010).

P. Oerebech, K. Sigurjonsson and T. L. McDorman, 'The 1995 United Nations Straddling and Highly Migratory Fish Stocks Agreement: Management, Enforcement and Dispute Settlement' (1998) 13 International Journal of Marine and Coastal Law 119.

P. Okowa, 'Environmental Dispute Settlement: Some Reflections on Recent Developments' in M. D Evans (ed), *Remedies in International Law: The Institutional Dilemma* (Hart Publishing 1998).

P. Okowa, 'Issues of Admissibility and the Law on International Responsibility' in M. D Evans (ed), *International Law* (3rd edn Oxford University Press 2010).

S. F. Oldfield, 'The Evolving Role of CITES in Regulating the International Timber Trade' (2013) 22 RECIEL 291.

L. Oppenheim, *International Law – A Treatise: Vol. I – Peace,* edited by Lauterpacht (7th edn H. Longmans, Green and Co. 1948).

N. Oral, 'Protection of Vulnerable Marine Ecosystems in Areas Beyond National Jurisdiction: Can International Law Meet the Challenge?' in A. Strati, M. Gavouneli and N. Skourtos (eds), *Unresolved Issues and New Challenges to the Law of the Sea: Time Before and Time After* (Martinus Nijhoff Publishers 2006).

M. A. Orellana, 'The Swordfish Dispute between the EU and Chile at the ITLOS and the WTO' (2002) 71 Nordic Journal of International Law 55.

—— 'The Law on Highly Migratory Fish Stocks: ITLOS Jurisprudence in Context' (2004) 34 Golden Gate University Law Review 459.

F. Orrego Vicuña, *The exclusive economic zone: regime and legal nature under international law* (Cambridge University Press 1989).

—— *The Changing International Law of High Seas Fisheries* (Cambridge University Press 1999).

—— 'The International Law of High Seas Fisheries: From Freedom of Fishing to Sustainable Use' in O. S. Stokke (ed), *Governing High Seas Fisheries: The Interplay of Global and Regional Regimes* (Oxford University Press 2001).

—— 'Le droit de la pêche en haute mer: développement et conservation des ressources' in Mélanges offerts à Laurent Lucchini et Jean-Pierre Quéneudec (ed), *La mer et son droit* (Pedone 2003).

I. Österdahl, 'International organizations – institutions and organs' in J. Klabbers and A. Wallendahl (eds), *Research Handbook on the Law of International Organizations* (Edward Elgar 2011).

F. Ottolenghi and others, 'Capture-based aquaculture: The fattening of eels, groupers, tunas and yellowtails' (FAO 2004).

B. H. Oxman, 'The duty to respect generally accepted international standards' (1991–1992) 24 New York University Journal of International Law and Politics 109.

—— 'Complementary Agreements and Compulsory Jurisdiction' (2001) 95 American Journal of International Law 277.

I. Papanicolopulu, 'From the North Sea to the Bay of Bengal: Maritime Delimitation at the International Tribunal for the Law of the Sea' (23 March 2012) http://www.ejiltalk.org/from-the-north-sea-to-the-bay-of-bengal-maritime-delimitation-at-the-international-tribunal-for-the-law-of-the-sea/ accessed 5 July 2015.

G. Parkes, 'Precautionary fisheries management: the CCAMLR approach' (2000) 24 Marine Policy 83.

E. C. M. Parsons, N. A. Rose and T. M. Telecky, 'The trade in live Indo-Pacific bottlenose dolphins from Solomon Islands: A CITES decision implementation case study' (2010) 34 Marine Policy 384.

S. Patel, 'The Convention on International Trade in Endangered Species and the Last Unicorn' (1995–1996) 18 Houston Journal of International Law 157.
D. Pauly, 'Anecdotes and the shifting baseline syndrome of fisheries' (1995) 10 TRENDS in Ecology and Evolution 430.
D. Pauly and others, 'Fishing down marine food webs' (1998) 279 Science 860.
D. Pauly and others, 'Towards sustainability in world fisheries' (2002) 418 Nature 689.
J. Pauwelyn, *Conflict of Norms in Public International Law: How WTO Law Relates to other Rules of International Law* (Cambridge Studies in International and Comparative Law, Cambridge University Press 2003).
J. Peel, 'A Paper Tiger which Dissolves in the Rain? The Future for Resolving Fisheries Disputes under UNCLOS in the Aftermath of the Southern Bluefin Tuna Arbitration' (2002) 3 Melbourne Journal of International Law 53.
M. J. Peterson, 'General Assembly' in T. G. Weiss and S. Daws (eds), *The Oxford Handbook on The United Nations* (Oxford University Press 2007).
J. Pfeil, 'Fisheries Jurisdiction Case (Spain v Canada)' Max Planck Encyclopedia of Public International Law www.mpepil.com.
E. K. Pikitch and others, 'Status, trends and management of sturgeon and paddlefish fisheries' (2005) 6 Fish and Fisheries 233.
T. J. Pitcher and others, 'An evaluation of progress in implementing ecosystem-based management of fisheries in 33 countries' (2009) 33 Marine Policy 223.
T. Polacheck, 'Tuna longline catch rates in the Indian Ocean: did industrial fishing result in a 90% rapid decline in the abundance of large predatory fish?' (2006) 30 Marine Policy 470.
—— 'Politics and independent scientific advice in RFMO processes: A case study of crossing boundaries' (2012) 36 Marine Policy 132.
A. Powers, 'Farming the Ocean' (2007–2008) 22 Natural Resources and Environment 45.
A. Proelß, 'Marine Mammals' Max Planck Encyclopedia of Public International Law www.mpepil.com.
M. Prost and P. K. Clark, 'Unity, Diversity and the Fragmentation of International Law: How Much Does the Multiplication of International Organizations Really Matter' (2006) 5 Chinese Journal of International Law 341.
R. S. V. Pullin, 'Aquaculture up and down the food web' in V. Christensen and J. MacLean (eds), *Ecosystem Approaches to Fisheries: A Global Perspective* (Cambridge University Press 2011).
J.-F. de Pulvenis Séligny, 'The marine living resources and the evolving law of the sea' (2010) 1 Aegean Review of the Law of the Sea and Maritime Law 61.
M. Ragazzi, *The concept of international obligations erga omnes* (Oxford University Press 2000).
R. Rayfuse, 'The Interrelationship between the Global Instruments of International Fisheries Law' in E. Hey (ed), *Developments in International Fisheries Law* (Kluwer Law International 1999).
—— 'Countermeasures and High Seas Fisheries Enforcement' (2004) 51 Netherlands International Law Review 41.
—— *Non-Flag State Enforcement in High-seas Fisheries* (Martinus Nijhoff Publishers 2004).
—— 'The Future of Compulsory Dispute Settlement Under The Law of the Sea Convention' (2005) 36 Victoria University of Wellington Law Review 683.
—— 'To Our Children's Children's Children: From Promoting to Achieving Compliance in High-seas Fisheries' (2005) 20 International Journal of Marine and Coastal Law 509.
—— 'Building sustainable high seas fisheries through certification processes: issues and perspectives' (2009) 35 Océanis 93.
R. Rayfuse and R. Warner, 'Securing a Sustainable Future for the Oceans Beyond National Jurisdiction: The Legal Basis for an Integrated Cross-Sectoral Regime for High Seas Governance for the 21st Century' (2008) 23 International Journal of Marine and Coastal Law 399.
C. Raymakers, 'International Trade in Sturgeon and Paddlefish Species – the Effect of CITES listing' (2002) 87 International Review of Hydrobiology 525.
—— 'CITES, the Convention on International Trade in Endangered Species of Wild Fauna and Flora: its role in the conservation of Acipenseriformes' (2006) 22 Journal of Applied Ichthyology 53.

M. Recharte Uscamaita and R. Bodmer, 'Recovery of the Endangered giant otter Pteronura brasiliensis on the Yavarí-Mirín and Yavarí Rivers: a success story for CITES' (2009) 44 Oryx 83.

R. Reeve, *Policing International Trade in Endangered Species* (Earthscan 2002).

—— 'Wildlife Trade, Sanctions and Compliance' (2006) 82 International Affairs 881.

—— 'The Convention on International Trade in Endangered Species of Wild Fauna and Flora (CITES)' in G. Ulfstein (ed), *Making treaties work: Human Rights, Environment and Arms Control* (Cambridge University Press 2007).

A. Rengifo, 'Protection of Marine Biodiversity: A New Generation of Fisheries Agreements' (1997) 6 RECIEL 313.

K. W. Riddle, 'Illegal, Unreported, and Unregulated Fishing: Is International Cooperation Contagious?' (2006) 37 Ocean Development and International Law 265.

A. Rieser, 'International Fisheries Law, Overfishing and Marine Biodiversity' (1996–1997) 9 Georgetown International Environmental Law Review 251.

K. Rigg, R. Parmentier and D. Currie, 'Halting IUU Fishing: Enforcing International Fisheries Agreements' (OECD Directorate for Food, Agriculture and Fisheries 2004).

V. Röben, 'Conference (Meeting) of States Parties' Max Planck Encyclopedia of Public International Law www.mpepil.com.

J. G. Robinson and K. H. Redford, 'The Use and Conservation of Wildlife' in J. G. Robinson and K. H. Redford (eds), *Neotropical Wildlife Use and Conservation* (The Chicago University Press 1991).

C. A. Roheim and J. G. Sutinen, 'Trade and Market-Related Instruments to Promote Sustainable Fishing Practices' International Trade and Sustainable Development Series Issue Paper 3 (Geneva 2006).

C. Romano, 'The Southern Bluefin Tuna Dispute: Hints of a World to Come...Like It or Not' (2001) 32 Ocean Development and International Law 313.

S. Rosenne, 'Establishing the ITLOS' (1995) 89 American Journal of International Law 806.

—— 'Reflections on Fisheries Management Disputes' in R. Casado Raigón and G. Cataldi (eds), *L'évolution et l'état actuel du droit international de la mer: Mélanges de droit de la mer offerts à Daniel Vignes* (Bruylant 2009).

D. R. Rothwell, 'Building on the Strengths and Addressing the Challenges: The Role of Law of the Sea Institutions' (2004) 35 Ocean Development and International Law 131.

—— 'The Contribution of ITLOS to Oceans Governance through Marine Environmental Dispute Resolution' in T. M. Ndiaye and R. Wolfrum (eds), *Law of the sea, environmental law and settlement of disputes: Liber amicorum Judge Thomas A. Mensah* (Nijhoff 2007).

D. R. Rothwell and T. Stephens, 'Illegal Southern Ocean Fishing and Prompt Release: Balancing Coastal and Flag State Rights and Interests' (2004) 53 International and Comparative Law Quarterly 171.

—— *The International Law of the Sea* (Hart Publishing 2010).

D. A. Russell and D. Vanderzwaag, 'Ecosystem and Precautionary Approaches to International Fisheries Governance: Beacons of Hope, Seas of Confusion and Illusion' in D. A Russell and D. Vanderzwaag (eds), *Recasting Transboundary Fisheries Management Arrangements in Light of Sustainability Principles: Canadian and International Perspectives* (Legal Aspects of Sustainable Development vol 8, Martinus Nijhoff Publishers 2010).

N. de Sadeleer, *Environmental Principles – From Political Slogans to Legal Rules* (Oxford University Press 2002).

—— 'Le rôle ambivalent des principes dans la formation du droit de l'environnement: l'exemple du principe de précaution' (Le droit international face aux enjeux environnementaux, Aix-en-Provence, 4–6 June 2009).

R. Salama, 'Fragmentation of International Law: Procedural Issues Arising in Law of the Sea Disputes' (2005) 19 Australian and New Zealand Maritime Law Journal 24.

F. Salerno, 'Treaties Establishing Objective Regimes' in E. Cannizzaro (ed), *The Law of Treaties Beyond the Vienna Convention* (Oxford University Press 2011).

P. H. Sand, 'Endangered Species, International Protection' Max Planck Encyclopedia of Public International Law www.mpepil.com.

—— 'Commodity or Taboo? International Regulation of Trade in Endangered Species' (1997) Green Globe Yearbook 19.
—— 'Whither CITES? The Evolution of a Treaty Regime in the Borderland of Trade and Environment' (1997) 8 European Journal of International Law 29.
—— 'Japan's 'Research Whaling' in the Antarctic Southern Ocean and the North Pacific Ocean in the Face of the Endangered Species Convention (CITES)' (2008) 17 RECIEL 56.
—— 'Le rôle des "conférences des parties" aux conventions internationales' (Le droit international face aux enjeux environnementaux, Aix-en-Provence, 4–6 June 2009).
—— 'Enforcing CITES: The Rise and Fall of Trade Sanctions' (2013) 22 RECIEL 251.
P. Sands, *Principles of international environmental law* (2nd edn Cambridge University Press 2003).
P. Sands and others, *Principles of international environmental law* (3rd edn Cambridge University Press 2012).
P. Sands, P. Klein and D. W. Bowett, *Bowett's law of international institutions* (6th edn Sweet & Maxwell 2009).
M. 't Sas-Rolfes, 'Assessing CITES: Four Case Studies' in J. Hutton and B. Dickson (eds), *Endangered Species-Threatened Convention: The Past, Present and Future of CITES* (Earthscan 2000).
J. E. Scanlon, 'CITES at Its Best: CoP16 as a 'Watershed Moment' for the World's Wildlife' (2013) 22 RECIEL 222.
H. N. Scheiber, 'Historical Memory, Cultural Claims, and Environmental Ethics: The Jurisprudence of Whaling Regulation' in H. N. Scheiber (ed), *Law of the Sea: The Common Heritage and Emerging Challenges* (Martinus Nijhoff Publishers 2000).
H. G. Schermers, 'The legal bases of international organization action' in R.-J. Dupuy (ed), *Manuel sur les organisations internationales – A Handbook on International Organizations* (2nd edn Martinus Nijhoff Publishers 1998).
H. G. Schermers and N. Blokker, *International institutional law: Unity within diversity* (5th edn Nijhoff 2011).
C.-C. Schmidt, 'Addressing Illegal, Unreported and Unregulated (IUU) Fishing: Paper prepared for the International Fisheries Compliance 2004 Conference, 29–30 September 2004, Brussels' (2004) http://www.oecd.org/greengrowth/fisheries/34029751.pdf accessed 4 July 2015.
G. J. Schram and A. Tahindro, 'Developments in Principles for the Adoption of Fisheries Conservation and Management Measures' in E. Hey (ed), *Developments in International Fisheries Law* (Kluwer Law International 1999).
W. E. Schrank, 'Is there any hope for fisheries management?' (2007) 31 Marine Policy 299.
N. Schrijver, *Sovereignty Over Natural Resources: Balancing Rights and Duties* (Cambridge Studies in International and Comparative Law, Cambridge University Press 1997).
T. Scovazzi, '"Dici Quodammodo Potest Pisces Exhauriri": Fishing in the Mare Liberum' in R. Casado Raigón and G. Cataldi (eds), *L'évolution et l'état actuel du droit international de la mer: Mélanges de droit de la mer offerts à Daniel Vignes* (Bruylant 2009).
B. Sepúlveda-Amor, 'Contribution of the International Court of Justice to the development of the international law of the sea' (2010) 1 Aegean Review of the Law of the Sea and Maritime Law 5.
A. Serdy, 'Accounting for Catch in Internationally Managed Fisheries: What Role for State Responsibility?' (2010) 15 Ocean and Coastal Law Journal 23.
—— 'Postmodern International Fisheries Law, or We Are All Coastal States Now' (2011) 60 International and Comparative Law Quarterly 387.
D. Shelton, 'Soft law' in D. Armstrong (ed), *Routledge Handbook of International Law* (Routledge 2009).
Y. Shigeta, 'Obligation to Protect the Environment in the ICJ's Practice: To What Extent Erga Omnes?' (2012) 55 Japanese Yearbook of International Law 176.
M. R. Simpson, C. M. Miri and C. Busby, 'Assessment of Thorny Skate (Amblyraja radiata Donovan, 1808) in NAFO Divisions 3LNO and Subdivision 3Ps' SCR Doc. 08/43 (June 2008) http://archive.nafo.int/open/sc/2008/scr08-043.pdf accessed 4 July 2015.
I. Sinclair, *The Vienna Convention on the Law of Treaties* (2nd edn Manchester University Press 1984).

M. B. Sky, 'Getting on the List: Politics and Procedural Maneuvering in CITES Appendix I and II Decisions for Commercially Exploited Marine and Timber Species' (2009–2010) 10 Sustainable Development Law and Policy Review 35.

H. Slim, 'Les pavillons de complaisance' (Le pavillon – Colloque international, Institut Océanographique de Paris, 2–3 March 2007 (Pedone 2008)).

M. J. Smith and others, 'Assessing the impacts of international trade on CITES-listed species: Current practices and opportunities for scientific research' (2011) 144 Biological Conservation 82.

C. Soanes and A. Stevenson (eds), *Concise Oxford English Dictionary* (11th edn Oxford University Press 2008).

J. Sommer, 'Environmental Law-Making by International Organizations' (1996) 56 Zeitschrift für ausländisches öffentliches Recht und Völkerrecht 628.

B. K. Sovacool, 'A Game of Cat and Fish: How to Restore the Balance in Sustainable Fisheries Management' (2009) 40 Ocean Development and International Law 97.

K. I. Stergiou and V. Christensen, 'Fishing down food webs' in V. Christensen and J. MacLean (eds), *Ecosystem Approaches to Fisheries: A Global Perspective* (Cambridge University Press 2011).

J. D. Stevens and others, 'The effects of fishing on sharks, rays, and chimaeras (chondrichthyans), and the implications for marine ecosystems' (2000) 57 ICES Journal of Marine Science 476.

P. Stoett, 'To trade or not to trade?: The African elephant and CITES' (1996–1997) 52 International Journal of Marine and Coastal Law 567.

O. S. Stokke, 'Governance of High Seas Fisheries: The Role of Regime Linkages' in D. Vidas and W. Ostreng (eds), *Order for the oceans at the turn of the century* (Kluwer Law International 1999).

—— 'The Interplay of International Regimes: Putting Effectiveness Theory to Work' (Lysaker 2001). FNI Report 14 http://www.fni.no/doc&pdf/FNI-R1401.pdf accessed 6 July 2015.

—— 'Trade Measures and the Combat of IUU Fishing: Institutional Interplay and Effective Governance in the Northeast Atlantic' (2009) 33 Marine Policy 339.

P. Stoll and F. Schorkopf, *WTO – World Economic Order, World Trade Law* (Martinus Nijhoff Publishers 2006).

C. D. Stone, 'Too Many Fishing Boats, Too Few Fish: Can Trade Laws Trim Subsidies and Restore the Balance in Global Fisheries?' in K. P. Gallagher and J. Werksman (eds), *The Earthscan Reader on International Trade and Sustainable Development* (Earthscan 2002).

A. Stroud, 'Essay: A Review of the Role of the CITES Secretariat in the Implementation of the Non-Detriment Finding Requirement' (2005–2006) 30 William and Mary Environmental Law and Policy Review 661.

P. Sturma, 'The Responsibility of International Organizations and their Member States' in M. Ragazzi (ed), *Responsibility of international organizations: Essays in Memory of Sir Ian Brownlie* (Martinus Nijhoff Publishers 2013).

S. Suhre, 'Misguided Morality: The Repercussions of the International Whaling Commission's Shift from a Policy of Regulation to One of Preservation' (1999–2000) 12 Georgetown International Environmental Law Review 305.

R. Sumaila and W. W. L. Cheung, 'Vulnerability and Sustainability of Fish Stocks Worldwide: With Emphasis on Fish Stocks within the Commonwealth' in R. Bourne and M. Collins (eds), *From Hook to Plate: The State of Marine Fisheries: A Commonwealth Perspective* (2009).

U. R. Sumaila, J. Adler and H. Keith, 'Global scope and economics of illegal fishing' (2006) 30 Marine Policy 696.

J. Swan, 'Regional Fishery Bodies and Governance: Issues, Actions and Future Directions' FAO Fisheries and Aquaculture Circular 959 (Rome 2000).

—— 'Fishing Vessels operating under open registers and the exercise of flag State responsibilities: Information and options' FAO Fisheries Circular 980 (Rome 2002).

—— 'International Action and Responses by Regional Fishery Bodies or Arrangements to Prevent, Deter and Eliminate Illegal, Unreported and Unregulated Fishing' FAO Fisheries and Aquaculture Circular 996 (Rome 2004).

—— 'Port State Measures to Combat IUU Fishing: International and Regional Developments' (2006–2007) 7 Sustainable Development Law and Policy Review 38.

T. Swanson, 'Developing CITES: Making the Convention Work for all the Parties' in J. Hutton and B. Dickson (eds), *Endangered Species-Threatened Convention: The Past, Present and Future of CITES* (Earthscan 2000).

A. K. Sydnes, 'Regional Fishery Organizations: How and Why Organizational Diversity Matters' (2001) 32 Ocean Development and International Law 349.

Y. Takei, 'Assessing Flag State Performance in Legal Terms: Clarifications of the Margin of Discretion' (2013) 28 International Journal of Marine and Coastal Law 97.

C. J. Tams, *Enforcing Obligations Erga Omnes in International Law* (Cambridge University Press 2005).

Y. Tanaka, 'The Changing Approaches to Conservation of Marine Living Resources in International Law' (2011) 71 Zeitschrift für ausländisches öffentliches Recht und Völkerrecht 291.

—— *The International Law of the Sea* (Cambridge University Press 2012).

R. G. Tarasofsky, 'Regional Fisheries Organizations and the World Trade Organization: Compatibility or Conflict?' (TRAFFIC International 2003).

W. W: Taylor, M. G. Schechter and L. G. Wolfson, *Globalization: Effects on Fisheries Resources* (Cambridge University Press 2007).

E. J. Techera and N. Klein, 'Fragmented governance: Reconciling legal strategies for shark conservation and management' (2011) 35 Marine Policy 73.

P. Thomas, 'Mushrooms and the Future of CITES' (2005) 30 Endangered Species Bulletin 24.

B. H Thompson, 'Tragically difficult: the obstacles to governing the commons' (2000) 30 Environmental Law 241.

E. Thorson, 'IELP White Paper on Export Quotas and Non-Detriment Findings' (2006).

R. Torpy, 'If Criminal Offenses Were Added to CITES, Would Nations Be Better Able to Restrict International Trade in Endangered Species and Protect Biodiversity?' (2012) 9 Brazilian Journal of International Law 57.

T. Treves, 'The Law of the Sea "System" of Institutions' (1998) 2 Max Planck Yearbook of United Nations Law 325.

—— 'Le Tribunal International du Droit de la Mer dans la pléiade des juridictions internationales' in O. Delas and others (eds), *Les juridictions internationales: complémentarité ou concurrence* (Bruylant 2005).

—— 'Do fisheries organisations and agreements provide an appropriate framework for sustainable management of deep-sea fisheries in the high seas?' (2009) 35 Océanis 71.

—— 'Cross-fertilization between different international courts and tribunals: the Mangouras case' in H. P Hestermeyer and others (eds), *Coexistence, Cooperation and Solidarity: Liber Amicorum Rüdiger Wolfrum* (Martinus Nijhoff Publishers 2011).

A. Trouwborst, 'The Precautionary Principle and the Ecosystem Approach in International Law: Differences, Similarities and Linkages' (2009) 18 RECIEL 26.

H. Tudela, 'La protection du milieu marin et la pratique de la pêche devant le Tribunal international du droit de la mer' in O. Lecucq and S. Maljean-Dubois (eds), *Le rôle du juge dans le développement du droit de l'environnement* (Bruylant 2008).

H. Tuerk, 'The Waning Freedom of the Seas' in R. Casado Raigón and G. Cataldi (eds), *L'évolution et l'état actuel du droit international de la mer: Mélanges de droit de la mer offerts à Daniel Vignes* (Bruylant 2009).

T. Tyler, 'Saving Fisheries on the High-seas: The Use of Trade Sanctions To Force Compliance with Multilateral Fisheries Agreements' (2006–2007) 20 Tulane Environmental Law Journal 43.

G. Ulfstein, 'Dispute resolution, compliance control and enforcement in international environmental law' in G. Ulfstein (ed), *Making treaties work: Human Rights, Environment and Arms Control* (Cambridge University Press 2007).

—— 'Treaty Bodies' in D. Bodansky, J. Brunnée and E. Hey (eds), *Oxford Handbook of International Environmental Law* (Oxford University Press 2007).

United Nations, Agenda 21 – The United Nations Programme of Action from Rio (1992).

A. A. Vagelli, 'The unfortunate journey of Pterapogon kauderni: A remarkable apogonid endangered by the international ornamental fish trade, and its case in CITES' (2008) 18 SPC Live Reef Fish Information Bulletin 17.

G. van Bohemen, 'High Seas Fisheries Management: Reflections on Experience with Regional Fisheries Management Organisations in the South Pacific' in D. Vidas and P. J. Schei (eds), *The World Ocean in Globalisation: Climate Change, Sustainable Fisheries, Biodiversity, Shipping, Regional Issues* (Martinus Nijhoff Publishers 2011).

J. M. van Dyke, 'Giving teeth to the environmental obligations in the LOS Convention' in A. G. Oude Elferink and D. R. Rothwell (eds), *Oceans Management in the 21st Century: Institutional Frameworks and Responses* (Martinus Nijhoff Publishers 2004).

S. van Osch, 'Save our Sharks: Using International Fisheries Law Within Regional Fisheries Management Organizations to Improve Shark Conservation' (2011–2012) 33 Michigan Journal of International Law 383.

W. van Reenen, 'Rules of Reference in the New Convention on the Law of the Sea, in particular in Connection with the Pollution of the Sea by Oil from Tankers' (1981) 3 Netherlands Yearbook of International Law 3.

D. Vanderzwaag, 'The Precautionary Principle and Marine Environmental Protection: Slippery Shores, Rough Seas, and Rising Normative Tides' (2002) 33 Ocean Development and International Law 165.

—— 'Canadian Struggles with Freedoms of Navigation and Fisheries: a Short "Legal Cruise"' in E. Franckx and P. Gautier (eds), *The Exercise of Jurisdiction over Vessels: New Developments in the Fields of Pollution, Fisheries, Crimes at Sea and Trafficking of Weapons of Mass Destruction* (Bruylant 2010).

E. W. Vierdag, 'The Time of the 'Conclusion' of a Multilateral Treaty: Article 30 of the Vienna Convention on the Law of Treaties and Related Provisions' (1988) 59 British Yearbook of International Law 75.

G. Vigneron, 'Compliance and International Environmental Agreements: A Case Study of the 1995 United Nations Straddling Fish Stocks Agreement' (1997–1998) 10 Georgetown International Environmental Law Review 581.

—— 'The Most Recent Efforts in the International Community to Implement the 1995 United Nations Straddling Fish Stocks Agreement' (1999) 10 Colorado Journal of International Environmental Law and Policy 225.

A. C. J. Vincent and others, 'The role of CITES in the conservation of marine fishes subject to international trade' (2014) 15 Fish and Fisheries 563.

P. Vincent, *Droit de la mer* (Larcier 2008).

J. E. Viñuales, 'The Contribution of the International Court of Justice to the Development of International Environmental Law: A Contemporary Assessment' (2008) 32 Fordham International Law Journal 232.

R. Vita and A. Marin, 'Environmental impact of capture-based bluefin tuna aquaculture on benthic communities in the western Mediterranean' (2007) 38 Aquaculture Research 331.

B. Vukas, 'Treaties, Third-Party Effect' Max Planck Encyclopedia of Public International Law www.mpepil.com.

C. Walters and J.-J. Maguire, 'Lessons for stock assessment from the northern cod collapse' (1996) 6 Reviews of Fish Biology and Fisheries 125.

H. Walther, *L'affaire du "Lotus" ou De l'abordage hauturier en droit pénal international* (Les éditions internationales 1928).

C. A. Ward-Paige and others, 'Recovery potential and conservation option for elasmobranchs' (2012) 80 Journal of Fish Biology 1844.

D. Warner-Kramer, 'Control Begins at Home: Tackling Flags of Convenience and IUU Fishing' (2004) 34 Golden Gate University Law Review 497.

R. Warner, *Protecting the Oceans Beyond National Jurisdiction* (Martinus Nijhoff Publishers 2009).

R. Watson and D. Pauly, 'Systematic distortion in world fisheries catch trends' (2001) 414 Nature 534.

R. Watson, U. R: Sumaila and D. Zeller, 'How much fish is being extracted from the oceans and what is it worth?' in V. Christensen and J. MacLean (eds), *Ecosystem Approaches to Fisheries: A Global Perspective* (Cambridge University Press 2011).

G. J. W. Webb, 'Are All Species Equal? A Comparative Assessment' in J. Hutton and B. Dickson (eds), *Endangered Species-Threatened Convention: The Past, Present and Future of CITES* (Earthscan 2000).

D. G. Webster, 'The irony and the exclusivity of Atlantic bluefin tuna management' (2011) 35 Marine Policy 249.

B. Weeber, K. Thomas and C. Dorey, 'New Zealand orange roughy (Hoplostethus atlanticus)' (September 2010) http://www.greenpeace.org/new-zealand/Global/new-zealand/P3/publications/oceans/2010/Orange%20Roughy.pdf accessed 4 July 2015.

P. Weil, 'Towards relative normativity in international law?' (1983) 77 American Journal of International Law 413.

E. B. Weiss, 'Invoking State Responsibility in the Twenty-First Century' (2002) 96 American Journal of International Law 798.

J. Werksman, 'The Conference of Parties to Environmental Treaties' in J. Werksman (ed), *Greening International Institutions* (Earthscan 1996).

N. D. White, *The law of international organisations* (2nd edn Juris Publishing – Manchester University Press 2005).

—— 'Separate but Connected: Inter-Governmental Organizations and International Law' (2008) 5 International Organizations Law Review 175.

J. B. Wiener, 'Precaution' in D. Bodansky, J. Brunnée and E. Hey (eds), *Oxford Handbook of International Environmental Law* (Oxford University Press 2007).

A. Wiersema, 'Adversaries or Partners? Science and the Precautionary Principle in International Wildlife Treaty Regimes' (2008) 11 Journal of International Wildlife Law & Policy 211.

—— 'The New International Law-makers? Conferences of the Parties to Multilateral Environmental Agreements' (2009) 31 Michigan Journal of International Law 231.

—— 'Uncertainty and Markets for Endangered Species under CITES' (2013) 22 RECIEL 239.

M. Wijkman, 'Managing the global commons' (1982) 36 International Organization 511.

W. Wijnstekers, *The Evolution of CITES* (9th edn CIC – International Council for Game and Wildlife Conservation 2011).

D. M. Wilkinson, 'The Use of Domestic Measures to Enforce International Whaling Agreements: A Critical Perspective' (1989) 17 Denver Journal of International Law and Policy 271.

A. Willock, 'Uncharted Waters/Implementation Issues and Potential Benefits of Listing Toothfish in Appendix II of CITES' (TRAFFIC 2002)

—— 'Administrative and Monitoring Implications of listing and down-listing of commercially-exploited aquatic species, including the implications of Annex 4 of Resolution Conf. 9.24' Background paper to CITES Workshop on Introduction from the Sea Issues, 30 November – 2 December 2005, Geneva (Switzerland) http://www.cites.org/eng/news/meetings/ifs-05/IFS05-TRAFFIC-paper.pdf accessed 4 July 2015.

A. Willock, M. Burgener and A. Sancho, 'First Choice or Fall Back? An Examination of Issues Relating to the Application of Appendix III of CITES to Marine Species' (TRAFFIC 2004).

C. Wold and E. Thorson, 'The Application of CITES to Atlantic Bluefin Tuna and Other Marine Species' (International Environmental Law Project 2010).

C. Wolf, P. Hübner and J. Lüthy, 'Differentiation of sturgeon species by PCR-RFLP' (1999) 32 Food Research International 699.

R. Wolfrum, 'IMO Interface with the Law of the Sea Convention' in M. H. Nordquist and J. N. Moore (eds), *Current Maritime Issues and the International Maritime Organization* (Kluwer Law International 1999).

—— 'Article XX GATT [Introduction]' in R. Wolfrum, P.-T. Stoll and A. Seibert-Fohr (eds), *WTO – Technical Barriers and SPS Measures* (Martinus Nijhoff Publishers 2007).

—— 'Enforcing Community Interests through International Dispute Settlement: Reality or Utopia' in U. Fastenrath and others (eds), *From Bilateralism to Community Interest: Essays in Honour of Bruno Simma* (Oxford University Press 2011).

R. Wolfrum and N. Matz, 'The Interplay of the United Nations Convention on the Law of the Sea and the Convention on Biological Diversity' (2000) 4 Max Planck Yearbook of United Nations Law 445.

—— *Conflicts in international environmental law* (Beiträge zum ausländischen öffentlichen Recht und Völkerrecht, Springer 2003).

World Bank and FAO, *The Sunken Billions – The Economic Justification for Fisheries Reform* (2009).

World Commission on Environment and Development, *Our Common Future (usually referred to as the Brundtland Report)* (Oxford University Press 1987).

World Ocean Review, *Living with the oceans* (Maribus, Future Ocean 2010).

—— *The Future of Fish – The Fisheries of the Future* (Maribus, Future Ocean 2013).

A. Wuertz and others, 'Identification of caviar from increasing global aquaculture production — Dietary capric acid as a labelling tool for CITES implementation in caviar trade' (2009) 298 Aquaculture Research 51.

N. Yagi, 'High Seas Fisheries Today: Challenges and Remedies Under the Global Economy' in D. Vidas and P. J. Schei (eds), *The World Ocean in Globalisation: Climate Change, Sustainable Fisheries, Biodiversity, Shipping, Regional Issues* (Martinus Nijhoff Publishers 2011).

A. Yankov, 'The International Tribunal for the law of the Sea: Its Place Within the Dispute Settlement System of the UN Law of the Sea Convention' (1997) 37 Indiana Journal of International Law 356.

—— 'Reflagging of fishing vessels: a critical assessment of its impact on the enforcement of fishing regulations and the responses thereto' in P. Ehlers, E. Mann-Borgese and R. Wolfrum (eds), *Marine Issues: From a Scientific, Political and Legal Perspective* (Kluwer Law International 2002).

M. Yeater and J. Vasquez, 'Demystifying the Relationship Between CITES and the WTO' (2001) 10 RECIEL 271.

S. Yee, "Member Responsibility' and the ILC Articles on the Responsibility of International Organizations: Some Observations' in M. Ragazzi (ed), *Responsibility of international organizations: Essays in Memory of Sir Ian Brownlie* (Martinus Nijhoff Publishers 2013).

M. A. Young, 'Protecting Endangered Marine Species: Collaboration Between the Food and Agriculture Organization and the CITES Regime' (2010) 11 Melbourne Journal of International Law 1.

—— *Trading Fish, Saving Fish: The Interaction between Regimes and International Law* (Cambridge Studies in International and Comparative Law, Cambridge University Press 2011).

O. R. Young, 'Fragmentation or interaction: the WTO, fisheries subsidies, and international law' (2009) 8 World Trade Review 477.

S. Young, 'Contemporary Issues of the Convention on International Trade in Endangered Species of Wild Fauna and Flora (CITES) and the Debate Over Sustainable Use' (2003) 14 Colorado Journal of International Environmental Law and Policy 167.

J. Zander, *The Application of the Precautionary Principle in Practice* (Cambridge University Press 2010).

D. Zeller and D. Pauly, 'Good news, bad news: Global fisheries discards are declining, but so are total catches' (2005) 6 Fish and Fisheries 156.

K. Zemanek, *The legal foundations of the international system: general course on public international law* (Recueil des Cours vol 266, Martinus Nijhoff Publishers 1997).

—— 'New Trends in the Enforcement of erga omnes Obligations' (2000) 4 Max Planck Yearbook of United Nations Law 1.

W. M. von Zharen, 'Ocean Ecosystem Stewardship' (1998) 23 William and Mary Environmental Law and Policy Review 1.

J. Ziemer, *Das gemeinsame Interesse an einer Regelung der Hochseefischerei: Dargestellt am Beispiel des Fish Stocks Agreement* (Veröffentlichungen des Walther-Schücking-Instituts für Internationales Recht an der Universität Kiel vol 128, Duncker & Humblot 2000).

Treaties

'Charter of the United Nations' 1945, 1 *United Nations Treaty Series* xvi.

Constitution of the Food and Agriculture Organization of the United Nations 1945.

'International Convention for the Regulation of Whaling' (ICRW) 1946, 161 *United Nations Treaty Series* 72.

'Agreement for the establishment of the General Fisheries Commission for the Mediterranean' (GFCM Agreement) 1949, 126 *United Nations Treaty Series* 237.

'Convention on Fishing and Conservation of the Living Resources of the High Seas' 1958, 559 *United Nations Treaty Series* 285.

'Convention on the High Seas' 1958, 450 United Nations Treaty Series 11.

'International Convention for the Conservation of Atlantic Tunas' (ICCAT Convention) 1966, 673 *United Nations Treaty Series* 63.

'Vienna Convention on the Law of Treaties' (VCLT) 1969, 1155 *United Nations Treaty Series* 331.

'Convention on International Trade in Endangered Species of Wild Fauna and Flora' (CITES) 1973, 993 *United Nations Treaty Series* 243.

'Convention on Future Multilateral Cooperation in the Northwest Atlantic Fisheries' (NAFO Convention) 1978, 1135 *United Nations Treaty Series* 369.

'Convention on the Conservation of Migratory Species of Wild Animals' (CMS) 1979, 1651 *United Nations Treaty Series* 333.

'Convention on the Conservation of Antarctic Marine Living Resources' (CCAMLR Convention) 1980, 1329 *United Nations Treaty Series* 47.

'Convention on the Future Multilateral Cooperation in North-East Atlantic Fisheries' (NEAFC Convention) 1980, 1285 *United Nations Treaty Series* 129 (revised 2006, 'New Convention' http://www.neafc.org/system/files/Text-of-NEAFC-Convention-04.pdf)

'United Nations Convention on the Law of the Sea' (UNCLOS) 1982, 1833 *United Nations Treaty Series* 397.

'Vienna Convention on the Law of Treaties between States and International Organizations or between International Organizations' 1986 A/CONF.129/15 (not yet in force).

'Agreement on the Conservation of Populations of European Bats' 1991, 1863 *United Nations Treaty Series* 101.

'United Nations Framework Convention on Climate Change' 1992, 1771 *United Nations Treaty Series,* 107.

'Agreement for the Establishment of the Indian Ocean Tuna Commission' (IOTC Agreement) 1993, 1927 *United Nations Treaty Series* 329.

'Convention for the Conservation of Southern Bluefin Tuna' (CCSBT Convention) 1993, 1819 *United Nations Treaty Series* 360.

'FAO Agreement to Promote Compliance with International Conservation and Management Measures by Fishing Vessels on the High Seas: Compliance Agreement' (Compliance Agreement) 1993, 2221 *United Nations Treaty Series* 91.

'General Agreement on Tariffs and Trade' (to be read together with GATT 1947) (GATT 1994) 1994, 1867 *United Nations Treaty Series* 187.

'Agreement establishing the World Trade Organization' (WTO Agreement) 1994, 1867 *United Nations Treaty Series* 154.

'Understanding on Rules and Procedures Governing the Settlement of Disputes' 1994, 1869 *United Nations Treaty Series* 401.

'Agreement for the Implementation of the Provisions of the United Nations Convention on the Law of the Sea of 10 December 1982 Relating to the Conservation and Management of Straddling Fish Stocks and Highly Migratory Fish Stocks' (UNFSA) 1995, 2167 *United Nations Treaty Series* 3.

'Agreement on the Conservation of African-Eurasian Migratory Waterbirds' 1995.

'Convention on the Conservation and Management of Highly Migratory Fish Stocks in the Western and Central Pacific Ocean' (WCPFC Convention) 2000, 2275 *United Nations Treaty Series* 43.

'Agreement on the Conservation of Albatrosses and Petrels' 2001, 2258 *United Nations Treaty Series* 257.
'Convention on the Conservation and Management of Fishery Resources in the South East Atlantic Ocean' (SEAFO Convention) 2001, 2221 *United Nations Treaty Series* 189.
'Convention for the Strengthening of the Inter-American Tropical Tuna Commission Established by the 1949 Convention Between the United States of America and the Republic of Costa Rica' (Antigua Convention) 2003.
'Southern Indian Ocean Fisheries Agreement' (SIOFA Agreement) 2006.
'Agreement on the Conservation of Gorillas and Their Habitats' 2007, 2545 *United Nations Treaty Series* 55.
'Agreement on Port State Measures to Prevent, Deter and Eliminate Illegal, Unreported and Unregulated Fishing' (Port State Measures Agreement) 2009.
'Convention on the Conservation and Management of High Seas Fishery Resources in the South Pacific Ocean' (SPRFMO Convention) 2009.

Case-Law

Arbitral Awards

Arbitration between the United States and the United Kingdom relating to the rights of jurisdiction of United States in the Bering's sea and the preservation of fur seals, Award, 15 August 1893, 2007 XXVIII Reports of International Arbitral Awards 263.
Southern Bluefin Tuna (New Zealand v. Japan, Australia v. Japan), Arbitral Award, 4 August 2000, Arbitral Tribunal, (2006) XXIII Reports of International Arbitral Awards 1.
The Objection by the Russian Federation to a Decision of the Commission of the South Pacific Regional Fisheries Management Organisation, Findings and Recommendations, 5 July 2013, Review Panel – Permanent Court of Arbitration.

GATT Panel/WTO Dispute Settlement Body

Canada — Measures Affecting Exports of Unprocessed Herring and Salmon (United States v. Canada), GATT Panel Report adopted, 22 March 1988, BISD 35S/98.
Restrictions on Imports of Tuna (Mexico v. United States), GATT Panel Report circulated not adopted, 3 September 1991, DS21/R – 39S/155, (1991) 30 ILM 1594.
Restrictions on Imports of Tuna (European Economic Community (EEC) and the Netherlands v. United States), GATT Panel Report circulated not adopted, 16 June 1994, DS29/R, (1994) 33 ILM 839.
European Communities v United States of America), WTO Dispute Settlement Body – Appellate Body Report Adopted, 13 February 1998, WT/DS26/AB/R and WT/DS48/AB/R.
United States – Import Prohibition of Certain Shrimp and Shrimp Products (India, Malaysia, Pakistan and Thailand v United States of America), WTO Dispute Settlement Body – Appellate Body Report Adopted, 6 November 1998, WT/DS58/AB/R.
Measures Affecting Asbestos and Asbestos-Containing Products (Canada v. European Communities), WTO Dispute Settlement Body – Appellate Body Report Adopted, 12 March 2001, WT/DS135/AB/R.

United States – Import Prohibitions of Certain Shrimp and Shrimp Products (Recourse to Article 21.5 of the DSU by Malaysia), WTO Dispute Settlement Body – Panel Report, 15 June 2001, WT/DS58/RW.

United States – Import Prohibitions of Certain Shrimp and Shrimp Products (Recourse to Article 21.5 of the DSU by Malaysia), WTO Dispute Settlement Body – Appellate Body Report Adopted, 22 October 2001, WT/DS58/AB/RW

Permanent Court of International Justice/International Court of Justice

Case concerning certain German interests in Polish Upper Silesia (Germany v. Poland), Judgment, 25 May 1926, Series A Publications of the Permanent Court of International Justice 1926 4.

Case of the "S.S. Lotus" (France v. Turkey), Judgment, 7 September 1927, Series A Publications of the Permanent Court of International Justice 4.

Factory at Chorzów (Germany v. Poland), Judgment, 13 September 1928, Series A Publications of the Permanent Court of International Justice 1928 4.

Reparation for injuries suffered in the service of the United Nations, Advisory Opinion, 11 April 1949, I.C.J. Reports 1949 174.

Case concerning rights of nationals of the United States of America in Morocco (France v. United States of America), Judgment, 27 August 1952, I.C.J. Reports 1952 176.

Case concerning the right of passage over Indian territory (Portugal v. India), Preliminary objections, 26 November 1957, I.C.J. Reports 1957 125.

Certain Expenses of the United Nations (Article 17, Paragraph 2, of the Charter), Advisory Opinion, 20 July 1962, I.C.J. Reports 1962 151.

South West Africa cases (Ethiopia v South Africa; Liberia v. South Africa) – second phase. Judgment, 18 July 1966, I.C.J. Reports 1966 6.

Federal Republic of Germany/Denmark; Federal Republic of Germany/Netherlands), Judgment, 20 February 1969, I.C.J. Reports 1969 3.

Case concerning the Barcelona Traction, Light and Power Company, Limited (Belgium v. Spain) – second phase, Judgment, 5 February 1970, I.C.J. Reports 1970 3.

Legal consequences for states of the continued presence of South Africa in Namibia (South West Africa) notwithstanding Security Council Resolution 276 (1970), Advisory Opinion, 21 June 1971, I.C.J. Reports 1971 16.

Fisheries Jurisdiction Case (Germany v. Iceland), Judgment, 25 July 1974, I.C.J. Reports 1974, 174.

Fisheries Jurisdiction Case (United Kingdom v. Iceland), Judgment, 25 July 1974, I.C.J. Reports 1974 3.

Nuclear Tests Case (Australia v. France), Judgment, 20 December 1974, I.C.J. Reports 1974 253.

Nuclear Tests Case (New Zealand v. France), Judgment, 20 December 1974, I.C.J. Reports 1974 457.

Interpretation of the Agreement of 25 March 1951 between the WHO and Egypt, Advisory Opinion, 20 December 1980, I.C.J. Reports 1980 73.

Case of the delimitation of the maritime boundary in the Gulf of Maine Area (Canada/United States of America), Judgment, 12 October 1984, I.C.J. Reports 1984 246.

Case concerning the Continental Shelf (Libyan Arab Jamahiriya/Malta), Judgment, 3 June 1985, I.C.J. Reports 1985 13.

Military and Paramilitary Activities in and against Nicaragua (Nicaragua v. United States of America), Judgment, 27 June 1986, I.C.J. Reports 1986 14.

Maritime Delimitation and Territorial Questions between Qatar and Bahrein (Qatar v Bahrain), Jurisdiction and Admissibility, 1 July 1994, I.C.J. Reports 1994 112.

Request for an examination of the situation in accordance with paragraph 63 of the Court's judgment of 20 December 1974 in the Nuclear Tests (New Zealand v. France) case (New Zealand v. France), Order, 22 September 1995, I.C.J. Reports 1995 288.

Legality of the Threat or Use of Nuclear Weapons, Advisory Opinion, 8 July 1996, I.C.J. Reports 1996 226.

Gabcikovo-Nagymaros Project (Hungary/Slovakia), Judgment, 25 September 1997, I.C.J. Reports 1997 7.

Fisheries Jurisdiction (Spain v. Canada), Judgment on Jurisdiction, 4 December 1998, I.C.J. Reports 1998 432.

Legal consequences of the construction of a wall in the occupied Palestinian territory, Advisory Opinion, 9 July 2004, I.C.J. Reports 2004 136.

Case concerning armed activities on the territory of the Congo (new application: 2002) (Democratic Republic of the Congo v. Rwanda), Judgment on jurisdiction and admissibility, 3 February 2006, I.C.J. Reports 2006 6.

Maritime Delimitation in the Black Sea (Romania v. Ukraine), Judgment, 3 February 2009, I.C.J. Reports 2009 61.

Case concerning Pulp Mills on the River Uruguay (Argentina v. Uruguay), Judgment, 20 April 2010, I.C.J. Reports 2010 14.

Questions relating to the obligation to prosecute or extradite (Belgium v. Senegal), Judgment, 20 July 2012, I.C.J. Reports 2012 422.

Whaling in the Antarctic (Australia v. Japan; New Zealand intervening), Judgment, 31 March 2014, ICJ.

International Criminal Tribunal for the Former Yugoslavia

Prosecutor v. Dusko Tadic, Judgment, 15 July 1999, ICTY IT-94-1-A.

International Tribunal for the Law of the Sea

The M/V "Saiga" (Saint Vincent and the Grenadines v. Guinea), Judgment on prompt release, 4 December 1997, ITLOS case n°1.

The M/V "Saiga" (Saint Vincent and the Grenadines v. Guinea), Judgment, 1 July 1999, ITLOS case n°2.

Southern Bluefin Tuna cases (New Zealand v. Japan; Australia v. Japan), Provisional Measures, 27 August 1999, ITLOS cases n°3 & 4.

Case concerning the Conservation and Sustainable Exploitation of Swordfish Stocks in the South-Eastern Pacific Ocean (Chile/European Union), Order of removal from the list, 16 December 2009, ITLOS case n°7.

The MOX Plant Case (Ireland v. United Kingdom), Provisional measures, 3 December 2001, ITLOS case n°10.

The "Volga" Case (Russian Federation v. Australia), Judgment on prompt release, 23 December 2002, ITLOS case n°11.

Dispute concerning delimitation of the maritime boundary between Bangladesh and Myanmar in the Bay of Bengal (Bangladesh/Myanmar), Judgment, 14 March 2012, ITLOS case n°16.

Responsibilities and obligations of States sponsoring persons and entities with respect to activities in the Area (Request for Advisory Opinion submitted to the Seabed Disputes Chamber), Advisory Opinion, 1 February 2011, ITLOS case n°17.
M/V "Virginia G" (Panama/Guinea-Bissau), Judgment, 14 April 2014, ITLOS case n°19.
Request for an Advisory Opinion Submitted by the Sub-Regional Fisheries Commission (SRFC), Advisory Opinion, 2 April 2015, ITLOS case n°21.

MIX
Papier aus verantwortungsvollen Quellen
Paper from responsible sources
FSC® C105338
FSC
www.fsc.org

If you have any concerns about our products,
you can contact us on
ProductSafety@springernature.com

In case Publisher is established outside the EU,
the EU authorized representative is:
Springer Nature Customer Service Center GmbH
Europaplatz 3, 69115 Heidelberg, Germany

Printed by Libri Plureos GmbH
in Hamburg, Germany